Lecture Notes in Artificial Intelligence   13469

Subseries of Lecture Notes in Computer Science

More information about this subseries at https://link.springer.com/bookseries/1244

Pablo García Bringas · Hilde Pérez García ·
Francisco Javier Martínez de Pisón ·
José Ramón Villar Flecha ·
Alicia Troncoso Lora · Enrique A. de la Cal ·
Álvaro Herrero · Francisco Martínez Álvarez ·
Giuseppe Psaila · Héctor Quintián ·
Emilio Corchado (Eds.)

Hybrid Artificial Intelligent Systems

17th International Conference, HAIS 2022
Salamanca, Spain, September 5–7, 2022
Proceedings

 Springer

Editors
Pablo García Bringas ⓘ
University of Deusto
Bilbao, Spain

Hilde Pérez García ⓘ
University of León
León, Spain

Francisco Javier Martínez de Pisón ⓘ
University of La Rioja
Logroño, La Rioja, Spain

José Ramón Villar Flecha ⓘ
University of Oviedo
Oviedo, Spain

Alicia Troncoso Lora ⓘ
Data Science and Big Data Analytics Lab
Pablo de Olavide University
Sevilla, Spain

Enrique A. de la Cal ⓘ
Department of Computer Science
University of Oviedo
Oviedo, Spain

Álvaro Herrero ⓘ
Applied Computational Intelligence
University of Burgos
Burgos, Burgos, Spain

Francisco Martínez Álvarez ⓘ
Universidad Pablo de Olavide
Seville, Spain

Héctor Quintián ⓘ
Department of Industrial Engineering
University of A Coruña
Ferrol, Spain

Giuseppe Psaila ⓘ
DIGIP
University of Bergamo
Dalmine, Bergamo, Italy

Emilio Corchado ⓘ
University of Salamanca
Salamanca, Spain

ISSN 0302-9743   ISSN 1611-3349 (electronic)
Lecture Notes in Artificial Intelligence
ISBN 978-3-031-15470-6   ISBN 978-3-031-15471-3 (eBook)
https://doi.org/10.1007/978-3-031-15471-3

LNCS Sublibrary: SL7 – Artificial Intelligence

This Springer imprint is published by the registered company Springer Nature Switzerland AG
The registered company address is: Gewerbestrasse 11, 6330 Cham, Switzerland

Preface

This volume of Lecture Notes on Artificial Intelligence (LNAI) includes accepted papers presented at the 17th International Conference on Hybrid Artificial Intelligence Systems (HAIS 2022), held in the beautiful city of Salamanca, Spain, during September 5–7, 2022.

HAIS has become a unique, established, and broad interdisciplinary forum for researchers and practitioners who are involved in developing and applying symbolic and sub-symbolic techniques aimed at the construction of highly robust and reliable problem-solving techniques, bringing the most relevant achievements in this field.

The hybridization of intelligent techniques, coming from different computational intelligence areas, has become popular because of the growing awareness that such combinations frequently perform better than the individual techniques such as neuro-computing, fuzzy systems, rough sets, evolutionary algorithms, agents and multiagent systems, and so on.

Practical experience has indicated that hybrid intelligence techniques might be helpful to solve some of the challenging real-world problems. In a hybrid intelligence system, a synergistic combination of multiple techniques is used to build an efficient solution to deal with a particular problem. This is, thus, the setting of the HAIS conference series, and its increasing success is the proof of the vitality of this exciting field.

The HAIS 2022 International Program Committee selected 43 papers, which are published in these conference proceedings, yielding an acceptance rate of about 65%.

The selection of papers was extremely rigorous in order to maintain the high quality of the conference and we would like to thank the Program Committee for their hard work in the reviewing process. This process is very important in creating a conference of high standard and the HAIS conference would not exist without their help.

The large number of submissions is certainly not only a testimony to the vitality and attractiveness of the field but an indicator of the interest in the HAIS conferences themselves.

HAIS 2022 enjoyed outstanding keynote speeches by distinguished guest speakers: Prof. Ajith Abraham, Director of Machine Intelligence Research Labs (MIR Labs), Prof. Guy De Tré head of the research group on Database, Document, and Content Management (DDCM) at Ghent University (Belgium), and Felix Barrio General Director at INCIBE (Spain).

HAIS 2022 has teamed up with "Neurocomputing" (Elsevier) and the "Logic Journal of the IGPL" (Oxford University Press) for a suite of special issues, including selected papers from HAIS 2022.

Particular thanks go as well to the conference's main sponsors, Startup Olé, the CYL-HUB project financed with NEXT-GENERATION funds from the European Union, the Ministry of Labor and Social Economy, the Recovery, Transformation and Resilience Plan, and the State Public Employment Service, channelled through the Junta de Castilla y León, the BISITE research group at the University of Salamanca, the CTC research group at the University of A Coruña, and the University of Salamanca. They jointly contributed in an active and constructive manner to the success of this initiative.

We want to thank all the contributing authors, members of the Program Committee, and the Local Organizing Committee for their hard and highly valuable work. Their work has helped to contribute to the success of the HAIS 2022 event.

We thank the team at Springer for their help and collaboration during this demanding publication project.

September 2022

Pablo García Bringas
Hilde Pérez García
Francisco Javier Martínez de Pisón
José Ramón Villar Flecha
Alicia Troncoso Lora
Enrique A. de la Cal
Álvaro Herrero
Francisco Martínez Álvarez
Giuseppe Psaila
Héctor Quintián
Emilio Corchado

Organization

General Chair

Emilio Corchado   University of Salamanca, Spain

International Advisory Committee

Ajith Abraham	Machine Intelligence Research Labs, Europe
Antonio Bahamonde	University of Oviedo, Spain
Andre de Carvalho	University of São Paulo, Brazil
Sung-Bae Cho	Yonsei University, Korea
Juan M. Corchado	University of Salamanca, Spain
José R. Dorronsoro	Autonomous University of Madrid, Spain
Michael Gabbay	Kings College London, UK
Ali A. Ghorbani	University of New Brunswick, Canada
Mark A. Girolami	University of Glasgow, Scotland
Manuel Graña	University of País Vasco, Spain
Petro Gopych	Universal Power Systems USA-Ukraine LLC, Ukraine
Jon G. Hall	The Open University, UK
Francisco Herrera	University of Granada, Spain
César Hervás-Martínez	University of Córdoba, Spain
Tom Heskes	Radboud University Nijmegen, The Netherlands
Dusan Husek	Institute of Computer Science, Czech Academy of Sciences, Czech Republic
Lakhmi Jain	University of South Australia, Australia
Samuel Kaski	Helsinki University of Technology, Finland
Daniel A. Keim	University of Konstanz, Germany
Marios Polycarpou	University of Cyprus, Cyprus
Witold Pedrycz	University of Alberta, Canada
Xin Yao	University of Birmingham, UK
Hujun Yin	University of Manchester, UK
Michał Woźniak	Wroclaw University of Technology, Poland
Aditya Ghose	University of Wollongong, Australia
Ashraf Saad	Armstrong Atlantic State University, USA
Fanny Klett	German Workforce Advanced Distributed Learning Partnership Laboratory, Germany
Paulo Novais	Universidade do Minho, Portugal

Rajkumar Roy	The EPSRC Centre for Innovative Manufacturing in Through-life Engineering Services, UK
Amy Neustein	Linguistic Technology Systems, USA
Jaydip Sen	Tata Consultancy Services Ltd., India

Program Committee Chairs

Pablo García Bringas	University of Deusto, Spain
Hilde Pérez García	University of León, Spain
Francisco Javier Martínez de Pisón	University of La Rioja, Spain
José Ramón Villar Flecha	University of Oviedo, Spain
Alicia Troncoso Lora	Pablo Olavide University, Spain
Enrique A. de la Cal	University of Oviedo, Spain
Álvaro Herrero	University of Burgos, Spain
Francisco Martínez Álvarez	Pablo Olavide University, Spain
Giuseppe Psaila	University of Bergamo, Italy
Héctor Quintián	University of A Coruña, Spain
Emilio Corchado	University of Salamanca, Spain

Program Committee

Alfredo Cuzzocrea	University of Calabria, Italy
Álvaro Herrero Cosio	University of Burgos, Spain
Álvaro Michelena Grandío	University of A Coruña, Spain
Anca Andreica	Babes-Bolyai University, Romania
Andreea Vescan	Babes-Bolyai University, Romania
Andrés Blázquez Colino	University of Salamanca, Spain
Angel Arroyo	University of Burgos, Spain
Antonio Dourado	University of Coimbra, Portugal
Antonio Jesús Díaz Honrubia	Polytechnique University of Madrid, Spain
Arkadiusz Kowalski	Wrocław University of Technology, Poland
Borja Sanz	University of Deusto, Spain
Camelia Serban	Babes-Bolyai University, Romania
Carlos Cambra	University of Burgos, Spain
Carlos Carrascosa	Polytechnic University of Valencia, Spain
Carlos Pereira	ISEC, Portugal
Damian Krenczyk	Silesian University of Technology, Poland
Daniel Urda	University of Burgos, Spain
David Iclanzan	Sapienta Hungarian University of Transylvania, Romania
Diego P. Ruiz	University of Granada, Spain
Dragan Simic	University of Novi Sad, Serbia
Eiji Uchino	Yamaguchi University, Japan

Eneko Osaba	TECNALIA Research & Innovation, Spain
Enol García González	University of Oviedo, Spain
Enrique De La Cal Marín	University of Oviedo, Spain
Enrique Onieva	University of Deusto, Spain
Esteban Jove	University of A Coruña, Spain
Federico Divina	Pablo de Olavide University, Spain
Fermin Segovia	University of Granada, Spain
Fidel Aznar	University of Alicante, Spain
Francisco Martínez-Álvarez	Pablo de Olavide University, Spain
Francisco Zayas-Gato	University of A Coruña, Spain
Francisco Javier Martínez de Pisón Ascacíbar	University of La Rioja, Spain
Georgios Dounias	University of the Aegean, Greece
Giorgio Fumera	University of Cagliari, Italy
Giuseppe Psaila	University of Bergamo, Italy
Gloria Cerasela Crisan	University of Bacau, Romania
Héctor Quintián	University of A Coruña, Spain
Hugo Sanjurjo-González	University of Deusto, Spain
Ignacio Turias	University of Cádiz, Spain
Igor Santos	Mondragon Unibertsitatea, Spain
Iker Pastor-López	University of Deusto, Spain
Ioana Zelina	Technical University of Cluj Napoca, Baia Mare, Romania
Ioannis Hatzilygeroudis	University of Patras, Greece
Javier De Lope	Polytechnic University of Madrid, Spain
Jorge García-Gutiérrez	University of Seville, Spain
Jose Dorronsoro	Autonomous University of Madrid, Spain
Jose Garcia-Rodriguez	University of Alicante, Spain
Jose Alfredo Ferreira Costa	Federal University of Rio Grande do Norte, Brazil
Jose Luis Calvo-Rolle	University of A Coruña, Spain
José Luis Casteleiro-Roca	University of A Coruña, Spain
José Luis Verdegay	University of Granada, Spain
Jose M. Molina	University Carlos III of Madrid, Spain
Jose Manuel Lopez-Guede	University of the Basque Country, Spain
José María Armingol	University Carlos III de Madrid, Spain
José Ramón Villar	University of Oviedo, Spain
Juan Pavón	Complutense University of Madrid, Spain
Juan Humberto Sossa Azuela	CIC-IPN, Mexico
Juan J. Gude	Universidad de Deusto, Spain
Lidia Sánchez-González	University of León, Spain
Luis Alfonso Fernández Serantes	FH-Joanneum University of Applied Sciences, Austria

Manuel Castejón-Limas	University of León, Spain
Manuel Graña	University of the Basque Country, Spain
Míriam Timiraos Díaz	University of A Coruña, Spain
Nashwa El-Bendary	Arab Academy for Science, Technology, and Maritime Transport, Egypt
Noelia Rico	University of Oviedo, Spain
Nuño Basurto	University of Burgos, Spain
Pablo García Bringas	University of Deusto, Spain
Paula M. Castro	University of A Coruña, Spain
Peter Rockett	University of Sheffield, UK
Petrica Pop	Technical University of Cluj-Napoca, Baia Mare, Romania
Qing Tan	Athabasca University, Canada
Robert Burduk	University of Wroclaw, Poland
Ruben Fuentes-Fernandez	Complutense University of Madrid, Spain
Sean Holden	University of Cambridge, UK
Theodore Pachidis	International Hellenic University, Greece
Urszula Stanczyk	Silesian University of Technology, Poland

HAIS 2022 Organizing Committee Chairs

Emilio Corchado	University of Salamanca, Spain
Héctor Quintián	University of A Coruña, Spain

HAIS 2022 Organizing Committee

Álvaro Herrero Cosio	University of Burgos, Spain
José Luis Calvo Rolle	University of A Coruña, Spain
Ángel Arroyo	University of Burgos, Spain
Daniel Urda	University of Burgos, Spain
Nuño Basurto	University of Burgos, Spain
Carlos Cambra	University of Burgos, Spain
Esteban Jove	University of A Coruña, Spain
José Luis Casteleiro Roca	University of A Coruña, Spain
Francisco Zayas Gato	University of A Coruña, Spain
Álvaro Michelena	University of A Coruña, Spain
Míriam Timiraos Díaz	University of A Coruña, Spain

Contents

Deep Learning

Evolutionary Computation

HAIs Applications

Image and Speech Signal Processing

Optimization Techniques

Bioinformatics

A Comparison of Machine Learning Techniques for the Detection of Type-4 PhotoParoxysmal Responses in Electroencephalographic Signals

Fernando Moncada Martins[1]([✉]) [iD], Víctor Manuel González[1] [iD],
Beatriz García[2] [iD], Víctor Álvarez[3] [iD], and José Ramón Villar[3] [iD]

[1] Electrical Engineering Department, University of Oviedo, Oviedo, Spain
{UO245868,vmsuarez}@uniovi.es
[2] Neurophysiology Department, University Hospital of Burgos, Burgos, Spain
bgarcialo@saludcastillayleon.es
[3] Computer Science Department, University of Oviedo, Oviedo, Spain
villarjose@uniovi.es

Abstract. Photosensitivity is a neurological disorder in which the patients' brain produces different types of abnormal electrical responses, known as Photoparoxysmal Responses (PPR), to specific visual stimuli, potentially triggering an epileptic seizure in extreme cases. The diagnosis of this condition is based on the manual analysis and detection of these discharges in their electroencephalogram. This research focuses on comparing different Machine Learning techniques for the automatic detection of Type-4 PPR (the most extreme PPR) in a real EEG dataset, after the transformation using Principal Component Analysis. Different two-class and one-class classifiers are tested, and the best performing methods for Type-4 PPR detection are 2C-KNN and DL-NN. Obtained results are compared with those achieved from a previous research, resulting in a performance increase of 15%. This system is currently in study with subjects at Burgos University Hospital, Spain.

Keywords: EEG · PPR detection · Photoparoxysmal responses · Photosensitivity · Epilepsy

1 Introduction

Electroencephalography (EEG) is a technique used for measuring the brain electrical activity by a set of electrodes placed on the scalp that measures the electrical discharges produced by the neurons. Because of this non-invasive quality,

This research has been funded by the Spanish Ministry of Economics and Industry, grant PID2020-112726RB-I00, by the Spanish Research Agency (AEI, Spain) under grant agreement RED2018-102312-T (IA-Biomed), and by the Ministry of Science and Innovation under CERVERA Excellence Network project CER-20211003 (IBERUS) and Missions Science and Innovation project MIG-20211008 (INMERBOT). Also, by Principado de Asturias, grant SV-PA-21-AYUD/2021/50994.

P. García Bringas et al. (Eds.): HAIS 2022, LNAI 13469, pp. 3–13, 2022.
https://doi.org/10.1007/978-3-031-15471-3_1

EEG is widely used for clinical diagnosing and monitoring of different neurological disorders, e.g. Alzheimer's disease [10,17] or epilepsy [3].

Photosensitivity is an abnormal sensitivity of the brain that provokes electrical epileptic discharges called Photoparoxysmal Responses (PPR) as a response to certain visual stimuli, like flashing lights or light reflections, response. The literature [18] has set 4 different types of PPR depending on the spreading and the waveform of the provoked discharge (see Fig. 1). The 4 types of PPR are named i) Type-1 PPR –with spikes in the occipital region–, ii) Type-2 PPR –showing spikes followed by a biphasic slow wave in occipital and parietal regions–, iii) Type-3 PPR –denoted by spikes followed by a biphasic slow wave in occipital and parietal regions and spread to frontal regions–, and iv) Type-4 PPR – characterized by generalized poly-spikes and waves–. Among them, Type-4 PPR is the most dangerous one, as it can lead to a real epileptic seizure. Besides, in real scenarios, the PPR characteristics varies from one individual to other and with several clinical variables –such as medical treatment, sleep quality, time of day, etc.–, making the identification even harder.

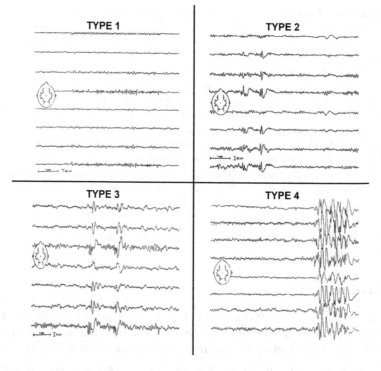

Fig. 1. The four types of PPR: **top-left** corresponds to Type-1 (spikes within the occipital rhythm); **top-right** corresponds to Type-2 (parieto-occipital spikes with biphasic slow wave); **bottom-left** corresponds to Type-3 (parieto-occipital spikes with biphasic slow wave and spread to the frontal region); **bottom-right** corresponds to Type-4 (generalized spikes and waves).

This study is a continuation of a previous study in [9], comparing different Machine Learning (ML) techniques for the automatic detection of Type-4 PPR in EEG recordings from patients who have been clinically diagnosed with photosensitivity disorder with different intensity. The final goal is to find the best method to detect the different types of PPR in order to design an automatic PPR detection system than can be used in clinical procedures.

The structure of this study is as follows. The next section focuses on some preliminaries knowledge and the related work the automated PPR detection. Section 3 describes the modelling alternatives considered in this research, detailing the different methods. Section 4 gives details of the experimentation set-up that has been carried out, while Sect. 5 includes all the obtained results and the discussion on them. The final section draws the conclusion of this research.

2   Preliminaries and Related Work

Photosensitivity evaluation is diagnosed using the Intermittent Photic Stimulation (IPS) [8,12], where the patient -monitored with EEG- is stimulated with flashing lights at different frequencies, increasing first and decreasing later, so to detect the frequency thresholds where the individual is sensitive. The detection and analysis of PPR is carried out manually by the clinical neurophysiologists and nurses, who perform a visual inspection of the patient's whole EEG recording searching these phenomena while taking into account their clinical context.

To our best knowledge, no automated method for the detection of PPR using the standard stimulation system has been developed so far. [14] designed a PPR detection method by analysing the brain response provoked by a flashing stimulation, but following a different stimulation pattern from the standard IPS used in clinical diagnosis. There are other recent studies that analyse the photosensitivity and epilepsy based directly on generalized seizures: in [11], a detection method based on the fluctuation from a high-frequency and a low-frequency components in each EEG channel is proposed; [16] applied the Extreme Gradient Boost technique for the classification of seizures, while a channel-independent Long Short-Term Memory Network is used in [2]; the combined information extracted from both EEG and electrocardiogram (ECG) signals is used in [20] in a multi-modal Neural Network; in [3], K-Nearest Neighbours and Artificial Neural Networks are used for the detection of ictal discharges and inter-ictal states; [19] proposed an EEG single-channel analysis applying three types of visibility graphs to represent different EEG patterns. Other studies make use of additional and different biometric measures for the same purpose, such as ECG [5,6,15], electromyograms (EMG) [1,21] or magnetoencefalograms (MEG) [13].

In our previous study [9], which was published in the Neural Computing and Applications journal, a proof of concept about Type-4 PPR detection was proposed. The PPR detection task was performed by two classifiers: an unsupervised 1C-KNN and a supervised 2C-KNN. 1C-KNN classified the EEG windows as normal or abnormal and 2C-KNN classified only the abnormal windows as PPR or not. For that purpose, the EEG recordings were divided into two sets:

the first one included all the EEG windows before the first PPR of the current subject labelled as normal an was used for the 1C-KNN training; the second one was formed by normal and PPR windows from the other subjects and was used for the 2C-KNN training. This process is shown in Fig. 2.

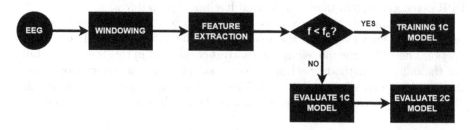

Fig. 2. When a EEG window comes from frequencies smaller than the stimulation frequency value at which the first PPR appears (fc, the cut frequency), the window is preserved for the training of the one-class models; otherwise, the window is labeled as normal or as anomaly. In this latter case, the two-class classifier labels then the anomaly window as PPR or not.

This ML procedure was used in the analysis of the effects of Virtual Reality in photosensitivity. The exposure of photosensitive and epilepsy patients to digital activities and environments has become a major concern with the huge proliferation of this technology in the last years [4].

3   Type-4 PPR Detection Using ML

The process we propose are designed following the steps shown in Fig. 3. For this study, we focused in channel Fz because is one of the EEG channels where PPRs express themselves best.

Firstly, sliding windows of 1-s length and 90% overlapping were extracted from the raw Fz channel and preprocessed (mean subtraction and Band-Pass Butterworth Filter in the range 1–45 Hz). Due to the high imbalance of the original data (only 3% of EEG windows are PPR), undersampling is computed to reduce it, reaching a PPR ratio of 10%. Then, a dimensional reduction step is performed, where a set of transformations and features are extracted from the EEG windows and Principal Component Analysis (PCA) algorithm is used to reduce the dimensions even more. The resulting components are grouped into clusters and finally analysed and classified by ML algorithms as normal or Type-4 PPR window.

This procedure is detailed in the next sections: the dimensional reduction process is detailed in Subsect. 3.1 and the following clustering of the reduced data and the selection and configuration of the ML models are explained in Subsect. 3.2.

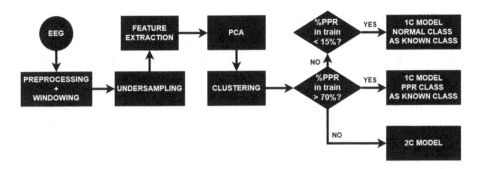

Fig. 3. The workflow of the designed experiment. The original EEG data are preprocessed and windowed and the imbalance of the data is reduced by using undersampling. The features are extracted from each EEG window. PCA and K-means are computed for dimensional reduction and clustering respectively. Depending on the proportion of PPR windows within each cluster, a different classification model is chosen.

3.1   Dimensional Reduction

Depending on the sampling rate used in the EEG recordings, the windows could include a large number of samples, which could make the tasks of the classifiers more difficult. Moreover, finding variables that present significant differences between the normal brain activity and the PPR windows could make the detection easier. Because of that, this step is focused in dimensional reduction of the processed EEG signal. A total of 32 features from temporal, statistical and spectral domains are extracted from each window.

- **Temporal Domain**: Sum of Absolute Values, Maximum Difference, Sum of Absolute Differences, Total Energy, Absolute Energy, Area Under the Curve, Entropy and Autocorrelation.
- **Statistical Domain**: Kurtosis, Skewness, Standard Deviation, Variance, Maximum, Minimum, Mean of Absolute Deviation, Root Mean Square.
- **Spectral Domain**: Fundamental Frequency, Maximum Frequency, Median Frequency, Maximum Power Spectrum, Centroid, Decrease, Spread, Distance, Skewness, Entropy, Kurtosis, Positive Turning, Roll-Off, Roll-On, Variation, Bandwith, Human Range Energy.

A cross-correlation vector between each feature and the real labels allows to find the most relevant and representative of Type-4 PPR features, which turned out to be Spectral Distance, Maximum Difference or Standard Deviation, among others. Then, a cross-correlation matrix between all features allows to analyze their independence: the correlation value is extremely high between the best features, which means that the represent the Type-4 PPR equally well.

Then, PCA algorithm is computed to reduce these 32 features into a even smaller set. PCA is the most widely used dimension-reducing technique [7]. This method extracts uncorrelated linear combinations from the initial dataset called Principal Components, which have the maximum variance possible of the original data.

3.2   Clustering and Classification

Clustering is a ML technique that divides samples into different groups based on their similarity. After the feature extraction and the dimensional reduction performed by PCA, the new transformed set presents higher distinctions between normal and PPR windows, so grouping the data into clusters allows to distinguish better between the different classes. This step is performed by K-means algorithm after dividing the data into training set and test set. Given the training data, K-means algorithm is used to create the appropriate number of clusters (following the "elbow rule"), and testing data are then associated to the most suitable cluster.

Once the data are divided into the clusters, the proportion of PPR samples within the training data of each cluster is calculated. Depending on this value, the training and testing data of each cluster are used in a different Machine Learning technique following these rules:

- If the proportion of PPR samples within the cluster is lower than 15%, i.e., the normal class clearly predominates the cluster, a one-class (1C) classifier will be used and the normal class will be considered as the known class, training the model only with the normal samples.
- On the contrary, if the proportion of PPR samples within the cluster is higher than 75%, it will mean that the PPR class clearly predominates the cluster and a 1C classifier where the PPR class will be considered as the known class will be used, training the model only with the PPR samples.
- If the proportion of PPR samples within the cluster is in the range 15%–75%, a two-class (2C) classifier will be used and the model will be trained with all the cluster's training data.

The Machine Learning algorithms selected for this study were one-class K-Nearest Neighbours (1C-KNN) and one-class Support Vector Machine (1C-SVM) as the one-class classifiers; and two-class K-Nearest Neighbours (2C-KNN), two-class Support Vector Machine (2C-SVM), Random Forest (RF) and Neural Network with Dense Layer as the hidden layer (DL-NN) as the two-class classifiers.

4   Materials and Methods

This section introduces the data set used in this research and describes the design of the experiments. The whole process was implemented in Python.

4.1   Data Set Description

The dataset used in this research has been gathered from Burgos University Hospital. It includes ten anonymized EEG recorded sessions from different photosensitive patients recorded with the hospital's own equipment. Among these patients, four did not present Type-4 PPR but many PPR from the other categories, so they were excluded from this study.

Each session consisted of a continuous recording while the first half of the conventional IPS procedure was performed: the frequency of the stimulation was only increased in the range 1–50 Hz, but the half corresponding to the descending frequencies half was not included. The duration of each session varies in the range 3–5 min. The EEG signals were recorded at a sampling rate of 500 Hz from 19 electrodes placed following the 10–20 standardized system, as shown in Fig. 4.

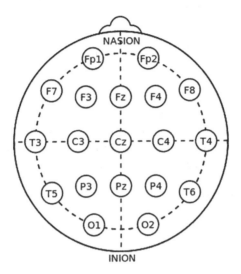

Fig. 4. Position of the 19 scalp electrodes used to record EEG signals according to the international 10–20 system of electrode placement, where Nasion is located at the center of the frontonasal area and the Inion in located at the posterioinferior part of the skull.

The selected EEG recordings were manually labelled by visual analysis, marking every PPR that can be categorized as Type-4. The proportion of both class instances within the dataset is showed in Table 1.

Table 1. Distribution of instances for each class within the original dataset.

N° Windows	N° Normal Windows	N° PPR Windows
20130	19510 **96'92%**	620 **3'08%**

4.2 Experimentation Design

For this experiment, a leave-one-out 10-fold Cross-Validation scheme is used (90% train + 10% test) with two different sets of 9 and 12 components created by PCA algorithm after reducing the set of 32 extracted features (corresponding to the 90% and 95% of the variance respectively). Both sets are used for the following process independently.

Once the data of each fold are divided into the respective fold's clusters, the proportion of PPR samples within the training data of each cluster is calculated, the corresponding classification algorithms are executed using the cluster's data and each classifier is trained and evaluated with the train data and the test data associated to its cluster, following the rules previously explained.

For each classifier, different parameter values are also tested:

- For KNN algorithm, the K number of nearest neighbours tested are 3, 5, 7, 9 and 11.
- For SVM algorithm, the parameters to be tested are the regularization parameter C and the kernel coefficient γ. The values tested are $C = [0.1, 1, 10, 100, 1000]$ and $\gamma = [1, 0.1, 0.01, 0.001, 0.0001]$
- For DL-NN, the N number of neurons of the hidden layer tested are 10, 20, 30, 40 and 50. The number of neurons of input and output layers have been fixed to the number of input variables and 1, respectively.
- For RF, the T number of trees tested are 100, 200, 300, 400 and 500; and as for the depth of the trees, the number of L levels tested are 5, 6, 7 and 8.

To measure the performance of the different classification techniques, the Accuracy (*Acc*), the Sensitivity (*Sens*), and the Specificity (*Spec*) measurements will be calculated.

5   Results and Discussion

The results gathered in this research consist of the Acc, Sens and Spec values of all classifiers applied in each cluster into which each fold is divided and with each of the proposed parameter values. Since there are multiple combinations of algorithms with their results, Table 2 shows the best Acc, Sens and Spec values obtained in each fold by each classifier: the upper table corresponds to the one-class classifiers applied in the class-dominant clusters and the lower table corresponds to the two-class classifiers applied in the balanced clusters.

The shown results are just one example of the best-performing case. Analyzing the overall results of all classifiers, on the one hand, the results from the one-class classifiers reveal that both of these methods are not the most suitable ones for this procedure: 1C-KNN method presents high *Sens* but low *Spec* values, which means that it tends to label the new data as outliers from the known normal class; on the contrary, 1C-SVM method presents high *Spec* and low *Sens* values, which means that it cannot distinguish the PPR class from the known normal class. The higher *Acc* values from 1C-SVM are due to the data imbalance.

On the other hand, the two-class algorithms present *Acc* and *Sens* values in the range 60%–80% and *Spec* values in the range 70%–90% for various parameter values. These results are higher than the one-class ones, which means that they

Table 2. Best Type-4 PPR detection results of each ML algorithm proposed in each fold. The upper table corresponds to the one-class classifiers (1C-KNN and 1C-SVM), while the lower table corresponds to the two-class classifiers (2C-KNN, 2C-SVM, RF, DL).

P_i	1C-KNN			1C-SVM		
	Acc	Sens	Spec	Acc	Sens	Spec
P_1	0.5175	0.8462	0.4255	0.7339	0.3077	0.8750
P_2	0.3634	0.5353	0.4276	0.7716	0.1087	0.8750
P_3	0.5846	0.8542	0.5306	0.6648	0.1667	1.0000
P_4	0.3208	0.6389	0.5206	0.6038	0.1042	1.0000
P_5	0.4910	0.7667	0.6280	0.8840	0.1667	1.0000
P_6	0.4590	0.7321	0.4245	0.7680	0.2500	1.0000
P_7	0.4605	0.6571	0.4367	0.7503	0.2387	1.0000
P_8	0.4732	0.6727	0.6796	0.7943	0.3824	0.5833
P_9	0.4621	0.8125	0.3377	0.8048	0.3125	0.9176
P_{10}	0.4192	0.6489	0.5523	0.7399	0.2426	1.0000

P_i	2C-KNN			2C-SVM			RF			DL-NN		
	Acc	Sens	Spec	Acc	Sens	Spec	Acc	Sens	Spec	Acc	Sens	Spec
P_1	0.7089	0.6279	0.8056	0.7342	0.6279	0.8611	0.7342	0.6279	0.8611	**0.7722**	**0.6977**	**0.8611**
P_2	0.7500	0.6786	0.7917	0.7500	0.6071	0.8333	0.7763	0.6071	0.8750	**0.7763**	**0.6786**	**0.8333**
P_3	0.7703	0.7576	0.7805	0.7973	0.6970	0.8780	0.8243	0.6970	0.9268	**0.8243**	**0.6970**	**0.9268**
P_4	**0.8028**	**0.8636**	**0.7755**	0.8028	0.7273	0.8367	0.8028	0.7727	0.8163	0.7465	0.7727	0.7347
P_5	**0.8250**	**0.7561**	**0.8974**	0.7875	0.6829	0.8974	0.8375	0.7073	0.9744	0.8000	0.7561	0.8462
P_6	0.8219	0.8276	0.8182	0.7671	0.7931	0.7500	**0.8356**	**0.8276**	**0.8409**	0.7808	0.8276	0.7500
P_7	**0.7711**	**0.7692**	**0.7727**	0.7831	0.6923	0.8636	0.7590	0.7179	0.7955	**0.7711**	**0.7692**	**0.7727**
P_8	0.6944	0.6061	0.7692	0.6944	0.6970	0.6923	0.7361	0.6364	0.8205	**0.7361**	**0.6667**	**0.7949**
P_9	**0.8462**	**0.8649**	**0.8333**	0.8571	0.8378	0.8704	0.8352	0.8378	0.8333	0.8352	0.8649	0.8148
P_{10}	**0.7831**	**0.6857**	**0.8542**	0.7831	0.6571	0.8750	0.7711	0.0286	0.8750	0.7711	0.7143	0.8125

are able to distinguish both classes better. Moreover, the parameter values that produce the best detection performance for each classifier are the following ones: $K = 3$–5 for 1C-KNN, $\gamma = 0.0001$ for 1C-SVM, $K = 9$ for 2C-KNN, $C = 1$ and $\gamma = 0.01$ for 2C-SVM, $T = 100$–200 and $L = 5$–6 for RF, and $N = 20$–30 for DL-NN.

If the class imbalance of the data is taken into account, which makes *Spec* measure easier to raise than *Sens*, the best performing methods for Type-4 PPR detection are 2C-KNN and DL-NN.

6 Conclusions and Future Work

In this research, we compared different ML algorithms for Type-4 PPR detection. For this purpose, a set of 32 features from time, spectral and statistical domains was extracted from EEG windows and PCA technique was applied for dimensional reduction. Then, the instances were divided into clusters and a certain set of classifiers was used in each cluster according to the proportion of PPR samples within the cluster: 1C-KNN and 1C-SVM were applied in those clusters in which one class was predominant; otherwise, 2C-KNN, 2C-SVM, RF and DL-NN were used.

The ML results show that two-class classifiers can detect Type-4 PPR better than one-class classifiers, and among them, the best detection algorithms are 2C-KNN and DL-NN. Furthermore, comparing this procedure with the one tested in our previous research [9], the Type-4 PPR detection performance has been increased by around 15%, but it is not as high as expected and there is still room for improvement.

Due to the lack of a large and appropriate EEG dataset, the results are not good enough by employing classic ML algorithms. Improving the results of this research would be possible by artificially increasing the available dataset through data augmentation techniques, which allow creating more EEG recordings with Type-4 PPRs from those currently available. Re-evaluating the procedure proposed in this study with the new data may allow determining whether a plausible solution can be found using classical ML.

In addition, Deep Learning techniques can also be tested along with the data augmentation technique, such as the classification of time series using LSTM-RNN (Recurrent Neural Networks) or an autoencoder. All of these proposals represent future research work.

References

1. Beniczky, S., Conradsen, I., Henning, O., Fabricius, M., Wolf, P.: Automated real-time detection of tonic-clonic seizures using a wearable EMG device. Neurology **90**, e428–e434 (2018). https://doi.org/10.1212/WNL.0000000000004893
2. Chakrabarti, S., Swetapadma, A., Pattnaik, P.K.: A channel independent generalized seizure detection method for pediatric epileptic seizures. Comput. Methods Programs Biomed. **209**, 106335 (2021). https://doi.org/10.1016/j.cmpb.2021.106335. https://linkinghub.elsevier.com/retrieve/pii/S0169260721004090
3. Choubey, H., Pandey, A.: A combination of statistical parameters for the detection of epilepsy and EEG classification using ANN and KNN classifier. SIViP **15**(3), 475–483 (2020). https://doi.org/10.1007/s11760-020-01767-4
4. Fisher, R.S., et al.: Visually sensitive seizures: an updated review by the epilepsy foundation. Epilepsia **63**(4), 739–768 (2022). https://doi.org/10.1111/epi.17175
5. Jahanbekam, A., et al.: Performance of ECG-based seizure detection algorithms strongly depends on training and test conditions. Epilepsia Open **6**, 597–606 (2021). https://doi.org/10.1002/epi4.12520
6. Jeppesen, J., et al.: Seizure detection based on heart rate variability using a wearable electrocardiography device. Epilepsia **60**, 2105–2113 (2019). https://doi.org/10.1111/epi.16343
7. Jolliffe, I.T.: Principal Component Analysis. Encyclopedia of Statistics in Behavioral Science, 2nd edn, vol. 30 (2002). https://doi.org/10.2307/1270093
8. Kasteleijn-Nolst Trenite, D.: Photosensitivity and epilepsy. In: Mecarelli, O. (ed.) Clinical Electroencephalography, pp. 487–495. Springer, Cham (2019). https://doi.org/10.1007/978-3-030-04573-9_29
9. Moncada, F., et al.: Virtual reality and machine learning in the automatic photoparoxysmal response detection. Neural Comput. Appl. (2022). https://doi.org/10.1007/s00521-022-06940-z

10. Morrison, C., Rabipour, S., Taler, V., Sheppard, C., Knoefel, F.: Visual event-related potentials in mild cognitive impairment and Alzheimer's disease: a literature review. Curr. Alzheimer Res. **16**(1), 67–89 (2018). https://doi.org/10.2174/1567205015666181022101036. http://www.eurekaselect.com/166483/article
11. Omidvarnia, A., Warren, A.E., Dalic, L.J., Pedersen, M., Jackson, G.: Automatic detection of generalized paroxysmal fast activity in interictal EEG using time-frequency analysis. Comput. Biol. Med. **133**, 104287 (2021). https://doi.org/10.1016/j.compbiomed.2021.104287
12. Rubboli, G., Parra, J., Seri, S., Takahashi, T., Thomas, P.: EEG diagnostic procedures and special investigations in the assessment of photosensitivity. Epilepsia **45**(5), 35–39 (2004). https://doi.org/10.1111/j.0013-9580.2004.451002.x
13. Soriano, M.C., et al.: Automated detection of epileptic biomarkers in resting-state interictal meg data. Front. Neuroinform. **11**, 43 (2017). https://doi.org/10.3389/fninf.2017.00043
14. Strigaro, G., Gori, B., Varrasi, C., Fleetwood, T., Cantello, G., Cantello, R.: Flash-evoked high-frequency EEG oscillations in photosensitive epilepsies. Epilepsy Res. **172**, 106597 (2021). https://doi.org/10.1016/j.eplepsyres.2021.106597
15. Ufongene, C., Atrache, R.E., Loddenkemper, T., Meisel, C.: Electrocardiographic changes associated with epilepsy beyond heart rate and their utilization in future seizure detection and forecasting methods. Clin. Neurophysiol. **131**, 866–879 (2020). https://doi.org/10.1016/j.clinph.2020.01.007
16. Vanabelle, P., Handschutter, P.D., Tahry, R.E., Benjelloun, M., Boukhebouze, M.: Epileptic seizure detection using EEG signals and extreme gradient boosting. J. Biomed. Res. **34**, 228 (2020). https://doi.org/10.7555/JBR.33.20190016
17. Vecchio, F., et al.: Classification of Alzheimer's disease with respect to physiological aging with innovative EEG biomarkers in a machine learning implementation. J. Alzheimer's Disease **75**, 1253–1261 (2020). https://doi.org/10.3233/JAD-200171
18. Waltz, S., Christen, H.J., Doose, H.: The different patterns of the photoparoxysmal response - a genetic study. Electroencephalogr. Clin. Neurophysiol. **83**, 138–145 (1992). https://doi.org/10.1016/0013-4694(92)90027-F
19. Wang, L., Long, X., Arends, J.B., Aarts, R.M.: EEG analysis of seizure patterns using visibility graphs for detection of generalized seizures. J. Neurosci. Methods **290**, 85–94 (2017). https://doi.org/10.1016/j.jneumeth.2017.07.013
20. Yang, Y., et al.: A multimodal AI system for out-of-distribution generalization of seizure detection (2021). https://doi.org/10.1101/2021.07.02.450974
21. Zibrandtsen, I.C., Kidmose, P., Kjaer, T.W.: Detection of generalized tonic-clonic seizures from ear-EEG based on EMG analysis. Seizure **59**, 54–59 (2018). https://doi.org/10.1016/j.seizure.2018.05.001

Smartwatch Sleep-Tracking Services Precision Evaluation Using Supervised Domain Adaptation

Enrique A. de la Cal[1(✉)], Mirko Fáñez[1], M. Dolores Apolo[2],
Andrés García-Gómez[2], and Víctor M. González[3]

[1] Computer Science Department, Faculty of Geology, University of Oviedo,
Oviedo, Spain
delacal@uniovi.es, mirko@mirkoo.es
[2] University of Extremadura, 09001 Badajoz-Cáceres, Spain
mdapolo@unex.es
[3] Electrical Engineering Department, EPI, University of Oviedo, Gijón, Spain
vmsuarez@uniovi.es

Abstract. In 2021 a 6-week clinical pilot was deployed to evaluate the effect of the Therapeutic Horse Riding Treatment (THRT) on the state of health of the recruited patients. Among all the clinical measures deployed to evaluate the effectiveness of the intervention, this work is focused on the sleep quality assessment. All the patients recruited have been worn with a WearOS® smartwatch 24 hours per day for the whole timeline, having the native sleep tracking app activated (SleepApp), with the further smartwatch 24 hours per day for the whole timeline, having the native sleep tracking app activated (SleepApp), with the further goal of analysing the effect of the intervention on the sleep quality of each patient. Nevertheless, before analysing this effect, it is essential to evaluate the accuracy of the sleep information computed by the SleepApp.

Thus, the main goal of this work is to estimate the accuracy of the sleep stages labels were obtained with the study smartwatch SleepApp. The proposed method is based on the supervised domain adaptation paradigm, where the source domain is the sleep dataset (SrcDataset) obtained with the smartwatch used in the pilot (SrcWatch), and the target domain is a dataset (TarDataset) available in a public repository, obtained as a combination of data from another smartwatch model (TarWatch) and the sleep stages obtained using PolySomnoGraphy (PSG) are considered as ground truth.

Finally, several classification machine learning models will be trained with the SrcDataset to obtain a sleep stages classification model tested on the TarDataset to evaluate the precision of the SleepApp sleep stages labels. The results show that although the best model trained with the SrcDataset sports a pretty good CV performance, the results obtained validating the former model but with the TarDataset are quite poor.

© Springer Nature Switzerland AG 2022
P. García Bringas et al. (Eds.): HAIS 2022, LNAI 13469, pp. 14–26, 2022.
https://doi.org/10.1007/978-3-031-15471-3_2

1   Introduction

In 2021 a clinical pilot trial was deployed on two groups of 10 people diagnosed with arthritis to evaluate the effect of the Therapeutic Horse Riding Treatment (THRT) on the state of health of the recruited patients. Both groups followed a programme of 12 sessions over six weeks, receiving the experimental group 12 THRT sessions, and the control group an equivalent Standardised Physical Exercise (SPE) programme. The effectiveness of the intervention program will be evaluated in terms of two groups of clinical measures: i) a first group of clinical measures, pain, joint mobility and balance, obtained at specific milestones during the pilot timeline, and ii) an actigraphy analysis through a multisensor WearOS® marketed smartwatch, triaxial accelerometer (3DACC) and Heart Rate (HR) sensor based on Photoplethysmography (PPG), worn 24 hours/day by all the patients during the whole pilot horizon.

Among all the clinical measures that can be obtained through an actigraphy analysis, this work focuses on the patient's sleep quality assessment. In the medical setting, traditional PolySomnoGraphy (PSG) has long been cited as the "gold standard" for sleep assessment, but sleep actigraphs are more affordable than polysomnographs, and their use has advantages, particularly in the case of extensive field studies [13]. However, actigraphy cannot be considered a substitute for polysomnography as it does not provide measures for brain activity (EEG), eye movements (EOG), muscle activity (EMG) or heart rhythm (ECG) [1]. But a PSG is not suitable for a long-term study such as this one. Thus, the native sleep tracking app installed in the smartwatches (SleepApp) has been activated, collecting sleep information from the 20 patients during the whole pilot duration, 23 hours approximately per day, to analyse the intervention's effect on the sleep quality of each patient. However, before analysing this effect, it is essential to evaluate the quality of the sleep information computed by the SleepApp. Therefore, the analysis of the effect of THRT on the patients' sleep quality will be tackled in further studies.

Thus, the main goal of this work is to estimate the accuracy of the sleep stages labelling obtained with the SleepApp built-in smartwatch. The proposed method is a supervised domain adaptation method, where:

- the source domain is the sleep dataset (SrcDataset) obtained with the smartwatch (SrcWatch) used in the pilot, and
- the target domain is a dataset (TarDataset) available in a public repository [15], obtained with a combination of another multisensor smartwatch data (TarWatch) and a PSG recorded simultaneously.

The SrcDataset includes the typical raw features used by smartwatches sleep Apps [3,5,15]: 3DACC components plus HR, and the sleep stages labels[1] (W, N1, N2, N3 and REM) while the TarDataset includes the same features as the

---

[1] Sleep can be split in five stages [12]: Wake, N1, N2, N3, and REM (Rapid Eye Movement). Stages N1 to N3 have considered non-rapid eye movement sleep, each progressively going into a deeper sleep. Sleep is arranged in 30-s epochs, and each of these epochs is assigned a specific sleep stage.

SrcDataset but labelled with the PSG considered as ground truth. Finally, several classification machine learning models will be trained with the SrcDataset to obtain a sleep stages classification model that will be tested on the TarDataset in order to evaluate the precision of the SleepApp sleep stages labels obtained with the SleepApp (see Fig. 1).

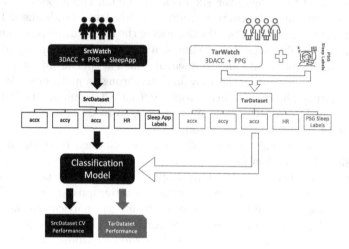

Fig. 1. The main idea

The structure of the paper is as follows. The following section deals with the description of the proposal. Section 3 describes the details of the pilot, the used datasets, and the experimental setup and shows and discusses the obtained results. Finally, conclusions and future work are drawn.

2   The Proposal

This proposal needs to be fed with two datasets: i) the SrcDataset that includes the RAW features accx, accy, accz, HR, and the sleep stages estimated with the smartwatch's (now on SrcWatch) native SleepApp that will be evaluated, and on the other side ii) the TarDataset that includes the same RAW features collected with a smartwatch (now on TarWatch) worn by the participants undergoing a PSG.

Figure 2 shows the complete procedure of the method proposed in this work. In the following subsection each step will be detailed.

Fig. 2. The general schema of the proposal

2.1   Step1: RAW Signals Preprocessing

During the pre-processing steps, the motion artifacts will be removed, and both datasets will be resampled to a homogeneous sampling rate for all the raw features to simplify the rest of the steps. Besides, the four RAW features time series will be windowed in 30-s samples with an overlapping step of 15-s. 30 s is the common window size used in the PSG technology as well the sleep analysis literature [3,15].

Raw Signals Motion Artifact Removal. Photoplethysmography (PPG) is an optical technology that uses LED as a Light Source (LS) and PhotoDetector (PD) as a receiver to capture the reflected LS calculating volumetric changes in blood to estimate HR [8]. However, as PPG signals are not robust due to Motion Artifacts (MA), produced by misalignment and wrong distance between the LS/PD and the user's wrist skin. Usually, a PPG sensor is built together with a 3DACC in a multisensor device like a smartwatch. Thus, PPG motion artifact removal techniques can be split into two [8]: i) those based on the 3DACC components, and ii) the ones based on the PPG peaks. We propose a new motion artifacts removal algorithm based on the Heart Rate Recovery (HRR) concept. It is supposed that the HR cannot increase or decrease faster than some gradient. Thus, we have estimated a logarithmic HRR-Threshold Curve (HRR-TC) that correlates Recovery Time (secs) and HR Variation (bpm) (see Fig. 3a). The source data to build the curve have been taken from a clinical trial on HRR with 40727 participants [14], where Table 1 shows the obtained average results:

Table 1. HRR average results

	Recovery time (sec)	HR variation (bpm)	HRR-TC HR variation (bpm)
HRR10	10	18.4 ± 7.7	26.1
HRR20	20	24.2 ± 8.9	33.1
HRR30	30	28.6 ± 9.6	38.2
HRR40	40	31.9 ± 10.1	42.0
HRR50	50	34.3 ± 10.4	44.7

In order to decide whether an HR window of 30-s (W) includes or not a MA, the HRR-TC is evaluated for the absolute value of the Recovery time (Rec_t_W = |HR_Max_t-HR_Min_t|) secs) between the maximum (HR_Max_t, HR_Max) and minimum (HR_Min_t, HR_Min) points of W. And if the absolute difference between the HR_Max and HR_Min (Var_HR_W=|HR_Max-HR_Min|) bpm) overpasses the HRR-TC evaluated on Rec_t_W secs, window W should be removed from the dataset as well as the corresponding ACC components windows. Thus, this curve was evaluated for the whole 30-s HR windows obtained after the windowing process. For example, in Fig. 3b, the Var_HR_W between the maximum and minimum bpm values is 55 bpm in Rec_t_W of 15.10 s. If the HRR-TC is evaluated for 15.10 s, a HRR-Threshold of 30.47 bpm is obtained, so as 55 bpm is higher than 30.47 bpm, this window should be removed from the dataset, as well as the three ACC components windows.

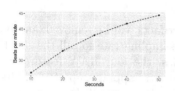 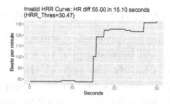

(a) HRR threshold curve (HRR-TC)   (b) Example of window with an artifact

Fig. 3. Motion artifacts identification

Raw Signals Normalization. In order to reduce the over-fitting of the learned models is important to normalise the time series. Thus, we have considered the work of [11], where five popular normalisation techniques (such as decimal scaling, median, min-max, vector and z-grade) for higher-order neural networks (HONNs) training are analysed, obtaining the best results for the vector and decimal scaling alternatives. Therefore, we have selected by its simplicity the vector normalisation technique (see Eq. 1).

$$N_i = \frac{T_i}{\sqrt{\sum_{j=1}^{k} T_j^2}} \tag{1}$$

2.2   Step2: Features Computing

Table 2 shows the calculated features for each normalised 30-s window which includes the 3DACC x, y and z axes (a_x, a_y, a_z), heart rate (HR) and accelerometer transformations MAGACC (Eq. 2), AoM (Eq. 3) and Arm Angle (Eq. 4) (the last 2 calculated in 3-s sub-windows).

$$MAGACC_t = \sqrt{ax_t^2 + ay_t^2 + az_t^2}, \text{ with } t \text{ epoch time} \tag{2}$$

$$AoM_t^{t+3} = max(MAGACC_t^{t+3}) - min(MAGACC_t^{t+3}), \text{ with } t \text{ in secs} \tag{3}$$

$$ArmAngle_t^{t+3} = \frac{180}{\pi} * arctan(\frac{MEAN_t^{t+3}(a_z)}{\sqrt{MEAN_t^{t+3}(a_x)^2 + MEAN_t^{t+3}(a_y)^2}}) \tag{4}$$

2.3   Steps 3 and 4: Models Training and Domain Adaptation

Before carrying out the domain adaptation process to the target dataset, a conservative strategy to obtain the optimal model has been designed (see Fig. 4 [In Black]): two kinds of source models have been trained and validated taking as input the pre-processed SrcDataset:

- First, one model trained and tested with the data of one patient has been obtained (Best1PModel), and
- A Leave One Out (LOO) user-centred validation strategy has been deployed with a selection of S patients' data to obtain a more general model (Best-LOOModel).

Table 2. Table of features. Each feature applied to several measures m. 45 transformations in total are calculated.

Feature	Equation	Applied to m	Description
Mean	$\frac{1}{N}\sum_{n=1}^{N} m(n)$	$a_x, a_y, a_z,$ MAGACC, AoM, ArmAngle, HR	Sum of window values divided by window size N
Standard deviation	$\sqrt{\frac{1}{N-1}\sum_{n=1}^{N}(m(n)-MEAN_m)^2}$	$a_x, a_y,$ a_z,MAGACC, AoM,ArmAngle, HR	Amount of dispersion in window m values
Correlation	$\frac{1}{N-1}\sum_{n=1}^{N}\frac{(a1_n-MEAN_{a1})*(a2_n-MEAN_{a2})}{STD_{a1}*STD_{a2}}$	a_x, a_y, a_z	Statistical relationship between 2 different acc. axes $a1$ and $a2$
Kurtosis	$\frac{E[(m-MEAN_m)^4]}{(STD_m)^2}-3$	$a_x, a_y, a_z,$ MAGACC	Standardized fourth central moment of a distribution minus 3
Crest factor	$\frac{max(m)}{\sqrt{\frac{1}{N-1}\sum_{n=1}^{N}m(n)^2}}$	$a_x, a_y, a_z,$ MAGACC	Quantification of measure m peaks
Skewness	$E\left[\left(\frac{m-MEAN_m}{STD_m}\right)^3\right]$	$a_x, a_y, a_z,$ MAGACC	Asymmetry value of measure m about its mean
Zero crossing	$\lvert(2\leq n\leq N)\wedge(m(n)*m(n-1)<0)\rvert$	$a_x, a_y, a_z,$ MAGACC	Amount of measure m sign changes
Entropy	$-\sum_k (p_m(k)*ln(p_m(k)))$	$a_x, a_y, a_z,$ MAGACC	Measure of the information content of measure m. p_m is the probability mass function of m
Band energy	$\frac{\sum_{n=1}^{N} m'(n)}{\sum_{n=1}^{N} m(n)}$	$a_x, a_y, a_z,$ MAGACC	The energy E of normalised subband signal m' (0.25-3Hz) normalised by total E
Spectral flux	$\sum_{n=2}^{N}(m(n)-m(n-1))^2$	$a_x, a_y, a_z,$ MAGACC	Measure of how quickly the power spectrum of the signal m changes

The Best1PModel has been obtained after a comparison of several well-known ML classification algorithms validated with a CV schema (It will be detailed in Materials and methods Sect. 3.1). After computing the Best1PModel, it will be deployed on the TarDataset to check if the labels of the SrcDataset allow to obtain a model suitable for the TarDataset. The same procedure will be run for the BestLOOModel with the TarDataset in case the performances of Best1PModel and BestLOOModel were quite similar (see Fig. 4 [In Blue]).

3 Numerical Results

3.1 Materials and Methods

The SrcDataset. The pilot of the SrcDataset lasted six weeks between September 2021 and November 2021, collecting data from 20 patients using TicWatch® E2 devices with a 3DACC and PPG sensors at 10 Hz and 1 Hz, respectively.

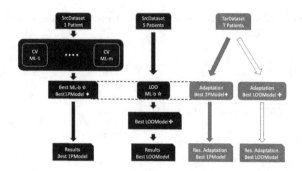

Fig. 4. Models train/test (in black) and adaptation (in blue). (Color figure online)

After the pilot deployment, 16 patients out of the 20 finalised the experiment due to different personal and/or health issues. Therefore, the study dataset described in Sect. 2 has been collected from this group of patients suffering from any arthritis disease, according to the inclusion and exclusion criteria stated in the trial registered in [2]. After analysing the complete SrcDataset, just the smartwatches of 4 patients out of the 16 kept significant sleep labels (10 h at least per sleep stage), so these 4 patients were selected for this experiment.

The TarDataset. TarDataset was collected at the University of Michigan [15], from a study group of 32 subjects worn with an Apple® Watch, and monitoring sleep concurrently via PSG. Triaxial acceleration and PPG HR were recorded from the Apple® Watch at 50 Hz and 1 Hz, respectively. In current work, 10 hours per sleep stage will be selected by merging all participants' data in order to get dataset size equivalent to SrcDataset.

Sleep Architectures Definition. Four sleep architectures have been selected (see Table 3) in order to check the performance of the models proposed in different contexts:

Table 3. Sleep stages architectures

Number of sleep stages	Sleep architecture
2	WAKE(W) and SLEEP(N1, N2, N3 and REM)
3	WAKE(W), NREM(N1, N2 and N3) and REM
4	WAKE(W), N1, NREM(N2 and N3) and REM
5	WAKE(W), N1, N2, N3 and REM

3.2 Experimentation Set up

The experimentation is divided in three steps: i) the first step is devoted to obtaining the Best1PModel, taking as train/test dataset the SrcDataset of the patient #1 out of the 4. Six classical machine learning algorithms, taking as input the 45 features of each 30-s window and as output the four sleep architectures described before, have been selected in order to compare the performance of our proposal: RandomForestClassifier [4], DecisionTree [7], KNN and MLP (Multi-Layer Perceptron) with three configurations [9] considering the hidden layer size suggestion by the geometric pyramid rule [10]. A 5×2 cross-validation has been carried out for each ML algorithm and sleep architecture; ii) after obtaining the Best1PModel, this model will be validated on the TarDataset to check the accuracy of the sleep labels obtained from the SrcWatch SleepApp; and iii) finally, taking the best ML algorithm obtained in the first step, a LOO user-centred validation will be carried out on the 4-Patients SrcDataset, and it will be considered whether or not to validate this model on the T-Patients TarDataset, according to the results of the Best1PModel with the TarDataset.

In addition, the performance measures used in all the experiments included in this section are: sensitivity, specificity and, as a way to summarise the results, the geometric mean of sensitivity and specificity (GM).

3.3   Numerical Results

RAW Signals Pre-processing. Table 4 shows the number of 30-s windows included in the complete original datasets considered in this study, before and after the Motion Artifacts removal. It can be stated that the MA algorithm detected more anomalies in the SrcDataset (-1.692%) than in the TarDataset (-0.426%). Thus, it can be concluded that data gathered from the Apple® Watch (TarDataset) are more accurate and clean of artifacts than the one collected from the TicWatch® (SrcDataset). After the removal process, 10 h of 30-s windows (1200 windows) have been randomly selected for each sleep stage for the experiments in the current work (1/4-Patients SrcDataset/TarDataset).

Table 4. Number of 30-s windows after the motion artifact removal

	W	N1	N2	N3	REM	Total	%Change
SrcDataset before MA removal	102645	16086	68368	67746	25780	280625	-1.692%
SrcDataset after MA removal	101235	15803	67110	66449	25319	275916	
1-Patient SrcDataset	1200	1200	1200	1200	1200	6000	
4-Patients SrcDataset	4800	4800	4800	4800	4800	24000	
TarDataset before MA removal	3573	1839	23397	6819	10714	46342	-0.426%
TarDataset after MA removal	3533	1785	23339	6816	10672	46145	
T-Patients TarDataset	1200–4800	1200–4800	1200–4800	1200–4800	1200–4800	6000–24000	

Results for 1-Patient SrcDataset 5x2CV Experiments. For the shake of simplicity, just the mean and standard deviation (Std) of the 1-Patient Src-Dataset 5×2CV experiments obtained with the selected ML algorithms with all the sleep architectures[2] have been included in Table 5. It can be stated that the

---

[2] For obvious reasons, 5 and 2-stages experiments do not include results for the Hierarchical Machine Learning alternatives (*-H).

RF model outperforms the remaining ML techniques for all the sleep architectures. Even thought the primary goal of this work is not to propose a competitive sleep staging algorithm, the results obtained are comparable with the baseline results from the literature [6]. The 5-stages results (see boxplots in Fig. 5) state that RF outperforms the remaining alternatives for each sleep stage. Nevertheless, the confusion matrix for the best CV fold (see Fig. 6) shows a good sensitivity for the WAKE stage but a moderate one for the remaining stages

Table 5. Summary results for 1-Patient SrcDataset 5 × 2CV experiments

ML technique	Measure	5 sleep stages		4 sleep stages		3 sleep stages		2 sleep stages			
		Mean	Std	Mean	Std	Mean	Std	Mean	Std	Mean	Std
RF	Sens	**0.66**	0.11	0.72	0.08	0.75	0.10	**0.89**	0.01	0.76	0.08
	Spec	**0.91**	0.01	0.91	0.02	0.87	0.03	**0.89**	0.01	0.90	0.02
DT	Sens	0.46	0.20	0.54	0.18	0.66	0.15	0.87	0.03	0.63	0.14
	Spec	0.86	0.05	0.85	0.05	0.83	0.06	0.87	0.03	0.85	0.05
KNN	Sens	0.54	0.11	0.61	0.08	0.67	0.11	0.84	0.04	0.67	0.09
	Spec	0.88	0.02	0.87	0.03	0.83	0.05	0.84	0.04	0.86	0.04
NN_6	Sens	0.29	0.24	0.71	0.10	0.43	0.16	0.65	0.15	0.52	0.16
	Spec	0.82	0.14	0.90	0.01	0.72	0.12	0.65	0.15	0.77	0.11
NN_15	Sens	0.30	0.17	0.35	0.32	0.47	0.17	0.69	0.09	0.45	0.19
	Spec	0.83	0.09	0.79	0.22	0.73	0.10	0.69	0.09	0.76	0.13
NN_30	Sens	0.32	0.15	0.37	0.16	0.49	0.11	0.71	0.09	0.47	0.13
	Spec	0.83	0.06	0.79	0.09	0.75	0.08	0.71	0.09	0.77	0.08
RF-H	Sens	-	-	0.41	0.12	0.75	0.11	-	-	0.58	0.12
	Spec	-	-	0.80	0.05	0.88	0.02	-	-	0.84	0.04
NN-H_6	Sens	-	-	0.35	0.29	0.41	0.35	-	-	0.38	0.32
	Spec	-	-	0.78	0.20	0.71	0.27	-	-	0.75	0.24
NN-H_15	Sens	-	-	0.38	0.18	0.46	0.23	-	-	0.42	0.21
	Spec	-	-	0.79	0.10	0.73	0.15	-	-	0.76	0.13
NN-H_30	Sens	-	-	0.41	0.20	0.50	0.17	-	-	0.46	0.19
	Spec	-	-	0.80	0.09	0.75	0.08	-	-	0.78	0.09

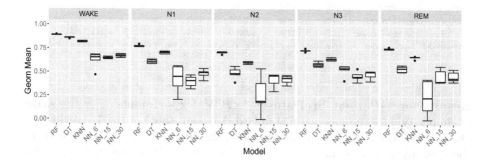

Fig. 5. 5-stages performance (GM) for 1-Patient SrcDataset 5 × 2CV experiments

(from N2 = 56% to N1 = 64%). Thus, the RF model for four sleep architectures will be selected as the winner ML technique for the 4-Patients LOO experiments and the domain adaptation learning step.

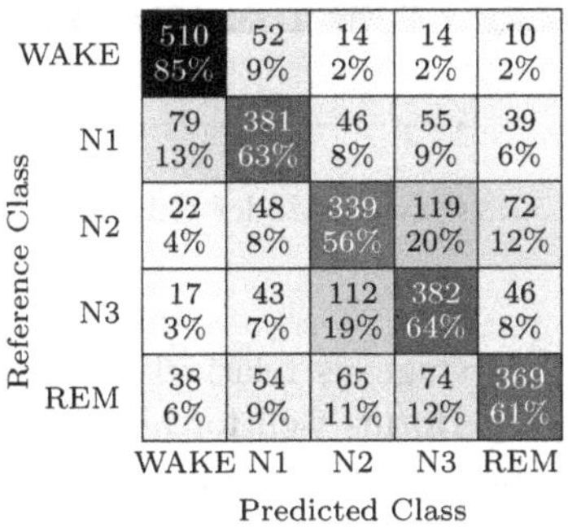

	WAKE	N1	N2	N3	REM
WAKE	510 85%	52 9%	14 2%	14 2%	10 2%
N1	79 13%	381 63%	46 8%	55 9%	39 6%
N2	22 4%	48 8%	339 56%	119 20%	72 12%
N3	17 3%	43 7%	112 19%	382 64%	46 8%
REM	38 6%	54 9%	65 11%	74 12%	369 61%

Reference Class (rows), Predicted Class (columns: WAKE N1 N2 N3 REM)

Fig. 6. Confusion matrix for 5-stages RF 5×2CV results (best CV fold)

Table 6. Summary of LOO validation with 4-Patients SrcDataset running the RF technique

		WAKE		N1		N2		N3		REM		Sens	Spec	GM
		Sens	Spec	Sens	Spec	Sens	Spec	Sens	Spec	Sens	Spec			
5-stages	Mean	0.70	0.90	0.20	0.86	0.34	0.75	0.27	0.84	0.20	0.84	0.34	0.84	0.54
	Std	0.11	0.01	0.03	0.05	0.12	0.10	0.12	0.07	0.03	0.02	0.08	0.05	0.06
		WAKE		N1		N2-N3				REM				
4-stages	Mean	0.71	0.87	0.20	0.86	0.54	0.64	-	-	0.22	0.85	0.42	0.81	0.58
	Std	0.13	0.02	0.02	0.02	0.06	0.03	-	-	0.05	0.03	0.07	0.03	0.04
		WAKE		N1-N2-N3						REM				
3-stages	Mean	0.72	0.87	0.63	0.60	-	-	-	-	0.26	0.84	0.54	0.77	0.64
	Std	0.11	0.02	0.02	0.04	-	-	-	-	0.02	0.02	0.05	0.03	0.04
		WAKE		SLEEP										
2-stages	Mean	0.75	0.84	0.84	0.75	-	-	-	-	-	-	0.80	0.80	0.80
	Std	0.11	0.05	0.05	0.11	-	-	-	-	-	-	0.08	0.08	0.08
	Mean	0.72	0.87	0.47	0.77	0.44	0.70	0.27	0.84	0.23	0.84	0.52	0.80	0.65
	Std	0.12	0.03	0.03	0.06	0.09	0.07	0.12	0.07	0.03	0.02	0.07	0.05	0.06

Results for 4-Patients SrcDataset LOO Experiments. The winner technique obtained in the previous step has been used in LOO experiments. In this case, data from 4 patients from the SrcDataset have been used in a LOO user-centred experiment[3] obtaining the results included in Table 6. The sensitivity

---

[3] Trained with 3 patients' data and tested with 1 patient's data, alternating the patients' 4 times.

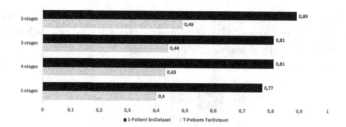

Fig. 7. Comparison of the $5 \times 2\text{CV}$ RF Model *GM* for 1-Patient SrcDataset vs T-Patients TarDataset

for the N1, N2, N3 and REM stages is relatively poor in all the sleep architectures, while the wake stage performs well for all the sleep architectures. Since this model sports terrible generalisation results, it will not be deployed on the TarDataset.

Domain Adaptation from the 1-Patient SrcDataset RF Model to the TarDataset. Figure 7 shows the GM for the RF model obtained after the $5 \times 2\text{CV}$ run with 1-Patient SrcDataset, compared with the same model validated with the T-Patients TarDataset. The performance of the RF model deployed on the T-Patients TarDataset is relatively low for every sleep architecture, with a GM under 0.5 in all the cases. So, it can be concluded that data and sleep stages labels collected from the SrcWatch do not allow learning a good model to be deployed on the ground-truth TarDataset labelled with the PSG.

4   Conclusions and Future Work

This work presents a methodology based on domain adaptation learning to check the accuracy of the sleep-tracking service of a marketed smartwatch (SrcWatch) deployed during a clinical pilot. The dataset obtained from the SrcWatch (SrcDataset) allows us to learn a sleep classification model that will be validated with another dataset (TarDataset) based on a combination of an Apple® Watch and a PSG considered ground truth for the sleep staging. However, even if the model obtained based on RF through $5 \times 2\text{CV}$ from the SrcDataset performs well for 1-Patient SrcDataset (All sleep architectures Mean GM = 0.82), the performance after deployed it on the TarDataset is relatively poor (GM = 0.44). Besides, a LOO user-centred experiment with the RF model using the SrcDataset has been carried out, obtaining low accuracy (GM = 0.65), so this model has not been deployed on the target dataset (TarDataset). Thus, it can be concluded that the accuracy of the sleep stages labels obtained with SrcWatch SleepApp for the clinical pilot is not quite good, so it is necessary to look for suitable alternatives.

One of the future issues to be deployed is carrying out the opposite process to the one presented in this work: training a model with the TarDataset and validating it with the SrcDataset. Other pending work is applying different anomaly detection algorithms (ADA) to obtain a cleaner signal and training the ADA with a clean dataset in a controlled environment.

Acknowledgement. This work has been supported by Ann Kern-Godal's Memorial Fund for Horse-Assisted Therapy under the grant "Clinical trial and data collection, multicenter, randomised trial, with experimental group, control group, and blind evaluation by third parties, in order to determine the effect of Therapeutic riding, on a group of people with Arthritis" (Grants Programme 2020). Also, the research has been funded by the Spanish Ministry of Economics and Industry, grant PID2020-112726RB-I00, by the Spanish Research Agency (AEI, Spain) under grant agreement RED2018-102312-T (IA-Biomed), and by the Ministry of Science and Innovation under CERVERA Excellence Network project CER-20211003 (IBERUS) and Missions Science and Innovation project MIG-20211008 (INMERBOT). Also, by Principado de Asturias, grant SV-PA-21-AYUD/2021/50994. By European Union's Horizon 2020 research and innovation programme (project DIH4CPS) under the Grant Agreement no 872548. And by CDTI (Centro para el Desarrollo Tecnológico Industrial) under projects CER-20211003 and CER-20211022 and by ICE (Junta de Castilla y León) under project CCTT3/20/BU/0002. Finally, as this is the first publication coming from the AINISE Pilot trial, the author's team would like to express our appreciation to the group of selfless collaborators that undertook all the essential tasks to make this pilot possible, led by Georgina Dieste.

References

1. Ancoli-Israel, S., et al.: The SBSM guide to actigraphy monitoring: clinical and research applications. Behav. Sleep Med. **13**(sup 1), S4–S38 (2015). https://doi.org/10.1080/15402002.2015.1046356
2. Apolo Arenas, D., De La Cal Marín, E.A., Garcia-Gomez, A., Hernández Rodríguez, I.: Clinical, Data Collection, Multicenter, Randomized Trial to Determine the Effect of Therapeutic Riding Compared With Physical Activity on a Group of People With Arthritis (ID: NCT05068050). Technical report (2021). https://clinicaltrials.gov/ct2/show/NCT05068050
3. Beattie, Z., et al.: Estimation of sleep stages in a healthy adult population from optical plethysmography and accelerometer signals. Physiol. Meas. **38**(11), 1968–1979 (2017)
4. Breiman, L.: Random forests. Mach. Learn. **45**(1), 5–32 (2001)
5. Fedorin, I., Slyusarenko, K., Lee, W., Sakhnenko, N.: Sleep stages classification in a healthy people based on optical plethysmography and accelerometer signals via wearable devices. In: 2019 IEEE 2nd Ukraine Conference on Electrical and Computer Engineering, UKRCON 2019 - Proceedings, pp. 1201–1204 (2019)
6. Grandner, M.A., Lujan, M.R., Ghani, S.B.: Sleep-tracking technology in scientific research: looking to the future. Sleep **44**(5), 1–3 (2021)
7. Hastie, T., Tibshirani, R., Friedman, J.: The Elements of Statistical Learning Data Mining, Inference, and Prediction. Springer Series in Statistics. Technical report (2001)
8. Ismail, S., Akram, U., Siddiqi, I.: Heart rate tracking in photoplethysmography signals affected by motion artifacts: a review. EURASIP J. Adv. Signal Process. **2021**(1), 1–27 (2021). https://doi.org/10.1186/s13634-020-00714-2

9. Kingma, D.P., Ba, J.L.: Adam: a method for stochastic optimization. In: 3rd International Conference on Learning Representations, ICLR 2015 - Conference Track Proceedings. International Conference on Learning Representations, ICLR (2015)
10. Masters, T.: Practical Neural Network Recipies in C++ (1993)
11. Panigrahi, S., Behera, H.S.: Effect of normalization techniques on univariate time series forecasting using evolutionary higher order neural network. Int. J. Eng. Adv. Technol. **3**(2), 280–285 (2013)
12. Patel, A.K., Reddy, V., Araujo, J.: Physiology, Sleep Stages (2022). https://www.ncbi.nlm.nih.gov/books/NBK526132/
13. Tang, N.K., Harvey, A.G.: Correcting distorted perception of sleep in insomnia: a novel behavioural experiment? Behav. Res. Ther. **42**(1), 27–39 (2004)
14. Van de Vegte, Y.J., Van der Harst, P., Verweij, N.: Heart rate recovery 10 seconds after cessation of exercise predicts death. J. Am. Heart Assoc. **7**(8), e008341 (2018)
15. Walch, O., Huang, Y., Forger, D., Goldstein, C.: Sleep stage prediction with raw acceleration and photoplethysmography heart rate data derived from a consumer wearable device. Sleep **42**(12), 1–19 (2019)

Tracking and Classification of Features in the Bio-Inspired Layered Networks

Naohiro Ishii[1(✉)], Kazunori Iwata[2], Naoto Mukai[3], Kazuya Odagiri[3], and Tokuro Matsuo[1]

[1] Advanced Institute of Industrial Technology, Tokyo 1400011, Japan
nishii@acm.org, matsuo@aiit.ac.jp
[2] Aichi University, Nagoya 453-8777, Japan
kazunori@aichi-u.ac.jp
[3] Sugiyama Jyogakuen University, Nagoya 464-8622, Japan
{naoto,kodagiri}@sugiyama-u.ac.jp

Abstract. Machine learning, deep learning and neural networks are extensively applied for the development of many fields. Though their technologies are improved greatly, they are often said to be opaque in terms of explainability. Their explainable neural functions will be essential to realization in the networks. In this paper, it is shown that the bio-inspired networks are useful for the explanation of tracking and classification of features. First, the asymmetric network with nonlinear functions is created based on the bio-inspired retinal network. They have orthogonal properties useful for the tracking of features compared to the conventional symmetric networks. Second the extended asymmetric networks are derived, which generate sparse coding for classification in the orthogonal subspaces. The sparse coding is realized in the extended layered asymmetrical networks. Finally, we classified Reuters collections data applying the explainable processing steps, which consist of the linear discriminations and the sparse coding with nearest neighbor relation for classification.

Keywords: Asymmetric network · Extended asymmetric layered networks · Explainability of functions in layered networks · Sparse coding · Tracking and classification for features spaces

1 Introduction

Recently, there has been a great deal of excitement and interesting in deep neural networks and sparse coding, because they have achieved breakthrough results in areas as machine learning, computer vision, neural computations and artificial intelligence [1–4]. However, due to the large complexity and highly nested structure of these deep networks and sparse coding, it is expected to be transparent, understandable and explainable in the multilayered structures [3]. Realizability of these functions is expected to describe in terms of variables of the network responses. Then, we can make clear how input variables have been changed by processing rules to the output decisions. In this paper, it is

© Springer Nature Switzerland AG 2022
P. García Bringas et al. (Eds.): HAIS 2022, LNAI 13469, pp. 27–38, 2022.
https://doi.org/10.1007/978-3-031-15471-3_3

shown that the bio-inspired networks are useful for the explanation of tracking and classification of features. First, the asymmetric network with nonlinear functions is created based on the bio-inspired retinal network. They have orthogonal properties useful for the tracking of features compared to the conventional symmetric networks [11]. Second the extended layered asymmetric networks are derived, which generate sparse coding in the orthogonal subspaces. The sparse coding is realized in the extended layered asymmetrical networks. Finally, experiments using Reuters collections as the real problems are performed to classify their data using linear discrimination and sparse coding for classification with nearest neighbor data, which consists of the linear discriminations and the sparse coding with the nearest neighbor relation [10].

2   Bio-Inspired Neural Networks

2.1   Background of Asymmetric Neural Networks Based on the Bio-Inspired Network

In the biological neural networks, the structure of the network, is closely related to the functions of the network. Naka et al. [8] presented a simplified, but essential networks of catfish inner retina as shown in Fig. 1. Visual perception is carried out firstly in the retinal neural network as the special processing between neurons.

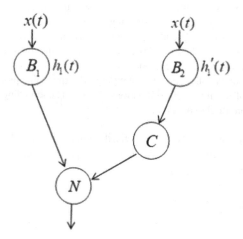

Fig. 1. Asymmetric network with linear and squaring nonlinear pathways

Visual perception is carried out firstly in the retinal neural network as the special processing between neurons. The following asymmetric neural network is extracted from the catfish retinal network [8]. The asymmetric structure network with a quadratic nonlinearity is shown which composes of the pathway from the bipolar cell B to the amacrine cell N and that from the bipolar cell B, via the amacrine cell C to the N [8, 9]. The asymmetric network plays an important role in the movement perception as the fundamental network. It is shown that N cell response is realized by a linear filter, which is composed of a differentiation filter followed by a low-pass filter. Thus, the asymmetric network is composed of a linear pathway and a nonlinear pathway with the cell C, which works as a squaring function.

2.2 Model of Asymmetric Networks

To make clear the characteristics of the asymmetric network with Gabor functions, orthogonality is computed, which implies independent relations created in the asymmetric structure network [11]. A model of the asymmetric network is shown in Fig. 2, in which impulse response functions of cells are shown in $h_1(t)$ and $h_1'(t)$. The $(\ )^2$ shows a square operation.

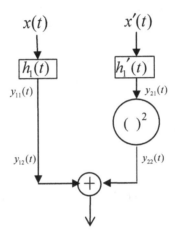

Fig. 2. Inner orthogonality of asymmetric network under stimulus condition

In Fig. 2, Gabor filters become to the following Eq. (1).

$$G_s(t') = \frac{1}{\sqrt{2\pi}\sigma}e^{-\frac{t'^2}{2\sigma^2\xi^2}}sin(t') \qquad G_c(t') = \frac{1}{\sqrt{2\pi}\sigma}e^{-\frac{t'^2}{2\sigma^2\xi^2}}cos(t') \qquad (1)$$

We assume here white noise $x(t)$, whose mean value is 0 and its power spectrum is p. The other white noise, $x'(t)$ is a high pass filtered one with zero mean and its power p'. The impulse response functions $h_1(t)$ and $h_1'(t)$ are replaced by the Gabor filters, $G_s(t')$ and $G_c(t')$ as shown in Eq. (1).

2.3 Tracking in the Asymmetric Networks

By applying the network in Fig. 3, the tracking characteristics for the input is experimented. The tracking output in the asymmetric network is described in Eq. (2). The output results for $y(t) = x(t)$ stimulus are shown in Fig. 4. The 2nd layered network is made of the iteration of the 1st layer network in Fig. 3, in which the tracking is well performed. The conventional energy model is shown in Fig. 5 [6] and in Eq. (3) under the same condition in Fig. 3. Then, the conventional neural network is insufficient for the exact tracking as shown in Fig. 6. But the tracking responses for y = 0.5 are well in the asymmetric and the symmetric networks as shown in Fig. 7.

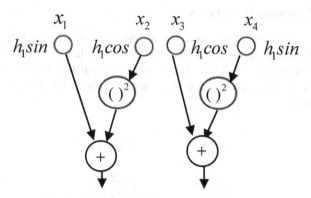

Fig. 3. Basic unit with orthogonal wavelet basis

$$y(t) = \int_0^T \left(\int_0^T h_1(\tau)sin(\tau)x_1(t-\tau)d\tau \right)dt + \int_0^T \left(\int_0^T h_1(\tau)cos(\tau)x_2(t-\tau)d\tau \right)^2 dt$$
$$+ \int_0^T \left(\int_0^T h_1(\tau)cos(\tau)x_3(t-\tau)d\tau \right)dt + \int_0^T \left(\int_0^T h_1(\tau)sin(\tau)x_4(t-\tau)d\tau \right)^2 dt$$

$$(2)$$

where the $h_1(t)$ is assumed to be Gaussian distribution $N(0, 1)$.

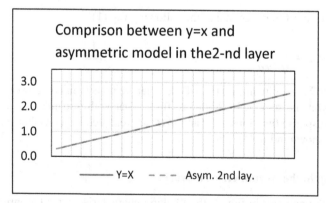

Fig. 4. Tracking characteristics for y = x in the asymmetric networks

$$y(t) = \int_0^T (\int_0^T h_1(\tau)sin(\tau)x_1(t-\tau)d\tau)^2 dt + \int_0^T (\int_0^T h_1(\tau)cos(\tau)x_2(t-\tau)d\tau)^2 dt$$

$$+ \int_0^T (\int_0^T h_1(\tau)cos(\tau)x_3(t-\tau)d\tau)^2 dt + \int_0^T (\int_0^T h_1(\tau)sin(\tau)x_4(t-\tau)d\tau)^2 dt$$

$$(3)$$

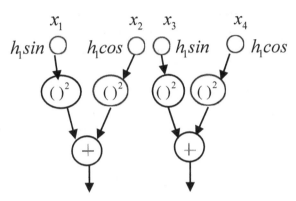

Fig. 5. Symmetric energy model

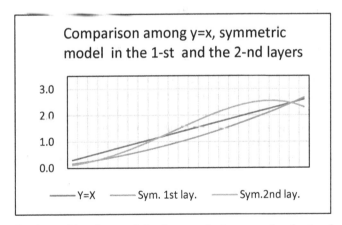

Fig. 6. Tracking characteristics for $y = x$ in the conventional network

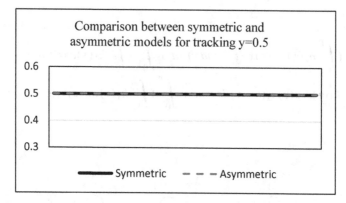

Fig. 7. Tracking characteristic for $y = 0.5$ in asymmetric and symmetric networks

2.4 Orthogonality in the Asymmetric Layered Networks

Rectification in the visual system [7] plays an important role for the generation of the orthogonality. Then, the half-wave rectification [7] is approximated in the following equation.

$$f(x) = \frac{1}{1 + e^{-\eta(x-\theta)}} \tag{4}$$

By Taylor expansion of Eq. (4) at $x = 0$, the Eq. (5) is derived as follows,

$$f(x)_{x=0} = f(\theta) + f'(\theta)(x - \theta) + \frac{1}{2!}f''(\theta)(x - \theta)^2 + \dots$$

$$= \frac{1}{2} + \frac{\eta}{4}(x - \theta) + \frac{1}{2!}(-\frac{\eta^2}{4} + \frac{\eta^2 e^{-\eta 0}}{2})(x - \theta) + \dots \tag{5}$$

The nonlinear terms, $x^2, x^3, x^4, \dots$ are generated in Eq. (5). Thus, the combination of Gabor function pairs is generated, in which the transformed network can be made of two layers of the extended asymmetrical network in Fig. 3 [11]. Then, the characteristics of the extended asymmetric network have pathways with higher order nonlinearities. The inputs in the 2nd layer is from responses of the 1st layer which are wavelets of the product of Gaussian and trigonometric functions of Eq. (1). These wavelets are shown in the following,

$$Ae^{-\frac{t^2}{2\sigma^2\xi^2}}\sin(t), \ A^2(e^{-\frac{t^2}{2\sigma^2\xi^2}})^2\sin^2(t), \ A^3(e^{-\frac{t^2}{2\sigma^2\xi^2}})^3\sin^3(t) \dots \tag{6}$$

$$Ae^{-\frac{t^2}{2\sigma^2\xi^2}}\cos(t), \ A^2(e^{-\frac{t^2}{2\sigma^2\xi^2}})^2\cos^2(t), \ A^3(e^{-\frac{t^2}{2\sigma^2\xi^2}})^3\cos^3(t) \dots \tag{7}$$

By applying the power reducing formula in the trigonometric functions, the following functions hold for the integer n to be odd,

$$\sin^n t = C_{1n} \sum_{k=0}^{(n-1)/2} \binom{n}{k} \cos((n - 2k)t), \quad \cos^n t = C_{1n} \sum_{k=0}^{(n-1)/2} (-1)^{(\frac{n-1}{2}-k)} \binom{n}{k} \sin((n - 2k)t), \tag{8}$$

while for the integer n to be even,

$$\sin^n t = C_{2n} + C_{1n} \sum_{k=0}^{(n-1)/2} \binom{n}{k} \cos((n-2k)t),$$

$$\cos^n t = C_{2n} + C_{1n} \sum_{k=0}^{(n-1)/2} (-1)^{(\frac{n-1}{2}-k)} \binom{n}{k} \sin((n-2k)t) \tag{9}$$

where C_{1n} and C_{2n} are constant value depending in Eqs. (8) and (9).

When n and m are odd, $(\sin^n t) \cdot (\cos^m t)$ is derived using Eq. (7).

$$C_{1n} C_{1m} \sum_k \sum_{k'} (-1)^{(\frac{(n-1)}{2}-k)} \binom{n}{k} \binom{m}{k'} \sin((n-2k)t) \cdot \cos((m-2k')t) \tag{10}$$

The correlation of the right term in Eq. (10) becomes

$$\int_{-\pi}^{\pi} \sin((n-2k)t) \cdot \cos((m-2k')t)dt$$

$$= \frac{1}{2} \int_{-\pi}^{\pi} \{\sin((n-m)-2(k-k'))t + \sin((n-m)+2(k-k'))t\}dt = 0 \tag{11}$$

Thus, the pair $\{(\sin^n t),(\cos^m t)\}$ becomes to be orthogonal. Similarly, the pair $\{(\sin^n t), (\cos^m t)\}$ becomes orthogonal in case of n to be odd and m to be even, and in case of n to be even and m to be odd. Only the pair $\{(\sin^n t), (\cos^m t)\}$ is not orthogonal in case of n to be even and m to be even. Thus, much orthogonal wavelets with the product of Gaussian and trigonometric function are generated in the 2nd layer. Further, much more orthogonal wave lets are generated in the 3rd layer. Then, the correlation between the wavelets $A^m (e^{-\frac{mt^2}{2\sigma^2 \xi^2}})^m \sin^m(t)$ and $A^l (e^{-\frac{lt^2}{2\sigma^2 \xi^2}})^l \cos^l(t)$ become zero using the linear function for Gaussian term.

3 Sparse Coding for Classification in the Extended Asymmetric Networks

3.1 Independence and Sparse Coding on the Orthogonal Subnetworks

Independence plays an important role for the classification scheme. Since orthogonal relations induce independence, the classification is expected in the extended asymmetric networks [11]. To process the outlier data from the group one in the subnetwork, additional independence is useful [5, 10], which is realized by the generation of the sparse coding. In Fig. 8, a schematic network of the mapping the lower dimensional data to the high dimensional one is shown. The data is assumed to be 4-dimensional one (x_1, x_2, x_3, x_4). To process simply and speedy the data, two 2-dimemsional data are generated as (x_1, x_2) and (x_3, x_4). The sub-data (x_1, x_2) is mapped to the 3rd dimensional

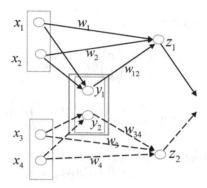

Fig. 8. Generation of independence through sparse code in the asymmetric networks

data as $y_1 = x_1 \cdot x_2$. Then, among the input data (x_1, x_2), the mapped one y_1 and the output z_1, the following linear equation holds.

$$[X] \cdot W = z_1 \tag{12}$$

where the input is shown in $[X] = \left[x_1, x_2, y_1\right]$, the weight matrix in $W = (w_1, w_2, w_{12})$.
This mapping is realized in the sparse coding scheme for classification.

3.2 Generation of Independent Basis Set via Sparse Coding Realization

Sparse coding realized in the extended asymmetric networks is made of the product operations of atoms and their negations. Here, atoms imply the output of the 1st layer of the asymmetric networks in Fig. 3, which are described in wavelet basis set $\{h_1 sin(x(t)), h_1^2 cos^2(x(t')), h_1 cos(x(t'')), h_1^2 sin^2(x(t'''))\}$. We assume here the data $\{x(t)\}$ is normalized or binary valued. Then, the atoms in the 1st layer asymmetric network become

$$\{h_1 sin(x(t)), h_1^2 cos^2(x(t')), h_1 cos(x(t'')), h_1^2 sin^2(x(t'''))\} \tag{13}$$

Further, we generate negation atoms

$$\{1 - h_1 sin(x(t)), 1 - h_1^2 cos^2(x(t')), 1 - h_1 cos(x(t'')), 1 - h_1^2 sin^2(x(t'''))\} \tag{14}$$

which are derived from the Eq. (5) in the nonlinear expansion of the half- rectification and the biological evidence of the (−) responses of the amacrine cell N in Fig. 1 [9].

New atoms are generated from the products of the combination of the atoms in Eqs. (13) and (14). For example, a new atom, $sin(x(t)) \cdot \{1 - cos^2(x(t'))\} \cdot sin^2(x(t'''))$ is generated as the combined atom using atoms in Eqs. (13) and (14). We call here the combined atom is realizable, when the atom exists in the expanded Eq. (5) of the layered asymmetric networks. The dictionary of atoms and combined atoms is made, which is described as the notation as D. First, the candidate sparse representation problem can be formulated as

$$\min_{\xi} \|\xi - D\xi\|_2 + \beta \|\xi\|_1 \tag{15}$$

where a signal data x can be represented in terms of a vector ξ and $\beta > 0$ is the parameter that controls the tradeoff between the sparsity and the reconstruction error. For the application of the sparse code to the classification problem, the second step is needed. Second, the candidate sparse code of the data x is different from the nearest neighbor data z in the classified results. We assume two candidate sparse codes $\alpha_1(x)$ and $\alpha_2(x)$. Then, the classification is performed correctly using a threshold element if the following equation is realized. For the classification, Eq. (12) is applied to the candidate sparse codes with nearest neighbor relation [10].

$$\begin{cases} \alpha_1(x)\omega_1 + \alpha_2(x)\omega_2 - \theta > 0 \\ \alpha_1(z)\omega_1 + \alpha_2(z)\omega_2 - \theta \leq 0 \end{cases} \tag{16}$$

where ω_1 and ω_2 are weights and θ is a threshold of the element. Equation (16) is solved if the determinant

$$\begin{vmatrix} \alpha_1(x) & \alpha_2(x) \\ \alpha_1(z) & \alpha_2(z) \end{vmatrix} \neq 0 \tag{17}$$

The non-zero determinant in Eq. (17) implies that the sparse codes, α_1 and α_2 are independent. Thus, the satisfied sparse code is obtained by two steps.

4 Application to Data Classification via Sparse Coding Realization in the Asymmetric Networks

We applied the asymmetric networks to the classification of the real data, which are generally characterized by the high dimensional attributes and their enormous volumes. These real data are expected to reduce the data to the lower dimensional data. We have developed the method to reduce to the lower dimensional data by the reduction techniques. To classify the real data in the threshold networks, hyperplanes and the subspace method developed here are applied. The classification steps in the asymmetric layered networks are shown in Fig. 9. As the real data, the part of the Reuters collections [12] was used, which consists of 'Cocoa data', 'Copper data' and 'Cpi data'. We used 16 reduced words from the original data of Cocoa, Copper and Cpi data, which are normalized. These reduced data, $(x_1, x_2, x_3, ..., x_{16})$ is transformed to the array in the asymmetric network as

$$h_1 sin(x_1) \ (h_1 cos(x_2))^2 \ h_1 cos(x_3) \ (h_1 sin(x_4))^2 \ h_1 sin(x_5) \ (h_1 cos(x_6))^2 ... \tag{18}$$

The problem is to classify each data into three data groups. We assume here the gaussian term in the impulse response function, h_1 in Eq. (18) to be 1 for simplicity. Processing steps for the classification are described in the following. First, by applying two steps processing in Fig. 9, the first data group, 'Cocoa data' are trained for the classification. In the 1st step in Fig. 9, the 'Cocoa data' is separated from 'Copper' and 'Cpi' data by the hyperplane in the linear discriminant analysis. The 'Cocoa data' is correctly classified to Cocoa group, which is described as Cocoa(+), while other Copper or Cpi data is misclassified as Copper data, which we call Cocoa(−). The Cocoa(−)

data is also processed via two steps in Fig. 9. The discrimination ratio of the Cocoa data by the two steps is shown in Fig. 10. In the 2nd step, sparse codes are generated for the misclassified Cocoa data, which the recover as the Cocoa data in the correct classification. Second, by applying two steps in Fig. 9, the second data group, 'Copper data' is also separated from 'Cocoa' and 'Cpi' data for the classification. Finally, by applying two steps in Fig. 9, the 3rd group, 'Cpi data' is separated from 'Cocoa' and 'Copper' data for the classification.

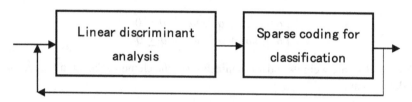

Fig. 9. Fundamental classification steps of each group data with two steps

The discrimination between 'Cocoa data group' and other groups is shown in Fig. 10 in which the blue line shows the value of the Cocoa(+) to be 0.76. The Cocoa(+) be in blue line come 1.0 by the sparse coding processing. The brown dotted line shows the Copper and Cpi data(−) to be 0.83 in the 1st layer, next to be 1 by the sparse coding processing. Thus, all the Cocoa data and other Copper, Cpi data are classified correctly.

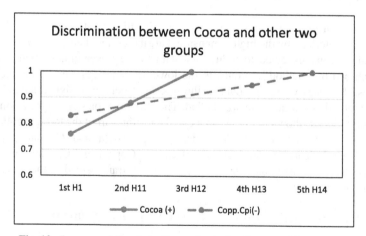

Fig. 10. Increase of discrimination ratios in the Cocoa and Copper data

Similar operations are performed in the 'Copper' and 'Cpi' data groups. The discrimination ratios are shown in Fig. 11. Thus, the correct classification among 'Cocoa', 'Copper' and 'Cpi' data groups are realized using linear discrimination and sparse coding processing with nearest neighbor relation in the extended asymmetrical networks.

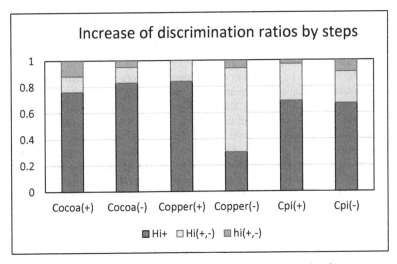

Fig. 11. Increase of discrimination ratios among 3 groups data by steps

5 Conclusion

Studies of machine learning, artificial intelligence and neural networks have progressed, and they are applied to a diverse range of fields. Although their models often are opaque in terms of explainability and realizability. Any intelligent functions are expected to create explainable design for the layered networks. In this paper, starting from the bio-inspired neural networks, It is shown that the asymmetrical network with nonlinear functions have superior characteristics for tracking moving inputs. Next, to explain the functions for the classification in the layered networks, the sparse coding realizable in the layered networks is proposed for the classification, which is realized in the structure of the asymmetric layered networks. The classification is experimented to verify two steps processing using real Reuters collection data, which show the exact classification with their steps.

References

1. LeCun, Y., Bengio, Y., Hinton, G.: Deep learning. Nature **521**, 436–444 (2015)
2. Kornblith, S., Lapusckin, S., Lee, H., Hinton, G.: Similarity of neural network representations revised. In: Proceedings of the 36th International Conference on Machine Learning, ICML-2019, pp. 1–16 (2019)
3. Samek, W., Lapusckn, S., Anders, C.J., Muller, K.-R.: Explaining deep neural networks and beyond: a review of methods and applications. Proc. IEEE **103**(3), 247–278 (2021)
4. Paiton, D.M., Frye, C., Lundquist, S.Y., Bowen, J.D., Zarcon, R., Olshausen, B.A.: Selectivity and robustness of sparse coding networks. J. Vis. **20**(12), 10, 1–28 (2020)
5. Fan, K.: On systems of linear inequalities. In: Kuhn, H.W., Tucker, A.W. (eds.): Linear Inequalities and Related Systems, pp. 99–156. Princeton University Press (1966)
6. Adelson, E.H., Bergen, J.R.: Spatiotemporal energy models for the perception of motion. J. Opt. Soc. Am. A 284–298 (1985)

7. Simonceli, E.P., Heeger, D.J.: A model of neuronal responses in visual area MT. Vis. Res. **38**, 743–761 (1996)
8. Naka, K.-I., Sakai, H.M., Ishii, N.: Generation of transformation of second order nonlinearity in catfish retina. Ann. Biomed. Eng. **16**, 53–64 (1988)
9. Sakai, H.M., Naka, K.-I.: Dissection of the neuron network in the catfish inner retina. I. Transmission to ganglion cells. J. Neurophysiol. **60**(5), 1549–1567 (1988)
10. Ishii, N., Torii, I., Mukai, N., Iwata, K., Nakashima, T.: Classification on nonlinear mapping of reducts based on nearest neighbor relation. In: 14th IEEE/ACIS International Conference on Computer and Information Science, pp.491–496. IEEE Computer Society (2015)
11. Ishii, N., Deguchi, T., Kawaguchi, M., Sasaki, H., Matsuo, T.: Orthogonal properties of asymmetric neural networks with gabor filters. In: Pérez García, H., Sánchez González, L., Castejón Limas, M., Quintián Pardo, H., Corchado Rodríguez, E. (eds.) HAIS 2019. LNCS (LNAI), vol. 11734, pp. 589–601. Springer, Cham (2019). https://doi.org/10.1007/978-3-030-29859-3_50
12. Reuters-21578 Text Categorization Collection. https://kdd.ics.edu/databases/reuters21578/reuters21578.html

Frailty Related Survival Risks at Short and Middle Term of Older Adults Admitted to Hospital

Guillermo Cano-Escalera[1], Manuel Graña[1(✉)], and Ariadna Besga[2,3]

[1] Computational Intelligence Group, University of the Basque Country (UPV/EHU), San Sebastian, Spain
manuelgrana@ehu.es
[2] Department of Neurology, Hospital Universitario de Araba, Vitoria, Spain
[3] Biomedical Research Centre in Mental Health Network (CIBERSAM) G10, Vitoria, Spain

Abstract. In the context of healthy aging research, frailty is characterized by a progressive decline in the physiological functions of multiple body systems that lead to a more vulnerable condition for the development of various adverse events, such as falls, hospitalization and mortality. In this paper, we assess the impact of frailty on survival of older adults at three time horizons: 1 month, 6 months, and 1 year, thus covering short and middle term survival. We consider frail and pre-frail patients recruited at hospital admission, according to the Fried phenotype. Cognitive status, comorbidities and pharmacology assessment was carried out at hospital admission and follow up survival was extracted from electronic health record (EHR). We compute Kaplan Meier estimates and log-rank test of survival probability functions at the selected censoring times for frail versus pre-frail patients. Additionally, we compute Cox regression for these censoring times over patient variables selected by independent sample t-test and logistic regression. We confirm that frailty predicts greater mortality at all censoring times. Significant hazard risks (HRs) identified in frail cohort were sex, age, weight, the occurrence of hospital readmission 30 days after discharge, sit and stand test scores, and the use of proton pump inhibitors, antiplatelet, quetiapine and paracetamol.

Keywords: Frailty · Mortality · Survival analysis

1 Introduction

World wide aging of the population pushes for healthy aging related research. Frailty [13] is a critical state strongly associated with aging that is characterized by a progressive decline in the physiological functions of multiple body systems leading to a condition of greater vulnerability for the development of various adverse events. Prevalence of frailty [7] in adults older than 65 years living in the community ranges from 4% to 17%. While prevalence of pre-frailty varies between 19% and 53% in different studies. Due to the clinical heterogeneity of

© Springer Nature Switzerland AG 2022
P. García Bringas et al. (Eds.): HAIS 2022, LNAI 13469, pp. 39–47, 2022.
https://doi.org/10.1007/978-3-031-15471-3_4

frailty, it is important to develop effective strategies for the provision of care to frail patients. Treatment of clinical frailty is one of the most important determinants for patients to progress correctly during their hospital stay [33]. In relation to interventions, attempts to manage the adverse consequences of frailty often focus on minimizing the risk of disability and dependence or treating underlying conditions and symptoms. In a complementary approach, the management of frailty implies the control the impact of possible stressors [1]. All improvements in care and prevention would reduce the economic costs because it has been shown that frailty is associated with an increase in the total costs of medical care [16].

In this paper we assess the increased mortality risk associated with frailty compared with pre-frailty status over the collected patient data. The study has been approved by the Ethics Committee of the Hospital Universitario de Alava. We also identify some patient variables associated with increased survival hazard risks (HR).

2   Materials and Methods

2.1   Study Design and Subjects

The patient recruitment process has been described elsewhere [4]. The patients were admitted to the services of internal medicine and neurology at the University Hospital of Alava (UHA). Frailty was evaluated by the Fried's frailty index [13] during the hospital stay. The frailty phenotype is based on a predefined set of five criteria, namely: involuntary weight loss, low mood, slow gait, poor grip strength, and sedentary behavior. The result is classified into three levels: robustness (none of the criteria), pre-frailty (one or two criteria) and frailty (three or more criteria). Partial noninvasive physical abilities and cognitive assessment was carried out also at admittance. Recruitment and tests requiring physical and cognitive involvement of the patient were carried out if and when the patient was stabilized and able. Patient recruitment was implemented in the period from September 2017 to September 2018. Inclusion and exclusion criteria have been detailed elsewhere [4]. This paper reports results over the entire sample, 501 considered frail and 213 pre-frail, and 17 patients were excluded as they were considered robust, using the Fried's frailty index. Mortality follow-up until January 2021 was performed by querying the institutional electronic health records (EHR). In this paper we consider 3 censoring times: 1 month, 6 months and 1 year after hospital discharge. Non-survivors were defined as all persons known to have died during follow-up period before the censoring time. Following open science practices, the anonymized dataset has been published in the Zenodo public repository (https://zenodo.org/record/5803234).

To assess the functional status of the patient, the following tests were applied: Short Physical Performance Battery (SPPB) test [15], Fried's frailty index (FFI) [13], and Barthel's index score (BIS) [22] measuring performance in activities of daily living. The FFI score has been used to classify patients into the robust, pre-frail, and frail categories. The nutritional state was assessed using the Mini

Nutritional Assessment-Short Form (MNA-SF)-as a mean to identify older subjects at risk of malnutrition before the apparition of severe changes in weight or serum or protein concentrations [19,28]. The Pfeiffer's Brief Screening Test for Dementia (PBSTD) [11,23] was applied for mental condition assessment. Additionally, we considered the number of falls in the previous month as a feature of frailty. The electronic health records (EHR) were accessed to extract sociodemographic data, survival data, clinical data such as comorbidities, pharmacological information, and survival data. Missing values were corrected by setting the default values after consultation with clinicians and double checking on the EHR. After data curation, variables with more than 20% cases with missing values were ignored to ensue processing.

2.2 Statistical Methods

We have used the Jasp package (https://jasp-stats.org) that is implemented in R but runs independently. The data was processed using Independent Samples T-Test, and later using Logistic Regression the variables $p > 0.05$. Data were also processed in Rstudio 1.2 and R 3.6.3 (www.r-project.org, accessed on 1 March 2022) using packages HSAUR2, Survival, and Survminer. We compute the Kaplan-Meier estimate of the survival function and its variance [2,12,25] at three censoring date (1 month, 6 months and 1 year). Log-rank test [12] assessed the statistical significant difference of the probability of survival curves of frail versus pre-frail patients. Significance threshold was 0.05. Separate analyses for the frail cohort (N = 501) and the pre-frail cohort (N = 213) were carried out. Cox's regression [8,12] was computed to assess the hazard risks (HR) of the variables at the 2 years. Independent multivariate Cox's regressions were carried out over variables. Proportional hazard assumption was tested for all Cox regression analyses by Schoenfeld residuals, and $p > 0.05$ discarded the null hypothesis.

3 Results

According to the Fried frailty scale, 68.54% of the sample is fragile, and 29.14% is pre-fragile, adding up to a total of 98% of the sample that is considered fragile or almost fragile. According to the SPPB, 64% of the sample presented a score of severe or moderate in terms of functional disability. Regarding the nutrition evaluated by the MNA-SF, 64% of the sample is at nutritional risk. However, the majority of patients (89%) were highly independent on the Barthel score, and 80% were almost normal on the Pfeiffer dementia scale. Table 1 summarizes the demographics of the entire dataset. Robust patients were excluded from the present analysis. The distribution of the frailty test outcomes is as follows: 64.44% of the patients were at a nutritional risk (MNA-SF score < 8). However, most of the patients (88.61%) were mildly to highly independent according to BIS, and 80.68% were almost normal in the Pfeiffer's scale of dementia.

Table 1. Demographics information of the cohort

Variable	Categories	N Total	%	Frail (N = 501) %	Pre-Frail (N = 213) %	p
Gender	Male	382	51.55	47.30	61.97	<0.001
	Female	359	48.45	52.70	38.03	<0.001
Weight (kg)	Mean (SD)	67.43 (13.60)		66.15 (13,87)	67.32 (13.72)	<0.001
Age (years)	Mean (SD)	84,37 (6.76)		85.52 (6.20)	84.26 (7.33)	<0.001
MS	Married	265	35.60	35.33	37.09	0.654
	Single	65	8.90	8.38	9.86	0.525
	Divorced	13	1.80	2.00	1.40	0.592
	Widowed	265	35.70	40.92	26.76	<0.001
	NA	133	18.00	13.37	24.89	<0.001
HD	Yes	501	67.61	71.85	61.50	0.585
	No	241	32.39	28.15	38.50	
NWS	Yes	419	56.54	68.46	31.92	<0.001
	No	322	43.46	31.54	68.08	
Living at	Own home	530	71.52	71.65	71.83	<0.001
	Alone	195	26.31	25.35	26.76	0.110
	Other's home	62	8.36	9.38	4.69	0.091
	Retirement house	64	8.63	9.98	3.28	0.009
Polypharmacy	Oligopharma <5	178	24.02	19.16	32.86	<0.001
	Moderate (5–9)	358	48.31	49.90	45.54	0.287
	Severe (>9)	205	27.67	30.94	21.60	0.011
Admission service	Neurology	81	10.93			
	Internal medicine	660	89.07			
Charlson comorbidity index		6.39 (2.08)				

Table 2. Log-rank test results of comparison of survival probability curves.

Cohort	1 month	6 month	1 year
	$p = 4e - 8*$	$p = 2e - 8$	$p = 1e - 8$

Note: *-strongly significant $p < 0.001$.

Figure 1 compares the survival probability Kaplan-Meier estimates of the frail cohort versus pre-frail cohort at selected censoring times of 1 month, 6 months, and 1 year. Log-rank test p values summarized in Table 2 confirm that frail patients had a significantly worse prognosis for survival at the 1 month, 6 months, and 1 year compared to pre-frail patients.

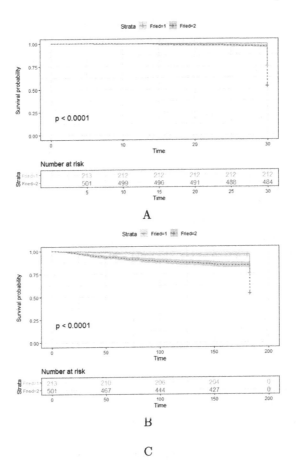

Fig. 1. Kaplan Meier survival curves for patients diagnosed with frail and pre-frail. Data is censored after 1 month (A), 6 months (B) and 1 year (C).

Table 3 contain the variables with significant HR for the frail cohort at 1 month, 6 months, and 1 year censoring times. Specific HRs in the frail cohort were readmissions before 30 days after discharge, scores of time to sit and stand up included in SPPB, and use of antiplatelet, quetiapine and paracetamol.

Table 3. Variables with significant ($p < 0.05$) high risk (HR, 95% Confidence Interval) found by Cox's regression carried out over variables. Frail cohort (N = 501), data censored at 1 month, 6 months and 1 year.

Variable	1 months		6 months		1 year	
	HR	p	HR	p	HR	p
Age	1.05 (1.01–1.09)	0.008	1.04 (1.01–1.08)	0.021	1.05 (1.01–1.09)	0.012
Gender	0.53 (0.36–0.78)	0.001	0.49 (0.33–0.73)	<0.001	0.45 (0.30–0.68)	<0.001
Weight	0.99 (0.97–1.00)	0.16	0.98 (0.97–1.00)	0.049	0.98 (0.97–1.00)	0.041
R30	0.65 (0.44–0.96)	0.03	0.52 (0.35–0.77)	0.001	0.52 (0.35–0.78)	0.001
SPPB-SUG	0.71 (0.52–0.96)	0.028	0.65 (0.47–0.89)	0.007	0.61 (0.45–0.84)	0.003
PPIs	0.68 (0.47–0.97)	0.035	0.64 (0.45–0.92)	0.016	0.61 (0.42–0.88)	0.008
Antiplatelet	1.61 (1.08–2.40)	0.021	1.61 (1.07–2.42)	0.021	1.77 (1.18–2.67)	0.006
Quetiapine	0.31 (0.12–0.79)	0.014	0.30 (0.11–0.77)	0.012	0.29 (0.11–0.76)	0.012
Paracetamol	1.34 (0.93–1.92)	0.114	1.45 (1.01–2.09)	0.044	1.50 (1.05–2.16)	0.028

4 Discussion

In a global context of population aging, there is a growing interest in improving the quality of care for the elderly, as well as an optimal management of health resources. Therefore, the prediction of risk factors in reference to patients considered frail is of increasing relevance. Although it is true that there is a lack of consensus in the literature regarding the definition of frailty and its categories, there are studies that have carried out research taking into account the categorization of frail and pre-frail [17,20] of the frailty phenotype of Fried. 98% of the sample of our study contains frail and pre-frail patients (68.54% frail, 29.14% pre-frail).

That frailty is a strong predictor of mortality [6] has been demonstrated in multiple systematic reviews [5]. Malnutrition has been shown to play a key role in the emergence of frailty [21]. Moreover, a combination of nutrition-related parameters has also been shown to be significantly associated with mortality in frail patients [18].

Our results show that the risk factors for presenting frailty are associated with a vulnerability in physical capacity, such as the need walking stick and falls. This relationship between falls and mobility and daily living problems in elderly subjects has already been demonstrated in a previous study. According to the WHO [30], falls are the second worldwide cause of death due to involuntary injuries and, they occur mostly in the elderly population group, increasing their incidence with age [3]. Lower resistance in the sit and get up test, a smaller calf circumference, and weight loss are quite significant frailty associated mortality risk factors. Our findings related to the loss of physiological capacity demonstrated in the MNA mobility, MNA weight loss 3 months and SPPB sit and stand up tests are in agreement with studies that indicate that weakness, slowness particularly in walking and limited capacity to perform routine physical activities often are the hallmark of the onset of frailty. In the cognitive aspect we have observed that depression, dementia or delirium are factors that elderly patients often present simultaneously.

Heart failure is considered another geriatric syndrome in the same way as dementia, falls and frailty [9]. It is a leading cause of morbidity and mortality worldwide, contributes to a substantial deterioration of quality of life for patients. The presence of frailty in chronic heart failure is associated with an 1.5-fold increased hazard for death and hospitalization [32]. In heart failure treatment, overdosing of diuretics is common, as doses often reflect requirements for acute compensation, which is two- to threefold the requirement of that in maintenance therapy. Hence, trial data demonstrate a positive correlation between mortality and diuretic use/dose [29]. A typical component of frailty, immobility, is widely considered to be a risk factor for venous thrombosis. In a prospective study, they observed that the risks of venous thrombosis were greater for people considered frail as pre-frail, thus coinciding with our result of a significant risk factor for venous thrombosis in pre-frail people with respect to survival. Cerebrovascular disease (CVA) is another disease that increases its incidence with age and constitutes a major cause of disability. In our study, it is observed that having presented CVA constitutes a risk factor for mortality in the pre-frail group. This may be related to the fact that a large part of people with previous cerebrovascular disease present functional limitations characteristic of the frail elderly and there are also several studies that have found that post-stroke fatigue, described as a subjective feeling that consists of a deep feeling of tiredness and a lack of mental and/or physical energy is one of the most commonly reported symptoms after stroke and is associated with a reduced quality of life and an increased risk of death [10,26].

In relation to pharmacology in the elderly patient, when multiple diseases coexist, a high number of drugs tends to be prescribed, with the risk of suffering adverse drug reactions and drug interactions and a higher risk of mortality. Furthermore, the risk increases with age as a consequence of the physiological changes of aging [14]. Among the drugs most prescribed, proton pump inhibitors (PPIs) indicated for gastric protection stand out. It is known that prolonged use of PPIs can cause serious systemic side effects [24]. Recent analyzes suggest that they can be inappropriately prescribed to 50% to 80% of patients and it has been observed that it is associated with an increased risk of cardiovascular disease and chronic kidney disease and, in these cases, with excess mortality [31]. Antiplatelet management is also particularly challenging in elderly patients due to the increased risk of both ischemic and hemorrhagic events, leading to higher mortality rates. Quetiapine is an antipsychotic medicament used for patients whose neuropsychiatric symptoms have not responded to other alternative pharma. Our findings are in agreement with literature reporting that quetiapine increases mortality by 2% [27].

5 Conclusions and Future Work

The objectives of this study were twofold: on the one hand, the evaluation of the increase in mortality due to frailty in an elderly population, and on the other, the identification of the high-risk variables that affect said mortality. Comparison of the survival probability curves confirms that frailty is a factor of higher mortality at 1 month, 6 months and 1 year. The study also confirms that drugs such as PPIs, antiplatelet, quetiapine and paracetamol are correlated with higher risk of mortality. It is worth noting that the results show that the frail and pre-frail categories correspond to clearly separated groups having different significant hazard risk variables.

References

1. Andreasen, J., Lund, H., Aadahl, M., Sørensen, E.E.: The experience of daily life of acutely admitted frail elderly patients one week after discharge from the hospital. Int. J. Qual. Stud. Health Well-Being **10**, 27370 (2015)
2. Barakat, A., Mittal, A., Ricketts, D., Rogers, B.A.: Understanding survival analysis: actuarial life tables and the Kaplan-Meier plot. Br. J. Hosp. Med. **80**(11), 642–646 (2019)
3. Berková, M., Berka, Z.: Falls: a significant cause of morbidity and mortality in elderly people. Vnitrni Lekarstvi **64**(11), 1076–1083 (2018)
4. Cano-Escalera, G., Graña, M., Irazusta, J., Labayen, I., Besga, A.: Survival of frail elderly with delirium. Int. J. Environ. Res. Public Health **19**(4), 2247 (2022)
5. Chang, S.-F., Lin, P.-L.: Frail phenotype and mortality prediction: a systematic review and meta-analysis of prospective cohort studies. Int. J. Nurs. Stud. **52**(8), 1362–1374 (2015)
6. Clegg, A., Young, J., Iliffe, S., Rikkert, M.O., Rockwood, K.: Frailty in elderly people. Lancet **381**(9868), 752–762 (2013)
7. Collard, R.M., Boter, H., Schoevers, R.A., Oude Voshaar, R.C.: Prevalence of frailty in community-dwelling older persons: a systematic review. J. Am. Geriatr. Soc. **60**(8), 1487–1492 (2012)
8. Cox, D.R.: Regression models and life-tables. J. Roy. Stat. Soc. Ser. B (Methodol.) **34**(2), 187–220 (1972)
9. Dharmarajan, K., Rich, M.W.: Epidemiology, pathophysiology, and prognosis of heart failure in older adults. Heart Fail. Clin. **13**(3), 417–426 (2017)
10. Elf, M., Eriksson, G., Johansson, S., von Koch, L., Ytterberg, C.: Self-reported fatigue and associated factors six years after stroke. PLoS ONE **11**(8), e0161942 (2016)
11. Erkinjuntti, T., Sulkava, R., Wikström, J., Autio, L.: Short portable mental status questionnaire as a screening test for dementia and delirium among the elderly. J. Am. Geriatr. Soc. **35**(5), 412–416 (1987)
12. Everitt, B.S., Hothorn, T.: A Handbook of Statistical Analyses Using R. CRC Press, Boca Raton (2010)
13. Fried, L.P., et al.: Frailty in older adults: evidence for a phenotype. J. Gerontol. A Biol. Sci. Med. Sci. **56**(3), M146-56 (2001)
14. Gómez, C., Vega-Quiroga, S., Bermejo-Pareja, F., Medrano, M.J., Louis, E.D., Benito-León, J.: Polypharmacy in the elderly: a marker of increased risk of mortality in a population-based prospective study (NEDICES). Gerontology **61**(4), 301–309 (2015)
15. Guralnik, J.M., et al.: A short physical performance battery assessing lower extremity function: association with self-reported disability and prediction of mortality and nursing home admission. J. Gerontol. **49**(2), M85–M94 (1994)

16. Hajek, A., et al.: Frailty and healthcare costs-longitudinal results of a prospective cohort study. Age Ageing **47**(2), 233–241 (2017)
17. Hanlon, P., Nicholl, B.I., Jani, B.D., Lee, D., McQueenie, R., Mair, F.S.: Frailty and pre-frailty in middle-aged and older adults and its association with multimorbidity and mortality: a prospective analysis of 493 737 UK biobank participants. Lancet Public Health **3**(7), e323–e332 (2018)
18. Jayanama, K., Theou, O., Blodgett, J.M., Cahill, L., Rockwood, K.: Frailty, nutrition-related parameters, and mortality across the adult age spectrum. BMC Med. **16**(1), 188 (2018)
19. Kaiser, M.J., et al.: Validation of the mini nutritional assessment short-form (MNA-SF): a practical tool for identification of nutritional status. JNHA J. Nutr. Health Aging **13**(9), 782 (2009)
20. Kidd, T., et al.: What are the most effective interventions to improve physical performance in pre-frail and frail adults? a systematic review of randomised control trials. BMC Geriatr. **19**(1), 184 (2019)
21. Laur, C.V., McNicholl, T., Valaitis, R., Keller, H.H.: Malnutrition or frailty? Overlap and evidence gaps in the diagnosis and treatment of frailty and malnutrition. Appl. Physiol. Nutr. Metab. **42**(5), 449–458 (2017). PMID: 28322060
22. Mahoney, F.I., Barthel, D.W.: Functional evaluation: the barthel index. Md. State Med. J. **14**, 61–65 (1965)
23. Pfeiffer, E.: A short portable mental status questionnaire for the assessment of organic brain deficit in elderly patients. J. Am. Geriatr. Soc. **23**(10), 433–441 (1975)
24. Porter, B., Arthur, A., Savva, G.M.: How do potentially inappropriate medications and polypharmacy affect mortality in frail and non-frail cognitively impaired older adults? A cohort study. BMJ Open **9**(5), e026171 (2019)
25. Rich, J.T., Neely, J.G., Paniello, R.C., Voelker, C.C.J., Nussenbaum, B., Wang, E.W.: A practical guide to understanding Kaplan-Meier curves. Otolaryngol. Head Neck Surg. **143**(3), 331–336 (2010)
26. Preto, L.S.R., Conceição, M.C.D., Amaral, S.I.S., Figueiredo, T.M., Sánchez, A.R., Fernandes-Ribeiro, A.S.: Fragilidad en ancianos que viven en la comunidad con y sin enfermedad cerebrovascular previa. Revista Científica de la Sociedad Española de Enfermería Neurológica **46**, 11–17 (2017)
27. Schneider, L.S., Dagerman, K.S., Insel, P.: Risk of death with atypical antipsychotic drug treatment for dementia: meta-analysis of randomized placebo-controlled trials. JAMA **294**(15), 1934–1943 (2005)
28. Vellas, B., Guigoz, Y., Garry, P.J., Nourhashemi, F., Bennahum, D., Lauque, S., Albarede, J.L.: The mini nutritional assessment (MNA) and its use in grading the nutritional state of elderly patients. Nutrition **15**(2), 116–122 (1999)
29. Wehling, M.: Morbus diureticus in the elderly: epidemic overuse of a widely applied group of drugs. J. Am. Med. Dir. Assoc. **14**(6), 437–442 (2013)
30. WHO. Falls. Technical report, WHO, April 2021
31. Xie, Y., Bowe, B., Yan, Y., Xian, H., Li, T., Al-Aly, Z.: Estimates of all cause mortality and cause specific mortality associated with proton pump inhibitors among us veterans: cohort study. BMJ **365** (2019)
32. Yang, X., et al.: Impact of frailty on mortality and hospitalization in chronic heart failure: a systematic review and meta-analysis. J. Am. Heart Assoc. **7**(23), e008251 (2018)
33. Murillo, A.Z., Herrero, Á.C.: Frailty syndrome and nutritional status: assessment, prevention and treatment. Nutr. Hosp. **36**(Spec No2), 26–37 (2019)

On the Analysis of a Real Dataset of COVID-19 Patients in Alava

Goizalde Badiola-Zabala[1(✉)], Jose Manuel Lopez-Guede[1,2], Julian Estevez[1,3], and Manuel Graña[1,3]

[1] Computational Intelligence Group, Basque Country University (UPV/EHU),
Leioa, Spain
goizalde.badiola@ehu.eus
[2] Department of Systems and Automatic Control,
Faculty of Engineering of Vitoria-Gasteiz, Basque Country University (UPV/EHU),
Nieves Cano 12, 01006 Vitoria-Gasteiz, Spain
[3] Department of Computer Science and Artificial Intelligence,
Faculty of Informatics, Basque Country University (UPV/EHU),
Paseo Manuel de Lardizabal 1, 20018 Donostia-San Sebastian, Spain

Abstract. COVID-19 has been spread to many countries all over the world in a relatively short period, largely overwhelmed hospitals have been a direct consequence of the explosive increase of coronavirus cases. In this dire situation, the demand for the development of clinical decision support systems based on predictive algorithms has increased, since these predictive technologies may help to alleviate unmanageable stress on healthcare systems. We contribute to this effort by a comprehensive study over a real dataset of covid-19 patients from a local hospital. The collected dataset is representative of the local policies on data gathering implemented during the pandemic, showing high imabalance and large number of missing values. In this paper, we report a descriptive analysis of the data that points out the large disparity of data in terms of severity and age. Furthermore, we report the results of the principal component analysis (PCA) and Logistic Regression (LR) techniques to find out which variables are the most relevant and their respective weight. The results show that there are two very relevant variables for the detection of the most severe cases, yielding promising results. One of our paper conclussions is a strong recommendation to the local authorities to improve the data gathering protocols.

Keywords: COVID-19 · Modeling · Machine learning

1 Introduction

On December 2019, several cases of pneumonia of unknown ethology were reported by the Authorities of the Republic of China to the WHO. These cases had been detected in the city of Wuhan, a city located in the Hubei province of China. After being confirmed that it was a new coronavirus called SARS-CoV-2,

© Springer Nature Switzerland AG 2022
P. García Bringas et al. (Eds.): HAIS 2022, LNAI 13469, pp. 48–59, 2022.
https://doi.org/10.1007/978-3-031-15471-3_5

the disease rapidly spread internationally, affecting most of the countries, and becoming a global pandemic with cases increasing exponentially.

Hospitals have been overloaded as a result of this issue, causing other diseases not to be treated [3], lack of staff and personal protective equipment. The focus on speeding up healthcare services is therefore essential to avoid bottlenecks in the Emergency Department (ED) that is the major point of entry to the hospital in Spain. After the massive spread mentioned above, there have been several investigations to improve healthcare services trying to prevent a second possible lockdown. Several research reports have shown that, although the use of Artificial Intelligence (AI) has been established in another context or industry, the introduction of AI to perform remote tasks that are traditionally performed by in-person clinical staff represents a significant milestone in healthcare operations strategy [8]. Since these algorithms are capable of improving patient care, minimize clinical burden, morbidity and mortality during the pandemic [1].

The structure of the paper is as follows, Sect. 2 gives some background information and a short review of the state-of-the-art of data analysis and prediction algorithms addressing the covid-19 pandemic. Section 3 describes the dataset, the variables and their pre-processing. Section 4 presents some discussion points in relation to the data analysis performed. Finally, Sect. 5 concludes the article.

2   Related Work

In an effort to address the pandemic response needs, researchers have begun to develop Machine Learning (ML) models to help health services deal with this critical situation. It is important to discriminate critical patients in need for being hospitalized from those who can be managed as an outpatient [10], creating a standardized protocol to that the situation can be properly handled and prevent health services for being overload. Several studies had proposed triage algorithms to give priority to those patients at higher risk [9]. Deep learning (DL) algorithms for triaging patients with suspected COVID-19 [16] had been used for disease control, as well as simple scoring systems [11]. Most of them take into account vital signs at the entrance of ED, temperature, shortness of breath, oxygen saturation, respiratory rate, systolic and diastolic blood pressure, and also age and gender variables [15].

For this purpose, several authors have utilised algorithms like PCA both for analysing the most relevant components of the input dataset and for data dimensionality reduction [2,7,12]. This method converts a group of potentially correlated features into a reduced number of independent ones denominated principal components, performed through the projection of the original data set to the reduced PCA space employing the eigenvectors of the correlation/covariance matrix [14]. Hence, Pinto et al. [12] aim was to identify low- and high-risk patients, by performing PCA combined with clustering. They achieved the classification of patients by their degree of severity. In the work of Gawriljuk et al. [4] the PCA algorithm was used to reduce feature dimensionality to then create multiple iterations of ML models as a potentially valuable prioritisation tool for SARS-CoV-2 antiviral drug discovery programs.

In addition, feature importance techniques play an important role in building predictive models [13], as they provide insight about the data and the model and the basis for dimensionality reduction and feature selection that can improve the efficiency and effectiveness of a predictive model on the problem. The relative scores referred may give an indication of which variables are indeed relevant in predicting the target variable, and vice versa, which characteristics are the least relevant [5].

3 Methods

3.1 Study Design

The ongoing project has been arisen because of the needs in the Health Care field that have been emerged as a result of the current pandemic. Specifically, obtaining predictive models based on AI to help health-care employees improving the management and treatment of each patients individually is the main focus of the study. The study makes special emphasis on the specific characteristics of the population of the Historical Territory of Alava. The current study was approved by BIOARABA Health Research Institute.

The data needed for this study come from the University Hospital of Alava (HUA). The internal and private software allows access to the database to help academics to conduct extensive and deep researches. In brief, hospital participation is voluntary and data are handled confidentially.

3.2 Ethical Approval and Patient Consent

All ethical standards and preoccupation for patients privacy have been respected during the performance of the study. The ethics committee validated the present study and granted an exemption for patient data collection informed consent due to the retrospective character of the study, and its anonymization. The project has counted with the collaboration of BIOARABA - HUA Txagorritxu, which has made the sources of information available to the project, properly anonymized, strictly fulfilling all the criteria established by its Ethics Committee are followed, since no personal information about the patient was included in the study.

3.3 Data Collection and Description

Demographic, vital, drug supply and laboratory data were required between October 2020 and March 2021. All the information provided to address future objectives proposed in the project is extracted from the following SQL tables maintained by the health care system providers:

- The main table, Admissions, contains data about the hospitalization (hospital department and section, date, reason, severity and hospital admission diagnosis) and demographic data as sex, date of birth, age and date of death, if the patient has deceased.

- In the Diagnosis table, all the information regarding the patient's medical diagnosis is found. This is represented by patient's unique identifier, gender, date of birth, age, date of death (if any), date of diagnostic and the diagnosis ICD10 code.
- ED table gathers information of patients requiring immediate attention in the emergency department (ED). Where patient's detailed information is grouped, such as identifier, sex, date of birth, age, date of death, circumstance and autopsy (if any), date of admission, priority and severity level (triage), admission reason, medical specialty, whether the patient is pregnant, and if so, the pregnancy week [6].
- Finally, the Laboratory table contains the test results. The following information can be found: patient identifier, sex, date of birth, date of death (if any), date of request, code and test description, results, hospital service and section requesting, diagnosis in text, type of test and type of result.

Since COVID-19 is a new virus lacking previous studies in its field, the collected data is the same to each and every patient in the hospital, i.e. data from patients with or without COVID-19 had been gathered. In order to separate COVID patients from others, results of a PCR test was used.

The dataset contains 6059 positive laboratory results on COVID-19 out of the 70314 patients requiring hospital services in that period. Thus, the maximum information for each patient was stored in the final dataset, and after removing the information of those patients with laboratory records but without any other physiological data, the number of patients decreased from 6059 to 4818.

Among all these variables collected during this period by the hospital, the aim was to focus on those considered most relevant for future predictions, i.e. age, sex, whether the patient was admitted to the intensive care unit (ICU), Triage level, weight, height and body mass index (BMI), and vital signs such as temperature, heart rate, Respiratory Rate, Glycaemic Index, Blood Oxygen Saturation, Diastolic Blood Pressure and Systolic Blood Pressure. Likewise, the final data set with the aforementioned variables contained 2422 samples.

The percentage of empty values of each variable was examined and reflected in the Table 1, and in order to interpolate the missing data while preserving as much information as possible, irregular physiological variables for each sampling period were all filled by shape-preserving piece-wise cubic spline interpolation. However, observing some variables with not enough values collected and not being possible to fill the empty values. According to the theory, it is not advisable to apply data imputation when the number of missing values is so high. Therefore it was concluded to remove Respiratory Rate, BMI, weight, height and glycaemic index, observing the high percentage of missing values shown in Table 1.

3.4 Attribute Analysis

The range and mean value of all variables is presented in Table 2. Vital sign completeness was obtained dividing the record of the specified vital sign with the number of visits. Demographic data are also included in the table, where

Table 1. Percentage of missing values for each vital sign variable.

Variables	Percentage of missing values	Number of missing values (n = 2422)
Age	0%	0
Gender	0%	0
ICU	0%	0
Triage	0%	0
Temperature	8.51%	206
Heart Rate	13.87%	336
Respiratory Rate	92.93%	2251
BMI	99.66%	2421
Weight	99.66%	2414
Height	99.66%	2414
Glycaemic Index	95.04%	2302
Blood Oxygen Saturation	22.54%	546
Diastolic Blood Pressure	16.47%	399
Systolic Blood Pressure	16.63%	403

the number of patients for each specified age range and gender are represented, both divided by patients admitted to ICU.

The age has been divided into several ranges for correct visualisation and compression, and it can be noticed that the range with the most people in the ICU is between 30 and 49 years. Nevertheless, the big difference is evident in respect to the older and younger age groups. The youngest constitute the smallest percentage of the whole, while those over 80 years old represent more than a fifth of the whole, it is clear that there is a strong tendency towards the older population. As for gender, there is no difference in those who are not admitted to the ICU, but there is a significant contrast between the two genders in the most seriously ill patients, where more than two thirds of ICU patients are men.

Focusing on the vital signs data, very little significant dispersion was observed regarding body temperature and a large dispersion in heart rate and both blood pressures, where in these last two variables the highest values are recorded in patients in the ICU. Without a doubt, the biggest difference and the greatest distortion can be evidently seen in the triage variable. Thus, it is a variable that is completely unbalanced between classes, as the urgent and standard classes represent almost the total percentage of the data. It should also be emphasised that no patient classified as non-urgent was admitted to the ICU, and conversely, all the immediate cases were admitted to the ICU.

3.5   Principal Component Analysis

Principal component analysis (PCA) was performed to recognise the underlying structure of the data. The scree plot (see Fig. 1) is a useful graphical technique

Table 2. Demographic and clinical characteristics among COVID-19 patients.

Characteristic		All (%) [range]	Mean ± SD	Non-ICU (%) [range]	Mean ± SD	ICU (%) [range]	Mean ± SD
Age	Age < 30	300(12.39)	18.3 ± 7.94	296 (12.65)	18.68 ± 7.92	4 (4.94)	28 ± 0
	30–49	526 (21.73)	40.21 ± 5.54	524 (22.39)	40.21 ± 5.52	2 (2.47)	39.5 ± 13.44
	50–59	350 (14.47)	54.49 ± 2.68	340 (14.53)	54.48 ± 2.70	10 (12.35)	55 ± 2.26
	60–69	327 (13.50)	64.57 ± 2.94	304 (12.99)	64.57 ± 2.94	23 (28.39)	64.65 ± 3.01
	70–79	424 (17.51)	74.41 ± 2.74	390 (16.67)	74.43 ± 2.75	34 (41.97)	74.15 ± 2.71
	Age > 80	494 (20.40)	86.55 ± 4.55	486 (20.77)	86.55 ± 4.57	8 (9.88)	88 ± 2.98
	All	2421 [0–101]	58.34 ± 22.42	2340 [28–90]	58.05 ± 22.59	81 [0–101]	67.32 ± 13.76
Gender	Female	1166 (48.16)		1143 (48.85)		23 (28.39)	
	Male	1255 (51.84)		1197 (51.15)		58 (71.60)	
Temperature, ºC		[33.3–39.7]	36.4915 ± 6.71	[33.3–39.7]	36.50 ± 1.82	[34.20–38.9]	36.35 ± 0.91
Heart Rate, beats/min		[33–180]	85.63 ± 17.53	[33–180]	85.60 ± 17.53	[40–124]	86.36 ± 17.63
Blood Oxygen Saturation, %		[22–100]	96.38 ± 3.47	[22–100]	96.45 ± 3.38	[77–100]	94.51 ± 5.12
Blood pressure, mm Hg	Systolic	[63–229]	132.15 ± 26.83	[63–225]	131.98 ± 26.75	[80–229]	137.15 ± 28.97
	Diastolic	[35–160]	77.41 ± 46.86	[35–135]	77.39 ± 47.55	[48–160]	77.95 ± 18.31
Triage	I - Immediate	4 (0.16)		0 (0)		4 (4.94)	
	II - Very urgent	186 (7.68)		155 (6.62)		31 (38.27)	
	III - Urgent	1272 (52.54)		1232 (52.66)		40 (49.38)	
	IV - Standard	943 (38.95)		937 (40.04)		6 (7.41)	
	V - Non-urgent	16 (0.67)		16 (0.68)		0 (0)	

to capture the PCs, where those PCs with the sharpest change in the cumulative variance slope should be taken into account as relevant components. The first principal component accounts for 40.93% of the total variance. Additionally, the second principal component represents 33% of the variability of the data. The remaining two variables share 19.43% and 5.83%, respectively. Indeed, the remaining features constitute a very marginal part of the variability (0.81%) and it would be probably negligible. The correlation matrix (see Fig. 2) shows the relationship between each variable and the PC, showing that the three variables most closely related to the three PCs are age, SBP, heart rate, and DPB respectively.

For the calculation of the weights of the main components, the Python sklearn library has been used to import the PCA module, and in the PCA method, seven number of components have been calculated (n_components=7).

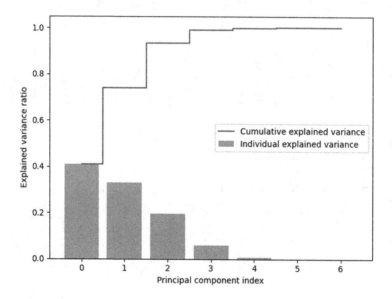

Fig. 1. Scree plot of 7 PCs

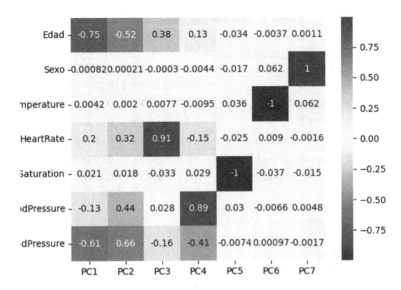

Fig. 2. Correlation matrix between variables and PC

3.6 Logistic Regression Feature Importance

Feature importance is the technique of assigning scores to input features to a predictive model that indicates the relative importance of each feature in predicting a target variable. The results may be interpreted by an expert in the field and could serve as a basis for collecting more or different data. The logistic regression model is a well known binary classifier In this case the classes have been separated into the less severe patients (belonging to triage number 1, 2 and 3) and the more severe classes (belonging to triage number 5 and 4), respectively. This is why the relevance coefficients are both positive and negative, positive scores indicate a feature predicting class 1, while negative scores indicate a feature predicting class 0. A bar chart is then created for the variable importance scores (see Fig. 3). Table 3 reproduces the exact values of the variable importance scores. The results suggest that age and heart rate are likely to be reliable predictors for more severe cases. Thus, most of the variance is covered by these two variables. The model constructed for the calculation of these weights has been obtained an accuracy of 0.937.

Table 3. Feature and score relations

	Feature	LR Score	PC	PC Score
0	Age	0.49531	PC0	0.40934
1	Sex	−0.17903	PC1	0.33
2	Temperature	−0.02779	PC2	0.19429
3	Heart Rate	0.30356	PC3	0.05807
4	Blood Saturation	−0.22305	PC4	0.00759
5	Diastolic Blood Pressure	−0.05858	PC5	0.004832
6	Systolic Blood Pressure	−0.15097	PC6	0.00169

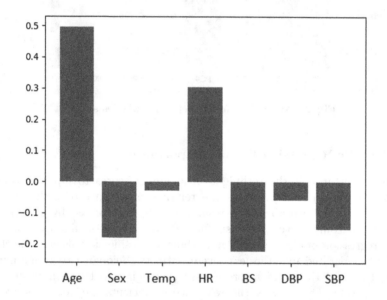

Fig. 3. Bar char with feature importance weights

4   Discussion

An important detection point for COVID-19 cases is the hospital ED. This is where the most severe cases requiring immediate attention should be detected. As it is a transmissible disease, precautionary measures should be taken. To this end, the triage method, which aims to identify the most seriously ill patients, has been taken into account. Early identification of patients, who are at high risk of being weak and who are likely to need hospitalization, would enable to accelerate delays in ED when there are multiple simultaneous admissions. It is a system for selecting and classifying patients on entry to the ED, based on their therapeutic needs and the resources available to care for them. In predictive algorithms, commonly, the number of instances in different classes is expected to be similar. For many real-life classification problems, however, the number of instances is potentially heavily skewed. For algorithms that try to predict the severity level of the patient at ED entry, as is in the case, the given dataset is unbalanced. This mismatch can lead to inaccuracies in model performance since training will be done with a high amount of very urgent and standard levels, and a insignificant amount of data from other levels. Therefore, corrective measures and procedures must be applied to deal with class imbalance, including data modification (undersampling or oversampling).

Apart from previous diagnoses, the great importance of variables such as the patient's BMI in predicting a subsequent worsening of the patient's condition has been highlighted. This data could be obtained either by calculating the BMI directly by the health worker or by the relation of height, weight and age of the subject. But unfortunately, as is clear from Table 1, our dataset does not have enough samples to take these measures into account. Despite the considerable relevance of these variables, less than 10% of samples have been recorded, and are not worth filling in with either the mean or interpolation. Consequently, the creation of a tool to standardise the previous comments of each patient is being considered for the future, creating a dataset that compiles the comorbidities of each patient, or even adding a series of flags with the diseases that are taken as relevant when predicting the gravity of the patient, and consider them as input variables in the model.

Considering all the above mentioned drawbacks a decision support system approach has been created through the creation of a LR as the nature of the data does not allowed to perform other types of prediction techniques, obtaining a very encouraging prediction accuracy results. The result of the PCA found that more than 98% of the data variance is represented by four features, several classification models could be created and compared to each other in order to select the one that best fits the given dataset. On the other hand, the results obtained through LR show that two variables are highly relevant for the detection of the most critical patients, representing almost 70% of the total variance.

5   Conclusion

The aim of this work was to create a detailed analysis of the data collected in order to develop predictive algorithms concerning the care of patients with covid-19 in the future. Following this exhaustive analysis, a forthcoming project is the implementation of a predictive tool for the level of severity of the patient at the entrance of the emergency department, where a set of demographic variables and vital signs could constitute the input variables in the model that will be developed. The main future objective is to detect as many of the most serious cases as possible in order to provide a direct and immediate service, to avoid future complications and to give priority to the patients who need the quickest attention.

However, a considerable difference has been detected in classes in the provided data, including the difference in the number of patients in the central levels of gravity, compared to the more extreme levels. There is also a difference in the age of the patients, where the majority of patients are elderly people and the vast minority belong to the younger ones. This balancing problem could be improved by using data balancing techniques to reduce the bias adding or reducing samples, and subsequently compare the accuracy of various classification models with different balancing techniques in order to achieve the optimal prediction result.

According the results obtained by creating the PCA and LR, where the PCA indicates that the most relevant variables are age, SBP, HR and DBP, and on the other hand LR indicating that age and HR are clear indicators of patient severity, a promising accuracy has been obtained that may lead to possible future implementations, resulting in a potential prediction of over 90% of cases.

Regarding the improvement of the acquisition and management of hospital data, it should be emphasised that as far as the human factor is concerned, the protocols for acquiring data on entry of each patient in the ED should be improved. Establishing guidelines on how to collect, enter and manage the data, as well as the frequency of collection, establishing the same protocol for different types of healthcare professionals and standardising the nomenclature of diagnoses. Furthermore, for admitted patients, direct recording of physiological sensors and storage without human intervention would improve the quality of the data. As far as the improvement of the data system is concerned, better administration and management of databases would be positive, such as being able to cross-reference tables of large quantities of data to detect the most relevant variables and improving access to past clinical histories, e.g. taking into account comorbidities. Future work, can be done creating algorithms to calculate the probability of being hospitalised or admitted to the ICU, as well as standardising patient treatment by performing patient clustering, even the possibility of predicting laboratory results. However, the main conclusion is that the data collecting protocols need to be improved in order to be able to achieve sound results.

Acknowledgments. The authors would like to express their gratitude to Fundación Vital for the financial support to the project "Aportaciones de Modelos Predictivos para COVID-19 basados en Inteligencia Artificial específicos para el Territorio Histórico de Alava - COVID19THA". In addition authors thank to the group "Nuevos desarrollos en salud" of Bioaraba and to Osakidetza-Servicio Vasco de Salud for their collaboration.

References

1. Aljaaf, A.J., Mohsin, T.M., Al-Jumeily, D., Alloghani, M.: A fusion of data science and feed-forward neural network-based modelling of COVID-19 outbreak forecasting in Iraq. J. Biomed. Inform. **118**, 103766 (2021)
2. Chung, H., et al.: Prediction and feature importance analysis for severity of COVID-19 in South Korea using artificial intelligence: model development and validation. J. Med. Internet Res. **23**(4), e27060 (2021)
3. Curigliano, G., et al.: Recommendations for triage, prioritization and treatment of breast cancer patients during the COVID-19 pandemic. The Breast **52**, 04 (2020)

4. Gawriljuk, V., et al.: Machine learning models identify inhibitors of SARS-COV-2. J. Chem. Inf. Model. **61**, 4224–4235 (2021)
5. Kakade, A., Kumari, B., Dholaniya, P.S.: Feature selection using logistic regression in case-control DNA methylation data of Parkinson's disease: a comparative study. J. Theor. Biol. **457**, 14–18 (2018)
6. Khaliq, O., Phoswa, W.: Is pregnancy a risk factor of COVID-19? Eur. J. Obstet. Gynecol. Reprod. Biol. **252**, 06 (2020)
7. Krysko, O., et al.: Artificial intelligence predicts severity of COVID-19 based on correlation of exaggerated monocyte activation, excessive organ damage and hyper-inflammatory syndrome: a prospective clinical study. Front. Immunol. **12** (2021)
8. Lai, L., et al.: Digital triage: novel strategies for population health management in response to the COVID-19 pandemic. Healthcare **8**(4), 100493 (2020)
9. Lastinger, L., Daniels, C., Lee, M., Sabanayagam, A., Bradley, E.: Triage and management of the ACHD patient with COVID-19: a single center approach. Int. J. Cardiol. **320**, 06 (2020)
10. Levenfus, I., Ullmann, E., Battegay, E., Schuurmans, M.: Triage tool for suspected COVID-19 patients in the emergency room: AIFELL score. Braz. J. Infect. Dis. **24**, 08 (2020)
11. Ng, J.J., Choong, A.M.T.L., Ngoh, C.L.Y.: A proposed scoring system for triage of patients who require vascular access creation in times of COVID-19. J. Vasc. Surg. **72**(3), 1150–1151 (2020)
12. Pinto, G.P., Vavra, O., Marques, S.M., Filipovic, J., Bednar, D., Damborsky, J.: Screening of world approved drugs against highly dynamical spike glycoprotein of SARS-CoV-2 using CaverDock and machine learning. Comput. Struct. Biotechnol. J. **19**, 3187–3197 (2021)
13. Rajbahadur, G.K., Wang, S., Ansaldi, G., Kamei, Y., Hassan, A.E.: The impact of feature importance methods on the interpretation of defect classifiers. IEEE Trans. Softw. Eng. **48**(7), 2245–2261 (2021)
14. Sayed, S., Elkorany, A., Sayed, S.: Applying different machine learning techniques for prediction of COVID-19 severity. IEEE Access **9**, 135697–135707 (2021)
15. Wallis, P., Gottschalk, S., Wood, D., Bruijns, S., De Vries, S., Balfour, C.: The cape triage score - a triage system for South Africa. S. Afr. Med. J. **96**, 53–6 (2006)
16. Wang, M., et al.: Deep learning-based triage and analysis of lesion burden for COVID-19: a retrospective study with external validation. Lancet Digital Health **2**, e506–e515 (2020)

Indoor Access Control System Through Symptomatic Examination Using IoT Technology, Fog Computing and Cloud Computing

Raúl López-Blanco[1](✉) , Ricardo S. Alonso[1,2] , Javier Prieto[1] ,
Sara Rodríguez-González[1] , and Juan M. Corchado[1,2]

[1] BISITE Research Group, University of Salamanca Edificio Multiusos I+D+i, Calle
Espejo 2, 37007 Salamanca, Spain
{raullb,ralorin,javierp,srg,corchado}@usal.es
[2] AIR Institute - Deep tech lab IoT Digital Innovation Hub, Salamanca, Spain
{ralonso,corchado}@air-institute.com
https://bisite.usal.es , https://air-institute.com

Abstract. The pandemic experienced in the last two years in the world has led people to be much more careful in their social relations, keeping their social distance and using hygienic prevention measures. However, when it is necessary to enter crowded closed environments, people feel insecure and are more afraid of contagion. This situation leads to the need for measures to control access to public places in order to prevent infection and to reinforce people's confidence. Various devices and solutions exist to control access, ranging from card-based identification to biometric sensors. However, they have shortcomings detected during the pandemic, such as the need to touch elements or the types of computing used, which can compromise security and/or response times. The solution proposed in this article integrates the best of these by incorporating facial recognition using neural networks, the presence or absence of a mask and medical Internet of Things (IoT) devices to monitor pulse, blood oxygen and body temperature. All this technology is used to check whether the person's access is safe for them and others. The data collection process in this system has proven to be efficient thanks to fog computing, which reduces latency times and prevents the user's data from being accessed by third parties while maintaining their privacy.

Keywords: Internet of Things · Access control · Fog computing · Cloud computing · COVID-19

1 Introduction

The rapid spread of the pandemic caused by COVID-19 [22] has shown that society was not prepared to control the transmission of an infectious disease. Moreover, it has shown that the mechanisms proposed to avoid situations such

© Springer Nature Switzerland AG 2022
P. García Bringas et al. (Eds.): HAIS 2022, LNAI 13469, pp. 60–72, 2022.
https://doi.org/10.1007/978-3-031-15471-3_6

as the one experienced worldwide have failed [2]. In fact, among the deaths in December 2020, the leading cause of death, above other diseases [15], is the SARS-CoV-2 virus.

In the search for a return to normality, measures such as social distance or ventilation in enclosed spaces are being implemented [32]. From these measures and the current pandemic phase in which the world finds itself, it is clear that it is necessary to control access to enclosed public spaces such as shopping centers or educational establishments [8]. Citizens feel insecure, especially in enclosed places [3], when it comes to socialising for fear of illness [29].

This work is proposed to try to solve the deficiencies found in some access control systems, some derived from the COVID-19 pandemic [33] and others due to their computing paradigms [12]. The proposal is based on the use of IoT sensors, which are able to collect biomedical data [6] from the user of the system and process it through fog computing to prevent this data from being compromised, using synchronisation with the cloud only if the user wants it.

In addition, the system uses Deep Learning models for facial recognition in particular deep neural networks (CNN) [28]. The implementation of this system aims to increase peace of mind and security of access to areas with a large influx of people [37]. The proposal aims to be an access control system capable of identifying the user [33] by different methods while collecting information on their state of health, with the aim of turning enclosed spaces into safe places [3].

In the rest of the article, the following sections can be found: Sect. 2 contains some related work on access control using IoT sensor technology. Section 3 describes the methodology of operation of this system for access to public and private enclosed spaces. Finally, the conclusions and future work are explained in Sect. 4.

2   Related Works

This section will describe some related work, including researches that uses advanced computing techniques combined with medicine, studies that apply IoT for medical purposes, and work that applies different technologies to access control.

The technological revolution of recent years and changes in legislation [20] have led to a very close collaboration between the medical sector and the technological innovation sector. The management of the large amount of data collected in medicine has been intervened with the help of Big Data technologies [27], and other technologies such as Blockchain through Smart Contracts [5].

The managed data has been used to perform more complex processing, such as Case-Based Reasoning [4], the application of Artificial Intelligence for medical image processing [31] or the use of Machine Learning for efficient processing of medical data [7]. Along with these technologies, different computing paradigms are applied, such as edge computing [30], fog computing [14] and cloud computing [24], which help to expand the fields of application of technology within medicine itself.

In fact, the application of fog computing makes this proposal a novelty compared to others that use cloud computing [21]. The use of this type of computing makes it possible to reduce latencies by being closer to the user [12] and increases the security of the user's data, which is very important to maintain with data as private as that concerning health [34].

Probably one of the most notable and tangible of the novelties introduced in medicine is the Internet of Things (IoT) [25]. This technology has been applied in very diverse fields such as the agricultural world [1] or medicine itself for the monitoring of environments such as, for example, temperature measurement in medical environments [6] or monitoring. Also for other cases such as the care of dependent patients [9] through elements such as smart bracelets [11].

In relation to COVID-19, one of the diseases that this development aims to detect, an attempt has been made to predict its evolution in order to detect the arrival of new waves of contagion through the use of Artificial Intelligence [19], as previously proposed by other studies [35]. Apart from predicting their behaviour, attempts have also been made to detect their presence in users of smartwatches [16] by means of biometric measurements made by them.

Biometric means of access such as fingerprints or cards were already present in places with restricted access due to their uniqueness and security [13]. With the advent of the pandemic, these have been displaced by the need to touch a physical element [33]. In the face of this, other methods of contactless identification [17], such as facial recognition, have become increasingly important [18].

From all the studies discussed above, it is clear that IoT technology and the accompanying information processing technologies (e.g., cloud computing and edge computing), are a growing trend in medicine. This technology, together with access controls, has taken on a new dimension and can monitor vital signs, which can be indicative of diseases such as COVID-19.

The present work aims at the preventive detection of COVID-19, due to the presence of symptoms present in the vital signs. These symptoms have been supported by scientific studies as effects related to the disease or their monitoring has been proven to be effective in preventing aggravation. The vital signs established to be measured have been heart rate [23], body temperature [36] and blood oxygen percentage [26].

3   Operation of the Control System

In this section we will study the workflow of the proposed access control system, its sensorics, its operation and the computing paradigms it uses. The proposed modular access control system is able to recognise the user in front of it through facial recognition, perform biomedical vital signs measurements and identify symptomatic signs of SARS-CoV-2 disease.

3.1 Facial Recognition Module

The face recognition module is able to recognize a system user in the situations described in Fig. 1.

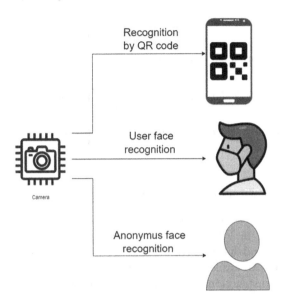

Fig. 1. User identification process.

From the photos taken by the recognition cameras, the system can perform three main recognition functionalities on the photo.

1. First it performs an analysis of the photo in search of a human face, this process is performed by a convolutional neural network (CNN) trained with a dataset of 140 000 faces and which has an architecture as shown in Table 1. Additionally this network has been trained with a dataset containing faces with and without mask which allows it to identify faces with mask, the training of this network has been supervised and based on labeled datasets.
2. When a face is detected, an attempt is made to identify it with one of the users registered in the system. This procedure is done by means of a convolutional neural network (CNN) trained with the photos of the users registered in the system that decomposes the features of each face into an array of features, with which each new photo in which a face is detected is compared.
3. In the case that there is no face, a QR code that can identify the user is searched, and if it is found, it is checked to ensure that it matches the one associated with the user's profile.

Table 1. ResNet-10 customized architecture.

Layers	Output size	ResNet-10
Convolutional_x_1	200×150	7×7, 64, stride 2
Convolutional_x_2	100×75	3×3 max pool, stride 2
		$\begin{bmatrix} 3 \times 3 \\ 3 \times 3 \end{bmatrix} \times 64$
Convolutional_x_3	50×38	$\begin{bmatrix} 3 \times 3 \\ 3 \times 3 \end{bmatrix} \times 128$
Convolutional_x_4	25×19	$\begin{bmatrix} 3 \times 3 \\ 3 \times 3 \end{bmatrix} \times 256$
Convolutional_x_5	13×10	$\begin{bmatrix} 3 \times 3 \\ 3 \times 3 \end{bmatrix} \times 512$

3.2 Steps of the Detection System

The detection and access control system follows a series of steps in its operation, which are described in Fig. 2. First, the user must disinfect its hands, using hydroalcoholic gel and latex gloves. Secondly, the user stands in front of the device, which has two tablets that act as cameras for facial recognition and screens to display instructions and recorded measurements.

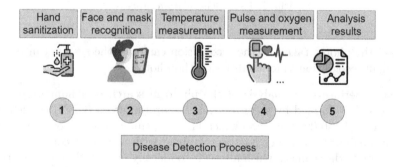

Fig. 2. Disease detection process.

All of these processes results in the recognition of a user and the system will then pass to the measurement phase, at which time the user will place its index finger on the system's pulse oximeter and the inside of its wrist on the temperature sensor and measurements will be taken for 30 s to obtain quality measurements on the constants temperature, pulse and blood oxygen.

After obtaining the results of the measurements the following rules are applied to identify if any of the parameters are out of the expected which may be an indicator of suffering from the disease. These rules have been defined by means of equations defined in chunks. An equation was established for each measured value, the function $f(o)$ Eq. 1 represents the alarm states for the values obtained by the oximetry; the function $g(t)$ (see Eq. 2) represents the values for the body temperature; in Eq. 3 shows the values for the heart rate.

$$f(o) = \begin{cases} A_2 \; if & o < O_1 \\ A_1 \; if \; O_1 \leq o < O_2 \\ A_0 \; if & o \leq O_2 \end{cases} \tag{1}$$

$$g(t) = \begin{cases} A_2 \; if & t \geq T_1 \\ A_1 \; if \; T_2 < t \leq T_3 \\ A_0 \; if & t \leq T_2 \end{cases} \tag{2}$$

$$h(r) = \begin{cases} A_2 \; if & r \geq R_1 \\ A_1 \; if & r < R_2 \\ A_1 \; if \; R_3 \leq r \leq R_1 \\ A_0 \; if \; R_2 \leq r < R_3 \end{cases} \tag{3}$$

$$alert = f(o) + g(t) + h(r) \tag{4}$$

The result obtained from Eq. 4 is the alert level to be displayed on the user interface so that the person in front of the system can see the collected measurements of their vital signs and if their access to the inside of the building can be carried out, this whole process is defined by the flowchart in the Fig. 3.

3.3   Medical Sensors

The chosen electronics include an industrial computer to serve as a local data processing system and two tablets that serve as camera and display. However, the most important sensors in this project are the temperature and pulse oximetry sensors. In order to solve the most accurate way possible to collect these data, sensors certified for use in medical-grade applications have been used.

For the pulse oximeter, a clamp capable of measuring cardiac pulse and oximetry has been used from the German manufacturer MedLab[1], which offers pulse oximeters oriented towards the development of medical-grade products. In addition, its wired connection allows the data to reach the intermediate computer in baud (Bd) making it impossible to expose the information.

The other critical point, the temperature sensor, has been chosen for its good performance in terms of measurement range. For the integration of this sensor, it has been placed on an Arduino nano, introducing a firmware capable of recalibrating it, and it has been covered with a plastic casing to avoid exposure to heat from the rest of the components.

---

[1] MedLab: http://www.medlab.eu/.

3.4   Fog Computing

Undoubtedly one of the differentiating elements of this system is the use of fog computing. The decision to use this type of computing for data processing comes from a considered decision, since the sensors chosen for the system are wired sensors that can only send the information they capture through the cable and are connected to an internal equipment of the machine.

This intermediate equipment is responsible for processing all the data collected by the sensors. It is also responsible for storing the data in a local database in encrypted form using the AES-256-CBC encryption algorithm so that any undesired access cannot know what the database contains.

In the system architecture shown in Fig. 4 there is only one case in which the data is sent to the cloud, this depends solely on the decision of the user who can choose to send its data to the cloud, which will allow it to see its measurement statistics in the system application and if it wishes it can give its data in an anonymised form for research purposes.

The approach of this architecture aims to address two very important fronts in the field of health and access control systems, which are data security [34] and latency reduction in its processing [12].

In short, the advantages that fog computing brings to the architecture of this device and that make it ideal for this system are the following:

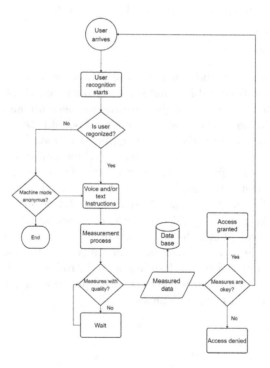

Fig. 3. System flow chart.

– Availability without a network. Users can use the device even if it is not connected to the network. In case the user decides to upload its data to the cloud, the device will save this action and execute it when connected. Compared to a cloud computing case, the system would be unusable if the connection is lost or does not exist.
– Lower latency. By bringing the computing closer to the users, the results obtained are immediate compared to systems containing fog computing that have to add network latency times to the computing times.
– Data security. The wired transport of data and execution within the device means that this type of sensitive data is not compromised by wireless networks, which is the case with cloud computing.

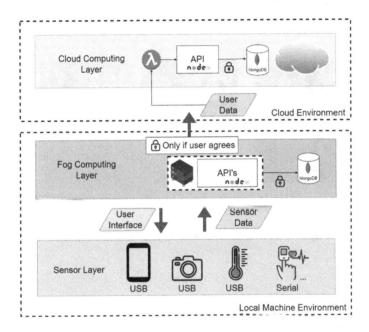

Fig. 4. Fog computing architecture.

3.5   Statistics Management and User Registration Module

The measurements taken by the access control system, as shown in Fig. 3 above, are recorded in the system database and stored in encrypted form to prevent user data from being compromised in the event of unwanted access to the system.

In order to maintain the privacy of the users, their data can be managed by the users themselves and they can decide between several options depending on the use they want to give them. The management of their data can be changed by them at will from the application or the next time they interact with this access control system.

Users can choose to keep their data only on each machine they have interacted with using its fog computing features whereby the sensors perform measurements and the data obtained is processed by the system's internal computer and stored in the local database in an encrypted form. Users can also know their health statistics in their application. In this case the data will be uploaded to the cloud and can be used in an anonymized form for research purposes against COVID-19.

The system also has a symptomatic surveillance interface that can be monitored by security personnel or healthcare personnel to control the access of patients whose biomedical values have been flagged by the system. This interface shows the values of temperature, oximetry, pulse, whether the user is wearing a mask or not and additionally shows the photo with which the user has been recognized, as shown in Fig. 5.

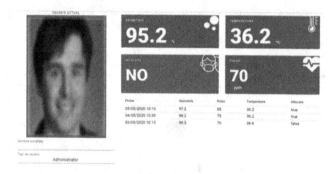

Fig. 5. User control system and latest measurements.

3.6 Accessibility Improvements

Accessibility has been another of the elements taken into account in order to meet the greatest possible functional diversities. Among the accessibility elements included are two screens at different heights for children and people with reduced mobility, voice instructions for the visually impaired and translations into more than five different languages. All these options make this access control system much more accessible and therefore more usable for everyone.

4   Conclusions and Future Work Lines

The conclusions drawn from this system have been obtained from experience with a prototype placed in an office center for a few weeks. It was found that users were initially reluctant to submit to the access control system, but when they saw that it was capable of identifying them and displaying their vital signs, they began to use it regularly.

Another of the milestones of this experimentation was the reduction in latency thanks to the fog computing architecture, which, as some studies have already pointed out [10], reduced data processing time by 10–20% compared to laboratory tests using cloud technologies. The summary of the general conclusions drawn from this study, which support the use of fog computing over cloud computing for this type of device, can be seen in Table 2.

Table 2. Advantages of using fog computing in the presented device.

	Fog computing	Cloud computing
Availability without a network	Yes	No
Time to obtain results	Computing time	Computation time + latency (10–20% more)
Security	Resistant to wireless network attacks	Vulnerable to DoS and MITM attacks

Future work with this symptomatic detection system could include the following lines:

- Implement a system that allows to recognize the COVID vaccination passport or other infectious diseases that may arise from which the population also needs to be vaccinated to access public places.
- To have a federated learning (FL) system that allows models capable of accurately predicting the presence of the disease, fed with data that do not need to leave the machine where they have been collected.
- This symptom based access control system could be scalable to many more diseases with similar symptomatology. Adding a Case-Based Reasoning (CBR) system could detect a wider range of diseases.

Acknowledgements. This work has been partialy supported by the project "XAI - Sistemas Inteligentes Auto Explicativos creados con Módulos de Mezcla de Expertos", ID SA082P20, financed by Junta Castilla y León, Consejería de Educación, and European Regional Development Fund (ERDF) funds.

References

1. Alonso, R.S., Sittón-Candanedo, I., Casado-Vara, R., Prieto, J., Corchado, J.M.: Deep reinforcement learning for the management of software-defined networks in smart farming. In: 2020 International Conference on Omni-layer Intelligent Systems (COINS), pp. 1–6. IEEE (2020)

2. Anderson, M., Mckee, M., Mossialos, E.: Covid-19 exposes weaknesses in European response to outbreaks (2020)
3. Asselmann, E., Borghans, L., Montizaan, R., Seegers, P.: The role of personality in the thoughts, feelings, and behaviors of students in Germany during the first weeks of the covid-19 pandemic. PloS one **15**(11), e0242904 (2020)
4. Blanco, X., Rodríguez, S., Corchado, J.M., Zato, C.: Case-based reasoning applied to medical diagnosis and treatment. In: Omatu, S., Neves, J., Rodriguez, J.M.C., Paz Santana, J.F., Gonzalez, S.R. (eds.) Distributed Computing and Artificial Intelligence. AISC, vol. 217, pp. 137–146. Springer, Cham (2013). https://doi.org/10.1007/978-3-319-00551-5_17
5. Casado-Vara, R., Corchado, J.: Distributed e-health wide-world accounting ledger via blockchain. J. Intell. Fuzzy Syst. **36**(3), 2381–2386 (2019)
6. Casado-Vara, R., De la Prieta, F., Rodriguez, S., Prieto, J., Corchado, J.M.: Cooperative algorithm to improve temperature control in recovery unit of healthcare facilities. In: Omatu, S., Mohamad, M.S., Novais, P., Díaz-Plaza Sanz, E., García Coria, J.A. (eds.) DCAI 2018. AISC, vol. 802, pp. 49–62. Springer, Cham (2020). https://doi.org/10.1007/978-3-030-00524-5_8
7. Castellanos-Garzón, J.A., Costa, E., Corchado, J.M., et al.: An evolutionary framework for machine learning applied to medical data. Knowl.-Based Syst. **185**, 104982 (2019)
8. Cheng, S.Y., Wang, C.J., Shen, A.C.T., Chang, S.C.: How to safely reopen colleges and universities during Covid-19: experiences from Taiwan. Ann. Internal Med. **173**(8), 638–641 (2020)
9. Costa, S.E., Rodrigues, J.J., Silva, B., Isento, J.N., Corchado, J.M.: Integration of wearable solutions in AAL environments with mobility support. J. Medical Syst. **39**(12), 1–8 (2015)
10. Farooqi, A.M., Alam, M.A., Hassan, S.I., Idrees, S.M.: A fog computing model for vanet to reduce latency and delay using 5G network in smart city transportation. Appl. Sci. **12**(4), 2083 (2022)
11. Fraile, J.A., Bajo, J., Corchado, J.M., Abraham, A.: Applying wearable solutions in dependent environments. IEEE Trans. Inf. Technol. Biomed. **14**(6), 1459–1467 (2010)
12. Guan, Y., Shao, J., Wei, G., Xie, M.: Data security and privacy in fog computing. IEEE Netw. **32**(5), 106–111 (2018)
13. Jain, A.K., Pankanti, S., Prabhakar, S., Hong, L., Ross, A.: Biometrics: a grand challenge. In: Proceedings of the 17th International Conference on Pattern Recognition, 2004. ICPR 2004, vol. 2, pp. 935–942. IEEE (2004)
14. Klonoff, D.C.: Fog computing and edge computing architectures for processing data from diabetes devices connected to the medical internet of things. J. Diab. Sci. Technol. **11**(4), 647–652 (2017)
15. Koh, H.K., Geller, A.C., VanderWeele, T.J.: Deaths from Covid-19. JAMA **325**(2), 133–134 (2021)
16. Mishra, T.: Pre-symptomatic detection of Covid-19 from smartwatch data. Nat. Biomed. Eng. **4**(12), 1208–1220 (2020)
17. Moisello, E., Malcovati, P., Bonizzoni, E.: Thermal sensors for contactless temperature measurements, occupancy detection, and automatic operation of appliances during the covid-19 pandemic: A review. Micromachines **12**(2), 148 (2021)
18. Mundial, I.Q., Hassan, M.S.U., Tiwana, M.I., Qureshi, W.S., Alanazi, E.: Towards facial recognition problem in Covid-19 pandemic. In: 2020 4th International Conference on Electrical, Telecommunication and Computer Engineering (ELTICOM), pp. 210–214. IEEE (2020)

19. Muñoz, L., Villarreal, V., Nielsen, M., Caballero, Y., Sittón-Candanedo, I., Corchado, J.M.: Artificial intelligence models and techniques applied to Covid-19: a review. Electronics **10**(23), 2901 (2021)
20. Parliment, E.: Regulation (eu) 2016/679 of the European parliament and of the council of 27 april 2016 on the protection of natural persons with regard to the processing of personal data and on the free movement of such data (2016). https://eur-lex.europa.eu/legal-content/EN/TXT/?qid=1532348683434& uri=CELEX%3A02016R0679-20160504
21. Pasquier, T., Bacon, J., Singh, J., Eyers, D.: Data-centric access control for cloud computing. In: Proceedings of the 21st ACM on Symposium on Access Control Models and Technologies, pp. 81–88 (2016)
22. Perez-Bermejo, M., Murillo-Llorente, M.T.: The fast territorial expansion of the covid-19 in Spain. J. Epidemiol., JE20200123 (2020)
23. Ponomarev, A., Tyapochkin, K., Surkova, E., Smorodnikova, E., Pravdin, P.: Heart rate variability as a prospective predictor of early Covid-19 symptoms. medRxiv (2021)
24. De la Prieta, F., Rodríguez-González, S., Chamoso, P., Corchado, J.M., Bajo, J.: Survey of agent-based cloud computing applications. Future Gener. Comput. Syst. **100**, 223–236 (2019)
25. Prieto, J., Amira, A., Bajo, J., Mazuelas, S., De la Prieta, F.: IoT approaches for distributed computing (2018)
26. Quaresima, V., Ferrari, M.: Covid-19: efficacy of prehospital pulse oximetry for early detection of silent hypoxemia. Crit. Care **24**(1), 1–2 (2020)
27. Sáez, J.A., Corchado, E.: Ksufs: a novel unsupervised feature selection method based on statistical tests for standard and big data problems. IEEE Access **7**, 99754–99770 (2019)
28. Schenkel, T., Ringhage, O., Branding, N.: A comparative study of facial recognition techniques: with focus on low computational power (2019)
29. Segal, S., Sharabany, R., Maaravi, Y.: Policymakers as safe havens: the relationship between adult attachment style, Covid-19 fear, and regulation compliance. Pers. Individ. Diff. **177**, 110832 (2021)
30. Sittón-Candanedo, I., Alonso, R.S., Rodríguez-González, S., García Coria, J.A., De La Prieta, F.: Edge computing architectures in industry 4.0: a general survey and comparison. In: Martínez Álvarez, F., Troncoso Lora, A., Sáez Muñoz, J.A., Quintlán, H., Corchado, E. (eds.) SOCO 2019. AISC, vol. 950, pp. 121–131. Springer, Cham (2020). https://doi.org/10.1007/978-3-030-20055-8_12
31. Srivastava, V., Purwar, R.: An extension of local mesh peak valley edge based feature descriptor for image retrieval in bio-medical images (2018). https://doi.org/10.14201/ADCAIJ2018717789
32. Sun, C., Zhai, Z.: The efficacy of social distance and ventilation effectiveness in preventing Covid-19 transmission. Sustain. Cities Soc. **62**, 102390 (2020)
33. Torres, C.M.C., Gomez, J.F.V., Carvallho, J.J., Trujillo, E.L., Tinjaca, N.B.: Implementation of industry 4.0 technologies in embedded systems for contagion mitigation and Covid-19 control in work areas. In: 2020 Congreso Internacional de Innovación y Tendencias en Ingeniería (CONIITI), pp. 1–6. IEEE (2020)
34. Winnie, Y., Umamaheswari, E., Ajay, D.: Enhancing data security in IoT healthcare services using fog computing. In: 2018 International Conference on Recent Trends in Advance Computing (ICRTAC), pp. 200–205. IEEE (2018)

35. Yigitcanlar, T., Butler, L., Windle, E., Desouza, K.C., Mehmood, R., Corchado, J.M.: Can building "artificially intelligent cities" safeguard humanity from natural disasters, pandemics, and other catastrophes? an urban scholar's perspective. Sensors **20**(10), 2988 (2020)
36. Zhang, L., Zhu, Y., Jiang, M., Wu, Y., Deng, K., Ni, Q.: Body temperature monitoring for regular Covid-19 prevention based on human daily activity recognition. Sensors **21**(22), 7540 (2021)
37. Zhang, M., et al.: Human mobility and Covid-19 transmission: a systematic review and future directions. In: Annals of GIS, pp. 1–14 (2022)

Data Mining and Decision Support Systems

Measuring the Quality Information of Sources of Cybersecurity by Multi-Criteria Decision Making Techniques

Noemí DeCastro-García[1,2](✉) (iD) and Enrique Pinto[2]

[1] Departamento de Matemáticas, Universidad de León, Campus de Vegazana S/n,
24007 León, Spain
ncasg@unileon.es
[2] Research Institute of Applied Sciences in Cybersecurity, Universidad de León,
Campus de Vegazana s/n, 24007 León, Spain
eping@unileon.es

Abstract. Companies and public institutions handle a large amount of data that could affect the quality of the information. This can produce that any decision-making process based on the data is not good enough.

In this article, we present a decision support software to assess the quality of the data of cybersecurity data sources integrating multi-criteria decision-making techniques. These methods let us rank the sources in terms of the quality of the information that they provide. First of them is the well known Weighted Sum Model. Since the evaluation of the cyber-security data sources has to be flexible to the institution's policies, we have adapted the application of the Analytic Hierarchy Process method. Also, we have included the computation of Goodman and Kruskal's coefficient to measure the concordance between the obtained rankings.

We have evaluated the data collected by the cyber data sources through the usual records received in a cybersecurity incident response team. The study was carried out over a real dataset containing $25,297,210$ cybersecurity event records, 27 sources, and 55 types of cybersecurity events. From the results, we have performed the diagnosis identifying those data aspects that can be improved.

Keywords: Data quality · Cybersecurity · Decision making systems · Diagnosis · Data sources

1 Introduction

The increasing number of cybersecurity incidents require that institutions to take advantage of analytical information to generate actionable knowledge. One of the most remarkable examples of the above situation is a CSIRT/CERT (Computer Security Incident Response Team/Computer Emergency Response Team). These

© Springer Nature Switzerland AG 2022
P. García Bringas et al. (Eds.): HAIS 2022, LNAI 13469, pp. 75–87, 2022.
https://doi.org/10.1007/978-3-031-15471-3_7

receive a massive amount of data related to cybersecurity events from multiple sources. However, if the sources' information quality is not adequate, the decisions that the CERTs propose could produce poor quality responses. In fact, ENISA highlights that the quality of actionable information at the disposal of a CERT team is one of the factors that has more effect on the success of an organization's security incident response team [8]. But these teams are usually more focused on responding to cybersecurity incidents than on the lesson learned that let us determine the data diagnosticity [7]. Then, it is necessary to provide them methods to evaluate the quality of the data (DQ) for each source, and for each cybersecurity event typology, to ensure that decisions and countermeasures taken by these teams are as appropriate as possible.

In this work, we present a decision support software that developes an empirical evaluation of sources of cybersecurity events using two Multi-Criteria Decision Making (MCDM) techniques. This ranks the set of cybersecurity sources in terms of the information quality that they provide. The main aim is to give CERTs a way to dynamically determine the sources' information quality and make decisions about the resources investment. The first MCDM method included in the tool is the weighted sum model (WSM). This is used for obtaining a general ranking of all the sources. For those cybersecurity event typologies where several data sources provide the data, we perform not only a WSM ranking but an Analytic Hierarchy Process (AHP). This allows experts to determine which data sources are most appropriate for each case, modifying the quality dimensions' importance easily without needing to fix an exact weight. Besides, in the cases that we have obtained both rankings, we have included Goodman and Kruskal's coefficient computation to measure concordance between the rankings obtained with WSM and AHP. In this way, if the coefficient is low or negative, it could be possible that the user has modified the weights of the quality dimensions only in one MCDM method. It could be helpful as an alert for the user. Also, it could help the expert to fix the weights in the WSM.

The case study was run over a real dataset provided by INCIBE under a confidentiality agreement. It is available in an anonimized form, [26]. We conclude that only one source has obtained a negative quality value. Then, it will be necessary to verify if this source provides some differentiating knowledge or it is better to eliminate it from the system. The rest of the values reveal the strengths and weakness of the information source system.

The article is organized as follows: In Sect. 2, we have included the background. In Sect. 3, we have described how the MCDM techniques are integrated. In Sect. 4, we have described all the experimental details of the work. In Sect. 5, we have developed the results and the discussion. Finally, the conclusions and references are given.

2   Background

We are going to review some concepts related to DQ issues in cybersecurity databases and the main results regarding the DQ model that will be used.

2.1 Related Work

We recall some definitions:

Definition 1. *A cybersecurity event is a cybersecurity change that may have an impact on organizational operations (including mission, capabilities, or reputation) [20].*

Be reminded that each type of cybersecurity event is called event typology. For example, IP Bot, ransomware, etc. These typologies can be categorized [18].

Definition 2. *A cybersecurity event record is the data that generates a cybersecurity event. Usually is collected by an engine in the form of a row with approximately 300–350 variables of different nature.*

Definition 3. *A cyber database is a datastore that collects cybersecurity events reports from several sources. The cyber databases contain a lot of unstructured information together with a high level of expert knowledge. The structure of the data of the security reports varies from machine-generated data to synthetic or artificial data.*

Taking advantage of analytical information in a CERT is a very recent research field, and minimal research is conducted to assess the data generated during the incident response investigations [3,4,11,19]. However, this kind of datastores usually has quality problems. In fact, the 70 % of threat intelligence sources are incomplete, and with a low quality of information, see [24] and [5]. For this reason, the study of DQ is one of the current challenges in cyber threat intelligence [28]. The question is how we can assess the DQ that is stored in a cybersecurity database.

Most of the existing research about DQ focused on determining the DQ dimensions that need to be assessed and how to do it. In [22] and [23], several dimensions of DQ are defined. In [1], six primary dimensions to assess DQ were proposed. With this reduction, the vision is more general and, therefore, applicable to more systems, although the approach remains somewhat static. Other studies introduce more dynamism to quality assessment: [29] establishes the concept of DQ depending on the context in which the data are used, and [16] performs the assessment depending on the data phase. Regarding more modern and complex storage systems, in [2], five main quality dimensions are established adapted to *big data* systems, and in [17], another model is proposed. It identifies three critical characteristics in the quality of the data: contextual adequacy, temporary adequacy, and operational adequacy. In addition, the family of standards *ISO/IEC 25000* [13], relating to the quality of software products and systems, and more specifically, the standard *ISO/IEC 25012 - Data Quality Model* [14], established a set of 15 standard quality dimensions. Besides, there is literature that points out more general frameworks or methodologies for DQ assessment [15]. Recently, in [31], a novelty methodology for measuring DQ dimensions and assessing the level of DQ is proposed.

2.2 DQ Model

The decision support software that is presented in this work is framed in a DQ model that lets us model the dynamics of the system of sources of information. A first version of the DQ model was presented in [6]. This has been extended, adding more quality dimensions and analyses that let us obtain the prescriptive diagnosis. Also, we have included the MCDM techniques.

Quality Dimensions. DQ dimensions that are included in the model are:

1. Quantity: count of received records.
2. Duplicity: number of duplicated records (replicated cases).
3. Completeness: number of records that have data in all significant fields.
4. Information level: average number of records that have useful (or mandatory) data fields.
5. Veracity: the degree to which the data can be considered reliable or true.
6. Unknown veracity: data whose reliability is unknown.
7. Relevance: the importance of the data. In the case of the cybersecurity event records, the severity of the event [4].
8. Frequency: how often the source updates the attribute values (based on the currency metric proposed in [12].
9. Consistency: the degree to which the data are good enough in terms of syntax format, type, range, etc.; that is, if the data need to be fit.
10. Price: cost of each record.
11. Exclusivity: indicates which sources provide data on each type of cybersecurity event. If one cybersecurity event is only provided by only one source, we will refer it as an exclusive cybersecurity event. Consequently, we will call non-exclusive events those that are collected by more than one source.
12. Diversity: indicates the number of event typologies for which the source provides data.

Note that the exclusivity, the diversity, and the price are not included in this case study.

Quality Measurement. We suppose that the assessment is applied over a real dataset containing N cybersecurity event records. Data come from different dynamic sources, $\mathcal{S}_j$ with $j = 1, \ldots, n$, which provide data on various cybersecurity event typologies $\mathcal{E}_l$, with $l = 1, \ldots, m$. Both n and m are variable over time. The data provided by each data source of each event typology are evaluated in each dimension in two different ways:

1. Absolute data: global raw information from the studied records.
2. Normalized data: raw data is normalized through a reference value. The normalization gives us a value $\mathcal{V}_{\mathcal{S}_j}^{\mathcal{E}_l}(\mathfrak{D}_i) \in [0, 1]$ for each quality dimension, event typology, and datasource.

3   Ranking of Sources by MCDM

We have integrated two MCDM methods: (1) Weighted sum model (WSM) that lets us obtain a general and non-exclusive event typology report ranking; (2) Analytic hierarchy process (AHP) that provides a non-exclusive event typology ranking.

3.1   Weighted Sum Model (WSM)

The first MCDM method that we have included in the tool is the simplest one: the weighted sum model (WSM) [9,30]. We have modeled the assessment process in the following way: once all the $\mathcal{V}_{S_j}^{\mathcal{E}_l}(\mathfrak{D}_i)$ values have been obtained, $\mathcal{V}_{S_j}^{\mathcal{E}_l}(\mathfrak{D}_i)$ is compared to tolerance thresholds $I_{i,l} = [a_{i,l}, b_{i,l}] \in [0,1]$ assigning a quality grade to each dimension in each event typology by the function described as follows:

$$\text{Degree}_{S_j}^{\mathcal{E}_l}(\mathfrak{D}_i) = \begin{cases} \text{High} & \text{if } \mathcal{V}_{S_j}^{\mathcal{E}_l}(\mathfrak{D}_i) > b_{i,l} \\ \text{Acceptable} & \text{if } \mathcal{V}_{S_j}^{\mathcal{E}_l}(\mathfrak{D}_i) \in I_{i,l} \\ \text{Low} & \text{if } \mathcal{V}_{S_j}^{\mathcal{E}_l}(\mathfrak{D}_i) < a_{i,l} \end{cases} \tag{1}$$

These thresholds $I_{i,l}$ are defined previously for each quality dimension and event typology. The reference values for the thresholds are flexible, and they can be defined and modified by the users. Otherwise, they remain by default. It is needed that these thresholds can be tuned at any time. For example, the experts can consider acceptable that the frequency in which a source provides information about ransomware is twelve hours. However, in an alert state, this frequency needs to be shorter (six hours or less).

The global quality value for each source is computed as follows:

$$\mathcal{V}_{S_j}^{\mathcal{E}_l} = w_+ \cdot \frac{\#\text{High}_{S_j}^{\mathcal{E}_l}(\mathfrak{D}_i)}{\#(\mathfrak{D}_i)} + w_0 \cdot \frac{\#\text{Acceptable}_{S_j}^{\mathcal{E}_l}(\mathfrak{D}_i)}{\#(\mathfrak{D}_i)} + w_- \cdot \frac{\#\text{Low}_{S_j}^{\mathcal{E}_l}(\mathfrak{D}_i)}{\#(\mathfrak{D}_i)} \tag{2}$$

where $\#\text{Degree}_{S_j}^{\mathcal{E}_l}(\mathfrak{D}_i)$ denotes the number of dimensions $\mathfrak{D}_i$ for S_j and $\mathcal{E}_l$ that have been assignated to Degree by (1). $\#(\mathfrak{D}_i)$ denotes the number of the assessed dimensions. The weights w_i can be modified by the user in terms to reward or penalyze the high and low level of the DQ in greater or lesser degree. By default, $w_+ = 1, w_0 = 0.5$ and $w_- = -1$. Then $\mathcal{V}_{S_j}^{\mathcal{E}_l} \in [-\frac{1}{2}, \frac{3}{2}]$.

3.2   Analytic Hierarchy Process (AHP)

The AHP allows us to structure knowledge in a similar way when the process is carried out by a human brain, helping experts modify the importance of the quality dimensions easily without needing to fix an exact weight. The hierarchical structure of the AHP framework consists of three levels. We have adapted these levels as well as the method in order to apply it to our case study. The structure is defined as follows:

1. *Level 1*: the goal is to rank the different sources in terms of the quality of the cybersecurity reports that they provide. That is, to order the sources in terms of the quality information.
2. *Level 2*: we take the set of the defined DQ dimensions as the criteria.
3. *Level 3*: the sources represent the alternatives.

The method performs the following computation steps for determining the final ranking of the alternatives concerning the overall goal:

1. Weight the dimensions: to evaluate the importance of each dimension $\mathfrak{D}_i$ against the others $\mathfrak{D}_k$, a pairwise comparison based on preference measurements by a 1-to 9-point Saaty's scale $\{1, 3, 5, 7, 9\}$ is needed to fill. This request is questioned to the users by the screen at the beginning of the execution of the script. In another case, it could be fixed by default with all ones. Then, a matrix of weights W is constructed. Its components take values from 1, if $\mathfrak{D}_i$ and $\mathfrak{D}_k$ are of equal importance, to 9 if $\mathfrak{D}_i$ is strongly more important than $\mathfrak{D}_k$.
2. Consistency level measurement of the matrix W: since real experts' preferences carry out the comparisons, it is needed to measure the consistency of the matrix W. This evaluation is conducted by the cconsistency matrix ratio (CR), and the consistency matrix index (CI) as the deviation or degree of consistency:

$$CI = \frac{\lambda_{max} - t}{t - 1} \text{ and } CR = \frac{CI}{RI} \tag{3}$$

where RI denotes the consistency index of a pairwise matrix generated randomly and λ_{max} is the largest eigenvalue for the matrix W. The random indexes (RI) chosen for the matrix consistency are extracted from [27]. If the matrix is consistent, we move to the next step. Otherwise, the tool turns back to the request of the comparisons of the dimensions.
3. Once the dimensions (criteria) have been weighted, the sources (alternatives) are compared. The source's assessment implies constructing a matrix for each dimension, $A(\mathfrak{D}_i)$, in which the sources are compared concerning this dimension. The eigenvalues of the matrices let us determine which is the best source in function of the dimensions and their importance. This evaluation in the tool is carried out automatically. With this goal, we first need to re-assign the grades of each source in each dimension. This process is performed as follows instead of the assignment given in Eq. (1):

$$\text{Degree}_{\mathcal{S}_j}^{\mathcal{E}_l}(\mathfrak{D}_i) = \begin{cases} \text{High} & \text{if } \mathcal{V}_{\mathcal{S}_j}^{\mathcal{E}_l}(\mathfrak{D}_i) \in [b_{i,l}, 1] \\ \textbf{High Medium} & \text{if } \mathcal{V}_{\mathcal{S}_j}^{\mathcal{E}_l}(\mathfrak{D}_i) \in [\frac{a_{i,l}+b_{i,l}}{2}, b_{i,l}) \\ \text{Low Medium} & \text{if } \mathcal{V}_{\mathcal{S}_j}^{\mathcal{E}_l}(\mathfrak{D}_i) \in [a_{i,l}, \frac{a_{i,l}+b_{i,l}}{2}) \\ \text{Low} & \text{if } \mathcal{V}_{\mathcal{S}_j}^{\mathcal{E}_l}(\mathfrak{D}_i) \in [0, a_{i,l}) \end{cases} \tag{4}$$

Once we have performed the above assignment, we have used the following rule for determining the component of the pairwise comparison for the sources for each dimension:

$$a_{jh}(\mathfrak{D}_i) = \begin{cases} 1 & \text{if } Colour(\mathcal{S}_j) = Colour(\mathcal{S}_h) \\ 3 & \text{if } (\mathsf{S}_j \wedge \mathsf{S}_h) \\ 5 & \text{if } (\mathsf{S}_j \wedge \mathsf{S}_h) \vee (\mathsf{S}_j \wedge \mathsf{S}_h) \\ 7 & \text{if } (\mathsf{S}_j \wedge \mathsf{S}_h) \vee (\mathsf{S}_j \wedge \mathsf{S}_h) \\ 9 & \text{if } (\mathsf{S}_j \wedge \mathsf{S}_h) \end{cases} \tag{5}$$

where a_{jh} is the component that results from comparing the source S_j with the source S_h under the dimension $\mathfrak{D}_i$.

4. Again, the consistency ratio is computed.
5. All the procedure described is carried out for each type of cybersecurity event typology.

For the described procedure, the source code presented in [21] has been adapted.

3.3 Concordance Between Rankings

For non-exclusive cybersecurity events, we have included the computation of the Goodman and Kruskal's coefficient [10] in order to obtain a measure of concordance between the rankings obtained with WSM and AHP. This nonparametric measure tells us how closely is the association that exists between two variables measured on an ordinal scale. Also, both variables to compare do not need to have the same categories in the ordinal variable. Goodman and Kruskal's gamma uses the following formula:$\gamma = \frac{N_c - N_d}{N_c + N_d}$ where N_c is the total number of pairs that rank the same (concordant pairs), and N_d is the number of pairs that do not rank the same (discordant pairs). Values range from −1 (negative association or perfect inversion) to 1 (positive association or perfect agreement). A value of zero indicates the absence of association. The closer you get to a 1 (or −1), the stronger the relationship.

4 Experimental Section

The study was run over a real dataset containing $N = 25,297,210$ cybersecurity event records. The data sample corresponds to a time window of approximately 24 h with $n = 27$ and $m = 55$. This dataset has been provided by INCIBE under a confidentiality agreement. For this reason, for this work, the sample has been anonymized. A Python script has been used to anonymize the data sample. The sample is available in [26].

The data sources in the sample are own, public or private. It is important to note that not all data sources contribute with the same amount of records, the number of variables, or type of information.

1. Own: $\mathcal{S}_{08}$, $\mathcal{S}_{11}$, $\mathcal{S}_{12}$, $\mathcal{S}_{13}$, $\mathcal{S}_{15}$, $\mathcal{S}_{16}$, $\mathcal{S}_{17}$, $\mathcal{S}_{18}$, $\mathcal{S}_{22}$, $\mathcal{S}_{25}$.
2. Public: $\mathcal{S}_{01}$, $\mathcal{S}_{02}$, $\mathcal{S}_{09}$, $\mathcal{S}_{21}$, $\mathcal{S}_{26}$, $\mathcal{S}_{27}$.
3. Private: $\mathcal{S}_{03}$, $\mathcal{S}_{04}$, $\mathcal{S}_{05}$, $\mathcal{S}_{06}$, $\mathcal{S}_{07}$, $\mathcal{S}_{10}$, $\mathcal{S}_{14}$, $\mathcal{S}_{19}$, $\mathcal{S}_{20}$, $\mathcal{S}_{23}$, $\mathcal{S}_{24}$.

The tests have been carried out with Windows 10, 1 CPU (Intel I5 CPU Model), and a RAM Memory of 8 GB have been used. The software is developed in Python 3.7. The source code is available in a Github [25] under GNU General Public License v3 (GNU GPLv3).

5   Results and Discussion

First of all, we note that the experimentation is carried out with the values of weights of WSM by default ($w_+ = 1, w_0 = 0.5, w_- = -1$). All the dimensions in the AHP are proposed with the same importance (value=1). Also, the thresholds are configurated by the experts of the CERT from which the dataset comes in terms of the cybersecurity event typologyes that are included in the case study.

In Table 1, the values of the general quality score ranking that we have obtained by the WSM are detailed ($\overline{X} = 0.4449, s = 0.2492, Me = 0.4444$). The worst source is S09 with a negative value of quality, and the best one is the source of S25. Considering that the maximum value of the quality score is 1.5, we can conclude that, in general, the level of the quality information in the system is not high enough. Further, from position 9, the quality of the sources does not achieve the value of 0.5. Then, it will be neccessary to analyze if the sources in these positions provide some significant value to the system, these can be improved in some dimension or, conversely, it is better to remove them from the system, more in those cases in which the CERT is paying for the information.

Table 1. General datasource quality ranking with WSM

Position	Datasource	Quality score	Position	Datasource	Quality score
1	S_25	**0.815**	10	S_21	0.444
2	S_22	0.778	11	S_19	0.417
2	S_11	0.778	12	S_08	0.408
2	S_13	0.778	13	S_20	0.389
2	S_16	0.778	14	S_27	0.333
3	S_15	0.715	15	S_10	0.264
4	S_26	0.667	16	S_02	0.25
5	S_17	0.648	17	S_24	0.18
6	S_18	0.584	18	S_05	0.167
7	S_23	0.567	19	S_14	0.167
8	S_06	0.527	20	S_01	0.0833
9	S_04	0.445	21	S_07	0.0
10	S_03	0.444	22	S_09	−0.056
10	S_12	0.444			

We can observe in Fig. 1 the average of the quality score by the nature of the source. Clearly, the sources that belong to the own CERT provides the information with the best quality. However, the average quality of the information of private sources is 0.3242. It is a low score, especially when such sources receive money for their information. It is crucial to analyze if they provide exclusive data about any event typology or provide some valuable knowledge. Then, let us now continue with the ranking of non-exclusive events. Here, is where the dimensions of exclusivity and diversity gain significance.

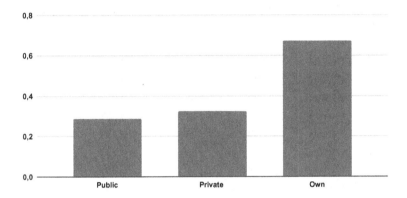

Fig. 1. Axis X: General quality score with WSM. The value of the weights and the criteria are fixed by default. Axis Y: type of source.

In Table 2, we can see the rankings obtained by the WSM and the AHP. We can observe that the ranking is the same in those events that are reported for a few sources. However, in those events that are provided for more sources (E28 and E53, for example), then we find more differences. The ranking obtained by the AHP method provides fewer ties; it is more concrete to choose the best source for each case. Regarding the importance of maintaining some private sources, we can observe that in those events typologies in which more sources collect data, the nature of the source with the best quality of information is private. This is the case in E28 (8 sources) and E53(7 sources). The best source for E54 (6 sources) is public, but the second one is private. But, it is also true that we could reduce the private sources to S_{06}, S_{19}, S_{23}. In a deeper analysis, we will need to check whether the rest of the private sources are actively involved in the system providing, for example, data about exclusive events.

Finally, although we have observed that the ranking provided by AHP is more effective, it is important to determine if the differences that are found between both rankings are significant. The values of Gamma and Kruskal's coefficients are all 1 except for E28 and E53 (0.760 and 0.889, respectively). This means that, although the ranking of AHP can be slightly different from that provided by WSM, both of them are consistent since the values of the Gamma coefficients in $E28$ and $E53$ are high enough to consider that there is a positive association between the rankings, although it is not perfect.

Table 2. Datasource quality ranking for not exclusive events

Event	Source	Nature of the source	Quality score with WSM	Position with WSM	Quality score with AHP	Position with AHP
E06	S_02	Public	0.556	1	0.5	1
	S_08	Own	0.556	1	0.5	1
E28	S_04	Private	0.333	4	0.161	2
	S_05	Private	0.333	4	0.098	5
	S_10	Private	0.389	3	0.129	4
	S_12	Own	0.444	2	0.152	3
	S_17	Own	0.167	6	0.086	7
	S_19	Private	0.667	1	0.233	1
	S_23	Private	0.333	4	0.087	6
	S_24	Private	0.222	5	0.054	8
E30	S_15	Own	0.611	1	0.567	1
	S_23	Private	0.389	2	0.433	2
E31	S_15	Own	0.556	2	0.433	2
	S_25	Own	0.778	1	0.567	1
E35	S_18	Own	0.778	1	0.567	1
	S_23	Private	0.556	2	0.433	2
E47	S_01	Public	−0.056	1	0.247	2
	S_02	Public	−0.056	1	0.236	3
	S_09	Public	−0.056	1	0.281	1
	S_24	Private	−0.056	1	0.236	3
E49	S_10	Private	0.444	2	0.253	2
	S_17	Own	0.444	2	0.253	2
	S_23	Private	0.889	1	0.495	1
E53	S_01	Public	0.222	3	0.179	2
	S_06	Private	0.611	1	0.247	1
	S_14	Private	0.167	4	0.109	5
	S_18	Own	0.389	2	0.143	3
	S_19	Private	0.167	4	0.109	5
	S_23	Private	0.167	4	0.13	4
	S_24	Private	−0.222	5	0.084	6
E54	S_03	Private	0.444	1	0.188	2
	S_05	Private	0.0	3	0.098	4
	S_06	Private	0.444	1	0.188	2
	S_07	Private	0.0	3	0.092	5
	S_20	Private	0.389	2	0.153	3
	S_21	Public	0.444	1	0.28	1

6   Conclusions and Future Work

This work presents a decision support software with which we perform an empirical assesment of the quality of the provided data by cybersecurity sources. The analysis includes the use of multi-criteria decision-making methods for the evaluation taking into account the importance of each evaluated dimension. From the results, we obtain a rank of sources depending of the quality of the data, an we make a diagnosis identifying some aspects that can be improved and the priority areas on which it is necessary to act.

In an extended work, the assessment of the quality of the information of the sources can be more extense. The implementation that we have designed includes the construction of reports to make deeper analyses, and obtain a prescriptive diagnosis about the situation.

Acknowledgements. This work is partially supported by the Spanish National Cybersecurity Institute (INCIBE) under contracts art.83, keys: X43 and X54.

References

1. Askham, N., Cook, D., Doyle, M., Fereday, H., Gibson, M., Landbeck, U., et al.: The six primary dimension for data quality assessment defining data quality dimensions: white paper dama (2013). https://www.whitepapers.em360tech.com/wp-content/files_mf/1407250286DAMAUKDQDimensionsWhitePaperR37.pdf
2. Cai, L., Zhu, Y.: The challenges of data quality and data quality assessment in the big data era. Data Sci. J. **14**, 1–10 (2015)
3. Carriegos, M.V., Fernández Díaz, R.Á.: Towards forecasting time-series of cybersecurity data aggregates. In: Herrero, Á., Cambra, C., Urda, D., Sedano, J., Quintián, H., Corchado, E. (eds.) CISIS 2019. AISC, vol. 1267, pp. 273–281. Springer, Cham (2021). https://doi.org/10.1007/978-3-030-57805-3_26
4. DeCastro-García, N., Muñoz Castañeda, A.L., Fernández-Rodríguez, M.: Machine learning for automatic assignment of the severity of cybersecurity events. Comput. Math. Methods **2020** (2020)
5. DeCastro-García, N., Muñoz Castañeda, A.L., Fernández-Rodríguez, M., Carriegos, M.V.: On detecting and removing superficial redundancy in vector database. Math. Prob. Eng. **2018**, 623–640 (2018)
6. DeCastro-García, N., Pinto, E.: A data quality assessment model and its application to cybersecurity data sources. In: Herrero, Á., Cambra, C., Urda, D., Sedano, J., Quintián, H., Corchado, E. (eds.) CISIS 2019. AISC, vol. 1267, pp. 263–272. Springer, Cham (2021). https://doi.org/10.1007/978-3-030-57805-3_25
7. Ertemel, A.: Consumer insight as competitive advantage using big data and analytics. Int. J. Comm. Finan. **1**, 45–51 (2015)
8. European Network and Information Security Agency: Actionable information for security incident response (2015). https://www.enisa.europa.eu/publications/actionable-information-for-security
9. Fishburn, P.: Additive utilities with incomplete product set: applications to priorities and assignments. Oper. Res. **15**, 537–542 (1967)
10. Goodman, L., Kruskal, W.: Measures of association for cross classification. J. Am. Stat. Assoc. **49**, 732–764 (1954)

11. Grispos, G., Glisson, W.B., Storer, T.: How good is your data? investigating the quality of data generated during security incident response investigations. In: Proceedings of the 52nd Hawaii International Conference on System Sciences, pp. 7156–7165 (2019)

12. Hinrichs, H.: Thesis disseration: Data quality management in data warehouse systems (2002)

13. International Organization for Standardization and International Electrotechnical Commission: Norma iso/iec 25000 (2018). https://iso25000.com/index.php/normas-iso-25000/iso-25012

14. International Organization for Standardization and International Electrotechnical Commission: Norma iso/iec 25012 (2018). https://iso25000.com/index.php/normas-iso-25000/iso-25012

15. Krogstie, J., Lindland, O.I., Sindre, G.: Defining quality aspects for conceptual models. In: Falkenberg, E.D., Hesse, W., Olivé, A. (eds.) Information System Concepts. IAICT, pp. 216–231. Springer, Boston, MA (1995). https://doi.org/10.1007/978-0-387-34870-4_22

16. Liu, L., Chi, L.N.: Evolutional data quality: a theory-specific view. In: Proceedings of the Seventh International Conference on Information Quality, pp. 292–304 (2002)

17. Merino, J., Caballero, I., Rivas, B., Serrano, M., Piattini, M.: A data quality in use model for big data. Future Gener. Comput. Syst. **63**, 123–130 (2016)

18. Ministerio de Interior: Guía nacional de notificación y gestión de ciberincidentes (2019). https://www.incibe-cert.es/sites/default/files/contenidos/guias/doc/guia_nacional_notificacion_gestion_ciberincidentes.pdf

19. Naseer, H., Maynard, S.B., Desouza, K.C.: Demystifying analytical information processing capability: the case of cybersecurity incident response. Decis. Supp. Syst. **143**, 113476 (2021). https://doi.org/10.1016/j.dss.2020.113476

20. National Institute of Standards and Technology (NIST): definition of cybersecurity event (2021). https://csrc.nist.gov/glossary/term/cybersecurity_event

21. Papathanasiou, J., Ploskas, N.: AHP. In: Multiple Criteria Decision Aid. SOIA, vol. 136, pp. 109–129. Springer, Cham (2018). https://doi.org/10.1007/978-3-319-91648-4_5

22. Pipino, L.L., Lee, Y.W., Wang, R.: Data quality assessment. Commun. ACM Dig. Libr. **45**, 211–218 (2002)

23. Piprani, B., Ernst, D.: A model for data quality assessment. In: Meersman, R., Tari, Z., Herrero, P. (eds.) OTM 2008. LNCS, vol. 5333, pp. 750–759. Springer, Heidelberg (2008). https://doi.org/10.1007/978-3-540-88875-8_99

24. Ponemon Institute LLC: The value of threat intelligence: Annual study of north american & united kingdom companies (2019). https://www.anomali.com/resources/whitepapers/2019-ponemon-report-the-value-of-threat-intelligence-from-anomali

25. Research Institute of Applied Sciences in Cybersecurity: Riasc data quality evaluation tool (2021). https://github.com/amunc/DataQualityEvaluation

26. RIASC: Data sample (2021). https://drive.google.com/drive/folders/1glEc9Wl1WsFrwQcvBU_cpmaQu-SiO6ck?usp=sharing

27. Saaty, T.: A scaling method for priorities in hierarchical structures. J. Math. Psychol. **15**, 234–281 (1977)

28. Schaberreiter, T., et al.: A quantitative evaluation of trust in the quality of cyber threat intelligence sources. In: Proceedings of the 14th International Conference on Availability, Reliability and Security. ARES 2019. Association for Computing Machinery, New York (2019)

29. Strong, D.M., Lee, Y.W., Wang, R.Y.: Data quality in context. Commun. ACM Dig. Libr. **40**, 103–110 (1997)
30. Triantaphyllou, E.: Multi-Criteria Decision Making: A Comparative Study. Kluwer Academic Publishers (now Springer), Dordrecht (2000). https://doi.org/10.1007/978-1-4757-3157-6_2
31. Valencia-Parra, A., Parody, L., Varela-Vaca, A.J., Caballero, I., Gómez-López, M.T.: Dmn4dq: when data quality meets dmn. Decis. Supp. Syst. **141**, 113450 (2021). https://doi.org/10.1016/j.dss.2020.113450

A Case of Study with the Clustering R Library to Measure the Quality of Cluster Algorithms

Luis Alfonso Pérez Martos[1]([✉]) [iD], Ángel Miguel García-Vico[2]([✉]) [iD],
Pedro González[3]([✉]) [iD], and Cristóbal José Carmona[3]([✉]) [iD]

[1] Department of Computer Science, University of Jaen, Jaén, Spain
lapm0001@gmail.com
[2] Andalusian Research Institute in Data Science and Computational Intelligence
(DaSCI), University of Granada, Granada, Spain
agvico@decsai.ugr.es
[3] Andalusian Research Institute in Data Science and Computational Intelligence
(DaSCI), University of Jaen, Jaén, Spain
{pglez,ccarmona}@decsai.ugr.es

Abstract. This pattern has been addressed by researchers, as it allows us to categorize unlabelled data, and its use is suitable for automatic data classification to reveal concentrations of data. While many classification methods have been proposed, there are no criteria on which methods are more suitable for a given dataset. This paper presents the *Clustering* library which contains a set of well-known clustering algorithms to cover two objectives: first, grouping data in a homogeneous way by establishing differences between clusters; and second, generating a ranking between algorithms and the attributes analyzed in the dataset. Finally, through the GUI we can run the experiment without knowing the library parameters.

Keywords: Unsupervised learning techniques · Clustering · R Package

1   Introduction

Clustering is the unsupervised classification of patterns into groups. Data clustering [1,2] is one of the main techniques within data mining. Its main aim is to partition a set of data objects according to some criteria, such that similar objects are grouped into the same cluster, and dissimilar objects are divided into different clusters. This allows extracting relevant knowledge about the behavior of the data. Some common application examples are market segmentation [3], social network analysis [4], or anomaly detection [5], amongst others. In all these examples, the aim of grouping similar objects is achieved by maximizing, or minimizing, some quality measures used in the distribution of the cluster data. In the literature, several measures can be found to validate the quality of the clusters. The first type of measure is based on external metrics, which involves

© Springer Nature Switzerland AG 2022
P. García Bringas et al. (Eds.): HAIS 2022, LNAI 13469, pp. 88–97, 2022.
https://doi.org/10.1007/978-3-031-15471-3_8

evaluating the results of an algorithm based on a pre-established structure. This is imposed on a dataset and reflects our intuition about the clustering structure of the dataset. The second type of measure is based on internal metrics, where the results of a clustering algorithm are evaluated in terms of the characteristics of the instances belonging to each cluster, e.g. the proximity matrix.

In the literature related to R implementations of clustering, we can find multiple algorithms and similarity measures. In our package, we have reviewed the Clustering Task View[1] and selected those algorithms that have been cited most often. The problem there is not much software that implements quality criteria on clustering to measure and analyze the quality of the different algorithms. Specifically, there are some problems associated with current libraries:

- It is not possible to work with a set of datasets, so it is difficult to compare several algorithms.
- It is not possible to standardize the input data for the algorithms.
- It is not focusing on measuring the quality of clusters.
- A limited number of libraries include a graphical interface.

This contribution presents a case study, using the developed library applied to a dataset, to bring new functionalities to the R community working with Clustering. The structure of this contribution is as follows. First, Sect. 2 presents the most important functionalities of the package. Section 3 presents a complete example of how the package works on a dataset and Sect. 4 has a graphical representation of the results. Finally, Sect. 5 outlines the conclusions and future directions.

2   The Clustering Package

Clustering library. It is a library that allows us to compare multiple clustering algorithms simultaneously while assessing the quality of the clusters extracted. The purpose of this library is the evaluation of a set of datasets to determine which attributes are the most suitable for obtaining clusters of interest. Therefore, evaluations of the clusters created, how they have been distributed, whether the distributions are uniform and how they have been categorized from the data can be performed. Finally, a GUI is also included to facilitate the study and analysis of the algorithms.

The algorithms included in the *Clustering* library are:

- agnes [7].
- clara [8].
- daisy [10].
- diana [9].
- fanny [11].
- fuzzy cm [12].

---

[1] https://cran.r-project.org/web/views/Cluster.html.

- fuzzy gk [14].
- fuzzy gg [13].
- hcluster [15].
- mona [16].
- pam [17].
- pvpick [18].
- pvclust [18].

The *Clustering* library implements a set of functionalities to address the issues raised by other packages.

- `clustering()`: It is the core function of the package. It is in charge of generating the `clustering` object. This object stores the results of the *Clustering* library execution. The class exports the well-known S3 methods print() and summary() that show the data structure without codification, and a summary with basic information about the dataset respectively. The library allows sorting and filtering operations for further processing of the results. The '[' operator makes use of the filter method of the dplyr library [19].
- External metrics. these methods are in charge of assessing the quality of the clusters extracted using the dataset's variables as target:
  - `best_ranked_external_metrics()`: The execution of this method gets the attributes with better behavior by algorithm, the measure of distance and number of clusters in a ranking way.
  - `evaluate_best_validation_external_by_metrics()`: This method groups the data by algorithm and distance measure, instead of obtaining the best attribute from the data set.
  - `evaluate_validation_external_by_metrics()`: It groups the results of the execution by algorithms.
  - `result_external_algorithm_by_metric()`: It is used for obtaining the results of a given algorithm grouped by the number of clusters.
- Internal metrics. It incorporates the same set of methods as mentioned above for external metrics:
  - `best_ranked_internal_metrics()`: The execution of this method gets the attributes with better behaviour by algorithm, measure of distance and number of clusters in a ranking way.
  - `evaluate_best_validation_internal_by_metrics()`: This method groups the data by algorithm and distance measure, instead of obtaining the best attribute from the data set.
  - `evaluate_validation_internal_by_metrics()`: It groups the results of the execution by algorithms.
  - `result_internal_algorithm_by_metric()`: It is used for obtaining the results of a given algorithm grouped by number of clusters.
- `plot_clustering()`: This method represents the results of clustering in a bar chart. The graph represents the distribution of the algorithms based on the number of partitions and the evaluation metrics employed, which can be internal or external.

- `export_external_file()`: It exports the results of external metrics in LATEX format, for integration into documents with that format.
- `export_internal_file()`: This method is used in order to export the results of the internal metrics.

3 A Case Study Using the *Clustering* Library on the Dataset of Deaths

At this point, we are going to work with the library to examine the real potential it has in evaluating algorithms, and internal and external quality measures as well as being able to work with a range of clusters that allows us to select the best algorithm from the configured data. The dataset we are going to use as an example is available on the web https://datos.gob.es/es/ under the name Deaths by most frequent causes, Autonomous Community, the autonomous city of death, period, and sex. This dataset includes information on Deaths by most frequent causes, community, the autonomous city of death, period, and sex. The attributes of the dataset are detailed below:

- Autonomous community and city of death: It is a nominal attribute with the name of the Autonomous regions of Spain. The values are: Andalucía, Aragón, Principado de Asturias, Illes Balears, Canarias, Cantabria, Castilla y León, Castilla-La Mancha, Cataluña, Ceuta, Comunitat Valenciana, Extremadura, Galicia, Comunidad de Madrid, Melilla, Región de Murcia, Comunidad Foral de Navarra, País Vasco, La Rioja.
- Most frequent causes: It is a nominal attribute with the most frequent types of deaths. The set of values for this attribute are: Cáncer de bronquios y pulmón, Cáncer de colon, Cáncer de mama, Cáncer de páncreas, Cáncer de próstata, Covid-19 Virus identificado, Covid-19 Virus no identificado (sospechoso), Demencia, Diabetes mellitus, Enf. crónicas de las vías respiratorias inferiores (ECVRI), Enfermedad de Alzheimer, Enfermedad hipertensiva, Enfermedades cerebrovasculares, Enfermedades isquémicas del corazón, Insuficiencia cardiaca, Insuficiencia renal, Neumonía, Resto enfermedades.
- Sex: It is a nominal attribute with the types of sex: Hombres, Mujeres.
- Period: It is a nominal attribute with the periods: 2020M01 2020M02 2020M03 2020M04 2020M05.
- Total: It attributes represent the number of occurrences. [1,2843].

From the set of metrics with which the package works, we are going to use Precision and Recall [20] as internal measures and Silhouette [6] as an internal measure. When working with Clustering it is necessary to indicate the number of clusters we are going to work with, that is why we are going to determine the optimal number of clusters using the elbow method as shown in the Fig. 1.

In our case, we will choose the range of 7 to 9 clusters. Table 1 shows the results obtained from the execution of the clustering() method.

In this table, *Algorithm* indicates the name of the algorithm, *Distance* represents the distance measurement employed (for methods with a single metric),

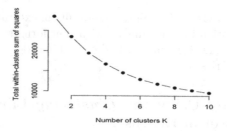

Fig. 1. Calculation of cluster number using the elbow method.

Table 1. Overview of clustering algorithms integrated into the *Clustering* library of CRAN.

Algorithm	Distance	Clusters	Data	Var	Time	Precision	Recall
clara	euclidean	7	dataset	1st	0.0171	0.6499	0.3157
clara	euclidean	7	dataset	2st	0.0185	0.3340	0.2979
clara	euclidean	7	dataset	3st	0.0208	0.3060	0.2410
clara	euclidean	7	dataset	4st	0.0213	0.2790	0.2027
clara	euclidean	7	dataset	5st	0.0451	0.0221	0.1653
clara	euclidean	8	dataset	1st	0.0178	0.6196	0.2483
clara	euclidean	8	dataset	2st	0.0199	0.5923	0.2198
clara	euclidean	8	dataset	3st	0.0205	0.2879	0.2013
clara	euclidean	8	dataset	4st	0.0209	0.2072	0.1848
clara	euclidean	8	dataset	5st	0.0515	0.0249	0.1731
clara	euclidean	9	dataset	1st	0.0198	0.7367	0.2516
clara	euclidean	9	dataset	2st	0.0219	0.4259	0.2042
clara	euclidean	9	dataset	3st	0.0240	0.2961	0.1905
clara	euclidean	9	dataset	4st	0.0252	0.2656	0.1571
clara	euclidean	9	dataset	5st	0.0485	0.0226	0.1544
kmeans_rcpp	–	7	dataset	1st	0.0492	0.5481	0.2684
kmeans_rcpp	–	7	dataset	2st	0.0524	0.5060	0.2677
kmeans_rcpp	–	7	dataset	3st	0.0527	0.2725	0.2590
kmeans_rcp	–	7	dataset	4st	0.0644	0.2349	0.2514
kmeans_rcp	–	7	dataset	5st	0.0787	0.0239	0.2343
kmeans_rcp	–	8	dataset	1st	0.0561	0.5806	0.2587
kmeans_rcp	–	8	dataset	2st	0.0570	0.5607	0.2480
kmeans_rcp	—	8	dataset	3st	0.0573	0.2696	0.2387
kmeans_rcp	–	8	dataset	4st	0.0601	0.2482	0.2170
kmeans_rcp	–	8	dataset	5st	0.0829	0.0245	0.2095
kmeans_rcp	–	9	dataset	1st	0.0463	0.6422	0.2363
kmeans_rcp	–	9	dataset	2st	0.0468	0.5026	0.2112
kmeans_rcp	–	9	dataset	3st	0.0475	0.2846	0.1998
kmeans_rcp	–	9	dataset	4st	0.0505	0.2550	0.1977
kmeans_rcp	–	9	dataset	5st	0.0745	0.0268	0.1

Clusters is the number of clusters used in that execution and *Data* is the dataset analyzed. For external metrics we must provide the structure of the dataset must be provided to compute them. The *Clustering* library tries to find the variable that provides the best partitioning of the data about the external metrics used. To achieve this, it selects each variable in the dataset as a target and calculates its associated external metrics. This is shown in Table 1 in column *Var*, which reflects which variable in the dataset has been used as the target. The remaining columns presented below, i.e., *Time*, *Precision*, and *Recall*, show the value of the external metrics employed concerning using that variable *Var* as the target.

Once the analysis of the data has been completed, the *Clustering* library has the necessary mechanisms to summarise the data by internal and external metrics, with the aim of:

– Find the optimal number of clusters for each algorithm.
– And to decide which variable best influences the results.

For this purpose, the *Clustering* library provides the following methods: best_ranked_external_metrics() and best_ranked_internal_metrics(). The results of the execution of these methods can be seen in the Tables 2 and 3. Highlight that the columns ending in *Attr* in the Tables 2 and 3 represent the variable in the dataset with the greatest influence on the metrics analysed.

Table 2. A summary of the external measurements sorted by algorithm, distance measure, and the number of clusters.

Algorithm	Distance	Clusters	Data	Var	Time	Precision	Recall	TimeAtt	PrecisionAtt	RecallAtt
clara	euclidean	7	dataset	1st	0.0171	0.0499	0.3157	4th	4th	1st
clara	euclidean	8	dataset	1st	0.0178	0.6196	0.2483	3th	3th	3st
clara	euclidean	9	dataset	1st	0.0198	0.7367	0.2516	2th	4th	4st
kmeans_rcpp	–	7	dataset	1st	0.0492	0.5481	0.2684	4th	4th	4st
kmeans_rcpp	–	8	dataset	1st	0.0561	0.5806	0.2587	2th	4th	3st
kmeans_rcpp	–	9	dataset	1st	0.0463	0.6422	0.2363	2th	4th	4st

Table 3. A summary of the internal measurements sorted by algorithm, distance measure, and the number of clusters.

Algorithm	Distance	Clusters	Data	Var	Time	Silhouette	TimeAtt	SilhouetteAtt
clara	euclidean	7	dataset	1st	2.110	0.19	4th	1st
clara	euclidean	8	dataset	1st	2.298	0.18	4th	1st
clara	euclidean	9	dataset	1st	2.194	0.20	4th	1st
kmeans_rcpp	–	7	dataset	1st	2.198	0.22	5th	1st
kmeans_rcpp	–	8	dataset	1st	2.308	0.23	4th	1st
kmeans_rcpp	–	9	dataset	1st	2.183	0.23	1th	1st

Finally, it should be noted that the *Clustering* library incorporates a series of methods that allow us to determine in certain situations which distance measure is the most appropriate concerning internal and external metrics. The main proposal is to reduce and facilitate the analysis and study of several algorithms for multiple data sets. The results can be seen in the Table 4 and Table 5 for the external metrics and Table 6 and Table 7 for the internal ones.

Table 4. Classification of the result by algorithm and distance measures through the *evaluate_best_validation_external_by_metrics*() method.

Algorithm	Distance	Clusters	Time	Precision	Recall	TimeAtt	PrecisionAtt	RecallAtt
kmeans_rcpp	–	9	0.0463	0.6422	0.2363	2nd	4th	4th
clara	euclidean	9	0.0198	0.7367	0.2516	2nd	4th	4th

Table 5. Results of the algorithms on external metrics through the *result_external_algorithm_by_metric*() method.

Algorithm	Distance	Clusters	Time	Precision	Recall	TimeAtt	PrecisionAtt	RecallAtt
kmeans_rcpp	–	9	0.0463	0.6422	0.2363	2nd	4th	4th
clara	euclidean	9	0.0198	0.7367	0.2516	2nd	4th	4th

Table 6. Classification of the result by algorithm and distance measures through the *evaluate_best_validation_internal_by_metrics*() method.

Algorithm	Distance	Clusters	Time	Silhouette	TimeAtt	SilhouetteAtt
kmeans_rcpp	–	8	2.308	0.23	4th	1st
clara	euclidean	9	2.194	0.20	4th	1st

Table 7. Results of the algorithms on internal metrics through the *result_internal_algorithm_by_metric*() method.

Algorithm	Distance	Clusters	Time	Silhouette	TimeAtt	SilhouetteAtt
kmeans_rcpp	–	8	2.308	0.23	4th	1st
clara	euclidean	9	2.194	0.20	4th	1st

From the results obtained, it can be determined for external measures, that the attribute of the dataset with which we achieve a categorization closest to a real categorization is the attribute *Period* for 9 clusters. For the internal measure, the appropriate number of clusters, we found with attribute *Autonomous community and city of death* using kmeans_rcpp as the algorithm, as reflected in the Table 7.

4   Graphical Distribution of Results

The *Clustering* library includes several methods that allow performing the query and sort operations as well as representing the distribution of data by cluster and algorithms as shown in Fig. 2.

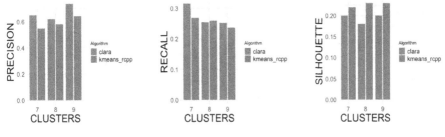

(a) Precision by number of clusters for each algorithm.

(b) Recall by number of clusters for each algorithm.

(c) Silhouette by number of clusters for each algorithm.

Fig. 2. Graphical representation of the external and internal metrics by several clusters for the algorithms indicated in the execution.

Until now, all operations performed on the *Clustering* library have been in console mode, but thanks to the `appClustering()` method, users will be able to interact via a GUI. method, users will be able to interact via a GUI. The interface design features a header, a side menu, and the main menu as can be seen in Fig. 3. With the header, the user can choose to view the data in numerical or graphical mode. On the left-hand side, you have the set of parameters with which the user can interact. Finally, the central menu presents the results depending on the mode chosen in the header (Fig. 4).

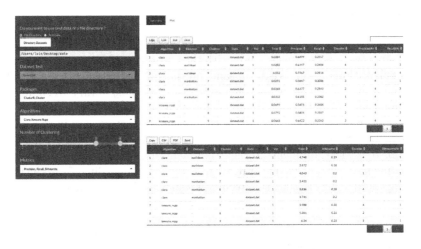

Fig. 3. Clustering app user interface.

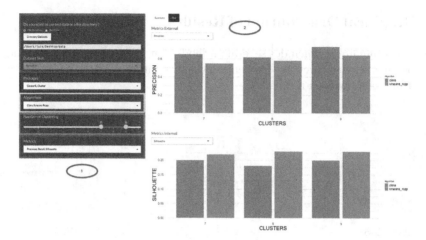

Fig. 4. Graphical visualisation of the results through the user interface of the Clustering package.

– Marked in red, the user can choose whether to work with test datasets or indicate a directory of dataset files to be processed.
– In blue, the libraries that implement the clustering algorithms mentioned throughout the paper can be selected. It is possible to mark all libraries or a subset of them. When a library is marked, all the algorithms implemented within the selected library are marked.
– In yellow, the algorithms implemented by the libraries are shown. Multiple algorithms can be selected.
– In green, it is possible to choose the desired number of clusters. It is possible to indicate ranges or select only one cluster by positioning the maximum and minimum on the same value.
– Finally, in violet, the evaluation metrics used when validating the clusters are chosen.

5   Conclusions

In this paper, we present a case study with the Clustering library, where we apply different algorithms as well as external and internal quality measures to analyze the clustering results. The advantages offered by the *Clustering* library are: analyzing one or multiple datasets using different algorithms, using multiple dissimilarity measures in the executions, working with a range of clusters, and incorporating quality metrics to analyze the most relevant attributes for the dataset, as well as being able to use the graphical interface that facilitates the use of the library without needing in-depth knowledge of R. As future work, the quality of the clusters is being improved using classification techniques such as Hyperrectangle with genetics (CHC).

Acknowledgments. This work was supported by the Spanish Ministry of Economy and Competitiveness under the project PID2019-107793GB-I00.

References

1. Kaur, M., Garg, S.: Survey on clustering techniques in data mining for software engineering. Int. J. Adv. Innov. Res. **5**(3), 238–243 (2014)
2. Steinbach, M., Karypis, G., Kumar, V.: A comparison of document clustering techniques. In: Proceedings of the International KDD Workshop On Text Mining, vol. 6 (2000)
3. Dolnicar, S.: A review of data-driven market segmentation in tourism. J. Travel Tour. Mark. **12**, 1–22 (2002)
4. Garg, N., Rani, R.: Analysis and visualization of Twitter data using k-means clustering. In: 2017 International Conference on Intelligent Computing and Control Systems (ICICCS), pp. 670–675 (2017)
5. Pandeeswari, N., Kumar, G.: Anomaly detection system in cloud environment using fuzzy clustering based ANN. Mobile Netw. Appl. **21**, 494–505 (2016)
6. Starczewski, A., Krzyżak, A.: Performance Evaluation of the Silhouette Index. In: Rutkowski, L., Korytkowski, M., Scherer, R., Tadeusiewicz, R., Zadeh, L.A., Zurada, J.M. (eds.) ICAISC 2015. LNCS (LNAI), vol. 9120, pp. 49–58. Springer, Cham (2015). https://doi.org/10.1007/978-3-319-19369-4_5
7. Lance, G., Williams, W.: A generalized sorting strategy for computer classifications. Nature **212**, 218 (1966)
8. Ramprasanth, H., Devi, A.: Outlier analysis of medical dataset using clustering algorithms. J. Anal. Comput. **15**, 1–9 (2019)
9. Kaufman, L., Rousseeuw, P.T.: Divisive analysis (program DIANA). In: Finding Groups in Data, pp. 253–279 (1990)
10. Kaufman, L., Rousseeuw, P.: Introduction. In: Finding Groups in Data, pp. 1–67 (1990)
11. Kaufman, L., Rousseeuw, P.: Fuzzy analysis (program FANNY). In: Finding Groups in Data, pp. 164–198 (1990)
12. Bezdek, J., Ehrlich, R., Full, W.: FCM: the fuzzy c-means clustering algorithm. Comput. Geosci. **10**, 191–203 (1984)
13. Gath, I., Geva, A.: Unsupervised optimal fuzzy clustering. IEEE Trans. Pattern Anal. Mach. Intell. **11**, 773–780 (1989)
14. Gustafson, D., Kessel, W.: Fuzzy clustering with a fuzzy covariance matrix. In: 1978 IEEE Conference on Decision and Control Including the 17th Symposium On Adaptive Processes, pp. 761–766 (1979)
15. R Core Team R: A Language and Environment for Statistical Computing. (R Foundation for Statistical Computing, 2021). https://www.R-project.org/
16. Kaufman, L., Rousseeuw, P.: Monothetic analysis (program MONA). In: Finding Groups in Data, pp. 280–311 (1990)
17. Kaufman, L., Rousseeuw, P.: Partitioning around medoids (program PAM). In: Finding Groups in Data, pp. 68–125 (1990)
18. Suzuki, R., Terada, Y., Shimodaira, H.: pvclust: hierarchical clustering with P-values via multiscale bootstrap resampling (2019). https://CRAN.R-project.org/package=pvclust, R package version 2.2-0
19. Wickham, H., François, R., Henry, L., Müller, K.: dplyr: a grammar of data manipulation (2021). https://CRAN.R-project.org/package=dplyr, R package version 1.0.5
20. Hanczar, B., Nadif, M.: Precision-recall space to correct external indices for biclustering. In: Proceedings of the 30th International Conference On Machine Learning, vol. 28, pp. 136–144

Comparing Clustering Techniques on Brazilian Legal Document Datasets

João Pedro Lima[1]([⊠])⬤ and José Alfredo Costa[2]⬤

[1] IMD, Universidade Federal do Rio Grande do Norte, Natal, RN 59078-900, Brazil
joaopedrodasilvalima@gmail.com
[2] DEE, Universidade Federal do Rio Grande do Norte, Natal, RN 59078-900, Brazil
alfredo@ufrnet.br

Abstract. The Brazilian justice system is one of the largest and most virtualized in the world. However, the country suffers from severe system congestion. Clustering is a machine learning technique that can increase the speed of the system by automating daily tasks in addition to helping to ensure legal certainty. However, choosing the right model for clustering is not a simple task. This work makes an empirical evaluation of six different approaches to the clustering of Brazilian legal documents. K-Means, Mini Batch K-Means and HDBSCAN algorithms were tested in different hyperparameters and together or not with the Kohonen's Map as a pre-clustering technique. The work also proposes a new NLP-oriented framework for the specific evaluation of textual clusters. Two databases of Brazilian legal documents were used. The results demonstrate that K-means and Mini Batch K-Means are the best choices, and the use of the Kohonen's map can increase the overall clustering performance.

Keywords: Clustering · Comparison · Natural Language Processing · Legal · Textual

1 Introduction

The digital revolution has accelerated the growth of data volume in all sectors of society, such as agriculture, finance, industry and legal systems. According to the recent Justice in Numbers 2021 report of Brazilian National Justice Council (CNJ) [5], the number of pending cases by the end of 2021 was a total of 75.4 millions, making the Brazilian Judicial System one of the biggest in the world.

The least two decades in Brazilian Judicial System were marked by a strong digitization process, with the creation of online platforms such as the PJe, responsible for implementing all the needs in a process lifetime. Due to this policy, today most of the cases are completely electronic (96.9% according to Justice in Numbers 2021). However, even in this scenario, the percentage of jammed cases (73%) is still a problem, showing that the migration to digital platforms is not enough to give the needed system's flow rate. These factors create the ideal

Supported by CAPES.

P. García Bringas et al. (Eds.): HAIS 2022, LNAI 13469, pp. 98–110, 2022.
https://doi.org/10.1007/978-3-031-15471-3_9

scenario for applying Machine Learning techniques, such as clustering, which can increase system's efficiency.

Clustering can be understood as the process of finding groups automatically on a dataset without any prior knowledge about its structure. There is no agreement on how a group should be defined, and several Machine Learning algorithms have tackled the problem with different assumptions [10].

In the legal domain, clustering is strongly related to the concept of mass trial and judicial security. Judicial Security is the guarantee that equal cases are treated equally, and mass trial is the process of giving a unique sentence simultaneously to a bunch of very similar cases. In both, a clustering algorithm can act as an assistive tool to the human responsible for the cases. It can also be a very useful tool to Information Retrieval and data organization.

Legal document clustering has already been explored in diverse research. [9] explores this technique to help lawyers find thematically similar amending acts in Polish Civil Code, [17] uses a fuzzy c-means approach to cluster similar argumentative sentences and determine what arguments are used in a case and [11] compares several text vectorization approaches in the task of clustering GDPR-like data protection texts in many languages.

However, working with clustering for legal texts is not a trivial task. Firstly, it inherits the traditional difficulties from the Natural Language Processing (NLP) world [1,24]. Texts are unstructured, non-numeric and ambiguous data, and representing they in a vector space is a hard problem in NLP applications. It also brings the difficulties from clustering [10]. As mentioned before, there is current no agreement on how a cluster should be defined, and as each algorithm makes its own premises about the data, choosing the right one for the application can be a tricky question to answer. It is also important to note that the mathematical optimal clusters are not necessarily the best ones from the human point of view. Lastly, it also includes several restrictions and requisites due to the Judicial System needs, such as model interpretability and transparency. In Brazil, these characteristics are especially related to the Resolution $N°$ 332, 21/08/2020 from CNJ.

In this paper, a research were conducted to empirically evaluate the performance of different clustering pipelines in the task of textual clustering of legal documents in Brazilian Portuguese. The research aims to answer what algorithm best fits the performance requirements for legal applications.

Algorithms such as K-Means and HDBSCAN were tested over a set of hyperparameters in two different ways: basic mode, where the algorithms operate normally over the data points, and second, with joint use with Kohonen's Map, where the algorithms cluster over the space reduced by a SOM neural network.

Two different databases were used: A 30,000 legal documents database from Tribunal de Justiça do Grande do Norte (TJRN) and a set of approximately 6,400 Brazilian Ordinary Laws.

The evaluation was made with a NLP-based framework that couples concepts of Topic Coherence Measures and Document Similarity. This approach aims to avoid the strong distributional assumptions of traditional cluster internal validity indices and focus on inherent aspects of the clustered texts. Human subjective evaluations have also been made to bring more robustness to the final conclusions.

The rest of this paper is structured as follows: Sect. 2 discuss the related works; Sect. 3 describes the clustering and NLP algorithms used; Sect. 4 describes the methodology and the proposed metric; Sect. 5 shows and discuss the results; Sects. 6 and 7 are, respectively, the conclusion and future work.

2   Related Work

In the Brazilian legal domain, [6] developed a clustering algorithm specifically for legal domain, extending the concept of "bag of words" to include domain specific knowledge to improve document retrieval.

The recent work in [21] compares different clustering techniques on a Brazilian legal dataset. The algorithms tested were Hierarchical, Lingo, K-means and Affinity Propagation, and, in addition to the external validity scores, the authors also reported several other performance characteristics of each model, helping to clarify vantages and disadvantages.

Similarly, [16] compared different approaches using the transformer architectures BERT, GPT-2 and RoBERTa on document clustering. The models are tested in its usual pretrained version and with fine tuning on specific legal domain texts, totaling six different approaches. The embedded texts are clustered by a K-Means algorithm, and the authors conclude that RoBERTa was the best choice.

The work in [13] makes a theoretical review of the main clustering algorithms on textual clustering. The work makes considerations about many clustering approaches characteristics, including susceptibility to noise, cluster granularity and efficiency.

[18] compares K-Means and Hierarchical Clustering methods using various distance metrics to measure similarity between articles. The authors report that the best performance metric for K-Means was 'correlation', while for the hierarchical method it was 'cityblock'.

3   Theoretical Basis

This section is a brief discussion of the tested algorithms and metrics used. For the following topics, consider $D = \{d_0, d_1, ..., d_n\}$ the set of documents in a corpus, $V = \{w_0, w_1, ..., w_m\}$ the vocabulary of this corpus, and $P = \{p_0, p_1, ..., p_n\}$, $p_i \in \mathbb{R}^k$ the respective vector representations of documents D.

3.1   Clustering Algorithms

K-Means. K-Means is the most widely used clustering algorithm and also one of the simplest in implementation. Its premise is to find k points $C = \{C_1, C_1, ..., C_k\}$ in the data space, each one representing a cluster, that minimizes the data's inertia objective function.

These points are randomly initialized and iteratively moved to optimal values, in a two-step process called Expectation-Maximization (EM). When fully converged, each data point x_i is assigned to a cluster y_i based on its Euclidean distance to the centroids:

$$y_i = \operatorname*{argmin}_{1 < j \leq k} \left\{ \|x_i - C_j\|^2 \right\}$$

K-Means has been widely studied over its history, and several upgrades [8,14] and variations [26] were created. One of its variations, that is included in the research as a separated algorithm, is the Mini Batch K-Means variation [22], that tries to speed up training by iterating only in a sample of the data.

Kohonen's Map. Kohonen's Map [12], also known as Self Organizing Map (SOM), is a type of unsupervised neural network that tries to represent a dataset's topological points distribution using a bidimensional grid of neurons.

Its premise is to move the neurons through the data space and enforce them into the data shape. The neurons can be initialized randomly or over the two main data's PCA components and are iteratively moved in each train epoch.

In summary, in each training epoch, each data point p_i pulls the closest neuron, which is called Best Matching Unit (BMU), to its direction with some learning rate α. Then, the pulled neuron will also pull its neighbors, but with α scaled down by a neighbor factor. This process continues until a maximum distance neighbor is reached and moved. The equation below shows update process of a neuron n_i from time step t to $t + 1$. x is the current point drawn from the dataset, $\alpha(t) \in (0, 1)$ is the current learning rate and $h(\cdot)$ is the scaling factor for BMU neighbors.

$$n_i^{t+1} = n_i^t + \alpha(t) h(i, BMU) \cdot (x - n_i^t)$$

By the end of training, the neurons's positions should summarize the data distribution over the space [7] and could be used to create an optimized 'smoothed' data representation, where each entry x is replaced by its final BMU vector. This final representation can be used to increase the performance of other clustering algorithms.

HDBSCAN. Hierarchical Density-Based Spatial Clustering of Applications with Noise (HDBSCAN) [15] is a clustering algorithm that extends the notions created by DBSCAN by implementing concepts of hierarchy and stability.

The DBSCAN algorithm is a density-based clustering approach that considers a cluster as a continuous area of constant density. In its context, 'density' is defined based on points proximity, as the number of points within a certain distance from a central point.

HDBSCAN follows the same cluster definition but implements several mechanisms to make the algorithm more resilient to noise and stable. Its implementation can be summarized into the following steps: (1) Transform the space according to the density, making the noisy points more prominent (2) Build the minimum spanning tree of the distances graph. (3) Construct a cluster hierarchy of connected components. (4) Condense the cluster hierarchy based on minimum cluster size. (5) Extract the stable clusters from the condensed tree.

3.2   Natural Language Processing Techniques

TF-IDF. TF-IDF [2] is a classic textual vectorization algorithm, which is based on two coefficients: text frequency (TF), and documental frequency (IDF), to build embed documents into numerical vectors.

The representation created by TF-IDF will be vectors of size $|V|$, where each dimension represents a word from the vocabulary.

The algorithm assigns a weight to each word j of the text i according to: $p_{ij} = \text{TF}(w_j, d_i) \times \log(\text{IDF}(w_j))$, where $\text{TF}(w_i, d_j)$ is the frequency of the word j in the text i, and $\text{IDF}(w_i)$ the inverse of the documentary frequency of the word throughout the corpus. The idea is that the weights can translate the importance of each word to the given document, based on its rarity.

Topic Coherence. Topic Coherence is a quality measurement metric brought from Topic Segmentation sub-field. In this context, a topic is understood as a set of words found by some topic segmentation algorithm, like {airplane, cloud, pilot}. The main goal of a topic coherence measurement is to perfectly mimic human evaluations about topic's quality. It uses probabilities about word co-occurrence drawn from a reference corpus to build a mathematical framework that allows an objective quantification of quality that strongly correlates with the human's opinion.

In [20] the authors explore the main characteristics of current philosophy and statistics about how a topic coherence should be built, and propose a unifying framework based on four main modules (or steps): Segmentation; Probabilities calculation; Confirmation Measurement and Aggregation.

Using this approach, the authors not only were able to formally unify all the existing metrics, but also built the foundations to create new ones.

4   Methodology

This research aims to evaluate the performance of several clustering pipelines on the task of Brazilian Legal Document Clustering. This section details how the research was carried on and its main decisions.

4.1   Models Description

A clustering pipeline is a sequence of algorithms responsible for clustering a dataset $X = \{x_1, x_2, x_3, \ldots, x_m\}$. A pipeline can be composed of as many layers as needed, each one doing a specific task, like dimensionality reduction and normalization. In this research, a constant structure is considered for a pipeline, composed of only two moving parts: a pre-clustering algorithm and a clustering algorithm.

The pre-clustering algorithm is responsible for optimizing the data for the clustering algorithm. Two approaches were tested: No pre-clustering (the data remains unchanged) and a 10×10 Kononen Map.

For clustering, the algorithms tested are K-Means, Mini Batch K-Means and HDBSCAN. Each one is trained multiple times in a grid of hyperparameters, that is described in Table 1. For K-Means models, the number of clusters (n_clusters) varies exponentially. In MB K-Means, two sample sizes (batch_size) are also tested. In HDBSCAN, the minimum cluster size and the distance metric used are varied. The randomness seed (random_state) was fixed to ensure reproducibility. The values were chosen to maximize the variability of the experiments.

Table 1. Tested models hyperparameters with respective python libraries

Algorithm	Hyperparameter grid	Python library
K-Means	n_clusters $= \{2, 4, 8, 16, 32\}$ random_state $= 214$	Scikit-Learn
MB K-Means	n_clusters $= \{2, 4, 8, 16, 32\}$ batch_size $= \{128, 1024\}$ random_state $= 214$	Scikit-Learn
HDBSCAN	min_cluster_size $= \{5, 15, 30, 100, 300\}$ distance_metric $= \{$'euclidean', 'cosine'$\}$ random_state $= 214$	hdbscan

4.2 Databases, Preprocessing and Embedding

Two databases were used in the research: TJRN30 and BrLaws.

TJRN30 [3] is a database of judicial transactions of the Tribunal de Justiça do Rio Grande do Norte (TJRN). This database contains a total of 30,000 documents of judicial movements distributed equally among 10 classes, each class representing a specific type of movement.

BrLaws is a database containing approximately 6,400 Brazilian Ordinary Laws' ementas dated from 1988 to 2020. An 'ementa' can be understood as a one-line summary of the law intentions and subjects and, in general, is a better option than the full law text for clustering applications. This database was built using web scraping techniques and has no labels associated with it.

All the databases pass through the same preprocessing steps: (1) The text has been converted to lowercase; (2) Were removed special symbols and punctuation; (3) The Portuguese stopwords have been removed; (4) Money values were substituted by a constant token MONEY; (5) All other numbers, including dates, laws number and personal document ids were substituted by a constant token NUMBER; (6) After tokenization, all tokens with document frequency less than 5% and bigger than 80% were removed.

The text is embedded using traditional TF-IDF, using the implementation available at Scikit-Learn, which is also responsible for some of the preprocessing steps above.

4.3   Clustering Evaluation Framework

The evaluation of clustering algorithms is one of the main steps in a model's life process. Along with the choice of algorithm, the choice of performance metric describes what is being expected from a cluster.

There are numerous ways to carry out this evaluation, which can be divided fundamentally between external evaluation and internal evaluation.

In the external evaluation [23], the groups are compared directly with predefined labels, and the indices calculate the degree of agreement between these two partitions. These predefined labels are, in most cases, human annotations on the data. The main external indices are Mutual Information, Rand Index, Purity and Homogeneity. However, this approach has serious drawbacks. Among the main ones, is the fact that it is inconsistent with the nature of the clustering task, which presupposes, above all, the nonexistence of annotated classes.

Internal indexes [4], on the other hand, do not need any information besides the partitioning itself. These indices measure the quality of clusters by measuring intrinsic characteristics of the partitioned data, such as distance between points in the same group, distance between points in different groups, density of clusters, etc. Among the main internal indices are Within-Cluster Sum of Squares (K-Means' Inertia), Silhouette Score, Sdbw, Density-Based Clustering Validation (DBCV), etc. These indexes are more suitable for real applications, as they guide the developer in a totally unsupervised way.

Like clustering algorithms, indexes also have assumptions about how clusters should be structured [25]. However, a cluster that achieves good internal indexes does not necessarily create meaningful, representative clusters that are useful to a human.

Furthermore, for the comparison of different algorithms in the same task, the choice of an index becomes difficult, because each algorithm can perform differently in each index. For example, K-Means can get better performance with silhouette score, while spectral clustering with Sdbw.

To avoid all these problems related to the use of conventional (internal and external) indexes, this article proposes an alternative framework for the specific evaluation of textual clusters.

In the textual cluster, the most important requirements are: (R1) That each cluster contains texts on a specific subject. (R2) That the clusters are not redundant. These requirements are not necessarily linked to any kind of assumption about the distribution of data or the topology of the clusters, but rather the meaning of the texts in each group.

The proposed assessment framework tries to correlate with the human cluster assessment process and is heavily based on the work [20]. It is composed of two metrics, the average individual coherence of the clusters, related with (R1) and the average similarity between the clusters, related with (R2). Both are calculated using a set of words representing each cluster. It can be understood as an intrinsic evaluation oriented to NLP. The process is detailed in the following sections.

Cluster Representative Words. The first step in calculating the proposed metric is to summarize the texts of each cluster i using a set of keywords W_i. The idea is that these words represent the subject that each cluster captures.

Finding words that define each cluster is not a trivial task, but a heuristic approach is to use the mean value of the word's TF-IDF values in a cluster as a metric of relevance. For this research, was considered a total of 20 representative words for each cluster (top-20 most relevant words).

Cluster Topic Coherence. The set of words representing each cluster can be understood as the main topic of that cluster. Assuming that the W_i sets are defined, a direct way to calculate the clustering quality is through Topic Coherence Metrics. Considering a coherence metric C_{metric}, the mean individual cluster quality is defined as $\bar{C} = \frac{1}{|W|} \sum C_{metric}(W_i, D)$

The metric used in the research was the default C_v implementation available at Gensim [19] library.

Cluster Similarity. To calculate intra-cluster similarity, the proposed approach is quite straightforward. For each pair i, j of clusters, the similarity between them is defined by $S(i,j) = |W_i \cap W_j|$, that is, the amount of representative shared between these two clusters.

Therefore, the average similarity between the clusters is the average similarity of the pairs $\bar{S} = \frac{2}{k(k-1)} \sum_{i<j}^{i,j \in \{1,2,...,k\}} S(i,j)$

4.4 Clustering Human Evaluation

As a complementary evaluation to the previously proposed framework, some models are also evaluated by humans. All models that created between 2 and 16 clusters go under a manual inspection of the 10 most representative words of each cluster, receiving two scores. The first represents individual cluster coherence, related to the requirement (R1), and the second represents the clusters' independence (no redundancy), related to the requirement (R2). Both scores range from 1 to 5, where 1 is 'very bad' and 5 is 'very good'.

5 Results and Discussion

Figure 1 contains the results of the experiments for each model in each database. The y axis represents the average coherence $\bar{C}$, the x axis represents the mean amount of shared representative words between the clusters $\bar{S}$. In addition to the metrics previously described, the number of clusters created by each model was also included in the plots.

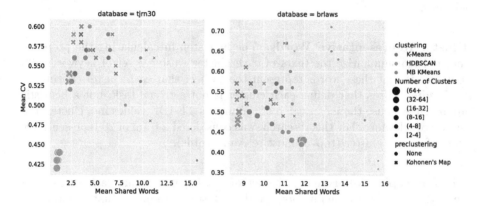

Fig. 1. Coherence score vs shared words

The best performing models are those closest to the upper left corner of the graph, with greater coherence and less redundancy. The K-Means variants (Normal K-Means and Mini Batch) generally perform better than HDBSCAN. This can be seen by the greater consistency of K-Means in approaching the upper left corner in both databases. Despite not being consistently good, HDBSCAN still has occasional high performance. On the TJRN30 database, although you can't easily declare an absolute winner, HDBSCAN is certainly among the top performing models.

The performance of K-Means and MB K-Means is very similar, and it is not possible to define a clear winner. Another clear fact extracted from the graph is the performance improvement provided by using the Kohonen Map as pre-clustering, since the best models are always those that benefited from this strategy.

From a assistive AI perspective, supposing that the clusters created will be used by some human actor, the ideal is that the number of clusters cannot be too high or too low. An interesting fact to note is that, by using this framework, the models with 'inadequate' number of clusters are naturally penalized.

Figure 2 contains the results of the human evaluation on some of the models, as mentioned in the previous section. A total of 66 models were evaluated. In this chart, the best models are those that approach the upper right corner.

The results are consistent with those obtained previously. K-means remains the best algorithm, although there is still no clear winner between its regular version and its MB version. On the other hand, the performance gap between HDBSCAN and the K-Means variants increased.

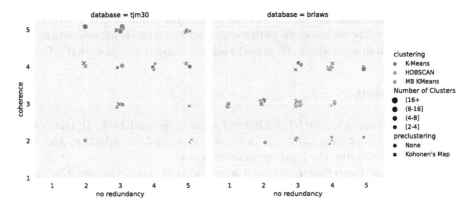

Fig. 2. Human scores

Another important aspect to take into consideration is the computational cost, as this can be a limiter aspect for a model adoption. Table 2 shows the mean execution time of each algorithm, as an approximation of this cost.

None of the models presented a prohibitive training time and, a priori, all could be used. However, it is important to note that K-Means performance is achieved by MB K-Means with a much shorter training time.

Table 2. Mean execution time (in seconds)

brlaws	HDBSCAN		KMeans		MB KMeans	
	None	Kohonen	None	Kohonen	None	Kohonen
	1.95	0.73	0.69	0.63	0.15	0.14
tjrn30	HDBSCAN		KMeans		MB KMeans	
	None	Kohonen	None	Kohonen	None	Kohonen
	1198.12	455.10	14.97	14.04	0.74	0.43

Table 3. Final scores.

Pre-clustering	Clustering	F. Score	H. Score	C. Cost	Total
None	K-Means	4	4	4	12
	MB K-Means	4	4	5	13
	HDBSCAN	2	2	2	6
Kohonen	K-Means	5	4	4	13
	MB K-Means	5	4	5	**14**
	HDBSCAN	3	3	3	9

Table 3 consolidates all the factors discussed into a final score. It gives a 1 to 5 score regarding the model's performance on the framework evaluation (F. Score), the human evaluation (H. Score) and the computational cost (C. Cost).

6   Conclusion

Working with clustering in NLP for legal texts is a hard task. It mixes the uncertainties of clustering, with the problems of textual vectorization and still imposes limitations due the legal application scope.

Today, many clustering techniques are available, and choosing the right one can be a hard question. This paper aims to evaluate different clustering techniques one the task of legal document clustering on Brazilian Portuguese Databases.

To avoid the strong premises of internal and external traditional clustering evaluation scores, this paper proposes a new framework for specifically evaluating textual clustering, implemented with concepts of topic coherence measurement. Manual human evaluation is also done.

The results show that the Mini Batch K-Means algorithm performs better than HDBSCAN and K-Means. It also shows that using a Kohonen's Map as a pre-clustering technique increases the final clustering performance.

7   Future Work

This work could be extended in many ways. Other pre-clustering techniques could be included, like dimensionality reduction algorithms. More clustering algorithms and more hyperparameters could be tested. In the evaluation, factors such as the percentage of the database clustered, especially considering HDB-SCAN ability to detect outliers, cluster size, and others structural characteristics could be included. Other Brazilian legal databases could also be tested.

References

1. Aguilar, J., Salazar, C., Velasco, H., Monsalve-Pulido, J., Montoya, E.: Comparison and evaluation of different methods for the feature extraction from educational contents. Computation **8**(2), 30 (2020)
2. Aizawa, A.: An information-theoretic perspective of tf-idf measures. Inf. Process. Manag. **39**(1), 45–65 (2003)
3. Araújo, D.C., Lima, A., Lima, J.P., Costa, J.A.: A comparison of classification methods applied to legal text data. In: Marreiros, G., Melo, F.S., Lau, N., Lopes Cardoso, H., Reis, L.P. (eds.) EPIA 2021. LNCS (LNAI), vol. 12981, pp. 68–80. Springer, Cham (2021). https://doi.org/10.1007/978-3-030-86230-5_6
4. Arbelaitz, O., Gurrutxaga, I., Muguerza, J., Pérez, J.M., Perona, I.: An extensive comparative study of cluster validity indices. Pattern Recogn. **46**(1), 243–256 (2013)
5. CNJ - Conselho Nacional de Justiça: Relatório Justiça em Números (2020)

6. de Colla Furquim, L.O., de Lima, V.L.S.: Clustering and categorization of Brazilian Portuguese legal documents. In: Caseli, H., Villavicencio, A., Teixeira, A., Perdigão, F. (eds.) PROPOR 2012. LNCS (LNAI), vol. 7243, pp. 272–283. Springer, Heidelberg (2012). https://doi.org/10.1007/978-3-642-28885-2_31
7. Costa, J.A.F., de Andrade Netto, M.L.: Clustering of complex shaped data sets via Kohonen maps and mathematical morphology. In: Dasarathy, B.V. (ed.) Data Mining and Knowledge Discovery: Theory, Tools, and Technology III, vol. 4384, pp. 16–27. International Society for Optics and Photonics, SPIE (2001)
8. Elkan, C.: Using the triangle inequality to accelerate k-means. In: Proceedings of the 20th International Conference on Machine Learning (ICML 2003), pp. 147–153 (2003)
9. Górski, L.: Towards legal change analysis: clustering of polish civil code amendments. In: ASAIL@ ICAIL (2019)
10. Hennig, C.: What are the true clusters? Pattern Recogn. Lett. **64**, 53–62 (2015)
11. Kawintiranon, K., Liu, Y.: Towards automatic comparison of data privacy documents: a preliminary experiment on gdpr- like laws. arXiv preprint arXiv:2105.10117 (2021)
12. Kohonen, T.: The self-organizing map. Proc. IEEE **78**(9), 1464–1480 (1990)
13. Liu, F., Xiong, L.: Survey on text clustering algorithm -research present situation of text clustering algorithm. In: 2011 IEEE 2nd International Conference on Software Engineering and Service Science, pp. 196–199 (2011)
14. Lloyd, S.: Least squares quantization in PCM. IEEE Trans. Inf. Theory **28**(2), 129–137 (1982)
15. McInnes, L., Healy, J., Astels, S.: hdbscan: hierarchical density based clustering. J. Open Source Softw. **2**(11), 205 (2017)
16. de Oliveira, R.S., Nascimento, E.G.S.: Brazilian court documents clustered by similarity together using natural language processing approaches with transformers. arXiv preprint arXiv:2204.07182 (2022)
17. Poudyal, P., Gonçalves, T., Quaresma, P.: Using clustering techniques to identify arguments in legal documents. In: ASAIL@ ICAIL (2019)
18. Rani, U., Sahu, S.: Comparison of clustering techniques for measuring similarity in articles. In: 2017 3rd International Conference on Computational Intelligence Communication Technology (CICT), pp. 1–7 (2017)
19. Rehurek, R., Sojka, P.: Gensim-python framework for vector space modelling. NLP Centre, Faculty of Informatics, Masaryk University, Brno, Czech Republic, vol. 3, no. 2 (2011)
20. Röder, M., Both, A., Hinneburg, A.: Exploring the space of topic coherence measures. In: Proceedings of the Eighth ACM International ConfERENCE on Web Search and Data Mining, pp. 399–408 (2015)
21. Sabo, I.C., Dal Pont, T.R., Wilton, P.E.V., Rover, A.J., Hübner, J.F.: Clustering of brazilian legal judgments about failures in air transport service: an evaluation of different approaches. In: Artificial Intelligence and Law, pp. 1–37 (2021)
22. Sculley, D.: Web-scale k-means clustering. In: Proceedings of the 19th International Conference on World Wide Web, pp. 1177–1178 (2010)
23. de Souto, M.C., Coelho, A.L., Faceli, K., Sakata, T.C., Bonadia, V., Costa, I.G.: A comparison of external clustering evaluation indices in the context of imbalanced data sets. In: 2012 Brazilian Symposium on Neural Networks, pp. 49–54. IEEE (2012)
24. Wang, Y., et al.: A comparison of word embeddings for biomedical natural language processing. J. Biomed. Inf. **87**, 12–20 (2018)

25. Xu, Q., Zhang, Q., Liu, J., Luo, B.: Efficient synthetical clustering validity indexes for hierarchical clustering. Expert Syst. Appl. **151**, 113367 (2020)
26. Yu, J.: General c-means clustering model. IEEE Trans. Pattern Anal. Mach. Intell. **27**(8), 1197–1211 (2005)

Improving Short Query Representation in LDA Based Information Retrieval Systems

Pedro Celard[1,2,3]($\boxtimes$)(iD), Eva Lorenzo Iglesias[1,2,3](iD),
José Manuel Sorribes-Fdez[1,2,3](iD), Rubén Romero[1,2,3](iD),
Adrián Seara Vieira[1,2,3](iD), and Lourdes Borrajo[1,2,3](iD)

[1] Computer Science Department, Universidade de Vigo, Escuela Superior de Ingeniería Informática, Campus Univ. As Lagoas, Ourense 32004, Spain
pedro.celard.perez@uvigo.es
[2] CINBIO - Biomedical Research Centre, Universidade de Vigo, Campus Univ. Lagoas-Marcosende, Vigo 36310, Spain
[3] SING Research Group, Galicia Sur Health Research Institute (IIS Galicia Sur), SERGAS-UVIGO, Vigo, Spain

Abstract. Incorporation of topic modeling techniques into Information Retrieval (IR) systems has been a promising area of research in the last years. Typically, queries submitted into IR systems are concise and made up using only the essential keywords. This leads to the formulation of short length queries, which have a negative impact on the LDA algorithm accuracy and relevant documents retrieval. This work presents a novel method to improve short query representation in information retrieval systems. The new technique (LDAW), modifies its representation based on the Latent Dirichlet Allocation (LDA) model. LDAW is tested with three biomedical corpora (TREC Genomics 2004, TREC Genomics 2005, and OHSUMED) and one legal cases corpus (FIRE 2017). Results prove that the application of the proposed method clearly outperforms the baseline methods (BM25 and non-modified LDA).

Keywords: Information retrieval · Short queries · Latent dirichlet allocation · Topic modeling · Text preprocessing

1 Introduction

In the last decade, content-based document management systems have gained great importance in the field of information systems. These are known as Information Retrieval (IR) systems and they allow the user access to a collection of documents in a flexible manner. The main goal of information retrieval systems is to offer a communication interface between the user and the data management system to obtain relevant information through this acquired knowledge.

Representing the text of the documents to apply automatic learning techniques for classification and clustering is an essential part of the content retrieval

© Springer Nature Switzerland AG 2022
P. García Bringas et al. (Eds.): HAIS 2022, LNAI 13469, pp. 111–122, 2022.
https://doi.org/10.1007/978-3-031-15471-3_10

process. The most widespread representation method is the Bag of Words, better known as BoW [13]. BoW represents a document with a vector where each element describes the weight or relevance of a word on that document taking the number of appearances as a measure. In many cases, this method ignores relevant results because they do not include any of the words featured in the query. This becomes a problem as document searches are usually based on a particular topic, not only specific words.

The use of topic modeling on large document collections has become one of the most popular fields of study in data sciences. It is largely supported by the technological advances of recent years. One of the most widely used techniques is the Latent Dirichlet Allocation (LDA), which offers the possibility of discovering a set of topics that gives shape to a collection of documents. LDA represents a document through a vector where each element symbolizes the influence of each topic found in the collection over the document [2]. Multiple authors have successfully implemented this technique.

Luo [10] presents a method to improve LDA topic modeling applied to sentiment classification accuracy by combining it with a Convolutional Neural Network Gated Recurrent Unit (CNN-GRU). In order to accomplish this, the LDA representation is fed to the CNN-GRU, this way it is able to learn from a latent space instead of a plain word space.

Celard *et al.* [4] present an LDA-based text pre-processing filter for Weka, a widely used text mining tool which includes a set of data mining methods. After performing a series of experiments using various classification algorithms, including SVM, k-NN and Naïve Bayes, they conclude that LDA is a viable text representation technique and that its feature reduction capability improves the performance of classification systems.

Aguilar *et al.* [1] reviews different text representation techniques including BM25 and LDA in and information retrieval environment. Regarding LDA, the authors conclude that it achieves the best results and obtains high similarity measures. They observed that LDA works best when a data set can be segregated in a high number of clusters, but still, LSA and BM25 outperforms LDA in document retrieval use case, making it necessary to improve its representation for this kind of works.

Gadelha *et al.* [5] apply different information retrieval techniques to automatically retrieve the traceability between textual software artifacts. They conclude that LDA is able to reliably reproduce the topics in the dataset, but the representation of each case using short text was not correctly representative. One of the main cause is that its keywords appeared in multiple topics at the same time, being unable to correctly build a discriminative representation without enough text.

The main objective of this work is to obtain a better representation of the user need by modifying the query rendered as an LDA vector in order to improve the retrieval performance in short text queries. The proposed method does not use additional data nor use a different algorithm, but changes the way the query vector is calculated based on the LDA algorithm using the word metrics of the corpus. This way, it could be combined with any other architecture, offering a way to improve other works in this field of study.

This paper is structured as follows. Section 2 shows an overview of the Information Retrieval systems and outlines the main components and techniques used in this work. Sections 3 and 4 describe the proposed technique and its evaluation procedure. In Sect. 5, the obtained results are discussed and analyzed. Finally, Sect. 6 presents the conclusions.

2   Material and Methods

2.1   Information Retrieval Systems

Considering the components of Information Retrieval systems (IR), an IR model can be defined as a set of related elements $[D, Q, F, R(q_i, d_j)]$. D represents a collection of documents, Q describes the needs of the user as queries, and F expresses the relationships between documents and queries. Finally, the ranking function $R(q_i, d_j)$ associates a position for a document $d_j \in D$ retrieved by a query $q_i \in Q$, thus sorting the collection of documents concerning a specific request.

Information Retrieval systems are capable of estimating the relevance of a document concerning a query [14]. In order to use them, the user must express his need as a query, and then the system collects this information and returns a list of elements that best meet the query requirements.

2.2   Latent Dirichlet Allocation

As studied in [4], Latent Dirichlet Allocation (LDA) is one of the most successful techniques in the field of topic modeling. It is a statistical model that implements the fundamentals of topic searching in a set of documents. LDA does not work with the meaning of each of the words, but assumes that when creating a document, intentionally or not, the author associates a set of latent topics to the text [2]. Based on this theory, an LDA model can be applied to an input collection of documents in order to infer their latent topics by using the so-called Gibbs Sampling algorithm. In this process, the algorithm iterates through the words and calculates the most representative words for each topic. The words can appear many times in the same document and be repeated in different documents at the same time. The algorithm is able to infer which topic best represents them in each iteration [2,12].

LDA is a statistical model that tries to capture this intuition. It considers that each document covers subjects in different proportions, and each word in a document corresponds to one of those subjects. After processing all the terms in the texts of the set of documents, the result obtained is a matrix where each document has a probability of belonging to each of the topics.

2.3   Relevance Estimation

The main goal of an IR system is to offer a solution to the need of the user. This need is represented as a written query, so the relevance of a document is defined as its ability to meet the need of the user. Several authors present works with models capable of providing the most effective ranking of documents. Two of the best-known algorithms are BM25 and k-Nearest Neighbours.

BM25 is a classification function widely used in information retrieval systems [3,9]. It takes into account the length of the document studied and the average length of the documents in the corpus in order to carry out a standardization, and thus avoiding that the length of the documents having too much influence on the calculation [15–17]. In order to optimize BM25 and use it as a ranking function, values for the internal parameters k_1 and b must be chosen. The implementation of Lucene BM25 [15] uses $k_1 = 1.2$ and $b = 0.75$ as default values.

K-Nearest Neighbours (k-NN) algorithm. The k-NN is based on the assumption that an element can be classified according to its closest k-elements, with k being an undefined integer. More formally, given a set of I elements represented in a R^d dimensional space with d dimensions, and a i_q query point, k-NN finds the k points $(i \in I)$ with the shortest distance to i_q.

3   Proposed Query Representation Method: LDAW

The proposed technique builds a new calculation method to better represent short queries in IR architectures to improve the relevance of the retrieved documents. As can be seen in Fig. 1, the employed architecture includes all essential parts of an IR system. Each document in the corpus is represented as a vector $d_D = [t_0, t_1, ..., t_K]$, as well as the queries, which are represented as $q_Q = [t_0, t_1, ..., t_K]$. The vector values indicate the probability that the text belongs to each of the K topics in the trained LDA model.

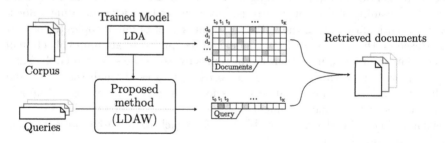

Fig. 1. IR system model overview.

To find the relevant documents for each query, a comparison between the vectors that represent the documents and the query is made. This work aims to improve the calculation of the vector that represents the query, to offer an improvement over the results obtained with the default vector returned by the LDA model.

3.1   LDAW Calculation Method

As explained before, a query is represented as a vector $q = [t_0, t_1, ..., t_K]$ where each element shows the probability of belonging to a topic. The proposed model called LDAW (Latent Dirichlet Allocation Weighting) modifies the calculation of this vector to obtain a better representation of the query considering the distribution of topics obtained when training the LDA model and different word metrics. A normalized vector with values between 0 and 1 ($||q||$) is the new representation obtained. As can be seen in Fig. 2, it is calculated through the sum of the original LDA vector represented as q_{LDA} and a set of new qW_w vectors, one for each of the words w of the query q.

$$||q|| = q_{LDA} + \sum_{w \in q} qW_w \tag{1}$$

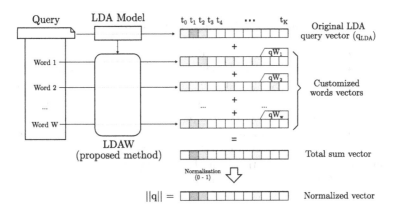

Fig. 2. Customized vectors for each term of the query.

For each word w of the query, a custom vector qW_w is created. The values of the vector are calculated using the distribution of topics of the original query represented as $t_k \in K$, the TF-IDF value of the word studied $w_{tf\text{-}idf}$, the significance of the word on the corpus w_{sign}, and the relevance of the word on the topic t_k representing the position of the vector $w_{t_k rel}$.

$$qW_{w_k} = w_{t_k rel} \cdot w_{sign} \cdot w_{tf\text{-}idf} \cdot t_k \tag{2}$$

Figure 3 shows how all these elements contribute to the creation of the new vector. The left hand of the image shows the relevance of the word on the distribution of the vocabulary for each particular topic. In the center of the image the significance of the word on the corpus vocabulary and its TF-IDF value are calculated. On the right hand of the image, the distribution of topics for the query according to the original LDA model can be found.

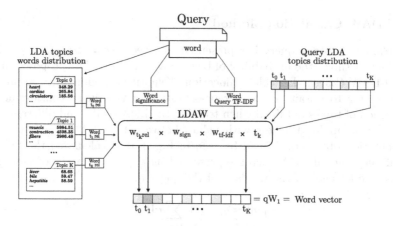

Fig. 3. Custom LDAW word vector calculation.

3.2   Relevance of the Word in Each LDA Topic

Each of the topics in the LDA model contains all the words of the vocabulary and a value indicating the relevance of each of the terms on the topic. To calculate the new vector of the query, the relevance of the word within the topic is used, representing this value as $w_{t_k rel}$. There is a big difference between the values of the most representative word of each topic. Using the value returned by LDA by default would cause an unwanted unbalance between each of the topics. To avoid this effect, $w_{t_k rel}$ is calculated by dividing the default LDA value by the sum of all the word values in the topic. In this way, the relevance is dependent on the individual value of the term within the topic itself.

$$w_{t_k rel} = \frac{w_{t_k}}{\sum\limits_{w \in t_k} w_{t_k}} \tag{3}$$

3.3   Relevance of the Word in the Corpus Vocabulary

The frequency of a term in the document only measures the number of occurrences of that term in the corpus, but it is not an indicative value of the amount of information it provides. Based on this assumption, it is assumed that the more times a word appears in the corpus, the less information it provides, and therefore the less relevant it is to the query. The relevance of each word in the query on the vocabulary is calculated comparing it to the most common term. To do this, we divide the number of repetitions of the most common term $|Cw|$ by the number of repetitions of the word studied $|w|$. This way the most common word will have a value equal to 1, while the more specific terms will have very high values, thus enhancing that word over the general calculation of the query vector.

$$w_{sign} = \frac{|Cw|}{|w|} \tag{4}$$

3.4   Relevance of the Word in the Query

When analysing a query, it is handled as a pseudo-document, thus allowing to calculate the TF-IDF value of each word. This value provides the relevance of the word using the frequency on the general vocabulary of the corpus (TF) and the inverse on the document (IDF), which in this case is the query made by the user.

When calculating the TF-IDF of a document, a value between 0 and 1 is obtained for each of the terms that make up the document. To avoid a multiplication by 0, that would entirely eliminate the relevance of the term for the general calculation, 1 is added to the TF-IDF value. This would mean that a term with a low TF-IDF value does not have its importance modified, while a high value could double it. To avoid problems of division by zero he includes the sum of 1 in the calculation. Yang [20] explains how the more terms a vocabulary contains, the higher the probability that the TF-IDF value reaches 0, and therefore the ability of an algorithm to extract the characteristics of a text is harmed. The solution is to add 1 to the denominator, so that the values do not reach 0, thus obtaining better results. The calculation of $w_{tf\text{-}idf}$ for the words in the query would look like this:

$$w_{tf\text{-}idf} = 1 + tf\text{-}idf(w) \tag{5}$$

3.5   Word Vector Calculation

Each of the values of the word customized vector is calculated as follows, where w represents a word of the query and k stands both for a vector element and an LDA topic.

$$qW_{w_k} = q_{LDA_k} + \left(\frac{w_{t_k}}{\sum\limits_{w \in t_k} w_{t_k}} \cdot \frac{|Cw|}{|w|} \cdot (1 + tf\text{-}idf(w)) \cdot q_{LDA_k} \right) \tag{6}$$

4   Evaluation

4.1   Data Sets

To carry out the experiments, a set of widely known in the information retrieval literature corpora are used (TREC 2004, TREC 2005, OHSUMED and FIRE 2017). These collections follow the standard representation of the TREC collection, with a colection of documents, a Queries file and a Qrels file. One of the tasks in TREC Genomics 2004 Track (TREC04) [8], and TREC Genomics 2005 Track (TREC05) [6] is to model the situation of a user in need of information using an information retrieval system to access biomedical scientific literature.

The OHSUMED collection, initially compiled by [7], is a subset of the MEDLINE database, which is a bibliographic database of important medical literature supported by the National Library of Medicine. The corpus was originally used at the TREC-9 Filtering Track. OSHUMED contains 348,566 references consisting of fields such as titles, abstracts, and MeSH descriptors from 279 medical journals published between 1987 and 1991.

FIRE stands for Forum of Information Retrieval and Evaluation. It's an India based organization for research on information retrieval [19]. The FIRE 2017 Precedence Retrieval Task [11] compiles a collection of legal documents used in Indian Supreme Court trials. The objective was to find relevant prior documents from a current case document.

Unlike previous corpora, FIRE 2017 does not include short queries. In order to carry out the retrieval process, a full text document is used as query. This helps to test the proposed method and proves its effectiveness over different type of data while it continues to offer better results than the base cases.

4.2   Evaluation Measures

In this work, a precision analysis is done for 5 (P@5), 10 (P@10), and 100 (P@100) returned documents. This measure is equal to the proportion of total documents returned with those that are relevant. In addition to the precision, the Mean Average Precision (MAP) is used to evaluate the results obtained. MAP is a standard measure used in many of the works related to Information Retrieval. The MAP applied to Q queries is the average of the results obtained in each query q [19].

The precision and the MAP are obtained through the statistical analysis tool "trec-eval" [8]. This tool has been officially developed for its use in many of the tasks organized by the Text REtrieval Conference (TREC). For its correct execution, the file with the qrels and a result file with the same structure explained above is provided as an input. Trec-eval returns a series of measurements, among which are those studied in this work.

4.3   Text Pre-processing

In written text semantics may cause multiple appearances of the same word in different variants, being necessary to pre-process them. Standard text pre-processing techniques are used in order to reduce the input feature size to train the models and duplicate documents are removed from the corpora. A predefined list of stop-words (common English words) is removed from the text, and a stemmer is applied.

4.4   Experiments Description

To check the feasibility of the vector calculation proposed in this work, the results obtained are compared with BM25 and the conventional LDA results. In order to carry out the experiments, as recommended in [18], BM25 parameter values are $k_1 = 2.0$ and $b = 0.8$. When training the LDA model, the chosen parameters values are 150 topics and 800 Gibbs sampling iterations. This values have been obtained in previous work [4]. The selected evaluation measures are Precision at 5, 10 and 100 documents and MAP (Mean Average Precision).

Performance comparisons have been carried in the following ways:

- **Base cases:** The results obtained with these configurations will be referred to as base line.
  - **BM25**: Base case using the BM25 model.
  - **Original LDA**: Base case using the unmodified LDA model.
- **Vector modification:** In Sect. 3.1 a new way of calculating the vector representative of the query made by the user is proposed. This new calculation consists of several elements that allow the study of the influence that each one of them has on the final result. The impact of each of the parts is made through a series of additive tests. In this way, the results obtained with the inclusion of their different parts can be studied.
  - **LDAWr**: Modification through the relevance of the word in each of the topics.
  - **LDAWrs**: Modification through the relevance of the word in each topic and the number of appearances of the word on the vocabulary.
  - **LDAWrst**: Modification through the relevance of the word in each topic, the number of appearances of the word on the vocabulary and the TF-IDF value of each word in the query.

5 Results and Discussion

Figure 4 shows the precision and MAP values obtained for each of the experiments performed taking into account the different variants of the proposed vector calculation. The inclusion of the word relevance of the topics (LDAWr) offers an improvement for all the corpus over BM25 and original LDA, with the exception of the OHSUMED, where similar results are obtained for LDA and LDAWr.

The significance of a word over the vocabulary of the corpus (LDAWrs) helps to obtain even better results than the base cases in all corpora. The same happens when the TF-IDF value of the word in the query is introduced in the calculation (LDAWrst). This last variations of the calculation offer similar MAP values, but individual precision results are influenced by the inclusion of the TF-IDF. LDAWrs gets better results than LDAWrst when only 5 documents are retrieved (P@5). On the contrary, LDAWrst gets higher values at P@10.

In the case of P@100, it offers an improvement over the base cases in corpus TREC04 and TREC05, while the results of OHSUMED and FIRE2017 do not offer a relevant difference. This is caused by the low number of relevant documents that these corpora offers for each query.

Taking into account the MAP value and precision, the proposed method is fully feasible. In all its variants the results are better than the base cases or very similar. Another important conclusion that should be mentioned is that the proposed method highly improves the results when it is used for document retrieval with short queries (TREC Genomics 2004, TREC Genomics 2005, and OHSUMED), but also delivers a significant improvement for long queries corpus as Fire 2017.

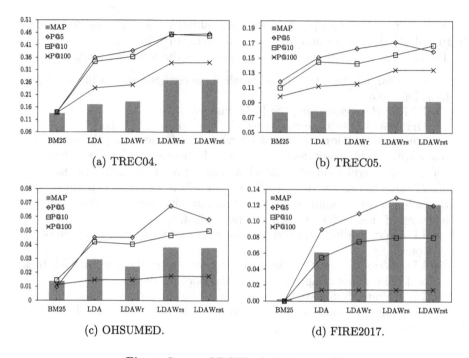

Fig. 4. Custom LDAW calculation results.

6 Conclusions

We have proposed a custom modification of the LDA vector to improve query representation in LDA-based ad-hoc retrieval, and evaluated different variants using several biomedical collections. Analyzing all the obtained results, it can be seen how the new calculation for the representative vector of the query presented in this work offers better results compared to the use of the BM25 model and the traditional LDA. A significant improvement of the values obtained after the modification of the vector using the feedback documents is also perceived. In general, the results are in line with expectations and support the theories discussed throughout this paper.

It has been concluded that the use of LDAWrs and LDAWrst offer better results, obtaining more relevant documents over short queries than the base cases. Analyzing the precision it is stated that the inclusion of the TF-IDF of each word in the query (LDAWrst) affects the vector in such a way that better results are obtained when the number of documents is close to 10 (P@10), while the non-inclusion of this value (LDAWrs) benefits the search for a smaller number of documents (P@5). The proposed method is not limited to information retrieval and query representation, as it could be beneficial to LDA models applied to short text as tweets, customer reviews, etc.

Moreover, the proposed technique makes use of different values related to each corpus: the relevance of the word in each LDA topic, and its relevance within the corpus vocabulary. These are calculated only once when training the LDA model. The only values that need to be computed for each query are the TF-IDF of its terms, the LDA vector itself (which must also be calculated in the original technique) and the final vector calculation, being it a simple multiplication. Taking these characteristics into account, and being a technique specialized in short queries, the proposed method adds very low computational load to the process.

For future work, taking into account the result documents as feedback could help obtaining more relevant documents for each query. In addition, existing parameter tuning and the inclusion of new ones could help achieving improved results. In addition, we plan to review different topic modeling techniques and methods to evolve the proposed vector calculation and improve its effectiveness and efficiency.

Acknowledgments. SING group thanks CITI (Centro de Investigación, Transferencia e Innovación) from University of Vigo for hosting its IT infrastructure. We also appreciate the support provided by Consellería de Educación, Universidades e Formación Profesional (Xunta de Galicia) under the scope of the strategic funding of ED431C2018 /55-GRC Competitive Reference Group. The funders had no role in study design, data collection and analysis, decision to publish, or preparation of the manuscript.

Pedro Celard is supported by a pre-doctoral fellowship from Xunta de Galicia (ED481A 2021/286).

References

1. Aguilar, J., Salazar, C., Velasco, H., Monsalve-Pulido, J., Montoya, E.: Comparison and evaluation of different methods for the feature extraction from educational contents. Computation **8**(2) (2020). https://doi.org/10.3390/computation8020030
2. Blei, D.: Probabilistic topic models. Commun. ACM **55**(4), 77–84 (2012). https://doi.org/10.1145/2133806.2133826
3. Bounhas, I., Soudani, N., Slimani, Y.: Building a morpho-semantic knowledge graph for Arabic information retrieval. Inf. Process. Manag. **57**(6) (2020). https://doi.org/10.1016/j.ipm.2019.102124
4. Celard, P., Vieira, A., Iglesias, E., Borrajo, L.: LDA filter: a latent dirichlet allocation preprocess method for weka. PLoS ONE **15**(11) (2020). https://doi.org/10.1371/journal.pone.0241701
5. Gadelha, G., Ramalho, F., Massoni, T.: Traceability recovery between bug reports and test cases-a Mozilla Firefox case study. Autom. Softw. Eng. **28**(2), 1–46 (2021). https://doi.org/10.1007/s10515-021-00287-w
6. Hersh, W., Cohen, A., Yang, J., Bhupatiraju, R.T., Roberts, P., Hearst, M.: Trec 2005 genomics track overview. In: TREC 2005 Notebook, pp. 14–25 (2005)
7. Hersh, W.R., Buckley, C., Leone, T.J., Hickam, D.H.: Ohsumed: an interactive retrieval evaluation and new large test collection for research. In: SIGIR, pp. 192–201 (1994)
8. Hersh, W., Bhupatiraju, R., Ross, L., Cohen, A., Kraemer, D., Johnson, P.: Trec 2004 genomics track overview. In: Proceedings of the Text REtrieval Conference, vol. 13 (01 2004)

9. Jian, F., Huang, J., Zhao, J., Ying, Z., Wang, Y.: A topic-based term frequency normalization framework to enhance probabilistic information retrieval. Comput. Intell. **36**(2), 486–521 (2020). https://doi.org/10.1111/coin.12248

10. Luo, L.X.: Network text sentiment analysis method combining LDA text representation and GRU-CNN. Pers. Ubiq. Comput. **23**(3–4), 405–412 (2019). https://doi.org/10.1007/s00779-018-1183-9

11. Mandal, A., Ghosh, K., Bhattacharya, A., Pal, A., Ghosh, S.: Overview of the fire 2017 irled track: information retrieval from legal documents. In: FIRE (2017)

12. Navarro, D., Griffiths, T., Steyvers, M., Lee, M.: Modeling individual differences using dirichlet processes. J. Math. Psychol. **50**(2), 101–122 (2006). https://doi.org/10.1016/j.jmp.2005.11.006

13. Nikolaos, T., George, T.: Document classification system based on hmm word map. In: Proceedings of the 5th International Conference on Soft Computing as Transdisciplinary Science and Technology, CSTST 2008, pp. 7–12. ACM, New York (2008). https://doi.org/10.1145/1456223.1456229

14. Ponte, J.M., Croft, W.: Language modeling approach to information retrieval. In: SIGIR Forum (ACM Special Interest Group on Information Retrieval), pp. 275–281 (1998). https://doi.org/10.1145/290941.291008

15. Perez-Iglesias, J., Perez-Agüera, J., Fernández, V., Feinstein, Y.: Integrating the probabilistic models bm25/bm25f into lucene (2009)

16. Robertson, S., Zaragoza, H.: The probabilistic relevance framework: Bm25 and beyond. Found. Trends Inf. Retr. **3**(4), 333–389 (2009). https://doi.org/10.1561/1500000019

17. rg Robertson, S.E., Walker, S.: Some simple effective approximations to the 2-poisson model for probabilistic weighted retrieval. In: SIGIR 1994, pp. 232–241. Springer, Heidelberg (1994). https://doi.org/10.1007/978-1-4471-2099-5_24

18. Robertson, S.E., Walker, S., Beaulieu, M., Willett, P.: Okapi at trec-7: automatic ad hoc, filtering, vlc and interactive track. Nist Spec. Publ. SP **500**, 253–264 (1999)

19. Tamrakar, A., Vishwakarma, S.: Analysis of probabilistic model for document retrieval in information retrieval. In: Proceedings - 2015 International Conference on Computational Intelligence and Communication Networks, CICN 2015, pp. 760–765 (2016). https://doi.org/10.1109/CICN.2015.155

20. Yang, Y.: Research and realization of internet public opinion analysis based on improved tf - idf algorithm. In: 2017 16th International Symposium on Distributed Computing and Applications to Business, Engineering and Science (DCABES), pp. 80–83 (2017). https://doi.org/10.1109/DCABES.2017.24

A New Game Theoretic Based Random Forest for Binary Classification

Mihai-Alexandru Suciu$^{(\boxtimes)}$ and Rodica Ioana Lung

Centre for the Study of Complexity, Babeş-Bolyai University, Cluj-Napoca, Romania
mihai-suciu@ubbcluj.ro, rodica.lung@econ.ubbcluj.ro
http://csc.centre.ubbcluj.ro/

Abstract. Decision trees and random forests are some of the most popular machine learning tools for binary classification, being used in many practical applications. Both methods provide a neighborhood for tested data during the prediction phase, and probabilities are usually computed based on the proportion of classes in those neighborhoods. The approach presented in this paper proposes replacing the prediction mechanism with one based on a probabilistic classifier based on the Nash equilibrium concept applied to the local data selected by the random forest classifier. Numerical experiments performed on synthetic data illustrate the behavior of the approach in a variety of settings.

Keywords: Decision trees · Random forests · Nash equilibria

1 Introduction

Decision trees are known to be among the most popular machine learning methods [19], and together with random forests [3] they are successfully used in applications in various domains such as medicine [4,9,18], chemistry [12], security [14], image classification [10], social networks [6], etc. Like many other efficient machine learning techniques, they are mainly used in two manners: either tailored for practical applications, highly adapted to their specificity, or in some standard form for comparison purposes with new methods. While tailoring the algorithm to a certain type of data for practical purposes is undoubtedly an essential task, it decreases the chances that a particular approach is actually tested outside the scope of its application.

This paper proposes a new general approach to interpreting information provided by decision trees and random forests. Both methods represent data in the form of subsets - partitions in the case of decision trees and sets of partitions in the case of random forests. These sets provide local information about data that can be further used to make predictions for test data that fit into that particular

This work was supported by a grant of the Romanian Ministry of Education and Research, CNCS - UEFISCDI, project number PN-III-P4-ID-PCE-2020-2360, within PNCDI III.

© Springer Nature Switzerland AG 2022
P. García Bringas et al. (Eds.): HAIS 2022, LNAI 13469, pp. 123–132, 2022.
https://doi.org/10.1007/978-3-031-15471-3_11

search space region. Local data provided by the random forests is aggregated, and a game-theoretic framework based on simple logistic regression is used to make predictions for the binary classification problem.

2   Decision Trees and Random Forests

The binary classification problem can be described as follows: given a data set **D** consisting of n points/instances $x_i \in \mathbb{R}^d$ and corresponding labels $y_i \in \{0, 1\}$, $i = 1, \ldots n$ we want to construct a model that predicts the class $\hat{y}_i$ for each x_i and further use this model to predict the class of new instances x from the data space that **D** belongs to [8].

Decision trees (DT) [2] split the data space into separate regions in a recursive manner by using some form of hyper-planes until the information in each region is considered pure and used to predict classes for all test data belonging to it. The model can be represented as a decision tree in which a node defines a new hyper-plane splitting its current region. A node becomes a leaf if data in the defining region is considered "pure" based on some indicator, related more or less to the fact that there is some majority of instances having the same class in that region. Various DT approaches use different purity indicators and splitting procedures. Examples of popular indicators used for these tasks are CART, the *entropy* or the *Gini index* [20].

Regardless of the splitting method/purity indicator used, a deep enough decision tree can correctly predict the class of every instance in the given data set **D**, while failing to correctly predict test instances because, like other classification models, in this manner they are *overfitting* the data. Thus, most improvements to DT aim to avoid this issue by finding a trade-off between the tree depth and some performance evaluation indicator and by adapting splitting methods and purity indices.

The bottom line is that DTs provide a partition of the training data that is further used for prediction. A tested instance x will be labeled based on the information provided by data in the corresponding tree leaf/region of the data space. Usually, the decision is made based on the proportion of labels in that particular data partition. Let DT be a decision tree build based on a data set **D** and x a tested value. Then the decision tree DT has partitioned **D** into data found in its leafs, denoted by $DT_1, \ldots, DT_m$, where m is the number of leafs of DT. Let $DT(x)$ be the data set corresponding to the leaf region of x, $DT(x) \subset D$. Typically the model would assign to x label y with a probability equal to the proportion of class y in $DT(x)$.

One way to enhance the performance of a decision tree is to employ other methods for analyzing the information from the leaves. In this paper, we propose the use of the FROG framework [17] to predict labels based on leaf data. FROG stands for Framework based on optimization and game theory for binary classification, and it is a game-theoretic framework for classification that aims to enhance the performance of probabilistic classification models by converting the classification problem into a multiplayer game and estimating model parameters to approximate the equilibrium of the game.

2.1 FROG

FROG constructs a game in which players are instances of the training data, their actions consist in choosing a class/label, and their payoffs are computed based on the F_1 score [5] adapted for their true class. Considering mixed strategy profiles for the game, an objective function having as a global minimum with value 0 the Nash equilibrium of the game is constructed and used to estimate the parameters of a probabilistic classification model.

Thus, if we consider logistic regression [8] the task is to find $\beta \in \mathbb{R}^{d+1}$ such that

$$\phi(x_i; \beta) = \frac{1}{1 + e^{-\beta x_i}} \tag{1}$$

can be used to predict the label y_i of x_i. In this model $\phi(x_i; \beta)$ represents the probability that x_i has label $y_i = 1$. β is estimated by maximizing the log likelihood function. A free term of 1 is added to x_i when computing the dot product βx_i.

Within FROG, instead of using the likelihood function, the Nash equilibrium of the game is approximated by minimizing an objective function constructed to compute game equilibria as global minima [11]. Thus, give a training data set $X \subset \mathbb{R}^{n \times d}$ and corresponding labels Y, the FROG game is defined as a triplet (N, C, U) where:

- N is the set of players, $N = \{1, \ldots, n\}$;
- C is the strategy profiles set of the game, consisting of the classes chosen by each player, $C = \{0, 1\}^n$. A strategy profile of the game $c \in C$ represents a possible classification of the data, where each player/instance i chooses class c_i;
- $U : C \to \mathbb{R}$ is the payoff function $U = (u_1, \ldots, u_n)$, where $u_i : C \to \mathbb{R}$ is the payoff function of player i computed as the F_1 score corresponding to the label y_i of x_i if player i has not chosen its correct label, and 1 otherwise

$$u_i(c) = \begin{cases} F_1(c; i) \text{ if } c_i \neq i \\ 1 \quad \text{ if } c_i = i \end{cases}. \tag{2}$$

The F_1 score

$$F_1(c; i) = \frac{2TP_{c,i}}{2TP_{c,i} + FP_c + FN_c}$$

[5] is computed as the ratio between the number of correctly classified instances with the desired label ($2TP_{c,i}$) divided by this number ($2TP_{c,i}$) to which we add the total number of incorrectly labeled instances ($FP + FN$). $TP_{c,i}$ counts how many instances in c have correctly chosen label i, and $FP_c + FN_c$ count how many instances are in c are incorrectly labeled.

A possible solution for the FROG game is the Nash equilibrium, a strategy profile of the game in which no player has a unilateral incentive for deviation, i.e. no player can improve its payoff by changing its strategy while all others maintain theirs. It is easy to show that the Nash equilibrium for the FROG game corresponds to the correct classification of the training data.

A mixed form Nash equilibrium represents a probability distribution over the strategies space indicating for each player the probability to choose a label such that it cannot improve its expected payoff by unilateral deviation. However, just the NE of the game cannot be used to make classification predictions. However, a probabilistic model that matches equilibrium probabilities would provide a classification model with "nice" properties such as stability against unilateral deviations.

The Nash equilibrium of a game can be computed as the minimum of a function on a polytope [11]. A function $v(\sigma)$ is defined as the sum of squared positive differences in unilateral deviations from σ to pure strategies for all players and all their pure strategies, where $\sigma \in [0,1]^n$ is a mixed strategy profile that can be described as a vector $\sigma = (\sigma_1, \sigma_2, \ldots, \sigma_n)$, with σ_i the probability that player/instance i chooses the label 1.

Within FROG, parameters of a probabilistic model are estimated to minimize function $v()$. Thus, in order to estimate parameter β in Eq. (1) we minimize $v(\phi(\beta x_1), \phi(\beta x_1), \ldots, \phi(\beta x_n))$ instead of optimizing the likelihood function. The function $v()$ is continuous, nonlinear, and challenging to minimize. In order to compute approximate optimum values, FROG uses the Covariance Matrix Adaption Evolution Strategy (CMA-ES) [7]. CMA-ES evolves the mean and covariance matrix of a normally distributed population of potential solutions to the problem; it is fast, reliable, and adaptive, with recommended parameter settings that makes it easy to use for different practical applications.

Thus, by using FROG, we assume that it may be possible to improve the performance of a DT that uses proportions of labels on the leaves for prediction unless all instances on a leaf have the same label, while the tested instance has a different one, in which case it is impossible to identify it correctly.

As a solution for this issue with regard to decision trees, as well as to avoid overfitting, Breiman proposed the use o Random Forests (RF) [3]. RFs combine decision trees grown on bootstrapped data on a randomly chosen subset of the features. Majority voting based on individual predictions in each tree is used to predict the class of a tested instance x. In [3] Breiman shows that random forests do not overfit as more trees are added, as they always converge. Also, they show that the accuracy of the random forests relies on the strength of the individual DT and their interdependence. However, in [1] disadvantages of using random forest are discussed, among which the possibility of bias in results due to the 'absent levels' during the training phase.

In this paper, we propose the enhancement of RFs by using FROG during the prediction phase by replacing the majority voting with the use of a FROG based probabilistic model trained on data provided by the decision trees in the forest. We call this method Random Forests - FROG (RF-FROG).

2.2 RF-FROG

While a typical decision tree offers a partition of the data space, a RF offers a set of partitions determined by the decision trees. For a given test instance x, the prediction is performed by majority voting based on information from the partitions sets it belongs to in each tree. Data in each leaf represents a neighborhood for the tested value x that varies from tree to tree depending on the bootstrapped sample and the features used to construct the tree. Aggregating corresponding data from all the decision trees construct a wider neighborhood offering a classifier relevant information for x extracted from all the trees. This information can be fed to FROG and a prediction be made based on the model provided by it.

The outline of RF-FROG is presented in Algorithm 1. RF-FROG takes as input the usual parameters of a random forest: training data $S \subset \mathbf{D}$, number of trees K, percent of the number of attributes sampled for splitting the trees p, maximum tree depth μ, and minimum purity level π. RF-FROG consists of two main steps: the forest is grown, and trees are preserved in the first step. In order to make a prediction for some test data in the second step, FROG is run for each tested instance on the data collected from the leaves in which that instance would fit from all the trees in the forest, and the prediction is made based on the underlying logistic regression model used by FROG.

In this respect RF-FROG acts similarly to a k-nearest neighbor method. As FROG is only used during the test phase, a small increase in computational complexity is determined by the size of the region in which a test instance is located, directly connected to the size of the forest trees. For $RF(xt)$ sets with less than 500 instances the increase is negligible.

Algorithm 1. RF-FROG

1: **Input**: training set $S \subseteq \mathbf{D}$, K, p, μ, π;
2: **Output**: predictions C for a (test) set T;
3: **for** k in $1 : K$ **do**
4:     $D_k \leftarrow$ sample of size N with replacement from S;
5:     $DT_k \leftarrow$ Decision tree based on D_k, p, μ, and π;
6: **end for**
7: **for each** $xt \in T$ **do**
8:     $RF(xt) = \cup_{t=1}^{K} DT_k(xt)$;
9:     Apply FROG on $RF(xt)$, i.e. use CMA-ES to mmpute

$$\beta = \operatorname*{argmin}_{\beta} v(\phi(\beta x_1), \ldots, \phi(\beta x_{|RF(xt)|}))^*$$

10:     Compute $c(xt_i) = \phi(\beta xt_i)$ (probability that xt_i has class 1);
11: **end for**
12: return $C = (c(xt_1), c(xt_2), \ldots, c(xt_{|T|})$
* $|\cdot|$ denotes the cardinality of a set.

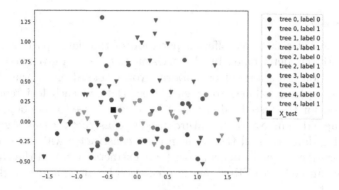

Fig. 1. $RF(xt)$ for Example 1. Labeled training instances of each tree from a Random Forest classifier for one test instance.

Example 1. Figure 1 illustrates $RF(x)$ for a tested instance x for a synthetic data set generated using the `make_classification` function from the `scikit-learn` [13] Python library with the following parameters: 200 instances, two attributes, seed 600; the data set is balanced (`weights` is 0.5) and the overlap for the two classes is 0.5 (`class_sep` parameter). The 200 instances are split into training and test sets. The Random Forest classifier has 5 trees. The entire data set and corresponding partition of data in leafs by using a single DT are depicted in Fig. 2. The Decision Tree has a maximum depth of 3 and uses entropy as the decision criterion when making splits.

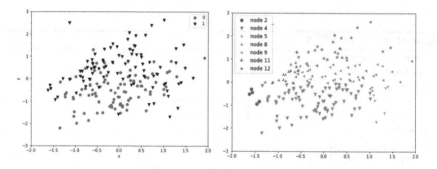

Fig. 2. Example 1: the left figure presents the original data with labeled instances and the right one presents the same instances separated in DT leaf nodes.

3   Numerical Experiments

Numerical experiments test if the use of FROG improves results reported by random forests on a set of synthetic data-sets generated with various properties in order to simulated different real-world situations. While the performance of a method on synthetic data-sets cannot guarantee its efficiency on real data, it provides a good indication about the behaviour of the method and where it may be applied [16].

Several test data sets with different degrees of difficulty are generated by using the make_classification function from scikit-learn with the following parameters: $D1 = \{100, 200, 500\}$ number of instances, $D2 = \{2, 5, 10, 15, 20\}$ the number of features, $D3 = \{0.1, 0.2, 0.5, 1\}$ the class separator, equal class weight (0.5), and seed 500. The obtained test data sets present different degrees of difficulty and provide a good test-bed for classification methods [16].

For each classification problem, we split the instances into K cross-validation folds at random to estimate the prediction error [8]. We use a stratified K-Folding approach to ensure that we obtain a balanced data set. For reproducibility we use scikit-learn to split the data into $K = 10$ folds using the StratifiedKFold function with seed 1. We fit the model on the training instances and report the area under the ROC curve (AUC) on the test instances for each fold.

We choose to report the area under the ROC curve (AUC) [8] as it is a performance measure of the accuracy of a classifier and measures the probability that a randomly selected positive sample is ranked higher than a randomly selected negative sample [5, 15]. If the tested model is able to classify all instances correctly, the AUC measure has a value of 1. A higher AUC value indicates a better result.

For the proposed approach, we generate random forests with the different parameters: $M1 = \{5, 10, 30, 50\}$ trees, $M2 = \{2, 3, 5, 10\}$ depth of a tree, and $M3 = \{0.5, 0.7, 1\}$ is the percentage of features kept from all features in the training set when subsampling the training instances for each tree.

Thus we obtain 48 variations of the classifier; each variant is tested on the 60 test data sets. We compare the performance of RF-FROG to the standard RF classifier that uses the same decision trees like the ones generated within RF-FROG. We report if there is a statistical difference between the two classifiers on the AUC values obtained for each test data set with each variation of the classifiers, Figs. 3a, 3b, and 3c. Statistical significance is assessed on the AUC values reported on the 10 folds by using a paired $t-test$ with $\alpha = 0.05$, testing the null hypothesis that AUC values reported by RF-FROG are lower than those reported by RF. A p-value lower than α indicates that the null hypothesis can be rejected.

Figure 3a presents the result for the variations that classify the test data sets with 100 instances. Each column presents results for a value of $D3$ parameter, and every four rows in each column show results for parameter $D2$. Each square of a 4×12 matrix represents the p-value; lighter squares represent lower values for the statistical test, whereas darker squares represent higher values. The lines of a 4×12 matrix present results for the parameter $M1$, the first four columns show results for the variation of $M2$ parameter when $M3$ parameter has the value 0.5, the next four columns show results for the variation of $M2$ when $M3$ has the value 0.7 and the last four columns in a 4×12 matrix show results for the variation of M2 when M3 has the value 1.

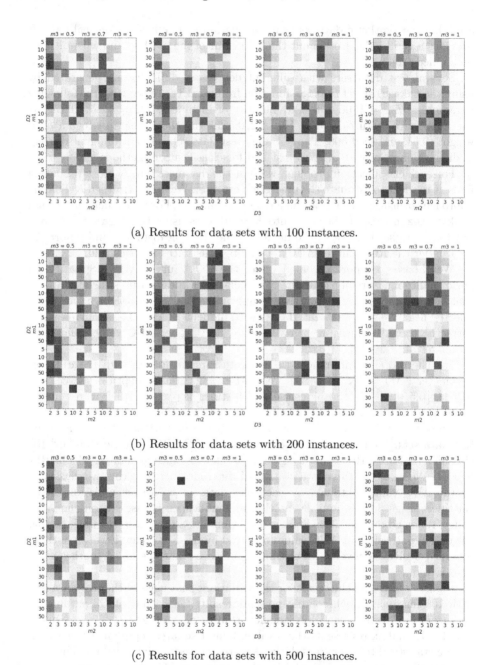

(a) Results for data sets with 100 instances.

(b) Results for data sets with 200 instances.

(c) Results for data sets with 500 instances.

Fig. 3. Results obtained on the synthetic data sets. Squares represent p-values, lighter shades correspond to lower values. A p-value below the significance level $\alpha = 0.05$ indicates that RF-FROG results may be considered significantly better than those reported by RF. Each square represents one configuration of the data set.

In 284 cases out of the 960 results presented in Fig. 3a, RF-FROG obtains results significantly better than RF, and in 4 out of 960 results, RF-FROG reports significantly worse results. RF-FROG is statistically indifferent to RF-based on AUC values in all other cases.

Figure 3b presents the result for the variations that classify the test data sets with 200 instances. In 353 cases out of the 960 results presented in Fig. 3b, RF-FROG obtains results significantly better than RF, and in 22 out of 960 results, RF-FROG obtains results significantly worse results. In all other cases RF-FROG is statistically indifferent to RF based on AUC values.

Figure 3c presents the result for the variations that classify the test data sets with 500 instances. Here, in 505 cases out of the 960 results presented in Fig. 3c, RF-FROG reports results significantly better than RF, and in 104 out of 960 results, RF-FROG reports results significantly worse results. RF-FROG is statistically indifferent to RF in all other cases based on AUC values.

4   Conclusions

A simple enhancement of the random forest classifier is presented in this paper. Data from the leaves of the RF trees is aggregated for each tested instance, and a game-theoretic approach based on optimization is used to compute the probability of belonging to a class. The game-theoretic approach requires using an advanced optimizer to compute the equilibria of a multiplayer game; in this case, CMA-ES is used. While results show significant differences between methods, it remains to be proven that including FROG in more advanced versions of RFs remains beneficial, and this is the scope of further work.

References

1. Au, T.C.: Random forests, decision trees, and categorical predictors: the "absent levels" problem. J. Mach. Learn. Res. **19**(1), 1737–1766 (2018)
2. Breiman, L., Friedman, J.H., Olshen, R.A., Stone, C.J.: Classification and Regression Trees. Wadsworth and Brooks, Monterey, CA (1984)
3. Breiman, L.: Random forests. Mach. Learn. **45**(1), 5–32 (2001). https://doi.org/10.1023/A:1010933404324
4. Czajkowski, M., Jurczuk, K., Kretowski, M.: Accelerated evolutionary induction of heterogeneous decision trees for gene expression-based classification. In: Proceedings of the Genetic and Evolutionary Computation Conference, pp. 946–954. GECCO 2021. Association for Computing Machinery, New York (2021). https://doi.org/10.1145/3449639.3459376
5. Fawcett, T.: An introduction to roc analysis. Pattern Recogn. Lett. **27**(8), 861–874 (2006). ROC Analysis in Pattern Recognition
6. Fazeen, M., Dantu, R., Guturu, P.: Identification of leaders, lurkers, associates and spammers in a social network: context-dependent and context-independent approaches. Soc. Netw. Anal. Min. **1**(3), 241–254 (2011). https://doi.org/10.1007/s13278-011-0017-9

7. Hansen, N., Müller, S.D., Koumoutsakos, P.: Reducing the time complexity of the derandomized evolution strategy with covariance matrix adaptation (CMA-ES). Evol. Comput. **11**(1), 1–18 (2003). https://doi.org/10.1162/106365603321828970

8. Hastie, T., Tibshirani, R., Friedman, J.: The Elements of Statistical Learning: Data Mining, Inference and Prediction, 2nd edn. Springer, New York (2009)

9. Lotte, F., et al.: A review of classification algorithms for EEG-based brain-computer interfaces: a 10 year update. J. Neural Eng. **15**(3), 031005 (2018). https://doi.org/10.1088/1741-2552/aab2f2

10. Ma, L., Li, M., Ma, X., Cheng, L., Du, P., Liu, Y.: A review of supervised object-based land-cover image classification. ISPRS J. Photogram. Remote. Sens. **130**, 277–293 (2017). https://doi.org/10.1016/j.isprsjprs.2017.06.001

11. McKelvey, R.D., McLennan, A.: Computation of equilibria in finite games. Handb. Comput. Econ. **1**, 87–142 (1996)

12. Mitchell, J.B.O.: Machine learning methods in chemoinformatics. Wiley Interdisc. Rev. Comput. Mol. Sci. **4**(5), 468–481 (2014). https://doi.org/10.1002/wcms.1183

13. Pedregosa, F., et al.: Scikit-learn: machine learning in Python. J. Mach. Learn. Res. **12**, 2825–2830 (2011)

14. Resende, P., Drummond, A.: A survey of random forest based methods for intrusion detection systems. ACM Comput. Surv. **51**(3), 1–36 (2018). https://doi.org/10.1145/3178582

15. Rosset, S.: Model selection via the AUC. In: Proceedings of the Twenty-First International Conference on Machine Learning, ICML 2004, p. 89. Association for Computing Machinery, New York (2004). https://doi.org/10.1145/1015330.1015400

16. Scholz, M., Wimmer, T.: A comparison of classification methods across different data complexity scenarios and datasets. Expert Syst. Appl. **168**, 114217 (2021). https://doi.org/10.1016/j.eswa.2020.114217

17. Suciu, M.-A., Lung, R.I.: Nash equilibrium as a solution in supervised classification. In: Bäck, T., Preuss, M., Deutz, A., Wang, H., Doerr, C., Emmerich, M., Trautmann, H. (eds.) PPSN 2020. LNCS, vol. 12269, pp. 539–551. Springer, Cham (2020). https://doi.org/10.1007/978-3-030-58112-1_37

18. Van, A., Gay, V.C., Kennedy, P.J., Barin, E., Leijdekkers, P.: Understanding risk factors in cardiac rehabilitation patients with random forests and decision trees. In: Proceedings of the Ninth Australasian Data Mining Conference - Volume 121, AusDM 2011, pp. 11–22. Australian Computer Society Inc, AUS (2011)

19. Wu, X., et al.: Top 10 algorithms in data mining. Knowl. Inf. Syst. **14**(1), 1–37 (2008). https://doi.org/10.1007/s10115-007-0114-2

20. Zaki, M.J., Meira, Jr, W.: Data Mining and Machine Learning: Fundamental Concepts and Algorithms. Cambridge University Press, Cambridge, 2 edn. (2020). https://doi.org/10.1017/9781108564175

Concept Drift Detection to Improve Time Series Forecasting of Wind Energy Generation

Tomás Cabello-López[✉][iD], Manuel Cañizares-Juan[iD],
Manuel Carranza-García[iD], Jorge Garcia-Gutiérrez[iD], and José C. Riquelme[iD]

Department of Computer Languages and Systems, University of Sevilla, 41012
Sevilla, Spain
tclopez@us.es

Abstract. Most of the current data sources generate large amounts of data over time. Renewable energy generation is one example of such data sources. Machine learning is often applied to forecast time series. Since data flows are usually large, trends in data may change and learned patterns might not be optimal in the most recent data. In this paper, we analyse wind energy generation data extracted from the Sistema de Información del Operador del Sistema (ESIOS) of the Spanish power grid. We perform a study to evaluate detecting concept drifts to retrain models and thus improve the quality of forecasting. To this end, we compare the performance of a linear regression model when it is retrained randomly and when a concept drift is detected, respectively. Our experiments show that a concept drift approach improves forecasting between a 7.88% and a 33.97% depending on the concept drift technique applied.

Keywords: Machine learning · Concept drift detection · Data streaming · Time series · Wind energy forecasting

1 Introduction

Today, a growing concern about climate change exists. Climate change has become a major problem for humankind [1]. It is produced by greenhouse gas emissions from human activities such as burning fossil fuels for energy [2]. Renewable energies represent the best alternative to the use of fossil fuels in power generation since they are an inexhaustible, cheap and exploitable source of energy everywhere in the world [3]. However, its increased use in the electricity mix involves an inherent risk of system instability due to renewable energies intermittency problems [4].

Energy production forecasting in advance might be a possible solution to renewable energy issues. In this context, regression models could provide future

Supported by the Spanish Ministry of Science and Innovation (PID2020-117954RB-C22) and the Andalusian Regional Government (US-1263341, P18-RT-2778).

P. García Bringas et al. (Eds.): HAIS 2022, LNAI 13469, pp. 133–140, 2022.
https://doi.org/10.1007/978-3-031-15471-3_12

generation and thus improve the estimation of other non-renewable energy sources needed to balance the power generation mix.

Although regression techniques have shown their potential for such forecasting problems, they are sensitive to abrupt changes in data distributions. Especially in situations where data flows are continuous. In this context, it is necessary to retrain the models to improve their results [5].

Concept drift techniques are based on recognising the moments in which a data stream shows significant changes in trend. Thus, when we detect them, we can retrain our model, performing a better adaptive and incremental training. Concept drift detection techniques are often applied as preprocessing for regression techniques. Jean Paul Barddal et al. [6] study an ensemble-based dynamic model using the ADWIN Concept Drifts detection technique. Jan Zenisek et al. [7], evaluate the benefits of using concept drifts detection techniques to help regression models perform preventive maintenance of industrial machinery. Elena Ikonomovska et al. [8], present a regression tree model for data stream processing using concept drifts detection techniques. Lucas Baier et al. [9] elaborate a strategy that uses simple machine learning models when concept drifts are detected and a general complex model otherwise.

Although concept drift is not a novel preprocessing of time series, our work is the first intent (to the best of our knowledge) to evaluate concept drift detection in a real context such as renewable energy forecasting.

In this paper, we have chosen Linear Regression (LR) model (a very well-known technique in the literature [10]) to address wind energy generation forecasting in Spain and evaluate its performance when it is used with and without concept drift detection techniques.

The content of this paper will be organised in the following sections as follows: all relevant information regarding the data, the models used and their comparison method in Sect. 2. The results obtained by the models and the limitations of the study are shown in Sect. 3. Finally, the main conclusions reached based on the results are in Sect. 4.

2   Materials and Method

2.1   Dataset

To train our models, we have used data about historical wind power generation from the API ESIOS [11]. Data was collected from 1st February 2022 to 1st April 2022 in 10-minute intervals. The total number of records is 7640.

Before introducing the data into a LR model, we must normalise it using the min-max technique. The models process the data using a sliding window strategy with size 4. Our training set thus consists of subsets of 4 records (in ten-minute intervals) previous to the wind power generation to be predicted.

2.2   Concept Drifts Detection Techniques

The techniques used for the detection of concept drifts have been Adaptive Windowing (ADWIN) [12], Kolmogorov-Smirnov Windowing Method (KSWIN)

[13] [14] and Page Hinkley [15]. The parameterization used for each technique has been determined by a grid search as can be seen in Table 1 which highlights the parameters finally selected.

Table 1. Grid search carried out in concept drift detection techniques. In bold, the best parameters selected.

Technique	Parameters	Values
ADWIN	delta	0.005, 0.001, 0.05, **0.2**, 0.7, 1, 0.15
KSWIN	alpha	**0.001**, 0.005, 0.05, 0.07, 0.5, 0.2, 0.7
	window_size	100, 150, 120, 110, 80, **72**, 160
	stat_size	30, **40**, 36, 35, 32, 42
PageHinkley	delta	0.005, 0.001, 0.002, 0.007, 0.01, **0.1**, 0.15
	lambda	50, **25**, 10, 5, 30, 70, 100
	alpha	0.99999, 0.999, 0.85, **0.7**, 0.9995, 0.9993

A Python library called scikit-multiflow has been used for the implementation of the Concept Drift detection techniques [16].

To predict the wind power generation and test the performance of the compared concept drift detection techniques we have used LR. The implementation of the LR model has been carried out through the sklearn library [17].

2.3   Comparison Procedure

In this paper, we compare LR performance with and without using concept drift detection techniques. To evaluate each option, we use Blocked Cross-Validation (BCV) method [18,19] which avoids test data leakage issues. BCV also allows to measure errors with lower variability and provide a more accurate estimate of the generalisation error than a hold-out validation procedure does.

In our experimentation, BCV splits the dataset in a number of folds N. N depends on the number of concept drifts detected. Thus, the whole dataset is divided in different blocks which corresponds with the folds used in BCV. Finally, every fold is used to validate a LR model using 80% of the data as training and 20% as test.

To evaluate the goodness of a a concept drift detection which provides N concept drift, we carried out a BCV with a LR model and a random distribution of N blocks within the dataset and compare LR performance with that obtained from the BCV obtained by the same LR but trained with the N concept-drift-derived block distribution. The measures used to estimate the accuracy of the LR models in the predictions are the mean absolute error (MAE) and the WAPE [20,21].

To reduce the random nature of the comparison, BCV on LR models have been repeated fifty times with random blocks. Then, the mean errors in the fifty executions where used to make the comparison with the results obtained from the concept-drift-derived blocks.

3   Results

This results has been obtained on a local computer with 16 gb of DDR4 8 GB × 2 (2666 MHz) RAM, an Nvidia GeForce RTX2060-6GB GDDR6 graphics card and an Intel Core i7-9750H processor. The concept drifts detected for each detection technique and the mean time in seconds that LR took to process each of the blocks can be seen in Table 2.

Table 2. Number of concept drifts detected for each detection technique.

Technique	Concept drifts	Mean time LR (seconds)
ADWIN	2	0.0016
KSWIN	253	0.0003
PageHinkley	185	0.0017

The results of a LR model with and without concept drifts detection according to several detection techniques can be seen in Table 3.

Table 3. MAE, WAPE results from training the model using concept drifts detectors and random folders. In bold, the best results by each technique.

Concept drift detector	Using detector		Not using detector	
	MAE	WAPE	MAE	WAPE
ADWIN-LR	**890.266**	**0.0077**	1348.324	0.012
KSWIN-LR	**128.386**	**0.0012**	139.366	0.0013
PageHinkley-LR	**126.975**	**0.0011**	140.86	0.0013

As we can see in Table 3, the use of a concept drift detector improves the mean performance of the LR model when it is applied in our case of study in all cases.

In the following figures, we can observe the mean MAE and WAPE obtained by our models along with the fifty random executions. The red line stands for the results obtained by LR when concept-drift-derived blocks were used to retrain. From now on, we will use CD (concept drift) model to refer to the model which uses the concept-drift-derived block and RB (random blocks) model to refer the model which uses the random blocks. We can see in Fig. 1 and Fig. 2 that the models which use ADWIN have very similar results ten out of the fifty times. Furthermore, three out of the fifty times the RB model obtains better results

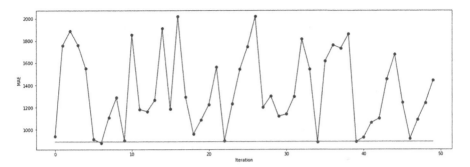

Fig. 1. MAE results using ADWIN. Red line stands for CD best value. (Color figure figure)

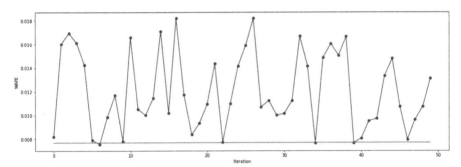

Fig. 2. WAPE results using ADWIN. Red line stands for CD best value. (Color figure figure)

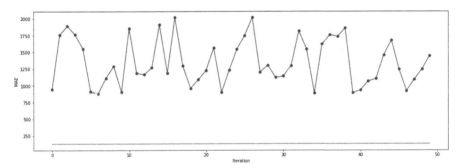

Fig. 3. MAE results using KSWIN. Red line stands for CD best value. (Color figure figure)

than the CD model. Even with better results in a few of iterations, the RB model provides a worse performance throughout the iterations being outperformed by CD model forty-seven times.

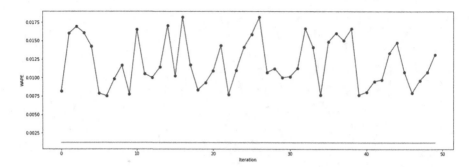

Fig. 4. WAPE results using KSWIN. Red line stands for CD best value. (Color figure figure)

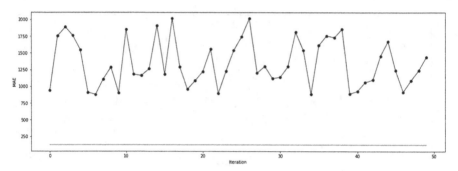

Fig. 5. MAE results using PageHinkley. Red line stands for CD best value. (Color figure figure)

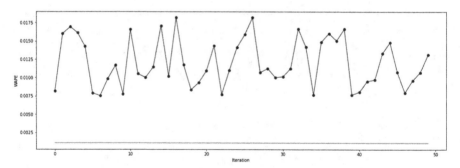

Fig. 6. WAPE results using PageHinkley. Red line stands for CD best value. (Color figure figure)

In the Fig. 3 and Fig. 4 the MAE and WAPE results for the KSWIN models are represented, in this case we can affirm that the CD model outperforms the RB model, being able to achieve much better results in all the iterations. Finally, the Fig. 5 and Fig. 6 show similar results for the PageHikley models. In these cases, the CD model manages to obtain better results than the RB model at every iteration.

Our results seem to confirm that LR model performs better predictions when the concept drifts are previously analysed. They encourage the use of concept drift detection as part of the data pre-processing stage when applying regression models for forecasting.

However, this paper has some limitations, it aims to be a practical exploration of the application of this set of techniques, so we don't attempt to define the best regression technique to deal with power generation forecasting issue but to confirm that concept drift detection can be helpful for regression problems. There are more techniques for time series forecasting with which we could obtain a more complete study. We have fixed the hyperparameters of the concept drifts detection methods through grid search, which does not prove us that there are not better configurations. Finally, we have only taken into account the last two months recorded by ESIOS, it will be interesting to see how the results evolve over time out of our interval of study.

4   Conclusions

In this paper, we analysed the wind energy generation data extracted from the Sistema de Información del Operador del Sistema (ESIOS) of the Spanish power grid. The study we performed indicated that the models improved their results between a 7.88% and a 33.97% by the prior detection of concept drifts in the data depending on the concept drift technique applied. The method of detecting concept drifts with which obtained the best result was the PageHinkley method.

Future work could compare the performance of the proposed model with other models, specially from deep learning area of expertise. This study could also be extended by trying to evaluate the performance of this combination of techniques for a longer time and with meteorological variables useful for the prediction of wind power generation.

References

1. Díaz Cordero, G.: El cambio climático. Ciencia y sociedad (2012)
2. Lizano, B.: Calentamiento Global: "la máxima expresión de la civilización petrofósil". Revista del CESLA. https://www.redalyc.org/articulo.oa?id=243329724003
3. Rafique, M.M., Bahaidarah, H.M., Anwar, M.K.: Enabling private sector investment in off-grid electrification for cleaner production: optimum designing and achievable rate of unit electricity. J. Clean. Prod. **206**, 508–523 (2019). https://doi.org/10.1016/j.jclepro.2018.09.123
4. Mills, A.D., Levin, T., Wiser, R., Seel, J., Botterud, A.: Impacts of variable renewable energy on wholesale markets and generating assets in the united states: a review of expectations and evidence. Renew. Sustain. Energy Rev. **120**, 109670 (2020). https://doi.org/10.1016/j.rser.2019.109670

5. Jaworski, M.: Regression function and noise variance tracking methods for data streams with concept drift. Int. J. Appl. Math. Comput. Sci. **28**(3), 559–567 (2018)
6. Barddal, J.P., Gomes, H.M., Enembreck, F.: Advances on concept drift detection in regression tasks using social networks theory. Int. J. Nat. Comput. Res. (IJNCR) **5**(1), 26–41 (2015)
7. Zenisek, J., Holzinger, F., Affenzeller, M.: Machine learning based concept drift detection for predictive maintenance. Comput. Ind. Eng. **137**, 106031 (2019). https://doi.org/10.1016/j.cie.2019.106031
8. Ikonomovska, E., Gama, J., Sebastião, R., Gjorgjevik, D.: Regression trees from data streams with drift detection. In: Gama, J., Costa, V.S., Jorge, A.M., Brazdil, P.B. (eds.) DS 2009. LNCS (LNAI), vol. 5808, pp. 121–135. Springer, Heidelberg (2009). https://doi.org/10.1007/978-3-642-04747-3_12
9. Baier, L., Kühl, N., Satzger, G., Hofmann, M., Mohr, M.: Handling concept drifts in regression problems – the error intersection approach. In: WI2020 Zentrale Tracks, pp. 210–224. GITO Verlag (2020). https://doi.org/10.30844/wi_2020_c1-baier
10. Ray, S.: A quick review of machine learning algorithms. In: 2019 International conference on machine learning, big data, pp. 35–39. cloud and parallel computing (COMITCon), IEEE (2019)
11. Api esios documentation. https://api.esios.ree.es/
12. Bifet, A., Gavalda, R.: Learning from time-changing data with adaptive windowing. In: Proceedings of the 2007 SIAM International Conference on Data Mining, pp. 443–448. SIAM (2007)
13. Raab, C., Heusinger, M., Schleif, F.-M.: Reactive soft prototype computing for concept drift streams. Neurocomputing **416**, 340–351 (2020)
14. Lima, M., Filho, T.S., de A. Fagundes, R.A.: A comparative study on concept drift detectors for regression. In: Britto, A., Valdivia Delgado, K. (eds.) BRACIS 2021. LNCS (LNAI), vol. 13073, pp. 390–405. Springer, Cham (2021). https://doi.org/10.1007/978-3-030-91702-9_26
15. Page, E.S.: Continuous inspection schemes. Biometrika **41**(1/2), 100–115 (1954)
16. Montiel, J., Read, J., Bifet, A., Abdessalem, T.: Scikit-multiflow: a multi-output streaming framework. J. Mach. Learn. Res. **19**(72), 1–5 (2018)
17. Buitinck, L. et al.: API design for machine learning software: experiences from the scikit-learn project. In: ECML PKDD Workshop: Languages for Data Mining and Machine Learning, pp. 108–122 (2013)
18. Bergmeir, C., Benítez, J.M.: On the use of cross-validation for time series predictor evaluation. Inf. Sci. **191**, 192–213 (2012). https://doi.org/10.1016/j.ins.2011.12.028. Data Mining for Software Trustworthiness
19. Bergmeir, C., Costantini, M., Benítez, J.M.: On the usefulness of cross-validation for directional forecast evaluation, computational statistics & data analysis. cFEnetwork: Ann. Comput. Financ. Econometr. **76**, 132–143 (2014). https://doi.org/10.1016/j.csda.2014.02.001
20. Shcherbakov, M.V., et al.: A survey of forecast error measures. World Appl. Sci. J. **24**(24), 171–176 (2013)
21. Hewamalage, H., Montero-Manso, P., Bergmeir, C., Hyndman, R.J.: A look at the evaluation setup of the m5 forecasting competition. arXiv preprint arXiv:2108.03588 (2021)

A Decision Support Tool for the Static Allocation of Emergency Vehicles to Stations

Miguel Ángel Vecina[2](✉)(iD), Joan C. Moreno[1](✉)(iD), Yulia Karpova[2](✉)(iD),
Juan M. Alberola[1](✉)(iD), Victor Sánchez-Anguix[2](✉)(iD), Fulgencia Villa[2](✉)(iD),
and Eva Vallada[2](✉)(iD)

[1] Valencian Research Institute for Artificial Intelligence, Universitat Politècnica de València, Camino de Vera, s/n, 46022 Valencia, Spain
{joamoteo,jalberola}@upv.es

[2] Instituto Tecnológico de Informática, Grupo de Sistemas de Optimización Aplicada, Ciudad Politécnica de la Innovación, Edificio 8g, Universitat Politècnica de València, Camino de Vera s/n, 46022 Valencia, Spain
{mivegar2,yukarkry,vicsana1,mfuvilju,evallada}@upv.es

Abstract. Medical emergencies require decision support tools for determining the allocation of emergency resources (e.g., ambulances) depending on different factors (e.g., requests, stations, etc.). In this sense, obtaining the optimal allocation is key to guaranteeing the effectiveness of the emergency services, which can be measured as the response time. Focused on this issue, we present a decision support tool for the ambulance location problem, which consists of determining the optimal allocation of emergency vehicles to prospective stations. This tool is evaluated with real data from the Valencian region to assess our proposal's performance.

Keywords: Decision making tool · Emergency vehicles · Optimization

1 Introduction

Any country's Emergency Medical Service (EMS) is responsible for stabilizing and pre-hospital transporting patients with medical urgency and emergency. This service aims to perform this task with the lowest response time. The response time is essential to measure the effectiveness of the service, evaluate its quality and, consequently, preserve the life and health of patients. Any EMS has a fleet of Emergency Medical Vehicles (EMVs) to carry out pre-hospital care. More specifically, to deal with emergencies, two types of EMV may be employed depending on the severity level of the scenario: Basic Life Support (BLS) and Advanced Life Support (ALS).

The rapid response of emergency medical services depends on many factors, such as the quantity and quality of available resources (e.g., personnel and vehicles), dispatch policies, the average speed of an ambulance, the demand for emergency services, and the location of the ambulance bases (i.e., dispatch bases).

© Springer Nature Switzerland AG 2022
P. García Bringas et al. (Eds.): HAIS 2022, LNAI 13469, pp. 141–152, 2022.
https://doi.org/10.1007/978-3-031-15471-3_13

The latter represents the waiting location for ambulances while they are idle. The importance of an appropriate initial area for dispatch bases is crucial, as it determines how quickly they may be able to respond to different emergencies. The rapid progress in the development of Geographic Information Systems (GIS) [17], Geographic Positioning Systems (GPS) [10], and the Internet of Things (IoT) may allow addressing the problem of locating medical emergency vehicles efficiently and effectively.

The beginning of GIS development dates back more than half a century, but its everyday use began not many years ago. The current application of GIS covers the fields of earth sciences, such as hydrology, forestry, and geology. It has also been extended to urban management, hospitality and tourism, food industries, and the management of medical services. In parallel, we have also seen a rise in artificial intelligence (AI) techniques in urban applications [22]. AI techniques may report advantages by leveraging the optimal use of data and spatial information. Incorporating AI algorithms into spatial decision support systems and GIS may support decision-making in complex and large problems that human decision-makers could not effectively tackle.

This article presents the integration and development of an optimization model for optimizing the locations of medical vehicles with a GIS-based decision support tool. This tool aims to provide decision-making support to the Emergency Medical Services. More specifically, the tool's purpose is to help quickly visualize the impact that decisions can have on the amount of population covered/serviced by emergency vehicles according to their dispatch locations. We employ real data from Valencia's region to illustrate the applicability of the optimization model and the decision support tool.

The rest of the paper is organized as follows. Section 2 shows a general background of models for emergency vehicles location. Section 3 shows the architecture of the decision support system. Section 4 describes the optimization model to allocate emergency vehicles to dispatch stations. In Sect. 5, we describe different experiments to evaluate the performance of several solvers for the problem instance and the model's effectiveness compared to the current solution implemented by human decision-makers. Finally, in Sect. 6, we draw some concluding remarks and future research lines.

2   Background

The problem of deciding the location of medical emergency vehicles has been studied in the literature for some decades and from different perspectives.

The development of Geographic Information Systems (GIS) has favored geo-referencing and visualization tools. These tools are valuable for deciding the location of medical emergency vehicles. For instance, one of the spatial techniques employed in this study is isochrones. An isochrone represents the area that is reachable within a given time from a certain point through specific transport (on foot, by bicycle, by car). Current technologies allow for the calculation of isochrones in real-time, provide data on the amount of the population within the area of an isochrone, and support basic geometric operations between isochrones. Therefore, an isochrone can provide an interesting tool for analyzing the area covered by emergency vehicles. While useful, the use of such isochrones for the problem of deciding the location of emergency vehicles has so far been very limited [15,18,19]. The optimization model and the decision support tool proposed in this paper heavily rely on the calculation and visualization of isochrones as a decision-making tool.

Optimization models are also important in the problem of deciding the dispatch bases for ambulances. The first models were static and did not consider possible fluctuations in the system [6]. These models rely on parameters and data known in advance and are not probabilistic. However, later models [3] try to introduce uncertainty in the formulation of the models (e.g., the probability that a vehicle may not be available, the probabilistic nature of response time, etc.). The development of the models aims not only to incorporate all the complexity of the prehospital care process but also to make the models more realistic.

The design of a real emergency system should consider the response time of the first call and that the response times for future emergencies are as adequate as possible. The location problem must be solved by considering an action's influence on future actions. This is why dynamic models arise that try to find an optimal system configuration avoiding dimensional, combinatorial and probabilistic complexity. Mathematical programming, queuing theory, and geometric optimization are some of the approaches from which the problem of the location of health emergency vehicles continues to be addressed [4,5]. As far as the authors are concerned, this is one of the first optimization models to be based on the use of isochrones.

3   Architecture

The decision support tool presented in this paper is a part of a larger project involving tools to support the operational decisions made when managing emergency vehicles. We have coined this suite of software as *REVES*. In this section, we describe the architecture for *REVES Map*, one of the key components of *REVES*, as well as the services offered to decision-makers.

Firstly, the reader should understand that an emergency system's operational management requires three major decisions. The first decision is related to the initial location of the emergency vehicles across different stations. This is known in the literature as the static ambulance location problem [7,8,12]. The second operational decision consists of, given an emergency, deciding what ambulance is allocated for the service and to what hospital the vehicle transport the patient. This is known as the ambulance dispatch problem [13,16,23]. Finally, another decision is the relocation of ambulances as they finish transportation requests. That is, whether or not the ambulance should return to its initial station or be allocated to a different station to cover other areas. This is known in the literature as the ambulance relocation problem [2,9,14]. The reader should be aware that, in this paper, we will focus on describing a decision support tool for the static ambulance location problem. However, *REVES* includes other tools to deal with the other two operational problems. Next, we describe the general architecture of *REVES Map*. Figure 1 illustrates the architecture proposed for the tool.

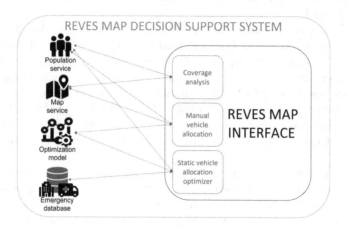

Fig. 1. General architecture of REVES Map.

The *REVES Map* decision support system employs a distributed architecture that separates the user interface from the business logic. We have employed a service architecture that decouples the user interface from prospective calculations. This architecture allows us to reuse services for other purposes and implement interfaces for different platforms if needed. As of today, there are several implemented services:

- **Emergency database service:** This service contains information about the number and type of available emergency vehicles in the area of interest, as well as the prospective stations for emergency vehicles. In addition, the database service also allows the user to record any allocation of vehicles to stations for future use and analysis.
- **Population service:** This service is responsible for estimating the population in a given area defined by a polygon. We base our estimations for this service on the Global Human Settlement Layer (GHSL) dataset. GHSL is a project by the European Commission which produces global spatial information and evidence-based analytics about the human presence on the planet. This information is available in free and open access population density maps that describe different features. The population service uses the *GHS-POP*[1] spatial raster dataset, which provides residential population estimates for the year 2015.

---

[1] https://ghsl.jrc.ec.europa.eu/ghs_pop2019.php.

– **Map service:** We employ a map service to show interactive maps, calculate isochrones, and make calculations such as routes. The interactive maps are based on *OpenStreetMap* [11], whereas the calculations are supported by *OpenRouteService*[2].
– **Optimization service:** We have designed an optimization service based on a Mixed-Integer Linear Programming (MILP) model. The model optimally allocates emergency vehicles to stations. We describe this service in more detail in Sect. 4. This service aims to efficiently allocate ambulances to dispatch stations based on available resources and effectively support human decision-making.

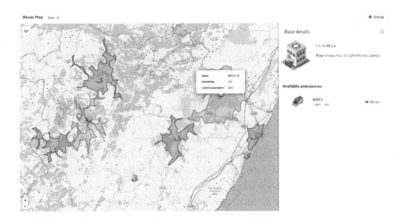

Fig. 2. Screenshot of REVES Map user interface showing the interactive map.

The user interface, which can be observed in Fig. 2, employs the services mentioned above to provide human decision-makers with tools to explore and analyze different static allocations of emergency vehicles to stations. As of today, the user interface supports the following functionalities:

– It allows the user to manually specify his/her allocation of emergency vehicles to stations. This task is supported using an interactive map like the one shown in the above figures.
– Moreover, the user interface lets users filter emergency vehicles by type or availability and set time values for the isochrones to increase or decrease the covered area.
– The user interface supports the visualization of the isochrones defined by current allocation and the population covered by isochrones. In addition to this, the user interface also shows intersections between isochrones and the population of those intersections. These features support a coverage analysis that can help human decision-makers to decide between different allocations.

---

[2] https://openrouteservice.org/.

– In addition, the user interface provides a link with the optimization service that allows the user to load the optimal allocation of vehicles to stations employing the model specified in Sect. 4.
– Finally, the user interface allows users to record allocations (manual or automatic) for later analysis.

4 Static Ambulance Allocation Model

This section describes the MILP model we designed to efficiently and effectively allocate emergency vehicles to dispatch bases.

4.1 Problem Description

The EMS of any country has a fleet of Advanced Life Support (ALS) and Basic Life Support (BLS) vehicles, identified as F^A and F^B, respectively. In addition, there is a set N of possible bases where vehicles can be parked to start their shift and where they return after attending the call. These stations are also known as dispatch bases o dispatch stations. Formally, the static ambulance allocation problem consists of placing each available Emergency Medical Vehicles (EMVs) in a base j belonging to the set N of possible bases. This assignment should cover the largest population of a geographical area. Our model divides the area into M health subareas that need to be covered. This subdivision reflects the reality of many health services like those employed in Valencia's region, where the province is divided into several separate health departments. Each health department (indexed by i) has h_i citizens to provide service. From each base j, an EMV can reach some population of each department i given a driving time (h_{ji}^A and h_{ji}^B, depending on whether the EMS is ALS or BLS). Due to the road network, it is impossible to cover the entire population in many cases. For this reason, this kind of problem can be formulated as a maximum coverage model [5]. Taking into account these ideas, we can formulate the following model.

4.2 Mathematical Model

As mentioned, the proposed decision support tool is composed of several services. One of those services is the optimization service, which optimally allocates emergency vehicles to prospective stations. This service is important for the human decision-maker as it provides an alternative allocation to the allocations designed by human decision-makers. We should point out that 159 prospective stations and 67 emergency vehicles (20 ALS and 47 BLS) are available in a region like Valencia. Thus, the search space becomes too massive for the cognitive capabilities of a human to handle. Hence, an optimization model can help to tackle this problem. Our proposed optimization model is based on a MILP formulation. This section describes the optimization model and explains each of its components.

Variables

The model has two type of variables

* Assignment of ambulances to bases:
  - $X_j^A = 1$, if the ALS ambulance is assigned to the base $j \in \mathcal{N}$; 0, otherwise
  - $X_j^B = 1$, if the BLS ambulance is assigned to the base $j \in \mathcal{N}$; 0, otherwise
* Level of coverage in each health department
  - D_i = uncovered population in the health department $i \in \mathcal{M}$
  - E_i = population covered more than once in the health department $i \in \mathcal{M}$

The formulation of the model is:

$$\min \sum_{i \in M} D_i \tag{1}$$

s.t.:

$$\sum_{j \in N} h_{ji}^A X_j^A + h_{ji}^B X_j^B + D_i - E_i = h_i \qquad \forall i \in M \tag{2}$$

$$\sum_{j \in N} X_j^A = F^A \tag{3}$$

$$\sum_{j \in N} X_j^B = F^B \tag{4}$$

$$X_j^A + X_j^B \leq 1 \qquad \forall j \in N \tag{5}$$

$$\sum_{j \cap N} X_j^A \geq f^A \qquad \forall j \in N \tag{6}$$

$$\sum_{j \in N} X_j^B \leq f^B \qquad \forall j \in N \tag{7}$$

$$\sum_{j \in k_s} (X_j^A + X_j^B) \leq 1 \qquad \forall k_s \in K, \quad s = 1, ..., v \tag{8}$$

$$\sum_{j \in l_s} (X_j^A + X_j^B) \leq 2 \qquad \forall l_s \in L, \quad s = 1, ..., w \tag{9}$$

$$X_j^A \in \{0, 1\} \qquad \forall j \in N \tag{10}$$

$$X_j^B \in \{0, 1\} \qquad \forall j \in N \tag{11}$$

$$D_i \geq 0 \qquad \forall i \in M \tag{12}$$

$$E_i \geq 0 \qquad \forall i \in M \tag{13}$$

The objective function (1) minimizes the population that remains uncovered in the province. Constraint (2) establishes that, for each health department, the whole of its population must be covered by the fleet of ALS and BLS. Constraints (3) and (4) set out that the total fleet of ALS and BLS, respectively, must be located. Constraints (5) ensure that a base can have a maximum of one ambulance. Constraints (6) and (7) require that in each department, there must be at least one number f^A of ALS and at most one number f^B of BLS, respectively. These last restrictions aim to avoid the concentration of ambulances in a few departments. For this case, the values chosen for the parameters f^A and f^B have been 1 and 10, respectively. Before explaining constraints (8) and (9), we must point out one of the problems that arise when performing the model: overlaps between isochrones. This implies that the same population will be overcovered. Both constraints control this situation, and thanks to the developed decision support tool, we can more easily identify these bases.

The zones with a high population density need to overlap to attend the emergencies on time. This is the purpose of Eq. (9), where L is the set of bases whose isochrones overlap, and the whose covered population is large. The constraint (8) avoids placing ambulances on bases with a high level of overlap between isochrones and with a population density. K is the set of bases whose isochrones overlap and the whose covered population is small.

5   Evaluation

This section defines the evaluation carried out to assess the efficiency of the developed model. First, we describe a set of computational experiments carried out to analyze the computational efficiency of the proposed model. Then, we describe an analysis carried out to assess the effectiveness of the solution proposed by our current model compared to the current solution implemented by human decision-makers in Valencia's region.

5.1   Computational Evaluation

As mentioned above, the first set of experiments aims to assess the most appropriate solver for our proposed model. That is, finding the solver that finds the optimal solution the fastest. Therefore, we designed a set of experiments to find the most appropriate solver. We selected three non-commercial solvers[3]: SCIP [1], CBC [21], and a CP-SAT solver [20].

To assess the adequateness of each solver for the proposed optimization model, we created a problem instance based on actual data from Valencia's province. The data for the problem was provided by the Primary Health and Emergency Service of Valencian region (SASUE). The problem instance allocates 20 advanced and 47 basic emergency vehicles to 159 prospective stations. The stations are scattered around Valencia's region and organized into 11 health departments, denoted from A to K. Table 1 shows the population of each of the departments.

The problem instance was solved with each of the solvers mentioned above. More specifically, the problem instance was solved 100 times per solver to capture stochastic differences in the execution of the solvers. Therefore, we solved the problem instance a total of 300 times. For each run, we recorded whether or not the solver found the optimal solution, the solution itself, and the execution time of the solver.

Taking as a reference the results obtained by CBC, we carried out a Dunn posthoc test ($\alpha = 0.05$) to identify differences between the results obtained by

---

[3] Non-commercial solvers were selected as a requirement of the public administration.

Table 1. Population of each health department in Valencia's province

A	B	C	D	E	F
176,288	344,019	194,397	277,280	195,000	284,060
G	H	I	J	K	Total
66,000	364,017	316,919	245,855	118,832	**2,582,667**

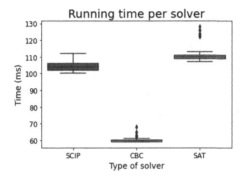

Fig. 3. Box and whisker diagram for the execution obtained with each solver

CBC and the other alternatives. The p-value of the comparison between CBC and SCIP was 5.13×10^{-17}. The p-value of the comparison between CBC and CP-SAT was 1.71×10^{-57}. Thus, the posthoc confirmed that the execution time obtained by CBC is statistically different from the execution time of SCIP and CP-SAT and lower than the times obtained by the other solvers. Thus, for the proposed instance, it is advisable to employ CBC. In all cases, the obtained response times is suitable for a real-time tool.

5.2 Model Evaluation

The model is applied to the real case of the province of Valencia. Before solving the problem, we studied whether, by placing a vehicle in each of the 159 possible bases and configuring an isochrone of 15 min, we could cover the entire population of the province of Valencia. In other words, we solve the problem without any resource constraints. We coin this solution as the ideal solution, as it represents an ideal scenario where resources are unlimited. This is the best solution that one could aspire to obtain. Table 2 shows, in the first row, the amount of population left uncovered per department in this ideal solution. Additionally, Fig. 4 (a) shows this ideal solution. In total, 0.5% of the population is not covered due to the road network. Table 2 compares the percentage of the uncovered population in each department of health according to the ideal solution (D_i^{min}, i.e., unlimited vehicles), the current assignment proposed by the EMS ($D_i^{current}$, i.e., the manual solution), and the configuration purposes by our mathematical model (D_i^{model}). The solution implemented by the EMS represents the baseline to improve, as it is the current solution provided by experts. The model solution halves the amount of population not covered with the current assignment. It also matches the ideal solution for some departments. Figure 4 graphically compares the solution provided for the case of unlimited resources (a), for the current distribution of EMV proposed by the EMS (b), and for the distribution proposed by the mathematical model implemented (c). We can appreciate that the solution provided by the model is closer to the ideal case with unlimited resources.

Table 2. Percentage of population not covered across health departments.

	A	B	C	D	E	F	G	H	I	J	K	Total
D_i^{min}	2.7	0.8	0.0	6.4	0.0	0.8	0.0	0.0	0.0	0.0	0.2	**0.5**
$D_i^{current}$	8.8	0.8	0.5	12.1	0.0	3.3	0.0	0.0	0.0	0.0	0.2	**1.5**
D_i^{model}	2.7	0.8	0.0	10.0	0.0	1.4	0.0	0.0	0.0	0.0	0.2	**0.7**

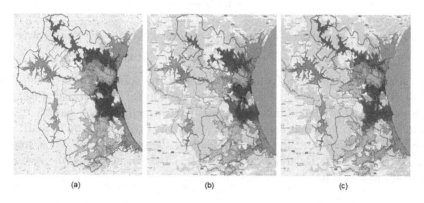

(a)   (b)   (c)

Fig. 4. Coverage of the province of Valencia according to the distribution of the EMV: a) Ideal coverage; b) Current coverage; c) Coverage proposed by the model

6   Conclusions

In this paper, we presented a decision support tool for allocating emergency vehicles. The tool is based on a service architecture that allows for implementing different interfaces and the reusing services. Apart from the graphical interface, which allows user interaction, the decision support system is composed of an optimization model based on mixed-integer linear programming. In detail, the tool presented in this paper deals with the static ambulance location problem, which determines the initial location of emergency vehicles. The decision support tool is composed of an optimization service that facilitates the decision-making process taken by humans since it optimally allocates emergency vehicles to prospective stations. In this sense, the optimization model can be solved by MILP solvers.

As REVES Map is intended to be used by the Primary Health and Emergency Service from the Valencian region, we used real data to evaluate our proposal's adequateness. According to our experiments, the performance of the CBC solver is better for our proposed model than the other solvers tested. The proposed model is efficient, and the optimal solution can be computed in real-time. In addition, we showed that the proposed optimization model improves the current solution employed by the regional Emergency Service. In future work, we plan to incorporate several other functionalities into the decision support tool. On the one hand, we plan to include analysis tools and algorithms for the ambulance dispatch problem, which refers to the problem of determining the ambulance and the hospital allocated for an emergency. On the other hand, we also plan to include tools and algorithms for the ambulance relocation problem, which considers the relocation of ambulances after finishing requests.

Acknowledgements. The authors are supported by Agència Valenciana de la Innovació under the project *IReves (Innovación en Vehículos de Emergencia Sanitaria)* (INNACC/2021/26) and by the Spanish Ministry of Science and Innovation under the project "OPRES-Realistic Optimization in Problems in Public Health" (PID2021-124975OB-I00), both partially funded with FEDER funds. Interested readers can visit http://ireves.upv.es. Part of the authors are supported by the Faculty of Business Administration and Management at Universitat Politècnica de València. Special thanks to *Servicio de Atención Sanitaria a Urgencias y Emergencias* and to *Servicios de Emergencias Sanitarias* of Comunitat Valenciana, and Dpt. of Coordination of Medical Transport for the information and data provided.

References

1. Achterberg, T.: SCIP: solving constraint integer programs. Math. Program. Comput. **1**(1), 1–41 (2009). https://doi.org/10.1007/s12532-008-0001-1
2. Andersson, T., Värbrand, P.: Decision support tools for ambulance dispatch and relocation. In: Mustafee, N. (ed.) Operational Research for Emergency Planning in Healthcare: Volume 1. TOE, pp. 36–51. Palgrave Macmillan UK, London (2016). https://doi.org/10.1057/9781137535696_3
3. Bélanger, V., Lanzarone, E., Nicoletta, V., Ruiz, A., Soriano, P.: A recursive simulation-optimization framework for the ambulance location and dispatching problem. Eur. J. Oper. Res. **286**(2), 713–725 (2020)
4. Bélanger, V., Ruiz, A., Soriano, P.: Recent optimization models and trends in location, relocation, and dispatching of emergency medical vehicles. Eur. J. Oper. Res. **272**(1), 1–23 (2019)
5. Bélanger, V., Ruiz, A., Soriano, P.: Recent Advances in Emergency Medical Services Management. Université Laval, Faculté des sciences de l'administration (2015)
6. Berlin, G.N., Liebman, J.C.: Mathematical analysis of emergency ambulance location. Socioecon. Plann. Sci. **8**(6), 323–328 (1974)
7. Erkut, E., Ingolfsson, A., Erdoğan, G.: Ambulance location for maximum survival. Nav. Res. Logistics (NRL) **55**(1), 42–58 (2008)

8. Gendreau, M., Laporte, G., Semet, F.: Solving an ambulance location model by tabu search. Locat. Sci. **5**(2), 75–88 (1997)
9. Gendreau, M., Laporte, G., Semet, F.: A dynamic model and parallel tabu search heuristic for real-time ambulance relocation. Parallel Comput. **27**, 1641–1653 (2001)
10. Grewal, M.S., Weill, L.R., Andrews, A.P.: Global Positioning Systems, Inertial Navigation, and Integration. Wiley, Hoboken (2007)
11. Haklay, M., Weber, P.: Openstreetmap: user-generated street maps. IEEE Pervasive Comput. **7**(4), 12–18 (2008)
12. Ingolfsson, A., Budge, S., Erkut, E.: Optimal ambulance location with random delays and travel times. Health Care Manag. Sci. **11**(3), 262–274 (2008). https://doi.org/10.1007/s10729-007-9048-1
13. Jagtenberg, C.J., van den Berg, P.L., van der Mei, R.D.: Benchmarking online dispatch algorithms for emergency medical services. Eur. J. Oper. Res. **258**(2), 715–725 (2017)
14. Jagtenberg, C.J., Bhulai, S., van der Mei, R.D.: An efficient heuristic for real-time ambulance redeployment. Oper. Res. Health Care **4**, 27–35 (2015)
15. Lam, S.S.W., Zhang, J., Zhang, Z.C., Oh, H.C., Overton, J., Ng, Y.Y., Ong, M.E.H.: Dynamic ambulance reallocation for the reduction of ambulance response times using system status management. Am. J. Emerg. Med. **33**(2), 159–166 (2015)
16. Lim, C.S., Mamat, R., Braunl, T.: Impact of ambulance dispatch policies on performance of emergency medical services. IEEE Trans. Intell. Transp. Syst. **12**(2), 624–632 (2011)
17. Longley, P.A., Goodchild, M.F., Maguire, D.J., Rhind, D.W.: Geographic Information Systems and Science. Wiley, Hoboken (2005)
18. Ong, M.E.H., et al.: Reducing ambulance response times using geospatial-time analysis of ambulance deployment. Acad. Emerg. Med. **17**(9), 951–957 (2010)
19. Peleg, K., Pliskin, J.S.: A geographic information system simulation model of EMS: reducing ambulance response time. Am. J. Emerg. Med. **22**(3), 164–170 (2004)
20. Perron, L.: Operations research and constraint programming at Google. In: Lee, J. (ed.) CP 2011. LNCS, vol. 6876, pp. 2–2. Springer, Heidelberg (2011). https://doi.org/10.1007/978-3-642-23786-7_2
21. Saltzman, M.J.: Coin-Or: an open-source library for optimization. In: Nielsen, S.S. (eds.) Programming Languages and Systems in Computational Economics and Finance. Advances in Computational Economics, vol. 18, pp. 3–32. Springer, Boston (2002). https://doi.org/10.1007/978-1-4615-1049-9_1
22. Sanchez-Anguix, V., Chao, K.M., Novais, P., Boissier, O., Julian, V.: Social and intelligent applications for future cities: current advances (2021)
23. Van Buuren, M., Jagtenberg, C., Van Barneveld, T., Van Der Mei, R., Bhulai, S.: Ambulance dispatch center pilots proactive relocation policies to enhance effectiveness. Interfaces **48**(3), 235–246 (2018)

Adapting K-Means Algorithm for Pair-Wise Constrained Clustering of Imbalanced Data Streams

Szymon Wojciechowski[1]([⊠]), Germán González-Almagro[2], Salvador García[2], and Michał Woźniak[1]

[1] Department of Systems and Computer Networks, Wrocław University of Science and Technology, Wrocław, Poland
szymon.wojciechowski@pwr.edu.pl
[2] Department of Computer Science and Artificial Intelligence (DECSAI) and DaSCI Andalusian Institute of Data Science and Computational Intelligence, University of Granada, Granada, Spain

Abstract. Contemporary man is addicted to digital media and tools supporting his daily activities, which causes the massive increase of incoming data, both in volume and frequency. Due to the observed trend, unsupervised machine learning methods for data stream clustering have become a popular research topic over the last years. At the same time, semi-supervised constrained clustering is rarely considered in data stream clustering. To address this gap in the field, the authors propose adaptations of k-means constrained clustering algorithms for employing them in imbalanced data stream clustering. In this work, proposed algorithms were evaluated in a series of experiments concerning synthetic and real data clustering and verified their ability to adapt to occurring concept drifts.

Keywords: Data streams · Pair-wise constrained · Clustering · Imbalanced data

1 Introduction

The development of technology and the increasing number of devices connected to one general-purpose network have made it possible to download and store massive amounts of constantly streaming data. Regardless of the information type, which might be readings from sensor mesh or news outlets [1], it will always be required to process it into a resource that the users can further utilize – knowledge. Therefore, it has become inevitable for the topic to be widely researched, leading to the development of data mining techniques for data streams.

Undoubtedly, one of the discussed challenges is to design algorithms capable of continuous data stream processing [2]. Therefore, the posed models must adjust to the constantly changing data characteristics, which phenomenon is named *concept drift* [3]. Moreover, models have to provide a prediction before

P. García Bringas et al. (Eds.): HAIS 2022, LNAI 13469, pp. 153–163, 2022.
https://doi.org/10.1007/978-3-031-15471-3_14

the next batch of data is available, which constrains the algorithm in terms of computational complexity.

The vast majority of the recent works focus on data stream classification, introducing various approaches for dealing with *concept drift*. Many of the proposed methods employ classifier ensembles [4] and dynamic selection of classifiers [5]. Moreover, various methods are based on concept drift detectors [6], and a prior probability estimation [7] which indicates when the data characteristics are changing.

Similarly, data streams were also researched in the context of clustering task [8]. Proposed algorithms are often adapting the existing methods to data stream processing [9]. At the same time, research on semi-supervised clustering with pair-wise constraints [10] has been mainly focused on the integration of constraints to classical algorithms [11] and methods based on optimization algorithms [12]. The adaptation of those algorithms to data streams was proposed only in a few recent works.

One of the mainly used techniques for this task is *C-DenStream* [13], which is adapting DBSCAN algorithm employing micro-clusters constructed on consecutive chunks. Another example is the *SemiStream* [14] algorithm, based on initial cluster initialization by *MPCk-means*, then detection of *o-clusters*, assigning them to *s-clusters* and *m-clusters*. Finally, the *CE-Stream* algorithm [15] extends the *E-Stream* algorithm, but it is limited to must-link constraints set.

Marking that there still is a lack of pair-wise constrained clustering algorithms for data streams, the main contribution of this work is a proposal for employing *COPk-means* and *PCk-means* algorithms for data streams clustering.

2   Algorithm

The data stream $\mathcal{DS}$ is a sequence of data chunks $\mathcal{DS} = \{DS_1, DS_2, \ldots, DS_k\}$. Each data chunk contains a set of samples described by a feature vector X for which the clustering algorithm $\kappa(X)$ assigns a label describing a cluster C. Additionally each chunk is also provided with two lists of pair-wise constraints ML and CL denoting *must-list* pairs and *cannot-link* pairs, respectively. Those extend a context of clustering with expert knowledge, which does not provide cluster labels directly, but only declares a relation between two samples.

The k-means clustering algorithm is an inductive learning method. The model is trained iteratively to minimize the intra-cluster and maximize the infra-cluster variance. In the first phase of the algorithm, k cluster centroids are initialized as points in feature space. Then, each observation is assigned to a cluster by finding the nearest centroid. After this procedure, the new centroids are calculated by averaging all assigned observations. The procedure is repeated until the model reaches convergence, which means that new centroids can not be further shifted. An additional stop condition, which guarantees algorithm execution in a feasible time, is the maximum number of maximum iterations. The algorithm pseudocode is presented in Algorithm 1.

Algorithm 1: Generic k-means clustering

INITIALIZE CENTROIDS ()
while *iteration* $<$ *max_iter* **do**
  | ASSIGN LABELS (X)
  | RECALCULATE CENTROIDS ()
  | **if** *convergence* **then**
  | | **break**
  | **end**
end

Constraints can be integrated into the k-means algorithm using one of two rules, defined as follows:

- *hard rule*: cluster labels assigned by the algorithm cannot be inconsistent with constraints. This rule is used by COPk-means (COPK) algorithm. Each label assignment is verified for feasibility against constraints. If one of the rules is violated, the algorithm terminates, leaving the rest of the samples unassigned.
- *soft rule*: cluster labels can be inconsistent with constraints, but constrain violation is penalized as an additional factor in minimized criteria. This rule is used by PCk-means (PCK) algorithm. The criteria minimized in the algorithm is defined as follows:

$$\mathcal{J}_{pckm} = \frac{1}{2} \sum_{x_i \in X} ||x_i - \mu_{l_i}||^2 +$$
$$\sum_{(x_i, x_j) \in \text{ML}} w_{ij} \mathbb{1}[l_i \neq l_j] + \sum_{(x_i, x_j) \in \text{CL}} \bar{w}_{ij} \mathbb{1}[l_i = l_j] \tag{1}$$

This approach assures that all samples are assigned to estimated clusters.

Another crucial part of the algorithm is selecting a proper cluster initialization method. The common methods are:

- *random* - which is very fast, but a poor initial match may lead to very slow convergence.
- *kmeans++* - allowing a better estimation of the initial centroids, but it introduces additional computation overhead.

For constrained clustering problems, centroids can also be initialized based on transitive closure of the must-link constraints, averaging the broadest spanning trees created during this procedure. The complexity of such a method depends on a backbone spanning-tree search algorithm.

Since the goal is to adapt the algorithm to process data streams, it is necessary to make changes to the initialization and modification of the centroids in the COPK and PCK algorithms. Authors propose two new algorithms namely COPK-S and PCK-S introducing modifications to described procedures.

Initialization. In sequential data chunks processing, the model created on the previous chunk has already determined the centroids that can be used in the next one. It can be assumed that those centroids are close to optimal, and, at the same time, reusing them reduces additional computational effort.

Assignment. Base algorithms assume that k clusters will be formed. However, sudden concept drift may lead to the complete disappearance of one of the clusters. It may turn out that no pattern will be assigned to a given centroid, which at the same time will prevent further shifting. Therefore, an additional modification is introduced to the proposed algorithms. Centroids to which no samples were assigned are omitted, but the same centroid will be reused for clustering in the next chunk.

3   Experiments

The proposed methods of adapting the k-means algorithms were tested in a series of experiments in order to find answers to the following research questions:

RQ1: What impact do the proposed modifications have on the ability to adapt the algorithm to the occurring concept drifts?

RQ2: What impact do the proposed modifications have on the clustering quality?

3.1   Research Protocol

The following experimental protocol was formulated to answer the research questions. Both synthetic streams – for which it was possible to observe the algorithm's operation with the defined behavior of the concept drift – and the real-life data were used, allowing for actual quality assessment of the proposed methods.

Synthetic data streams were generated with both the concept drift and the dynamic imbalance to the stream based on the artificial classification problem. Additionally, to create a set dedicated to clustering algorithms, two additional sets were created, in which the defined clusters were generated from the normal distribution at the given distribution means. This modification was aimed at creating a set dedicated to the clustering algorithm, bearing in mind that, at the same time, the objective of the algorithm is to determine the local concentration of observations correctly and not – as opposed to classification – to determine the correct labels.

There are no real data sets for the constrained clustering task with constraints provided by an actual expert. Therefore, it was necessary to generate both the synthetic and real streams based on a given set of labels for both the synthetic and real streams. Each set can introduce natural disproportion between the ML and CL, which is related to the number of classes and their volume. However, with a complete set of labels, it is possible to determine $\binom{n}{2}$ pairs of n samples and constraint type based on the similarity among their label. Therefore, to propose the method of generating the constraints, the percentage of possible sample connections is used, and it defines a number of pairs selected randomly and transformed into constraints.

The quality of clustering on selected data streams was assessed in a sequential protocol typical for this type of research. The incoming data are divided into chunks of equal size. The model is trained using the entire chunk, while the evaluation process follows the protocol for evaluating clustering algorithms. Additionally, for all tested methods, constraints lists are provided for each data chunk, and the set of labels itself is only used to compute the clustering metric.

3.2 Experimental Setup

Eight data streams were used for the experiments, four of which were synthetically generated, and four were prepared based on real data sets, obtained from MOA[1] and CSE[2]. A detailed description of the data sets is presented in the Table 1.

Selected data sets were divided into chunks of 200 samples. The real-life data streams were limited to the first 200 chunks, which preserved the original problem characteristics and provided a convenient length for reliable analysis.

The constraints were generated for each chunk using original problem labels. The ratio of generated constraints to all possible constraints was constant along the stream, but each data set was evaluated in three configurations of 1%, 2%, and 5% of possible constraints. The metric selected to measure and compare the clustering quality of evaluated algorithms was *Adjusted Rand Index* (ARI).

The tested algorithms and the experimental environment were implemented in Python programming language, using the *scikit learn* [16] and *stream-learn* libraries [17]. The repository allowing for the replication of presented results is available at Github[3].

3.3 Results

The graphs show ARI scores for both proposed methods and the reference approaches. In addition, for each method, the metric values are shown (a) for the entire data stream length and (b) averaged over all data chunks. This approach allows for a detailed analysis of the algorithm performance while enabling comparison in the context of the entire problem. Finally, all presented runs were smoothed using the cumulative sum.

Synthetic Datasets. Research on synthetic sets was carried out to study the algorithm behavior concerning the occurring drifts. The results obtained for synthetic data streams with concept drift are presented in Fig. 1.

One may observe that for the analyzed problem, the PCK-S achieved the best results. It should be underlined that in the case of *Blobs Gradual*, the clustering

---

[1] https://moa.cms.waikato.ac.nz/datasets/.

[2] https://www.cse.fau.edu/~xqzhu/stream.html.

[3] https://github.com/w4k2/pcsp.

Table 1. Summary of benchmark datasets.

Dataset	Source	Description
Blobs gradual	Synthetic	One normal distribution per class with a drifting mean value. Two separated clusters are intersecting, then returning to original concept
Blobs imbalance	Synthetic	One normal distribution per class with a drifting imbalance ratio. At the beginning static clusters have equal imbalance ratio, which is drifting towards 0.1 IR
Classification gradual	Synthetic	One distribution per class with two informative features. Mean of the cluster distributions is shifting over time
Classification imbalance	Synthetic	One distribution per class with two informative features. Recurring imbalance drift is occurring twice, varying between 0.05–0.95 IR
Forest covertype	MOA	Data contains the forest cover type obtained from US Forest Service (USFS) Region 2 Resource Information System (RIS)
Electricity	MOA	Data collected from Australian New South Wales Electricity Market
Airlines	MOA	Data contains flight arrival and departure details for all the commercial flights within the USA, from October 1987 to April 2008
KddCup99	CSE	Data created for network intrusion detection in simulated network. This data set was used for The Third International Knowledge Discovery and Data Mining Tools Competition

quality of the base PCK and for the proposition was equal. Moreover, all the algorithms performed poorly around the cluster intersection – which was expected behavior – but ARI significantly improved when 5% of constraints were provided. At the same time, attention should be paid to the massive variance of the obtained metric, indicating how high is its potential impact on model diversification. At the same time, despite the inferior quality of clustering employing the COPK, its significant improvement is noticeable for the COPK-S. However – addressing $RQ1$ – it cannot be unequivocally stated that the proposed methods allow for faster adaptation to the concept drift.

For the *Classification Gradual* problem, it can be seen that both PCK-S and COPK-S are better than the base algorithms and more stable – especially for the case where 2% of constraints were used. In the flows with the 1% of constraints used, an interesting behavior at the beginning of the stream is visible - improving overall model quality in the early run. It may suggest that the proposed initialization method causes some delay in achieving the algorithm's convergence, which stabilizes only after some time.

It is also essential to pay attention to the number of constraints on the tested algorithms. In all cases, increasing this parameter did not affect the quality of clustering by COPK and had a slight effect on COPK-S while significantly improving the PCK and PCK-S algorithms, leading to a stable result in *Classification Gradual* with 5% limitations.

For the *Blobs Gradual* problem, the PCK and PCK-S algorithms achieve the promising result using 5% constraints and minimize their variance while becoming insensitive to the emerging imbalance. It is also worth noting that, in contrast to the previously analyzed problem, it is possible to point to a much slighter decrease in the quality of PCK-S clustering at the time of imbalance. A similar relationship can be observed in *Classification Gradual*. In the scenario where the percentage of constraints is 0.5% and 1%, there is a sudden increase in the metric as the classes approach the prior equilibrium, followed by a rapid decrease. In the last case, PCK-S adjusts to imbalance the fastest, giving the metric the most stable value.

Real Datasets. The results of experiments carried out on real data streams are presented in Fig. 2. Results analysis is carried out to address *RQ2*. It is impossible to indicate the best clustering algorithm for the *forest covertype* and *electricity* sets where 1% and 2% restrictions were selected. Noticeable differences between the algorithms will appear only after considering the 5% constraints, which significantly improves the PCK and PCK-S algorithms. An interesting drop in the quality of the PCK clustering on the convtypeNorm set starts in chunk 75, which cannot be observed for PCK-S.

Significant differences can be observed in the case of the other two sets. It can be noted that for the results of *airlines*, all methods performed poorly, but there is an increase in the quality of clustering for COPK and COPK-S with an increasing number of constraints. Thus, this observation seems to be an exception to the rule observed in the previous sets. However, it should be noted that the metric only increased by 0.1 between the lowest and the highest percentage of constraints.

The most interesting observations can be made on the *KddCup99* set, for which COPK-S also turned out to be the best clustering algorithm. An interesting relationship can also be seen for this set, according to which adding constraints degenerate the quality of clustering. Most likely, it is related to the specific behavior of the algorithm, related to constraint feasibility [18].

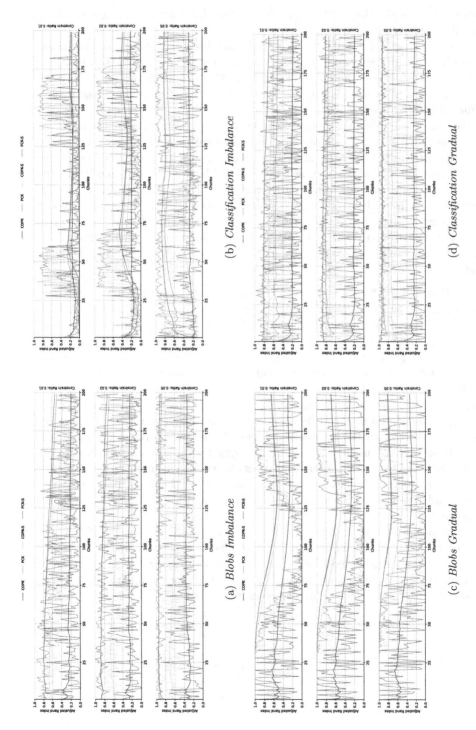

(a) *Blobs Imbalance*

(b) *Classification Imbalance*

(c) *Blobs Gradual*

(d) *Classification Gradual*

Fig. 1. Results for synthetic datasets.

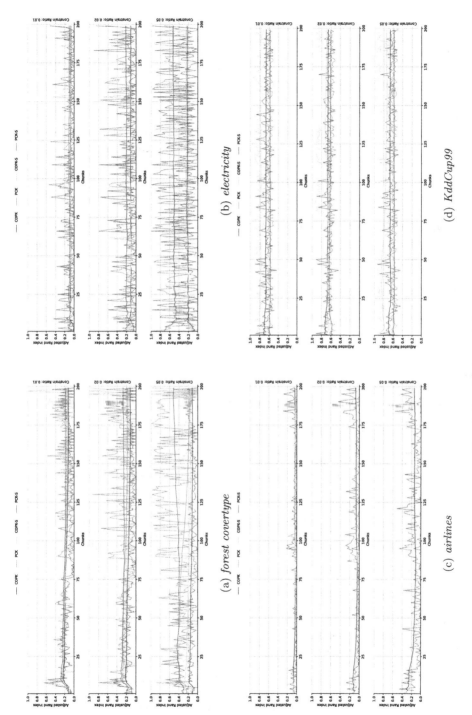

Fig. 2. Results for real datasets.

4   Conclusions

Two clustering algorithms for data streams: PCK-S and COPK-S, were presented in this work. Both algorithms are adopted to the task rarely discussed in the literature. The series of studies showed the ability to adapt the proposed methods to the existing concept drifts and obtain better results than in the case of the base algorithms.

The presented results provide the basis for further research. One of the aspects discussed in this work is the proposal to extend the proposed methods to be better adapted to the changing nature of the constraints. In some cases, the clustering quality showed significant variance throughout the run. An appropriate selection of constraints may be crucial for stabilizing the results.

In addition, the practical aspect of using the proposed methods should be emphasized in the context of other machine learning tasks, especially for active learning. Evaluating only the relation between two samples is easier for an expert than assigning them to imposed classes, which will reduce the cost of obtaining such knowledge.

Acknowledgement. This work was supported by the Polish National Science Centre under the grant No. 2017/27/B/ST6/01325 as well as by the statutory funds of the Department of Systems and Computer Networks, Wroclaw University of Science and Technology.

References

1. Ksieniewicz, P., Zyblewski, P., Choras, M., Kozik, R., Gielczyk, A., Wozniak, M.: Fake news detection from data streams. In: 2020 International Joint Conference on Neural Networks (IJCNN), pp. 1–8. IEEE, Glasgow, July 2020
2. Krawczyk, B., Minku, L.L., Gama, J., Stefanowski, J., Woźniak, M.: Ensemble learning for data stream analysis: a survey. Inf. Fusion **37**, 132–156 (2017)
3. Ramírez-Gallego, S., Krawczyk, B., García, S., Woźniak, M., Herrera, F.: A survey on data preprocessing for data stream mining: current status and future directions. Neurocomputing **239**, 39–57 (2017)
4. Bifet, A., Holmes, G., Pfahringer, B., Kirkby, R., Gavaldà, R.: New ensemble methods for evolving data streams. In: Proceedings of the 15th ACM SIGKDD International Conference on Knowledge Discovery and Data Mining, ser. KDD 2009, pp. 139–148. Association for Computing Machinery, New York (2009)
5. Zyblewski, P., Sabourin, R., Woźniak, M.: Preprocessed dynamic classifier ensemble selection for highly imbalanced drifted data streams. Inf. Fusion **66**, 138–154 (2021)
6. Guzy, F., Woźniak, M., Krawczyk, B.: Evaluating and explaining generative adversarial networks for continual learning under concept drift. In: International Conference on Data Mining Workshops (ICDMW), pp. 295–303 (2021)
7. Komorniczak, J., Zyblewski, P., Ksieniewicz, P.: Prior probability estimation in dynamically imbalanced data streams. In: 2021 International Joint Conference on Neural Networks (IJCNN), pp. 1–7. IEEE Shenzhen, July 2021
8. Silva, J.A., Faria, E.R., Barros, R.C., Hruschka, E.R., Carvalho, A.C.D., Gama, J.: Data stream clustering: a survey. ACM Comput. Surv. **46**(1), 1–31 (2013)

9. Cao, F., Estert, M., Qian, W., Zhou, A.: Density-based clustering over an evolving data stream with noise, pp. 328–339 (2006)

10. Davidson, I.: A survey of clustering with instance level constraints. ACM Trans. Knowl. Discov. Data (41) (2007)

11. González, S., García, S., Li, S.-T., John, R., Herrera, F.: Fuzzy k-nearest neighbors with monotonicity constraints: moving towards the robustness of monotonic noise. Neurocomputing **439**, 106–121 (2021)

12. González-Almagro, G., Luengo, J., Cano, J.-R., García, S.: Enhancing instance-level constrained clustering through differential evolution. Appl. Soft Comput. **108**, 107435 (2021)

13. Ruiz, C., Menasalvas, E., Spiliopoulou, M.: C-DenStream: using domain knowledge on a data stream. In: Gama, J., Costa, V.S., Jorge, A.M., Brazdil, P.B. (eds.) DS 2009. LNCS (LNAI), vol. 5808, pp. 287–301. Springer, Heidelberg (2009). https://doi.org/10.1007/978-3-642-04747-3_23

14. Halkidi, M., Spiliopoulou, M., Pavlou, A.: A semi-supervised incremental clustering algorithm for streaming data. In: Tan, P.-N., Chawla, S., Ho, C.K., Bailey, J. (eds.) PAKDD 2012. LNCS (LNAI), vol. 7301, pp. 578–590. Springer, Heidelberg (2012). https://doi.org/10.1007/978-3-642-30217-6_48

15. Sirampuj, T., Kangkachit, T., Waiyamai, K.: CE-stream : evaluation-based technique for stream clustering with constraints. In: The 2013 10th International Joint Conference on Computer Science and Software Engineering (JCSSE), pp. 217–222. IEEE, Khon Kaen, Thailand, May 2013

16. Pedregosa, F., et al.: Scikit-learn: machine learning in Python. J. Mach. Learn. Res. **12**, 2825–2830 (2011)

17. Ksieniewicz, P., Zyblewski, P.: Stream-learn-open-source Python library for difficult data stream batch analysis. arXiv:2001.11077 [cs, stat], January 2020

18. Davidson, I., Wagstaff, K.L., Basu, S.: Measuring constraint-set utility for partitional clustering algorithms. In: Fürnkranz, J., Scheffer, T., Spiliopoulou, M. (eds.) PKDD 2006. LNCS (LNAI), vol. 4213, pp. 115–126. Springer, Heidelberg (2006). https://doi.org/10.1007/11871637_15

Small Wind Turbine Power Forecasting Using Long Short-Term Memory Networks for Energy Management Systems

Esteban Jove[1], Santiago Porras[2(✉)], Bruno Baruque[3],
and José Luis Calvo-Rolle[1]

[1] CTC, CITIC, Department of Industrial Engineering, University of A Coruña,
Avda. 19 de febrero s/n, 15495 Ferrol, A Coruña, Spain
{esteban.jove,jlcalvo}@udc.es

[2] Departamento de Economía Aplicada, University of Burgos, Plaza Infanta Doña
Elena, s/n, 09001 Burgos, Burgos, Spain
sporras@ubu.es

[3] Departmento de Ingeniería Informática, University of Burgos, Avd. de Cantabria,
s/n, 09006 Burgos, Burgos, Spain
bbaruque@ubu.es

Abstract. The rising trend of energy prices during recent times, triggered by fuels shortage, wars and the consequent economic crisis, has derived on the promotion of green policies. In this context, the technologies to exploit wind energy have experienced a remarkable increase. Besides, the intermittent availability of wind energy can be complemented by energy management systems to reduce the energy bill. This work proposes a forecasting system to determine, in short-term, the power generated by small wind turbine using advanced automated learning techniques, such as the Long Short-Term Memory networks.

Keywords: LSTM · Deep learning · Time series forecasting · Energy Management System · Wind turbine

1 Introduction

The energy scene has presented several issues that has contributed to the promotion of clean and alternative energies. On the one hand, the great fossil fuels dependency resulted on the acceleration of climate change, with the consequent related concerns [4,20]. On the other hand, the energy prices experienced a significant increase during the last two years due to global pandemic, raw materials shortage and the Ukraine invasion, among others [9,11].

Under these circumstances of economic and climatic emergency, the lack of fossil fuels dependency and greenhouse gases emissions reduction plays a key role and both, national and regional governments, have made significant investments

© Springer Nature Switzerland AG 2022
P. García Bringas et al. (Eds.): HAIS 2022, LNAI 13469, pp. 164–174, 2022.
https://doi.org/10.1007/978-3-031-15471-3_15

to contribute to achieve a sustainable energy generation system, that could be able to face the climate change challenge [8,20].

Although governments focused their efforts to increase renewable electric power, this represents just the 15% of the global electric mix at the end of the first decade of 2000s, being the hydroelectric power the most used [18,19]. This value was raised up to a 22% by 2012 [20,24]. The annual report in [3] consideres that the wind power capacity increased from 17 GW in 2000 to 591 GW in 2018. Following this trend, an optimistic forecast estimates that by 2030 the share of wind generation will be almost 23% of the mix [15,17]. This value would represent five times the current share [7]. Regarding the wind generation in Spain, the installed power also presents a marked increase. It is estimated that the demand covered by this technology will be twice the current demand, going from 18.2% to 33.6% of the total share [1,16].

This kind of technology exploits the movement of air masses consequence of differential solar heating of the atmosphere [15]. The energy generated is proportional to the cube of wind speed, and since this value is strictly dependent of atmospheric variables, it can change within short time periods. This parameter can change from year to year, with the season, on a daily basis, or even in seconds [6,15].

This intermittency must be complemented through battery systems or a connected power grid. In this context, the possibility of having a reliable prediction of the power generated in the short term represents a really interesting tool to manage the energy [12,13]. This prediction would help manage, in advance, whether the power generated should be consumed, or stored in the battery depending on the state of charge and energy price.

This work presents an intelligent system to forecast the wind turbine power production in a bioclimatic house. The prediction is carried out using Long Short-Term Memory (LSTM) networks using an original dataset gathered by collecting the power generated during one year with 10 min of sampling rate. The proposal would be useful to implement Energy Management Systems. Then, it is presented the contribution of implementing a successful prediction system that estimates the energy generated in a wind turbine of a bioclimatic house. This contribution is the key for a more efficient energy management, ensuring both economic and energy savings.

The document is structured following this outline: after this section, the case study is described Sect. 2. Then, proposal is detailed in Sect. 3. The experiments and results carried out to achieve the forecasting are presented in Sect. 4. Finally, the last Section exposes the conclusions and future works derived from the proposal.

2   Case Study

2.1   Sotavento Galicia Building

The small wind turbine under study is located near a bioclimatic building founded by the Sotavento Galicia Foundation. This building has the aim of disseminating the use of different renewable energies, and promote their advantages with a low environmental impact. This bioclimatic house was built between the provinces of A Coruña and Lugo, in Galicia, is located in 43° 21′ N, 7° 52′ W, at a height of 640 m and is 30 km away from the sea. It is designed to supply electric power and also Hot Domestic Water (HDW) (see Fig. 1). Electricity is supplied by a wind turbine system of 1,5 kW, a photovoltaic field of 2,7 kW and also by the power network, while the HDW is supplied by a solar thermal system, a biomass system and a geothermal system.

Fig. 1. Scheme of electric and DHW installations

Focusing on the small wind turbine, which is the device whose behaviour will be predicted and it has been previously studied in [5,21], it is a BORNAY INCLIN 1500 model [2], with a three phase Permanent Magnet Synchronous Generator (PMSG) made of neodymium with blades and cover made of fiberglass and carbon fiber. This housing has a rudder in the backside to ensure the proper system orientation. The general features of the device are shown in Table 1.

2.2   Dataset Description

To carry out the proposal, the active power generated by the small wind turbine is measured and registered using a calibrated power meter during one year with a 10 min of sampling rate. The good interpretation of the dataset was ensured by means of a prior inspection that consisted of cleaning the duplicated measurements and null values. These values are considered not valid so, after discarding them, the 52.645 initial samples were reduced to 52.639.

Table 1. General characteristics of BORNAY INCLIN 1500 wind turbine

Mechanic features		
Number of blades	Radius	Material
2	1.43 m	Carbon fiber
Electric features		
Generator type	Nominal Power	Voltage
3-ph PMSG	1.5 kW	120 Vrms
Wind speed features		
Starter wind	Nominal speed	Maximum speed
3.5 m/s	12 m/s	14 m/s

3   Energy Management System

The energetic picture described in the Introduction section reveals two main issues that must taken into consideration to face the current economic and energetic challenges:

- The climate change derived in the excessive greenhouse emissions, which is a real threat to environment.
- The energy prices are increasing, resulting in more expensive bills, reducing the purchasing power of clients with the consequent economic crisis.

The Sotavento building was initially thought to tackle the first challenge through the use of green energies, but the second one needs a new proposal improve the economic efficiency. The main basis of this proposal lies on the fact that the energy companies establish different prices during one day. These values have a great fluctuation, with really expensive and cheap hours. Therefore, an intelligent system is proposed to predict, in advance, the energy generated by the wind power installation. With this information, an Energy Management System could take two possible decisions:

- If the energy price during the predicted period is high, the energy generated in the wind turbine is sent to the building.
- If the energy price during the predicted period is low, the energy generated in the wind turbine is stored in a battery and the building is supplied by the power network.

The resulting diagram would be the one shown in Fig. 2.

The technique chosen to carry out the prediction of the power generated by the wind turbine is the well known Long Short-Term Memory Network (LSTM). This technique, proposed by Hochreiter and Schmidhuber [14], was created to handle the long-term dependencies of Recurrent Neural Networks (RNN) [10,25]. This is done through the introduction of a gate into the cell, as detailed in [22,23].

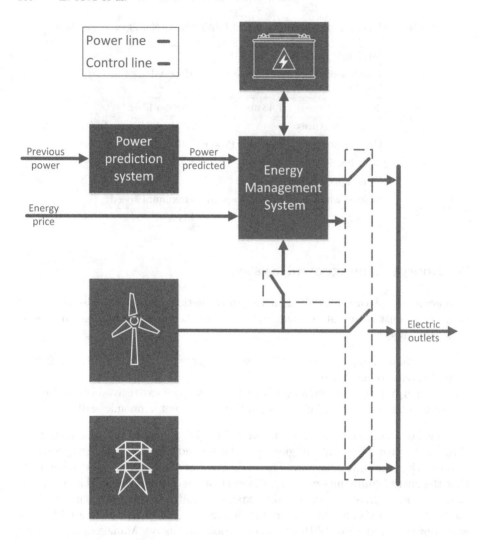

Fig. 2. Energy management system

This technique has been selected since is was applied to a wide range of problems, such as speech recognition, correlation analysis or trajectory prediction, among others [25]. The innovation of this proposal consists of the combination of an intelligent time series technique and the rules specified to manage the system in the proper way to ensure energy efficiency.

4   Experiments and Results

4.1   Experiments Setup

The proposal is evaluated using the procedure shown in Fig. 3 where an example with 14 samples represents the time series to be forecasted. The first five initial instances (blue squares) are used to fit the LSTM and the next three samples (yellow squares) are predicted. Then, the samples used to estimate the successive instances are shifted. Hence, samples from 4 to 8 (blue square in Test #2) are used to estimate the value of samples from 9 to 11. This procedure is repeated again to predict the last three values.

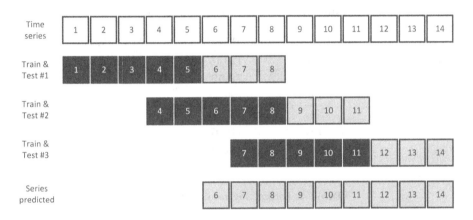

Fig. 3. Procedure followed to predict the time series (Color figure online)

Focusing on this particular case, the window width to carry out the prediction is 60 min. Furthermore, the window size of the forecasted values is tested at 30, 60 and 120 min. All these combinations are carried out without normalization and with a Z-Score normalization.

The forecasts achieved using LSTM are compared with a baseline prediction consisting of using the previous value as the future one. This means that the current value would be used to predict the power with 60 min in advance. Finally, the Mean Absolute Error (MAE) is the measure to determine the goodness of the proposal.

4.2   Results

The results obtained from the experiments detailed in the previous subsection are represented by months in Tables 2, 3, 4, 5, 6, 7, 8, 9, 10, 11, 12 and 13. The results achieved with LSTM that improve the baseline prediction are marked in bold.

Table 2. Forecasting results of January

Previous width (min)	Normalization	MAE LSTM (W)	MAE Baseline (W)
30	Yes	**31,41**	32,28
	No	**30,34**	
60	Yes	**32,16**	
	No	43,14	
120	Yes	**31,97**	
	No	37,14	

Table 3. Forecasting results of February

Previous width (min)	Normalization	MAE LSTM (W)	MAE Baseline (W)
30	Yes	120,15	112,86
	No	**79,59**	
60	Yes	120,39	
	No	123,56	
120	Yes	119,47	
	No	**106,05**	

Table 4. Forecasting results of March

Previous width (min)	Normalization	MAE LSTM (W)	MAE Baseline (W)
30	Yes	89,44	84,56
	No	**69,39**	
60	Yes	89,54	
	No	90,88	
120	Yes	89,62	
	No	**77,02**	

Table 5. Forecasting results of April

Previous width (min)	Normalization	MAE LSTM (W)	MAE Baseline (W)
30	Yes	89,44	84,56
	No	**69,39**	
60	Yes	89,54	
	No	90,88	
120	Yes	89,62	
	No	**77,02**	

Table 6. Forecasting results of May

Previous width (min)	Normalization	MAE LSTM (W)	MAE Baseline (W)
30	Yes	**27,31**	28,20
	No	**27,34**	
60	Yes	**27,72**	
	No	36,75	
120	Yes	**27,84**	
	No	33,44	

Table 7. Forecasting results of June

Previous width (min)	Normalization	MAE LSTM (W)	MAE Baseline (W)
30	Yes	27,81	27,42
	No	**22,08**	
60	Yes	27,57	
	No	29,37	
120	Yes	**27,39**	
	No	**22,70**	

Table 8. Forecasting results of July

Previous width (min)	Normalization	MAE LSTM (W)	MAE Baseline (W)
30	Yes	**38,18**	39,53
	No	**31,06**	
60	Yes	39,84	
	No	46,12	
120	Yes	39,65	
	No	36,72	

Table 9. Forecasting results of August

Previous width (min)	Normalization	MAE LSTM (W)	MAE Baseline (W)
30	Yes	**12,61**	15,94
	No	**15,54**	
60	Yes	**12,78**	
	No	21,17	
120	Yes	**13,12**	
	No	17,54	

Table 10. Forecasting results of September

Previous width (min)	Normalization	MAE LSTM (W)	MAE Baseline (W)
30	Yes	9,63	8,28
	No	11,25	
60	Yes	9,55	
	No	14,94	
120	Yes	9,50	
	No	13,21	

Table 11. Forecasting results of October

Previous width (min)	Normalization	MAE LSTM (W)	MAE Baseline (W)
30	Yes	28,17	26,97
	No	**23,63**	
60	Yes	28,50	
	No	28,52	
120	Yes	28,43	
	No	27,26	

Table 12. Forecasting results of November

Previous width (min)	Normalization	MAE LSTM (W)	MAE Baseline (W)
30	Yes	48,75	44,50
	No	**36,48**	
60	Yes	48,61	
	No	48,74	
120	Yes	48,95	
	No	45,03	

Table 13. Forecasting results of December

Previous width (min)	Normalization	MAE LSTM (W)	MAE Baseline (W)
30	Yes	87,08	84,48
	No	**62,32**	
60	Yes	87,94	
	No	**84,12**	
120	Yes	86,91	
	No	75,88	

5   Conclusions and Future Work

The results shows that the LSTM approach presents better results with at least one configuration in all months but September. This performance improvement would be a significant breakthrough to optimize an energy storage system, with the consequent energy and economic savings. Regarding the data normalization, it is not a significant trend about the best procedure, and a similar conclusion is achieved with the previous step width.

As future works, different paths could be followed. On the one hand, the use of techniques to analyze the seasonality of the time series could be considered in order to group the data prior to the training stage. A similar approach could be followed with the use of clustering algorithms. Finally, a deeper quantitative analysis of the results and the real implementation of this system could be carried out, even with the inclusion of atmospheric variables besides the previous power generated.

Acknowledgments. CITIC, as a Research Center of the University System of Galicia, is funded by Consellería de Educación, Universidade e Formación Profesional of the Xunta de Galicia through the European Regional Development Fund (ERDF) and the Secretaría Xeral de Universidades (Ref. ED431G 2019/01).

References

1. Informe sobre los elementos necesarios en la transición energética. https://www. aeeolica.org/uploads/Elementos_necesarios_para_la_Transicin_Energtica_FINAL. pdf (2017). Accessed 21 May 2020
2. Sotavento installation. http://web.archive.org/web/20080207010024/. http://www.808multimedia.com/winnt/kernel.htm (2020). Accessed 19 Mar 2020
3. Asociación Empresarial Eólica (AEE): Anuario eólico. la voz del sector 2019 (2019). https://www.aeeolica.org/images/Publicaciones/Anuario-Elico-2019.pdf
4. Aláiz-Moretón, H., Castejón-Limas, M., Casteleiro-Roca, J.L., Jove, E., Fernández Robles, L., Calvo-Rolle, J.L.: A fault detection system for a geothermal heat exchanger sensor based on intelligent techniques. Sensors **19**(12), 2740 (2019)
5. Baruque, B., Jove, E., Porras, S., Calvo-Rolle, J.L.: Small-wind turbine power generation prediction from atmospheric variables based on intelligent techniques. In: Herrero, Á., Cambra, C., Urda, D., Sedano, J., Quintián, H., Corchado, E. (eds.) SOCO 2020. AISC, vol. 1268, pp. 33–43. Springer, Cham (2021). https://doi.org/10.1007/978-3-030-57802-2_4
6. Baruque, B., Porras, S., Jove, E., Calvo-Rolle, J.L.: Geothermal heat exchanger energy prediction based on time series and monitoring sensors optimization. Energy **171**, 49–60 (2019)
7. British Petroleum: Renewable energy - wind energy (2018). https://www.bp.com/en/global/corporate/energy-economics/statistical-review-of-world-energy/renewable-energy.html.html#wind-energy
8. Casteleiro-Roca, J.L., et al.: Short-term energy demand forecast in hotels using hybrid intelligent modeling. Sensors **19**(11), 2485 (2019)
9. Chiah, M., Phan, D.H.B., Tran, V.T., Zhong, A.: Energy price uncertainty and the value premium. Int. Rev. Financ. Anal. **81**, 102062 (2022)

10. Choi, S.G., Cho, S.B.: Evolutionary reinforcement learning for adaptively detecting database intrusions. Logic J. IGPL **28**(4), 449–460 (2020)
11. Fáñez, M., Villar, J.R., De la Cal, E., González, V.M., Sedano, J.: Improving wearable-based fall detection with unsupervised learning. Logic J. IGPL **30**(2), 314–325 (2022)
12. Fernandez-Serantes, L.A., Casteleiro-Roca, J.L., Berger, H., Calvo-Rolle, J.L.: Hybrid intelligent system for a synchronous rectifier converter control and soft switching ensurement. Eng. Sci. Technol. Int. J. 101189 (2022)
13. Fernandez-Serantes, L.A., Casteleiro-Roca, J.L., Calvo-Rolle, J.L.: Hybrid intelligent system for a half-bridge converter control and soft switching ensurement. Revista Iberoamericana de Automática e Informática Ind. 1–5 (2022)
14. Hochreiter, S., Schmidhuber, J.: Long short-term memory. Neural Comput. **9**(8), 1735–1780 (1997)
15. Infield, D., Freris, L.: Renewable Energy in Power Systems. John Wiley & Sons (2020)
16. Jove, E., Casteleiro-Roca, J., Quintián, H., Méndez-Pérez, J., Calvo-Rolle, J.: Anomaly detection based on intelligent techniques over a bicomponent production plant used on wind generator blades manufacturing. Revista Iberoamericana de Automática e Informática Ind. **17**(1), 84–93 (2020)
17. Jove, E., Casteleiro-Roca, J.L., Quintián, H., Méndez-Pérez, J.A., Calvo-Rolle, J.L.: A new method for anomaly detection based on non-convex boundaries with random two-dimensional projections. Inf. Fusion **65**, 50–57 (2021). https://www.sciencedirect.com/science/article/pii/S1566253520303407
18. Leira, A., et al.: One-class-based intelligent classifier for detecting anomalous situations during the anesthetic process. Logic J. IGPL **30**(2), 326–341 (2020). https://doi.org/10.1093/jigpal/jzaa065
19. Lund, H.: Renewable energy strategies for sustainable development. Energy **32**(6), 912–919 (2007)
20. Owusu, P.A., Asumadu-Sarkodie, S.: A review of renewable energy sources, sustainability issues and climate change mitigation. Cogent Eng. **3**(1), 1167990 (2016)
21. Porras, S., Jove, E., Baruque, B., Calvo-Rolle, J.L.: Prediction of small-wind turbine performance from time series modelling using intelligent techniques. In: Analide, C., Novais, P., Camacho, D., Yin, H. (eds.) IDEAL 2020. LNCS, vol. 12490, pp. 541–548. Springer, Cham (2020). https://doi.org/10.1007/978-3-030-62365-4_52
22. Staudemeyer, R.C., Morris, E.R.: Understanding LSTM-a tutorial into long short-term memory recurrent neural networks. arXiv preprint arXiv:1909.09586 (2019)
23. Trojaola, I., Elorza, I., Irigoyen, E., Pujana-Arrese, A., Calleja, C.: The effect of iterative learning control on the force control of a hydraulic cushion. Logic J. IGPL **30**(2), 214–226 (2022)
24. Vega Vega, R., Quintián, H., Calvo-Rolle, J.L., Herrero, Á., Corchado, E.: Gaining deep knowledge of android malware families through dimensionality reduction techniques. Logic J. IGPL **27**(2), 160–176 (2019)
25. Yu, Y., Si, X., Hu, C., Zhang, J.: A review of recurrent neural networks: LSTM cells and network architectures. Neural Comput. **31**(7), 1235–1270 (2019)

CORE-BCD-mAI: A Composite Framework for Representing, Querying, and Analyzing Big Clinical Data by Means of Multidimensional AI Tools

Alfredo Cuzzocrea[1]([✉]) and Pablo G. Bringas[2]

[1] iDEA Lab, University of Calabria, Rende, Italy
`alfredo.cuzzocrea@unical.it`
[2] Faculty of Engineering, University of Deusto, Bilbao, Spain
`pablo.garcia.bringas@deusto.es`

Abstract. This paper introduces CORE-BCD-mAI, a *COmposite Framework for REpresenting, Querying, and Analyzing Big Clinical Data by means of multidimensional AI Tools*. The proposed framework combines *Artificial Intelligence* (AI) and *Big Data*, with the aim of accessing *big clinical datasets* for managing and decision-making purposes, in the *healthcare application scenario*. We present in detail methodologies supported by the CORE-BCD-mAI framework, along with its intrinsic anatomy. The paper also provides relevant research challenges deriving from the overall CORE-BCD-mAI proposal.

Keywords: AI-Supported Big Data Management · AI-Supported Big Data Analytics · AI-Supported Big Data Intelligence · Multidimensional AI tools · Big clinical data management and analytics

1 Introduction

Recently, a great deal of attention has been devoted to the emerging issue of coupling *Artificial Intelligence* (AI) and *Big Data* (e.g., [39]), both in the academic and the industrial research setting. Basically, the main idea here consists in devising AI techniques over big datasets, with the goal of successfully applying over such datasets a wide spectrum of AI methodologies like reasoning, automated learning, machine learning, natural language processing, and so forth. This, under the tight requirements of accessing, processing and managing big data, well-represented by the common *3V* (i.e., *Volume, Velocity, Variety*) *model* [6]. Here, one of the more annoying problems consists in the issue of ensuring *scalability* of AI techniques over massive-in-size, heterogeneous big datasets.

Big clinical data (e.g., [40]) represent a typical setting where the convergence of AI and Big Data provides clear benefits. Here, big clinical datasets are accessed for managing and decision-making purposes, in the *healthcare application scenario*. In such an application scenario, (big) data are very heterogenous in nature, beyond to be

© Springer Nature Switzerland AG 2022
P. García Bringas et al. (Eds.): HAIS 2022, LNAI 13469, pp. 175–185, 2022.
https://doi.org/10.1007/978-3-031-15471-3_16

of explosive size, and, as a consequence, *big data lake methodologies* appear to play a non-secondary role (e.g., [41]).

All considering, the so-delineated scientific setting opens the door to what we recognize under the terms: *AI-Supported Big Data Management, AI-Supported Big Data Analytics*, and *AI-Supported Big Data Intelligence*.

Along this line of research, this paper introduces **CORE-BCD-mAI**, namely "*A COmposite Framework for REpresenting, Querying, and Analyzing Big Clinical Data by means of multidimensional AI Tools*". CORE-BCD-mAI focuses the attention on research activities devoted to the definition and experimental evaluation of models, methodologies, techniques and algorithms for supporting (*i*) representation, (*ii*) querying, and (*iii*) analytics over **big clinical data** on the basis of **multidimensional AI**[1] **tools**. The paper also provides relevant research challenges deriving from the overall CORE-BCD-mAI proposal.

The remaining part of the paper is organized as follows. In Sect. 2, we provide motivations of our research. Section 3 provides methodologies and anatomy of the proposed CORE-BCD-mAI framework. In Sect. 4, we focus the attention on relevant research challenges related to the proposed CORE-BCD-mAI framework. Finally, Sect. 5 reports conclusions and future work of our research.

2 Motivations: Combining Multidimensional AI Tools and Big Clinical Data

CORE-BCD-mAI relies on two main components/concepts, namely big clinical data and multidimensional AI tools.

Clinical data (e.g., medical test data, patient record data, HVR[2]/ECG[3] data, omics data, clinical trial data, and so forth) are more and more an emerging instance of big data repositories, due to their intrinsic features such as volume, streaming nature and heterogeneity. This conveys in the so-called big clinical data (e.g., [24]). Among several properties, big clinical data are characterized by a strong integration with external data sources, such as census data, social data, pharmacovigilance data and so forth, which further enforces their (strong-)heterogeneity nature. The latter property opens the door to *multi-level analytics methodologies* that aim at analyzing (big) clinical data *at various level of granularity*, from the basic hospital level to the higher regional level and national level (e.g., epidemiological studies for healthcare-policy-oriented decision-making purposes).

Multidimensional AI tools refer to a collection of models, methodologies and analytics tools that aim at supporting big data representation, management and analytics on the basis of well-known multidimensional paradigms, being OLAP[4] [25, 26] the most significant model among them. OLAP allows us to represent and analyze data via fortunate *multidimensional metaphors* (e.g., dimensions, measures, levels, hierarchies, and so forth) by building very intuitive *multidimensional spaces* with analysis capabilities

---

[1] Artificial Intelligence.
[2] Heart Rate Variability.
[3] ElectroCardioGram.
[4] OnLine Analytical Processing.

well-over the capabilities of traditional SQL[5]-based analysis tools. Indeed, recently the research community has devoted a great deal of attention to the issue of addressing big data management and analytics issues via such paradigms (e.g., [27–29]).

Therefore, it is a matter of fact to recognize that representing, querying and analyzing big clinical data via big data techniques, and, specifically, by means of multidimensional AI tools, is an emerging research challenge, with also critical outcomes in the society.

In fact, it is well known that ICT[6] technologies can have a beneficial impact on reducing costs of national healthcare systems, by also allowing to devise suitable guidelines for driving future healthcare policies (e.g., [30]).

Indeed, nowadays there is a great attention on the challenge of effectively and efficiently managing and analyzing big clinical data, mostly before this has a critical impact on the national healthcare systems (and thus in the society). As a matter of fact, actually several (big) national projects in this sector exist, such as USA (e.g., [31]), Europe (e.g., [32]), and South Korea (e.g., [33]). On the other hand, most of today healthcare information systems are managed as (traditionally) legacy information systems, without taking advantages from the potentialities offered by big data technologies. In some sense, the main goal of CORE-BCD-mAI is fulfilling this gap.

By adopting big data technologies deeply, CORE-BCD-mAI is thus characterized by a high level of **originality**, especially due to the adaptation of multidimensional AI tools (which break the tradition with classical AI tools). On the other hand, the **perspective** of CORE-BCD-mAI is high, since the proposed representation, querying and analysis methodologies proposed within CORE-BCD-mAI can be further extended as to deal with other class of data, such as social data or government data, just to cite two well-known cases. The **methodology** pursued by CORE-BCD-mAI is traditional: design, and experimental testing on synthetic and real-life *big multidimensional clinical data sets*

3 CORE-BCD-mAI: Methodologies and Anatomy

The main research statement of CORE-BCD-mAI relies in arguing that big data techniques, and particularly multidimensional AI tools, are perfectly suitable to manage, process and support analytics over big clinical data, like also studied in several recent proposals in this field (e.g., [1–5]). First, a suitable *multidimensional data model* must be devised to this end, being the core component of all the CORE-BCD-mAI proposal. To this end, CORE-BCD-mAI introduces the so-called **big multidimensional NoSQL data model**, a specific data model that develops multidimensional structures on top of *NoSQL data* (e.g., OLAP data cubes [25]), since the latter are recognized as the most suitable solution to represent big clinical data (e.g., [34]).

Big data techniques are capable of integrating several classical Data Mining tools (like classification, clustering, frequent itemset mining, association rule mining, and so forth), with the strict requirement that these techniques must be *parallelizable*, in order to take advantages from the availability of rich 3V data [6] and the effectiveness of

---

[5] Structured Query Language.
[6] Information and Communication Technologies.

consolidated mining approaches (e.g., [7–11]). Therefore, we can safely consider big data techniques as an *enabling technology* for CORE-BCD-mAI goals, also given the success and the pervasivity of big data processing, management and mining tools and systems available at today.

The main goal of CORE-BCD-mAI thus consists in devising models, methodologies, technique and algorithms based on multidimensional AI tools for the *specific* case of big clinical data to be processed, i.e. HVR measurements and statistics, ECG traces, patient record information, omics data, clinical guidelines and so forth (including external data sources), with the final goal of supporting advanced big data analytics (including emerging *big data visualization methodologies* - e.g., [36]) for decision making purposes (e.g., early detection of severe pathologies). This allows us to devise, implement and experimentally assess specialized big data solutions supporting CORE-BCD-mAI goals, by also conveying in innovative best-practices to be considered in future research efforts. Most importantly, within the general CORE-BCD-mAI proposal, **critical research challenges** are identified and addressed, which constitutes the most relevant and significant outcome of the project.

With this goal in mind, CORE-BCD-mAI also pursues the definition and the prototypical implementation of a comprehensive *big multidimensional clinical data framework*, called CORE-BCD-mAI framework, that supports the various phases of the above-described data, information and decisional processing flows. To this end, CORE-BCD-mAI adopts well-known open-source big data processing technologies, as to take the possible-highest advantages from interoperability paradigms over freely-available open-source data representation, management and mining tools, as well as maximum diffusion and usage of the framework itself within the industrial and academic research communities.

Figure 1 provides the big picture of data, information and workflows of CORE-BCD-mAI, and their interconnections, where the different phases of the research activities are highlighted, namely: (*i*) *Big Multidimensional Clinical Data Representation*; (*ii*) *Big Multidimensional Clinical Data Querying*; (*iii*) *Big Multidimensional Clinical Data Analytics*; (*iv*) *Big Multidimensional Clinical Data Visualization*.

Looking at specific implementation-related aspects, the proposed framework relies on top of well-known *Cloud Computing architectures* (e.g., [16, 17]) where *Hadoop* [23] integrates various components of such a framework (namely: big multidimensional clinical data representation, querying, analytics, and visualization), mostly based on *MapReduce* [12], the most popular big data processing paradigm. This approach follows fortunate experiences where the Cloud Computing paradigm has already proven its effectiveness and efficiency in dealing with massive amounts of big data (e.g., [3, 5, 13, 21, 22]).

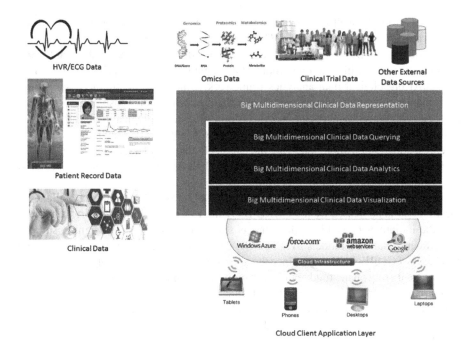

Fig. 1. CORE-BCD-mAI data/information/workflow structure.

4 CORE-BCD-mAI: Research Challenges

In our research, we have identified the following research challenges related to the proposed CORE-BCD-mAI framework:

- *Big Multidimensional Clinical Data Representation over Elastic NoSQL Databases*;
- *Compressed Big Multidimensional Clinical Data Querying*;
- *OLAP-Based Big Multidimensional Clinical Data Analytics*;
- *Big Multidimensional Clinical Data Visualization.*

In the following, we report our critical analysis on these challenges.

Big Multidimensional Clinical Data Representation over Elastic NoSQL Databases.
This challenge focuses on the relevant problem of effectively and efficiently representing big clinical data via *elastic paradigms* (e.g., [14, 15]), based on the big multidimensional NoSQL data model, which is one of most relevant outcomes of CORE-BCD-mAI. This approach is due to the following two motivations. First, it is well-understood that big clinical data may sensitively vary their size by several magnitude orders (for instance, this could be as following specific healthcare campaigns, e.g. for statistical purposes), thus alternative non-NoSQL data representation formats (e.g., relational data models) would be completely inadequate to capture this specific characteristic. Second, adopting multidimensional data models is the core strategy of CORE-BCD-mAI as to take

advantages from the well-known characteristics of the multidimensional paradigms, at all the various layers of the CORE-BCD-mAI framework (see Fig. 1).

Elastic paradigms are essentially realized via a strict integration with Cloud Computing architectures, like the one proposed as implementation blueprint of the CORE-BCD-mAI big clinical data analytics framework, which provide the necessary high-performance computing features to process big data while being also able to scale on data sets with possible-increasing-sizes. From a strict-technological point of view, big clinical data are stored in suitable *NoSQL databases* (e.g., [18]), which pursue the idea of representing data without marrying the classical relational data model while, instead, adopting a *document-oriented* (e.g., *MongoDB* [19]) or a *graph-oriented* (e.g., *Neo4j* [20]) schema-less data model.

How to devise multidimensional models on top of NoSQL data? This is a challenging question that is addressed by the CORE-BCD-mAI proposal. Indeed, apart from some sporadic initiatives (e.g., [35]), the latter is still an open problem. Designing suitable multidimensional data models over NoSQL data is a critical aspect that influences all the phases of CORE-BCD-mAI, since the upper-lying querying and analytics layers (see Fig. 1) heavily rely on the fundamental big multidimensional NoSQL data representation model implemented in CORE-BCD-mAI. Finally, it should be noted that this approach conveys in so-called big multidimensional clinical data, the fundamental data type for CORE-BCD-mAI (which, in turns, adheres to the big multidimensional NoSQL data model).

Compressed Big Multidimensional Clinical Data Querying. This challenge deals with the issue of devising models, techniques and algorithms for querying big multi-dimensional clinical data. In fact, retrieving the information necessary for supporting healthcare decision processes over large amounts of clinical data effectively and efficiently is of particular relevance in the context of CORE-BCD-mAI. At this level, effectiveness and efficiency play a key role with respect to the reliability of the entire framework. In this respect, the key idea consists in applying *data compression paradigms*, which already proven their effectiveness in the case of multidimensional data structures (e.g., [37]), as to balance efficiency with accuracy. In fact, in compression-based approaches, we admit to have less precision, which is perfectly tolerable for analytics purposes, by counterpartying it with "fast" *approximate answers*, which are critical for big data environments. This conveys in devising ad-hoc both big multidimensional data compression techniques as well as *approximate query answering algorithms* over compressed data structures (e.g., [38]).

As regards the specific problems investigated by the CORE-BCD-mAI proposal, the first, critical issue consists in devising partitioning strategies for multidimensional data structures, which drive the upper-lying compression data process. These strategies aim at obtaining a partitioned representation of the target multidimensional data structure in terms of so-called *buckets* (i.e., "bunches" of data) that are then compressed in order to obtain a compressed representation of the whole structure. The criterion of the partition strategy can be based on several alternatives, such as: (*i*) error-metrics based; (*ii*) similarity-based; (*iii*) space-bound based. The so-obtained compressed data structure is finally used for evaluating input queries, thus obtaining approximate answers. A similar approach can be also used to evaluate *OLAP queries*, which are particularly suitable to support big data analytics processes.

OLAP-Based Big Multidimensional Clinical Data Analytics. This challenge focuses the attention on the definition and testing of models, techniques and algorithms for supporting big data analytics over big multidimensional clinical data, based on the well-known OLAP paradigm. This means that the core analytics layer of the CORE-BCD-mAI framework adopts multidimensional data structures to represent the big data (e.g., OLAP data cubes) and flexible query answering techniques over such data structures (e.g., aggregate query evaluation) in order to finally support the underlying big data analytics processes. This problem, in fact, is directly related to the issue of devising approximate query answering techniques over compressed multidimensional clinical data. Indeed, it should be noted that approximate answers perfectly fulfill the goal of playing the role of basic components of major big data analytics processes over big multidimensional clinical data, for instance applied to aggregate queries. Just to mention a toy case study, a public healthcare government decision maker could be interested in accessing the evolution of the number of cancer patients (e.g., lung cancer) across a given time window (e.g., 2000–2010) in a given geographic zone (e.g., South France) via the CORE-BCD-mAI framework. Here, every point of the resulting data distribution is obtained as the (approximate) answer to a given OLAP aggregate query over big multidimensional clinical data. Here, numerical precision is not necessary, as, usually, in big data analytics we observe trends and patterns rather than exact values.

While quite a rich literature on OLAP analytics tools in traditional environments exists, the issue of applying these models, techniques and algorithms in emerging Cloud environments is still an open problem, just because making parallelizable these tools is not a trivial matter. For instance, to cite a simple case, some OLAP aggregate operators (e.g., SUM) are easily parallelizable over a Cloud node cluster, but some others (e.g., AVG) are not. On the contrary, to cite a more complex case, achieving parallel, Cloud-aware versions of multidimensional clustering over multidimensional data structures like OLAP cubes is an exciting research challenge that still needs a solution. Devising and making "computationally-feasible" OLAP analytics tools in Cloud environments, as to support complex big data analytics processes, is the main research issue that are investigated in the CORE-BCD-mAI proposal.

Big Multidimensional Clinical Data Visualization. This challenge regards the definition and testing of models, techniques and algorithms for supporting both visualization of big multidimensional clinical data and visualization of big data analytics results over big multidimensional clinical data. The goal is, thus, double-fold. These activities derive from the need for visualizing big multidimensional clinical data and the results of their processing, in order to provide user-friendly tools able to adequately support decision makers during the analysis of big multidimensional clinical data. It should be noted that the theme of big data visualization is relevant in the context of academic research but also in that of industrial research, and the case of the big multidimensional clinical data represents a very-interesting instance.

Looking in deeper details, supporting visualization of big multidimensional data (like OLAP data cubes) is an interesting open problem (with also powerful outcomes in real-life big data analytics tools). In the CORE-BCD-mAI proposal, in order to face-off this issue, the idea consists in devising *dimensionality-reduction big data visualization techniques* where complex, high-dimensional data structures can be visualized over lower-dimensional spaces (e.g., those feasible on typical access devices like desktop computers, laptops, smartphones, and so forth). In order to obtain this amenity, a possibility consists in exploiting the typical *hierarchical nature* of OLAP dimensions (e.g., [25]), and merging *multiple OLAP dimensions* into (shared) *visualizing OLAP dimensions* based on the *numerosity* of dimensional attributes (i.e., the number of members contained by a dimensional attribute) of the hierarchies. Basically, given two OLAP dimensions at different hierarchical levels, the one having a lower number of members in their dimensional attributes, said D_L, can be merged with the one having a higher number of members in their dimensional attributes, said D_H, by *re-distributing* the dimensional attributes of D_L as *children* of the dimensional attributes of D_H, thus obtaining one OLAP dimension only that "starts" with D_H and "continues" with D_L. The latter plays the role of visualizing OLAP dimension for D_H and D_L.

5 Conclusions and Future Work

In this paper, we introduced the CORE-BCD-mAI framework for combining big clinical data and multidimensional AI tools, along with relevant research challenges related to the investigated area.

The main perspective supported by the framework replies in combining multidimensional AI tools and big data, especially in the clinical setting. The framework exposes several point of research innovations in the investigated context, in both theory and practice. The conceptual architecture can behave, in future, as a relevant milestone for research efforts in the context of healthcare analytics methodologies, applications and systems.

Future work is mainly oriented towards making our proposed framework mora and more compliant with the emerging requirements dictated by big data trends (e.g., [42–50]).

References

1. Wei, H., et al.: Predicting health care risk with big data drawn from clinical physiological parameters. In: Huang, H., Liu, T., Zhang, H.-P., Tang, J. (eds.) SMP 2014. CCIS, vol. 489, pp. 88–98. Springer, Heidelberg (2014). https://doi.org/10.1007/978-3-662-45558-6_8
2. Sahoo, S.S.: Biomedical big data for clinical research and patient care: role of semantic computing. In: ICSC 2014, pp. 3–5 (2014)
3. Sahoo, S.S., et al.: Heart beats in the cloud: distributed analysis of electrophysiological 'Big Data' using cloud computing for epilepsy clinical research. JAMIA 21(2), 263–271 (2014)
4. Tsai, C.-F., Lin, W.-C., Ke, S.-W.: Big data mining with parallel computing: a comparison of distributed and MapReduce methodologies. J. Syst. Softw. 122, 83–92 (2016)
5. Forkan, A.R.M., Khalil, I., Atiquzzaman, M.: ViSiBiD: a learning model for early discovery and real-time prediction of severe clinical events using vital signs as big data. Comput. Netw. 113, 244–257 (2017)

6. Laney, D.: 3D data management: controlling data volume, velocity, and variety. Technical report, META Group (2001)

7. Fodeh, S., Zeng, Q.: Mining big data in biomedicine and health care. J. Biomed. Inform. **63**, 400–403 (2016)

8. Tzanis, G.: Biological and medical big data mining. IJKDB **4**(1), 42–56 (2014)

9. Wu, X., Zhu, X., Wu, G.-Q., Ding, W.: Data mining with big data. IEEE Trans. Knowl. Data Eng. **26**(1), 97–107 (2014)

10. Cuzzocrea, A.: Big data mining or turning data mining into predictive analytics from large-scale 3Vs data: the future challenge for knowledge discovery. In: Ait Ameur, Y., Bellatreche, L., Papadopoulos, G.A. (eds.) MEDI 2014. LNCS, vol. 8748, pp. 4–8. Springer, Cham (2014). https://doi.org/10.1007/978-3-319-11587-0_2

11. Pal, S.K., Meher, S.K., Skowron, A.: Data science, big data and granular mining. Pattern Recogn. Lett. **67**, 109–112 (2015)

12. Dean, J., Ghemawat, S.: MapReduce: simplified data processing on large clusters. Commun. ACM **51**(1), 107–113 (2008)

13. Cuzzocrea, A., Moussa, R.: A cloud-based framework for supporting effective and efficient OLAP in big data environments. In: CCGRID 2014, pp. 680–684 (2014)

14. Agrawal, D., Das, S., El Abbadi, A.: Big data and cloud computing: current state and future opportunities. In: EDBT 2011, pp. 530–533 (2011)

15. Lim, L.: Elastic data partitioning for cloud-based SQL processing systems. In: BigData Conference 2013, pp. 8–16 (2013)

16. Armbrust, M., et al.: A view of cloud computing. Commun. ACM **53**(4), 50–58 (2010)

17. Buyya, R., Yeo, C.S., Venugopal, S., Broberg, J., Brandic, I.: Cloud computing and emerging IT platforms: vision, hype, and reality for delivering computing as the 5th utility. Future Gener. Comput. Syst. **25**(6), 599–616 (2009)

18. Han, J., Haihong, E., Le, G., Du, J.: Survey on NoSQL database. In: IEEE ICPCA 2011, pp. 363–366 (2011)

19. Chodorow, K., Dirolf, M.: MongoDB – the definitive guide: powerful and scalable data storage, pp. I–XVII, 1–193. O'Reilly (2010). ISBN 978-1-449-38156-1

20. Webber, J.: A programmatic introduction to Neo4j. In: SPLASH 2012, pp. 217–218 (2012)

21. Lee, K.K.-Y., Tang, W.-C., Choi, K.-S.: Alternatives to relational database: comparison of NoSQL and XML approaches for clinical data storage. Comput. Methods Programs Biomed. **110**(1), 99–109 (2013)

22. Swenson, E.R., Bastian, N.D., Nembhard, H.B.: Data analytics in health promotion: health market segmentation and classification of total joint replacement surgery patients. Expert Syst. Appl. **60**, 118–129 (2016)

23. White, T.: Hadoop: The Definitive Guide. O'Reilly Media, Inc. (2009). ISBN 0596521979 9780596521974

24. Chrimes, D., Zamani, H.: Using distributed data over HBase in big data analytics platform for clinical services. Comput. Math. Methods Med. **2017**, 6120820:1–6120820:16 (2017)

25. Gray, J., et al.: Data cube: a relational aggregation operator generalizing group-by, cross-tab, and sub totals. Data Min. Knowl. Discov. **1**(1), 29–53 (1997)

26. Chaudhuri, S., Dayal, U.: An overview of data warehousing and OLAP technology. SIGMOD Rec. **26**(1), 65–74 (1997)

27. Cuzzocrea, A.: Scalable OLAP-based big data analytics over cloud infrastructures: models, issues, algorithms. In: ICCBDC 2017, pp. 17–21 (2017)

28. Cuzzocrea, A., Cavalieri, S., Tomarchio, O., Di Modica, G., Cantone, C., Di Bilio, A.: REMS.PA: a complex framework for supporting OLAP-based big data analytics over data-intensive business processes. In: ICEIS (1) 2019, pp. 223–230 (2019)

29. Cuzzocrea, A., De Maio, C., Fenza, G., Loia, V., Parente, M.: OLAP analysis of multi-dimensional tweet streams for supporting advanced analytics. In: SAC 2016, pp. 992–999 (2016)

30. Shahbaz, M., Gao, C., Zhai, L., Shahzad, F., Hu, Y.: Investigating the adoption of big data analytics in healthcare: the moderating role of resistance to change. J. Big Data **6**(1), 1–20 (2019). https://doi.org/10.1186/s40537-019-0170-y

31. Groves, P., Kayyali, B., Knott, D., Kuiken, S.V.: The 'Big Data' Revolution in Healthcare: Accelerating Value and Innovation. McKinsey Tech Rep (2016)

32. Habl, C., Renner, A.-T., Bobek, J., Laschkolnig, A.: Study on Big Data in Public Health, Telemedicine and Healthcare. European Commission Tech Rep (2016)

33. Nam, J., Kwon, H.W., Lee, H., Ahn, E.K.: National healthcare service and its big data analytics. Healthc. Inform. Res. **24**(3), 247–249 (2018)

34. Yang, E., et al.: A late-binding, distributed, NoSQL warehouse for integrating patient data from clinical trials. Database **2019**, baz032 (2019)

35. Chevalier, M., El Malki, M., Kopliku, A., Teste, O., Tournier, R.: Implementation of multidimensional databases with document-oriented NoSQL. In: Madria, S., Hara, T. (eds.) DaWaK 2015. LNCS, vol. 9263, pp. 379–390. Springer, Cham (2015). https://doi.org/10.1007/978-3-319-22729-0_29

36. Keim, D.A., Qu, H., Ma, K.-L.: Big-data visualization. IEEE Comput. Graph. Appl. **33**(4), 20–21 (2013)

37. Cuzzocrea, A., Serafino, P.: LCS-Hist: taming massive high-dimensional data cube compression. In: EDBT 2009, pp. 768–779 (2009)

38. Cuzzocrea, A.: Improving range-sum query evaluation on data cubes via polynomial approximation. Data Knowl. Eng. **56**(2), 85–121 (2006)

39. Tae, K.H., Roh, Y., Oh, Y.H., Kim, H., Whang, S.E.: Data cleaning for accurate, fair, and robust models: a big data - AI integration approach. In: DEEM@SIGMOD 2019, pp. 5:1–5:4 (2019)

40. Perez-Arriaga, M.O., Poddar, K.A.: Clinical trials data management in the big data era. In: Nepal, S., Cao, W., Nasridinov, A., Bhuiyan, M.D.Z.A., Guo, X., Zhang, L.-J. (eds.) BIGDATA 2020. LNCS, vol. 12402, pp. 190–205. Springer, Cham (2020). https://doi.org/10.1007/978-3-030-59612-5_14

41. Cuzzocrea, A.: Big data lakes: models, frameworks, and techniques. In: BigComp 2021, pp. 1–4 (2021)

42. Cuzzocrea, A., Mansmann, S.: OLAP visualization: models, issues, and techniques. In: Encyclopedia of Data Warehousing and Mining 2009, pp. 1439–1446 (2009)

43. Cuzzocrea, A., Furfaro, F., Masciari, E., Saccà, D., Sirangelo, C.: Approximate query answering on sensor network data streams. In: GeoSensor Networks, p. 49. CRC Press (2004)

44. Cuzzocrea, A., Wang, W.: Approximate range-sum query answering on data cubes with probabilistic guarantees. J. Intell. Inf. Syst. **28**(2), 161–197 (2007)

45. Bonifati, A., Cuzzocrea, A.: Storing and retrieving XPath fragments in structured P2P networks. Data Knowl. Eng. **59**(2), 247–269 (2006)

46. Cuzzocrea, A.: Overcoming limitations of approximate query answering in OLAP. In: IDEAS 2005, pp. 200–209 (2005)

47. Morris, K.J., Egan, S.D., Linsangan, J.L., Leung, C.K., Cuzzocrea, A., Hoi, C.S.: Token-based adaptive time-series prediction by ensembling linear and non-linear estimators: a machine learning approach for predictive analytics on big stock data. In: ICMLA 2018, pp. 1486–1491 (2018)

48. Audu, A.-R., Cuzzocrea, A., Leung, C.K., MacLeod, K.A., Ohin, N.I., Pulgar-Vidal, N.C.: An intelligent predictive analytics system for transportation analytics on open data towards the development of a smart city. In: Barolli, L., Hussain, F.K., Ikeda, M. (eds.) CISIS 2019.

AISC, vol. 993, pp. 224–236. Springer, Cham (2020). https://doi.org/10.1007/978-3-030-22354-0_21

49. Bellatreche, L., Cuzzocrea, A., Benkrid, S.: $\mathcal{F}$&$\mathcal{A}$: a methodology for effectively and efficiently designing parallel relational data warehouses on heterogenous database clusters. In: Bach Pedersen, T., Mohania, M.K., Tjoa, A.M. (eds.) DaWaK 2010. LNCS, vol. 6263, pp. 89–104. Springer, Heidelberg (2010). https://doi.org/10.1007/978-3-642-15105-7_8

50. Salman, M., Munawar, H.S., Latif, K., Akram, M.W., Khan, S.I., Ullah, F.: Big data management in drug-drug interaction: a modern deep learning approach for smart healthcare. Big Data Cogn. Comput. **6**(1), 30 (2022)

Generalized Fisher Kernel with Bregman Divergence

Pau Figuera[1]($\boxtimes$), Alfredo Cuzzocrea[2], and Pablo García Bringas[1]

[1] D4K Group, University of Deusto, Bilbao, Spain
pau.figueras@opendeusto.es, pablo.garcia.bringas@deusto.es
[2] iDEA Lab, University of Calabria, Rende, Italy
alfredo.cuzzocrea@unical.it

Abstract. The Fisher kernel has good statistical properties. However, from a practical point of view, the necessary distributional assumptions complicate the applicability. We approach the solution to this problem with the NMF (Non-negative Matrix Factorization) methods, which with adequate normalization conditions, provide stochastic matrices. Using the Bregman divergence as the objective function, formally equivalent solutions appear for the specific forms of the functionals involved. We show that simply by taking these results and plug-in into the general expression of the NMF kernel, obtained with purely algebraic techniques, without any assumptions about the distribution of the parameters, the properties of the Fisher kernel hold, and it is a convenient procedure to use this kernel the situations in which they are needed we derive the expression of the information matrix of Fisher. In this work, we have limited the study to the Gaussian metrics, KL (Kullback-Leibler), and I-divergence.

Keywords: Fisher kernel · Non-parametric · Bregman divergence · Non-negative Matrix Factorization

1 Introduction

Kernelization methods are crucial in classification problems of Machine Learning, determining the applicability of the SVM (Support Vector Machines) introduced by [25]. From a practical point of view, its study is justified as a discriminator when the observations are not separable. In this case, it is difficult to assign suitable labels $\mathcal{Y} = \{\mathcal{Y}_1, \mathcal{Y}_2, \dots\}$ for observed instances, and thus quality predictions. One consequence, from a quantitative point of view, is that the evaluation of the similarity with the scalar product does not provide good results. Kernelization is a transformation of the observations space (input space) to another (feature space), not necessarily of the same dimension. It is a dot product generalization if the problem is well defined (it satisfies the Representer Theorem: scalar products admit representation in a Hilbert space such that any linear combination of functions f and g obtained from observations is symmetric and

P. García Bringas et al. (Eds.): HAIS 2022, LNAI 13469, pp. 186–194, 2022.
https://doi.org/10.1007/978-3-031-15471-3_17

semidefinite positive [3], or Mercier's Theorem [19]: any dot product function f is non-negative).

Although there are several varieties of kernels, and from our point of view, the Fisher Kernel, introduced by [15], deserves special attention. It provides a metric for a probabilistic model, and a consistent estimator (the density converges in probability to the value of the parameter that generates the distribution) for the posterior (density of the data given the parameter) of the observed and unobserved instances of the statistical model selected to fit the data [24]. It has been early successfully applied to protein classification [14]. We omit many other works to overload this manuscript with unnecessary references.

A major formulation of the Fisher kernel is due to [12]. From approaches based on his work in which he introduces the PLSA (Probabilistic Latent Semantic Analysis), and in the framework of IR (Information Retrieval), for the co-occurrences that take place when d_i documents of a corpus and the w_j words of a theasaurus, the relative frequencies decompose as the product of mixtures $P(d_i, w_j) = P(d_i, z_k)P(z_k)P(w_j, z_k)$, and that is the symmetric formulation. In this case $P(d_i, z_k)$ and $P(w_j, z_k)$ follow multinomial distributions(of parameters $P(d_i, z_k) \sim \Psi$ and $P(w_j, z_k) \sim \Phi$, while $P(z_k)$ is a set of dummy variables with no statistical significance). A contribution to this approach is introduced in [5], assuming that the distributions $P(d_i, z_k)$ and $P(w_j, z_k)$ are only *iid* (independently identically distributed).

Nonparametric approaches admit two paths. The traditional is to smoothing data with a mixture of multivariate densities $\hat{f} = (1/p) \sum_p \prod_m w(x_q - x; h_q)$ (the function $\hat{f}$, and often named *kernel function* or *kernel smooth function*, which must not be confused with the functions that transform dot products from input to feature space discussed in this manuscript). Problems of this approach are the same as the multivariate nonparametric density estimation and related to the smoot parameter and the partition on the support. A classic works in this direction is [22] and more recently [27] and [11].

NMF techniques also allow obtaining a [18, 26] kernel. Under proper normalization conditions, the obtained matrices are stochastic [7], allowing us to relate the NMF to the Kernel Fisher (in this case, the parameters are the matrices into which it decomposes). This problem, closely related to our current research, has led us to formulate a nonparametric version based on the minimization of the KL divergence [10], and the asymptotic behavior of the decomposition (which always converges to the data matrix [9] with the aid of the *em* algorithm [2]. This statement, under Gaussian assumptions leads to more understandable and stable classifications [21] classifiers.

2   Statement of the Problem

For a set of m observed entities with n variables, and arranging the data as the matrix $\mathbf{X} \in \mathbb{R}^{m \times n}$ where the subscript i ($1 \leq i \leq m$) indicates the rows and j ($1 \leq j \leq n$) the columns, the Fisher kernel is defined as [13]

$$K(\mathbf{x}_i, \mathbf{x}'_{i'}) = U_\theta(\mathbf{x}_i), \mathcal{I}_F^{-1} U_\theta(\mathbf{x}_{i'}) \quad \text{for } i' \neq i \text{ being } U_\theta(\mathbf{x}_i) = -\frac{\partial}{\partial \theta} \log P(\mathbf{x}_i | \theta)$$

$$\mathcal{I}_F = E_\mathbf{x}[U_\theta U'_\theta] \tag{1}$$

$\mathbf{x}_i$ is a row vector of $\mathbf{X}$, and θ a distributional parameter. $U_\theta(\mathbf{x}_i)$ are the Fisher Scores, and $\mathcal{I}_F^{-1}$ the Fisher information Matrix.

If a function $Q(y | x)$ enough differentiable exists, the Fisher kernel is a consistent estimator, with Cramer Rao's lower bound [24], and therefore

$$\text{var}(\hat{\theta}) \geq \frac{1}{I(\hat{\theta})} \tag{2}$$

Parametric solutions refers to the problem of choice of a distributional parameter $\hat{\theta} \in \Theta$, and evaluates (1).

2.1   Non-parametric Approach

Nonparametric approaches can be considered in two ways. The most extended is to smooth the density with the aid of a mixture of densities. The multivariate *kdf* (kernel density function) is [17]

$$\hat{f}(\mathbf{x}) = \frac{1}{p} \sum_p \prod_n w(y_n - y_{np}; h_p) \tag{3}$$

and fitting $\hat{f}$ to f is necessity minimize the MISE (Mean Integrated Standard Error) [17]

$$\text{MISE}(\hat{f}) = \mathbb{E}\left\{ \int_\mathcal{D} (\hat{f} - f) dx \right\}^2 \tag{4}$$

being $\mathcal{D}$ the support on which the data are evaluated.

In our research, the main interest is based on the NMF algebra solutions, and restricting the problem to non-negative data, thus the constraint $\mathbf{X} \in \mathbb{R}_+^{m \times n}$ lets to make probabilistic sense if the transformation

$$[\mathbf{Y}]_{ij} = [\mathbf{X}]_{ij} \, \mathbf{D}_X^{-1} \mathbf{M}_X^{-1} \quad (i = 1, \ldots, m \text{ and } j = 1, \ldots, n) \tag{5}$$

with $\mathbf{D}_X$ and $\mathbf{M}_X$ diagonal matrices defined as $\mathbf{M}_X = \text{diag}\left(\sum_j x_{ij}\right)$ and $\mathbf{D}_X = \text{diag}(m)$, is done, and ensuring the change of $\mathbf{X} \in \mathbb{R}_+^{m \times n}$ to $\mathbf{Y} \in \mathbb{R}_{[0,1]}^{m \times n}$ is invariant under change of measurement scales. Then, exists nonnegative entries matrices $\mathbf{W}$ and $\mathbf{H}$ approximating $\mathbf{Y}$ such that [6]

$$[\mathbf{Y}]_{ij} = [\mathbf{W}]_{ik}[\mathbf{H}]_{kj} \quad \left([\mathbf{W}]_{ik} \in \mathbb{R}_+^{m \times K}, \, [\mathbf{H}]_{kj} \in \mathbb{R}_+^{n \times K}, \, k \in \mathbb{Z}_+\right) \tag{6}$$

minimizes some norm or divergence (formally, for vectors $\mathbf{a}$, $\mathbf{b}$, and $\mathbf{c}$, a divergence D is a map that not satisfies one of the distance axioms (i) $d(\mathbf{a}, \mathbf{b}) = d(\mathbf{b}, \mathbf{a})$

(symmetry); (ii) $d(\mathbf{a}, \mathbf{b}) = 0$ iff $\mathbf{a} = \mathbf{b}$ (identity); and (iii) $d(\mathbf{a}, \mathbf{b}) \leq d(\mathbf{a}, \mathbf{c}) + d(\mathbf{c}, \mathbf{b})$ (triangular inequality), usually symmetry).

Imposing the normalization conditions $\tilde{y}_{ij} = y_{ij} \sum_i y_{ij}$ on the columns of $\mathbf{Y}$ and $\mathbf{W}$, and the rows of $\mathbf{H}$, such that $\|\mathbf{y}\|_j = \|\mathbf{w}\|_k = \|\mathbf{h}\|_k = 1$, the product

$$[\widehat{\mathbf{Y}}]_{ij} = \frac{1}{m}[\widetilde{\mathbf{Y}}]_{ij} \tag{7}$$

$$= \frac{1}{mk}[\widetilde{\mathbf{W}}]_{ik}[\widetilde{\mathbf{H}}]_{kj} \tag{8}$$

is equivalent to the mixture of probabilities $P(\mathbf{w}|\, k = k')P(\mathbf{h}|\, k = k')$ [7].

The construction of a NMF kernel assumes a learned base $\mathbf{H}_\Phi$ for a function $\Phi(\cdot) : \mathbf{Y} \to \Phi(\mathbf{Y})$, hence [26]

$$\Phi(\mathbf{Y})\Phi(\mathbf{Y})' = \left(\mathbf{W}\mathbf{H}_\Phi\right)\left(\mathbf{W}\mathbf{H}_\Phi\right)' \tag{9}$$

$$= \mathbf{W}\mathbf{W}' \tag{10}$$

and

$$K(\mathbf{y}, \mathbf{y}) = \langle \mathbf{w}, \mathbf{w} \rangle \tag{11}$$

is the NMF kernel.

3   Non Parametric General Solutions

Bregman divergence is [4]

$$D_\phi(\mathbf{Y}\|\widehat{\mathbf{Y}}) = f(\mathbf{Y}) - f(\widehat{\mathbf{Y}}) - \langle \nabla f(\widehat{\mathbf{Y}}), \mathbf{Y} - \widehat{\mathbf{Y}} \rangle \tag{12}$$

and is equivalent to other norms and divergences, in the cases

$$D_\phi(\mathbf{Y}\|\widehat{\mathbf{Y}}) \sim d_{L2}(\mathbf{Y}, \widehat{\mathbf{Y}})$$
$$\sim \|\mathbf{y} - \widehat{\mathbf{y}}\|^2 \qquad (\text{if } f(\mathbf{y}) = \langle \mathbf{y}, \mathbf{y} \rangle) \tag{13}$$
$$D_\phi(\mathbf{Y}\|\widehat{\mathbf{Y}}) \sim D_{KL}(\mathbf{Y}\|\widehat{\mathbf{Y}}) \qquad \left(f(\mathbf{y}) = \sum \mathbf{y} \log \mathbf{y}\right) \tag{14}$$

A more complete table of equivalences is provided in the same paper.

In both cases it can be shown that the approximation error is arbitrarily small if $\widehat{\mathbf{Y}} \to \mathbf{Y}$ holds in probability, which occurs if $k \geq \min(m, n)$ [9].

Heuristically, it can be justified by considering the set of labels $\mathcal{Y} = (l_1, \ldots, l_p)$ which splits as $\mathcal{Y} = \{\mathcal{Y}_-, \mathcal{Y}_+\}$. If the available observations are represented by $\mathbf{Y}_t$, and adding a single new observation gives $\mathbf{Y}_n$. Both have ratios α and $(1 - \alpha)$. Furthermore, assuming the result of this new observation is known, which is written as a single-row matrix, we have

$$\bigcup_i \{\mathbf{Y}_t, \mathbf{Y}_n\} = \alpha \widehat{\mathbf{Y}}_t, (1 - \alpha) \widehat{\mathbf{Y}}_n \tag{15}$$

where the matrices with the symbol *hat* are their limit states. In this case, the equality between the matrices provides a correct classification of the first new observation. Repeating this process for all new observations, a partition into two classes is obtained. Proceeding in the same way for $\mathbf{Y}_+ = (l_2, \ldots, l_p)$ the result is obtained.

The general introduction of the Fisher kernel, for norms and divergences, gives the relation

$$I_F = E\Big[\frac{\partial^2}{\partial \theta^2} \log l([\widehat{\mathbf{Y}}]_{ij}|\theta)\Big] \tag{16}$$

$$= E(\mathcal{H}) \tag{17}$$

being $\mathcal{H}$ the Hessian of the likelihood. The introduction of a monotonically decreasing function J of the differences (or quotient) of entropies, with parameters θ and ϕ of the same class of densities, and reasoning as the classical work of Rao [20], and using the Jensen's difference, defined as $J(\theta, \phi) = H(\theta, \phi) - \lambda H(\theta) - \mu H(\phi)$, where H is an entropy, and λ y μ scalars such that $\lambda + \mu = 1$), and assuming that the parameter space is a sufficiently differentiable manifold, the expansion

$$J(\theta, \theta + d\theta) = J(\theta, \phi = \theta) + \frac{\partial}{\partial \theta} J(\theta, \phi = \theta) + \frac{1}{2!}\frac{\partial^2}{\partial \theta_i \partial \theta_j} J(\theta, \phi = \theta) + \cdots \tag{18}$$

vanishes the first two terms, if $\phi = \theta$, and

$$J(\theta, \theta + d\theta) \approx \frac{1}{2}\frac{\partial^2}{\partial \theta_i \partial \theta_j} J(\theta, \phi = \theta) \tag{19}$$

$$= g_{ij} \tag{20}$$

g_{ij} is the well-known geodesic distance.

For the case of the gaussian norm, g_{ij}, taking derivatives

$$g_{ij} = \frac{\partial^2}{\partial \mathbf{W} \partial \mathbf{H}} - \frac{1}{2}\|\mathbf{Y} - \mathbf{W}\mathbf{H}\|^2$$

$$= \mathbf{W}\mathbf{W}' \quad \Big([\mathbf{W}]_{ik} \leftarrow [\mathbf{W}]_{ik} \odot \Big(\frac{[\mathbf{Y}]_{ij}}{[\mathbf{W}\,\mathbf{H}]_{ij}}\,[\mathbf{H}]'_{kj}\Big)\Big) \tag{21}$$

and if the matrix $\mathbf{W}$ is orthogonal, provides the identity matrix. Within parenthesis, we indicate the solution of (6) with the suitable objective function ($L2$ norm). The kernel is

$$K(\mathbf{y}_i, \mathbf{y}'_i) = \langle \mathbf{w}, \mathbf{w}'_i \rangle \tag{22}$$

For the KL divergence

$$g_{ij} = \frac{\partial^2}{\partial \mathbf{W} \partial \mathbf{H}} \mathbf{Y} \log \frac{\mathbf{Y}}{\mathbf{W}\mathbf{H}}$$

$$= \mathbf{I} \tag{23}$$

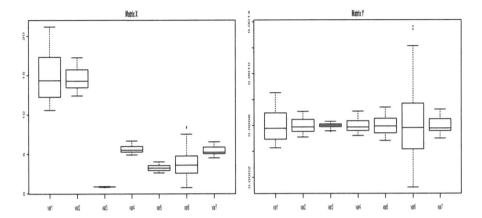

Fig. 1. Effect of transformation $\mathbb{R}_+ \to \mathbb{R}_{[0,1]}$ according (5). After the invariant change of scales, the most informative variables (those which has more variance) are not necessary the same.

In both cases, the kernel is given by (11), plug-in the obtained expression for W. Also it is immediate to see that the plug-inn is accomplished for I-divergence, defined as

$$D_{KL}(\mathbf{Y} \| \mathbf{W\,H}) = \sum_{ij} \left([\mathbf{Y}]_{ij} \log \frac{[\mathbf{Y}]_{ij}}{[\mathbf{W\,H}]_{ij}} - [\mathbf{Y}]_{ij} + [\mathbf{W\,H}]_{ij} \right) \qquad (24)$$

4   Examples

We illustrate the behavior of the Fisher Kernel with three data sets from the *UCI repository* [8]. We select attending non-presence of missing and suitability to the classification problem, which implies the existence of a categorical variable we identify with the labels $\mathcal{Y}$ with two or more levels. The data sets are *seed* (210×7), and *glass* (2146×10), all with no-negative real entries. The goal of the examples is to categorize new data, after training a piece of the given set (80% of the available data).

The transformation of non-negative real values to $[0,1]$ has effects on the information contained by the variables (variance), depending on its level (mean). Figure 1 illustrates this effect for the data set *seeds*. Also this transformation seems to have no effect on correlation matrices. Under the current state of the art it is an open question.

To obtain the results of Fig. 2, transformation (5) must first be performed. Then, depending on the criteria adopted, the Kernel is chosen. The non-parametric case assumes the simultaneous choice of distance/divergence and a value of K. The iterative process between the Formulas that define $\mathbf{W}$ and $\mathbf{H}$, and recalculating in each iteration the value of $\widehat{\mathbf{Y}}$, until obtaining a satisfactory approximation, gives the learned base $\mathbf{H}_\Phi$. The training phase uses the RLDA

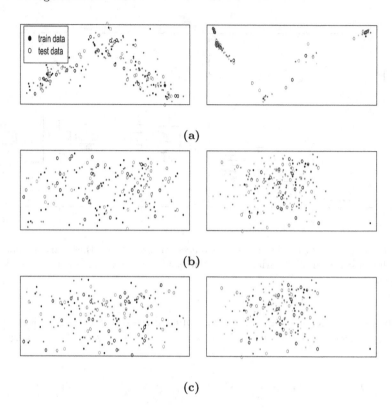

Fig. 2. (a) Shows the Gaussian parametric Fisher Kernel (left for dataset *seeds*, and right dataset *glass*. (b) Nonparametric Gaussian norm kernel, and (c) Nonparametric Kullback-Leibler kernel

(Rotational Linear Discriminant Analysis). A description of this methodology can be found in [23], and implemented in the *kfda* $\mathcal{R}$ package [16]. In essence is an iterative procedure with parameters μ (means of the data), and a rotation matrix $\mathbf{Q}$, randomly generated before the first iteration. Quality results are compared using several metrics in Table 1. They are (*i*) the confusion matrix, (*ii*), and (*iii*) F1-score.

K is also referred as the model components. In clustering applications, they are identified with [1, p. 22], and in the case of using these techniques in the information retrieval problem, with the latent variables [7]. In the case of using the KL divergence, it is accepted that the adjustment of these equations is a procedure equivalent to the EM (Expectation-Maximization) algorithm. This result is more general, and it is the limiting value of this algorithm, as Amari proved [2], calling it the *em* algorithm.

Table 1. Obtained results for the Fisher kernel for data sets *seeds* and *glass*. Param. corresponds to a Gaussian multivariate kernel. L2 NP FK is a Gaussian non-parametric Fisher kernel, and KL NP FK for a Kulback-Leibler divergence.

		Precision	F1 Score	Misclasif.
Seeds	Param	96.83%	.04	1.86%
	L2 NP FK	98.09%	.02	9.05%
	KL NP FK	89.95%	.09	10.48%
Glass	Param	95.31%	.04	3.17%
	L2 NP FK	97.66%	.02	1.87%
	KL NP FK	98.13%	.02	1.87%

5   Conclusion

The non-parametric Fisher kernel versions preserve statistical properties (it is a consistent estimator for the posterior, or distribution of the data conditioned on the distributional parameter). Also, it has the asymptotic property for classification error, which decreases arbitrarily, if the number of model components is suitable ($k \geq \min(m, n)$). However, solutions based on an iterative process, equivalent to the EM algorithm, is slow and limit the applicability, mainly for large datasets, and real-time results. On the other hand, according to the current state of the art in NMF techniques, there are open questions related to the information contained in the spaces $\mathbb{R}_+^{m \times n}$ and $\mathbb{R}_{[0,1]}^{m \times k}$ corresponding to input and feature spaces.

References

1. Aggarwal, C.C., Clustering, C.R.D.: Algorithms and Applications. CRC Press Taylor and Francis Group, Boca Raton (2014)
2. Amari, S.I.: Information geometry of the EM and EM algorithms for neural networks. Neural Netw. **8**(9), 1379–1408 (1995)
3. Aronszajn, N.: Theory of reproducing kernels. Trans. Am. Math. Soc. **68**(3), 337–404 (1950)
4. Banerjee, A., Merugu, S., Dhillon, I.S., Ghosh, J., Lafferty, J.: Clustering with Bregman divergences. J. Mach. Learn. Res. **6**(10), 1705–174 (2005)
5. Chappelier, J.-C., Eckard, E.: PLSI: the true fisher kernel and beyond. In: Buntine, W., Grobelnik, M., Mladenić, D., Shawe-Taylor, J. (eds.) ECML PKDD 2009. LNCS (LNAI), vol. 5781, pp. 195–210. Springer, Heidelberg (2009). https://doi.org/10.1007/978-3-642-04180-8_30
6. Chen, J.C.: The nonnegative rank factorizations of nonnegative matrices. Linear Algebra Appl. **62**, 207–217 (1984)
7. Ding, C., Li, T., Peng, W.: On the equivalence between non-negative matrix factorization and probabilistic latent semantic indexing. Comput. Stat. Data Anal. **52**(8), 3913–3927 (2008)

8. Dua, D., Graff, C.: UCI machine learning repository. University of California, Irvine, School of Information and Computer Sciences (2017). http://archive.ics.uci.edu/ml

9. Figuera, P., García Bringas, P.: On the probabilistic latent semantic analysis generalization as the singular value decomposition probabilistic image. J. Stat. Theory Appl. **19**, 286–296 (2020)

10. Figuera, P., Bringas, P.G.: A non-parametric fisher kernel. In: Sanjurjo González, H., Pastor López, I., García Bringas, P., Quintián, H., Corchado, E. (eds.) HAIS 2021. LNCS (LNAI), vol. 12886, pp. 448–459. Springer, Cham (2021). https://doi.org/10.1007/978-3-030-86271-8_38

11. Gudovskiy, D., Hodgkinson, A., Yamaguchi, T., Tsukizawa, S.: Deep active learning for biased datasets via fisher kernel self-supervision. In: Proceedings of the IEEE/CVF Conference on Computer Vision and Pattern Recognition, pp. 9041–9049 (2020)

12. Hofmann, T.: Learning the similarity of documents: an information-geometric approach to document retrieval and categorization. In: Advances in Neural Information Processing Systems, pp. 914–920 (2000)

13. Hofmann, T., Schölkopf, B., Smola, A.J.: Kernel methods in machine learning. Ann. Stat. **36**, 1171–1220 (2008)

14. Jaakkola, T., Diekhans, M., Haussler, D.: A discriminative framework for detecting remote protein homologies. J. Comput. Biol. **7**(1–2), 95–114 (2000)

15. Jaakkola, T.S., Haussler, D., et al.: Exploiting generative models in discriminative classifiers. In: Advances in Neural Information Processing Systems, pp. 487–493 (1999)

16. Kim, D.: KFDA: kernel fisher discriminant analysis (2017). https://CRAN.R-project.org/package=kfda, R package version 1.0.0

17. Langrené, N., Warin, X.: Fast and stable multivariate kernel density estimation by fast sum updating. J. Comput. Graph. Stat. **28**(3), 596–608 (2019)

18. Lee, H., Cichocki, A., Choi, S.: Kernel nonnegative matrix factorization for spectral EEG feature extraction. Neurocomputing **72**(13–15), 3182–3190 (2009)

19. Mercer, J.: XVI. Functions of positive and negative type, and their connection the theory of integral equations. Philos. Trans. R. Soc. Lond. Ser. A Containing Papers Math. Phys. Character **209**(441–458), 415–446 (1909)

20. Rao, C.R.: Differential metrics in probability spaces. Diff. Geom. Stat. Inference **10**, 217–240 (1987)

21. Salazar, D., Rios, J., Aceros, S., Flórez-Vargas, O., Valencia, C.: Kernel joint nonnegative matrix factorization for genomic data. IEEE Access **9**, 101863–101875 (2021)

22. Scholkopft, B., Mullert, K.R.: Fisher discriminant analysis with kernels. Neural Netw. Signal Process. IX **1**(1), 1 (1999)

23. Sharma, A., Paliwal, K.K.: Rotational linear discriminant analysis technique for dimensionality reduction. IEEE Trans. Knowl. Data Eng. **20**(10), 1336–1347 (2008)

24. Tsuda, K., Akaho, S., Kawanabe, M., Müller, K.R.: Asymptotic properties of the fisher kernel. Neural Comput. **16**(1), 115–137 (2004)

25. Vapnik, V.N.: An overview of statistical learning theory. IEEE Trans. Neural Netw. **10**(5), 988–999 (1999)

26. Zhang, D., Zhou, Z.-H., Chen, S.: Non-negative matrix factorization on kernels. In: Yang, Q., Webb, G. (eds.) PRICAI 2006. LNCS (LNAI), vol. 4099, pp. 404–412. Springer, Heidelberg (2006). https://doi.org/10.1007/978-3-540-36668-3_44

27. Zhou, Y., Shi, J., Zhu, J.: Nonparametric score estimators. In: International Conference on Machine Learning, pp. 11513–11522. PMLR (2020)

A HAIS Approach to Predict the Energy Produced by a Solar Panel

Ángel Arroyo[1]([✉])[iD], Hector Quintian[2][iD], Jose Luis Calvo-Rolle[2][iD],
Nuño Basurto[1][iD], and Álvaro Herrero[1][iD]

[1] Grupo de Inteligencia Computacional Aplicada (GICAP),
Departamento de Ingeniería Informática, Escuela Politécnica Superior,
Universidad de Burgos, Av. Cantabria s/n, 09006 Burgos, Spain
{aarroyop,nbasurto,ahcosio}@ubu.es
[2] Department of Industrial Engineering, University of A Coruña, CTC,
CITIC, Avda. 19 de febrero s/n, 15405 Ferrol, A Coruña, Spain
{hector.quintian,jlcalvo}@udc.es

Abstract. The current energy crisis coupled with concerns about climate change make renewable energies a priority. Solar energy being one of the most representative in Spain, being able to predict the behavior of a solar energy panel in the short term can be very useful to determine the coverage of this energy. To achieve this goal, four regression techniques in combination with a clustering algorithm have been applied, with the aim of predicting the solar energy generation of a panel during the year 2011 in the autonomous region of Galicia (Spain). A data set with continuous information has been used and it has been possible to validate the best way to combine regression and clustering techniques to achieve the best prediction.

Keywords: Regression · Neural networks · Solar energy · Renewable energies · Clustering

1 Introduction

Environmental care is currently not only a trend, but it is also an important issue for society and governments. Moreover, for obvious reasons, there is a clear trend, in which it is necessary to ensure care for the environment. In point of fact, no impact is very difficult or impossible. But nevertheless, aspects such as sustainability and the maximum possible reduction in environmental impact are very important [1]. In this sense, in terms of energy needs, renewable energies play a key role in contributing to a reduction in environment impact and emissions [2]. However, the impact of the power-plant implementation itself based on renewable sources has to be taken into account there is not usually any zero impact [3].

Due to it not being possible to achieve the null impact, even with the alternatives and use of renewable energies, there is a legal obligation to optimize and

© Springer Nature Switzerland AG 2022
P. García Bringas et al. (Eds.): HAIS 2022, LNAI 13469, pp. 195–207, 2022.
https://doi.org/10.1007/978-3-031-15471-3_18

plan installations with maximum efficiency [4]. Furthermore, the efficiency of the facilities must be measure in accordance with right ratios and criteria with the aim of ensuring the desired minimum impact [5].

Furthermore, when renewable energies or other possible sources are included in all types of buildings, even if the final use is not to generate energy, the management becomes much more difficult. For this reason, it is necessary to create tools to ensure the right handling of the different energy generation and consumption points.

Therefore, the Smartgrid concept [6], comes about for all of the aforementioned reasons, where it is necessary at least to measure the generation, the consumption, and of course, try to predict both of them, with the aim of taking decisions, and then to make the overall system more efficient in every way. In any case, it is a difficult task to match the generation with the demand regardless of the type of energy involved, always trying to buy the minimum amount of energy.

Given the current trend of the electric sector, and with the aim of optimizing the overall efficiency of the buildings and facilities, it is mandatory to have a trustworthy forecast to take the right decisions [7]. This prediction will contribute, for instance, to purchasing energy at the best price when necessary, store energy when it is more convenient, and so on.

During the modeling process, it is possible to take several different alternatives into account although the performance could be different. One of the reasons why the performance is not suitable is the non-linearity of the problem to be modeled. Intelligent systems are used in some different applications with very satisfactory results in general terms. Of course the non-linearity problem could be solved in many cases with the use of soft-computing techniques [8–12]. The problem when the system is non-linear could be the same, even by using simple intelligent systems. When it occurs, it is possible to divide the problem with clustering techniques such as K-means [13–19].

In this work the solar thermal facility of a bioclimatic house is modeled from several variables measured at the real building. The objective is to regress a dataset with information on the solar energy produced by a solar panel. To achieve this regression and to be able to predict this energy produced, several statistical techniques and Artificial Neural Networks (ANN) models have been applied, in combination with the well-known k-means clustering technique. A Hybrid Artificial Intelligent System (HAIS) has already been developed with the same objective of predicting the solar energy produced in [20]. In this paper, new ANN techniques are applied that offer better results and the combination of k-means with regression techniques is optimised, improving the results considerably.

Artificial Intelligence (AI) techniques have been applied in the past, as in [21] where one month's data from a solar collector is analysed using Phase Change Material. In [22] the author proposes the application of ANN to predict the three components of solar radiation: global horizontal, beam normal, and diffuse horizontal. None of the reviewed works cover in such an exhaustive way regression techniques with clustering with the objective of minimising the Mean Square Error (MSE) in the prediction of the solar energy produced.

The present research work is organized as follows. After the present introduction, the case under study is explained. Then, the applied techniques taken into account for the present research are detailed. The work is continued with the presentation of the results and their discussion. Finally, conclusions and future works are presented.

2   Case of Study

The proposal has been applied over the solar thermal facility, that is part of the systems installed within a real bioclimatic house. The physical system is described briefly bellow.

2.1   Sotavento Bioclimatic House

Sotavento bioclimatic house is a project of a demonstrative bioclimatic house of Sotavento Galicia Foundation (Fig. 1). The house is located within the Sotavento Experimental Wind Farm, which is a center of dissemination of renewable energy and energy saving. The farm is located between the councils of Xermade (Lugo) and Monfero (A Coruña), in the autonomous community of Galicia (Spain).

Fig. 1. Bioclimatic house

2.2   Bioclimatic House Facilities

The thermal and electrical facilities of the bioclimatic house have some renewable energy systems to complement these installations. Figure 2 schema describes the systems and components related with the thermal issues. The thermal installation includes 3 renewable energy sources (solar, biomass and geothermal) that supply the DHW (Domestic Hot Water) and the heating system. The electrical installation have two renewable energy sources (wind and photovoltaic) and one connection to the grid power, feeding the house lighting and power systems.

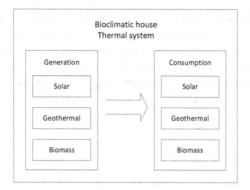

Fig. 2. Thermal facilities scheme

2.3   Solar Thermal System

Figure 3 shows the house solar thermal system, from the solar energy collection phase to the accumulation phase. The solid red line represents the part of the hydraulic circuit that flows hot liquid, while the dashed blue line represents the cold fluid. The collection zone of solar thermal system consists of solar panels (with an inclination of 19° in the North facade) that communicate with a solar accumulator through an ethyleneglycol closed hydraulic circuit.

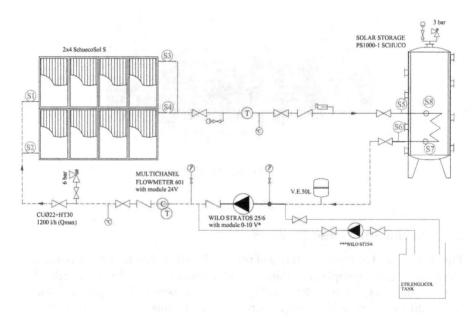

Fig. 3. Solar thermal system scheme

The hydraulic pump boosts the ethyleneglycol to solar panels inputs S1 and S2, which are arranged in a structure of two parallel blocks with four panels in series per block. The ethyleneglycol circulates inside the solar panels, absorbs the heat and sends it out through S3 and S4 with a temperature higher than the input. From there, the fluid is driven to the entrance of the parallel plate heat exchanger of the solar accumulator (S8), where the energy stored in the ethyleneglycol heats the water stored inside the solar accumulator. Once the fluid leaves the accumulator (S6), the hydraulic circuit is closed to send the fluid back to solar panels, boosting it with a circulation pump.

3   Techniques Applied

In order to analyse the data described in Sect. 3, four regression techniques (a linear statistic technique and three Artificial Neural Networks (ANN) techniques) combined with a clustering technique are applied.

3.1   Statistical Regression Techniques

Statistical regression techniques establish the relationship among different attributes in a dataset, by fixing a linear or non-linear (linear in this case study) equation linked to the input attributes.

Multiple Linear Regression. Linear Regression (LR) attempts to model the relationship between two or more explanatory variables and a response variable by fitting a linear equation to the dataset [23]. Every value of the predictor variable (x) is associated with a value of the criterions variables (y). The regression line for p explanatory variables (x_1, x_2, ..., x_n) is defined as follows:

$$U_y = \beta_0 + \beta_1 x_1 + \beta_2 x_2 + ... + \beta_n x_n \tag{1}$$

The fitted values b_0, b_1, ..., b_n estimate the parameters β_0, β_1, ...,β_p of the population regression line. The model is expressed as DATA = FIT + RESIDUAL, where the "FIT" term represents the expression $\beta_0 + \beta_1 x_1 + \beta_2 x_2 + ... + \beta_n x_n$. The "RESIDUAL" term represents the deviations of the observed values (y) from their means U_y, which are normally distributed with mean 0 and variance σ.

3.2   Artificial Neural Networks

The Artificial Neural Networks (ANN) are simplified models of natural neural systems. In this research work, the following ones were applied:

Extreme Learning Machine. Extreme Learning Machine (ELM) tries to overcomes some challenges faced by other techniques. ELM works for generalized Single-hidden Layer Feedforward Networks (SLFNs). The ELM concept is that the hidden layer of SLFNs need not be adjusted. Compared with other regression ANN techniques, ELM provides better generalization performance at a much faster learning speed and with least parametric setting [24]. ELM can be summed up as follows: the hidden layer of ELM need not be iteratively adjusted [25]; According to feedforward neural network theory [26], the training error and the norm of weights need to be minimized [27] and finally, the hidden layer feature mapping need to satisfy the universal approximation condition [27].

Support Vector Machine. A Support Vector Machine (SVM) [28] is a machine learning tool for classification and regression [29]. SVM regression is considered a nonparametric technique because it relies on kernel functions. In SVM regression, the set of training data includes predictor variables and observed response values. The goal is to find a function f(x) that deviates from y_n by a value no greater than ϵ for each training point x, and at the same time is as flat as possible.

Trilayered Neural Network. ANN is a commonly used Machine Learning technique that produces accurate results in regression tasks [30]. From the various types of ANNs, backpropagation ANN is applied in this study. The ANN structure consist of one or mores input(s), hidden layer and output. The ANN model is a feedforward, fully connected neural network for regression. The first fully connected layer of the neural network has a connection from the network input (predictor data), and each subsequent layer has a connection from the previous layer. Each fully connected layer multiplies the input by a weight matrix and then adds a bias vector [31]. An activation function follows each fully connected layer, excluding the last. The final fully connected layer produces the network's output, namely predicted response values. Among the different types of ANNs available with these characteristics, Trilayered Neural Network (TNN) has offered the best results in terms of Mean Square Error (MSE). TNN is a high model flexibility which increases with the first layer size, second layer size, and third layer size settings.

3.3 Clustering Technique

Cluster analysis [32] organizes data by grouping data samples according to a criterion distance. Two individuals in a valid group will be much more similar than those in distinct groups.

K-means. The k-means clustering algorithm [33], groups data samples into a predefined number of groups. It requires two input parameters: the number of clusters (k) and their initial centroids. Initially, each data sample is assigned to the cluster with the closest centroid. Once the clusters are defined, the centroids are recalculated and samples are reassigned. The algorithm to minimize the Sum Square Error (SSE) it can be defined:

$$SSE = \sum_{j=1}^{k} \sum_{x \epsilon G_i} p(x_i, c_j)/n \qquad (2)$$

where, k is the number of groups, p is the proximity function, c_j is the centroid of group j, and n is the number of samples. The Cityblock distance is the distance measure applied where each centroid is placed in the component-wise median of all the samples in the group. The distance from point x to each of the centroids is calculated as:

$$d_{st} = \sum_{j=1}^{n} |x_{sj} - x_{tj}| \qquad (3)$$

where j is an instance of the vector j.

4   Results and Discussion

The set of techniques and methods previously described are applied to the Solar Energy case study presented in Sect. 3. To achieve more accurate results, they are validated by the n-fold cross-validation [34] iterative process. Data partitions has been set to the value of 10 for all the experiments.

The regression process was carried out on the thermal power generated attribute of the panel solar. This attribute is set as the predictor attribute and the other six attributes are the criterion ones. Table 1 shows the distribution of samples for each of the twelve months.

Table 1. Number of samples for each of the three clusters per month.

Month	C1	C2	C3
1	3017	200	93
2	2777	296	234
3	2561	314	470
4	2586	93	49
5	1809	591	618
6	1939	565	555
7	2014	465	521
8	2002	501	496
9	1985	438	405
10	1378	339	377
11	1753	168	42
12	2653	306	35

For the twelve months in Table 1, one of the clusters groups the majority of the samples, while the other two are generally fairly balanced. The four regression techniques were applied to each of the clusters, yielding the results shown in Table 2. For each of the methods, the results produced are: mean execution time of the applied technique, Standard Deviation (Std) of this value, the mean MSE returned by the method and the Std of the MSE.

Table 2. Results for the three data clusters.

Tech	Time	Std Time	MSE	Std MSE	C
LR	1.49E−01	2.80E−01	1.17E−05	5.94E−08	C1
LR	3.78E−02	3.43E−03	1.18E−05	1.37E−07	C2
LR	2.52E−02	1.58E−03	2.89E−05	1.95E−07	C3
ELM	4.43E−02	1.03E−01	**7.15E−09**	2.23E−10	C1
ELM	1.80E−02	2.55E−03	1.76E−07	1.23E−08	C2
ELM	**8.33E−03**	1.46E−03	7.95E−07	4.56E−09	C3
SVM	5.85E−01	5.30E−01	5.00E−06	7.24E−07	C1
SVM	8.84E−01	7.29E−02	1.37E−05	1.32E−06	C2
SVM	6.01E−01	5.85E−02	1.48E−05	2.24E−06	C3
TNN	2.25	7.94E−01	5.08E−06	3.07E−06	C1
TNN	1.11E+01	4.14	2.28E−06	4.76E−07	C2
TNN	2.89	5.26E−01	1.18E−05	1.83E−06	C3

The best results (marked in bold) in Table 2, both in terms of runtime and minimum MSE corresponds to the ELM method. This one offers the best results and especially for clusters C1 and C3, which contain the least number of samples. SVM and TNN algorithms offer similar results in the approach of the minimum MSE, with LR being the worst performer.

Subsequently, the k-means technique was applied again, but in this step for each of the twelve months of 2011 year year.

Table 3 shows the MSE regression results of the four proposed techniques on the 36 datasets of Table 1.

Table 3 shows how the minimum MSE is clearly obtained by the ELM method, with the C2 in may standing out especially. The other three methods obtain very similar results, with TNN being perhaps the second best method and LR the worst of the four. Comparing these results with those obtained in Table 2, it can be seen that the grouping by months gives lower values in the calculation of the MSE, especially in the case of ELM.

Regarding the run times shown in Table 4, the fastest methods in calculating the MSE are LR and ELM, with the absolute best result being LR for C3 in the month of September. The slowest method is clearly TNN.

Table 3. MSE results per month

Month	Cluster	LR	ELM	SVM	TNN
1	1	1.82E−04	8.32E−07	5.26E−03	5.22E−05
1	2	6.20E−04	9.04E−06	2.55E−04	6.83E−04
1	3	8.85E−05	1.66E−07	7.51E−05	3.29E−04
2	1	1.84E−04	9.46E−07	5.20E−03	6.08E−05
2	2	4.10E−04	7.68E−06	1.77E−04	1.98E−04
2	3	1.53E−04	5.21E−08	5.73E−05	1.67E−05
3	1	2.10E−04	7.81E−07	1.91E−04	3.75E−05
3	2	3.74E−04	5.55E−06	2.08E−04	4.50E−04
3	3	8.88E−05	1.76E−08	1.61E−05	5.51E−06
4	1	2.36E−05	9.72E−08	1.67E−05	1.88E−06
4	2	1.42E−04	**3.59E−10**	2.57E−05	1.82E−05
4	3	1.93E−03	6.88E−06	6.63E−04	2.85E−04
5	1	2.57E−04	2.63E−06	1.33E−04	6.42E−05
5	2	2.10E−04	4.48E−06	1.25E−04	9.23E−05
5	3	7.15E−05	3.98E−08	2.83E−05	2.87E−05
6	1	2.11E−04	4.00E−06	1.79E−04	7.63E−05
6	2	2.05E−04	4.99E−06	1.36E−04	1.25E−04
6	3	8.31E−05	2.79E−08	3.43E−05	1.55E−05
7	1	2.38E−04	7.19E−06	2.60E−04	5.53E−05
7	2	2.68E−04	7.17E−06	1.52E−04	1.63E−04
7	3	7.64E−05	2.87E−08	2.61E−05	2.37E−05
8	1	2.59E−04	4.93E−06	2.60E−04	6.36E−05
8	2	2.56E−04	5.40E−06	1.05E−04	1.01E−04
8	3	8.21E−05	3.93E−08	2.80E−05	2.33E−05
9	1	1.98E−04	3.56E−06	1.60E−04	4.91E−05
9	2	2.98E−04	6.35E−06	1.40E−04	2.68E−04
9	3	1.24E−04	9.84E−08	6.40E−05	2.78E−05
10	1	3.85E−04	1.60E−06	4.85E−03	1.88E−04
10	2	3.35E−04	6.26E−06	1.40E−04	1.95E−04
10	3	1.01E−04	1.19E−07	1.51E−05	2.07E−05
11	1	3.23E−04	8.41E−07	3.88E−03	1.27E−04
11	2	6.41E−04	7.92E−06	3.95E−04	8.92E−04
11	3	3.72E−04	2.41E−07	1.14E−04	3.92E−04
12	1	2.36E−04	7.50E−07	3.99E−03	7.90E−05
12	2	3.98E−04	6.60E−06	1.56E−04	1.69E−04
12	3	1.85E−04	1.37E−06	8.68E−05	5.57E−05

Table 4. Time results per month

Month	Cluster	LR	ELM	SVM	TNN
1	1	1.22E−01	2.03E−02	3.18E+01	3.12
1	2	3.08E−03	8.30E−03	2.26E−02	7.32E−01
1	3	1.99E−03	6.72E−03	2.17E−02	5.92E−01
2	1	2.76E−03	8.04E−03	6.90E−02	2.65
2	2	8.06E−04	7.22E−03	2.29E−02	8.24E−01
2	3	3.52E−03	6.63E−02	2.21E−02	8.33E−01
3	1	3.76E−03	7.98E−03	5.99E−02	2.59
3	2	1.05E−03	6.82E−03	2.35E−02	7.36E−01
3	3	2.29E−03	7.29E−03	2.71E−02	1.28
4	1	9.86E−04	8.07E−03	5.81E−02	1.30
4	2	6.13E−03	1.87E−01	2.28E−02	3.72E−01
4	3	3.51E−03	7.61E−03	2.31E−02	7.21E−01
5	1	1.70E−03	8.29E−03	8.61E−02	2.33
5	2	1.41E−03	7.11E−03	3.27E−02	1.28
5	3	6.84E−04	7.19E−03	3.11E−02	1.40
6	1	1.29E−03	7.71E−03	4.16E−02	2.02
6	2	7.86E−04	7.14E−03	3.01E−02	1.31
6	3	7.31E−04	7.09E−03	2.80E−02	1.28
7	1	8.31E−04	7.44E−03	4.30E−02	2.02
7	2	1.21E−03	7.17E−03	2.62E−02	1.13
7	3	8.39E−04	7.07E−03	2.89E−02	1.21
8	1	5.97E−04	7.35E−03	4.15E−02	1.91
8	2	7.67E−04	6.87E−03	2.70E−02	1.12
8	3	6.72E−04	6.88E−03	2.59E−02	1.18
9	1	1.05E−03	7.44E−03	4.22E−02	1.90
9	2	1.73E−03	6.99E−03	2.57E−02	1.14
9	3	**3.97E−04**	6.97E−03	2.32E−02	1.13
10	1	2.42E−03	7.44E−03	6.75	1.62
10	2	5.87E−03	6.92E−03	2.11E−02	8.22E−01
10	3	2.99E−03	7.20E−03	2.22E−02	1.16
11	1	4.65E−04	7.13E−03	3.97E−02	1.93
11	2	9.14E−03	6.78E−03	2.28E−02	7.61E−01
11	3	1.09E−02	8.60E−03	3.63E−02	4.42E−01
12	1	1.60E−03	8.05E−03	4.49	2.56
12	2	7.53E−03	8.22E−03	2.40E−02	9.38E−01
12	3	6.30E−03	7.64E−03	1.93E−02	3.41E−01

It can be concluded that the ELM algorithm has proved to be the most optimal in the task of approximating the regression on the solar energy panel previously discussed, both when measuring the MSE and when considering the runtime execution which also achieves the best results in best cases. This fact makes the ELM algorithm to be considered as the most appropriate one in a possible future work on MV imputation.

5   Conclusions and Future Work

In this work the validity of four regression techniques in combination with a clustering technique were tested in order to predict the power generated in a solar panel.

Data from the year 2011 were used, in a first step the regression was performed over three data clusters and in a second over 36 clusters, 3 for each of the twelve months. Running the regression on each of the months significantly improves the results of running it over the whole year, allowing us to highlight some months, such as May, in which the results are particularly good, perhaps due to environmental conditions that do not change much. A remarkable fact is that regression techniques usually offer better results on clusters with few components, probably due to the high similarity of their samples.

Extreme Learning Machine stands out as the method with the best results, both in the calculation of the Mean Square Error and in the speed of execution time.

References

1. Kuwae, T., Hori, M. (eds.): Blue Carbon in Shallow Coastal Ecosystems. Carbon Dynamics, Policy, and Implementation, Springer, Singapore (2019). https://doi.org/10.1007/978-981-13-1295-3
2. Karunathilake, H., Hewage, K., Mérida, W., Sadiq, R.: Renewable energy selection for net-zero energy communities: life cycle based decision making under uncertainty. Renew. Energy **130**, 558–573 (2019)
3. Prakash, R., Bhat, I.K., et al.: Energy, economics and environmental impacts of renewable energy systems. Renew. Sustain. Energy Rev. **13**(9), 2716–2721 (2009)
4. Wei, M., Patadia, S., Kammen, D.M.: Putting renewables and energy efficiency to work: how many jobs can the clean energy industry generate in the US? Energy Policy **38**(2), 919–931 (2010)
5. Giacone, E., Mancò, S.: Energy efficiency measurement in industrial processes. Energy **38**(1), 331–345 (2012)
6. Amin, M.: Smart grid. Public Utilities Fortnightly (2015)
7. Potter, C.W., Archambault, A., Westrick, K.: Building a smarter smart grid through better renewable energy information. In: Power Systems Conference and Exposition, PSCE 2009, pp. 1–5. IEEE/PES, IEEE (2009)
8. Fontenla-Romero, O., Calvo-Rolle, J.L.: Artificial intelligence in engineering: past, present and future. DYNA **93**(4), 350–352 (2018)

9. Jove, E., et al.: Hybrid intelligent model to predict the remifentanil infusion rate in patients under general anesthesia. Logic J. IGPL **29**(2), 193–206 (2020). https://doi.org/10.1093/jigpal/jzaa046

10. Jove, E., Casteleiro-Roca, J.-L., Quintián, H., Méndez-Pérez, J.-A., Calvo-Rolle, J.L.: A new method for anomaly detection based on non-convex boundaries with random two-dimensional projections. Inf. Fusion **65**, 50–57 (2021). https://doi.org/10.1016/j.inffus.2020.08.011

11. García-Ordás, M.T., et al.: Clustering techniques selection for a hybrid regression model: a case study based on a solar thermal system. Cybern. Syst. 1–20 (2022). https://doi.org/10.1080/01969722.2022.2030006

12. Jove, E., et al.: Comparative study of one-class based anomaly detection techniques for a bicomponent mixing machine monitoring. Cybern. Syst. **51**(7), 649–667 (2020). https://doi.org/10.1080/01969722.2020.1798641

13. Casteleiro-Roca, J.-L., Calvo-Rolle, J.L., Méndez Pérez, J.A., Roqueñí Gutiérrez, N., de Cos Juez, F.J.: Hybrid intelligent system to perform fault detection on BIS sensor during surgeries. Sensors **17**(1), 179 (2017)

14. Quintián, H., Calvo-Rolle, J.L., Corchado, E.: A hybrid regression system based on local models for solar energy prediction. Informatica **25**(2), 265–282 (2014)

15. Casado-Vara, R., et al.: Edge computing and adaptive fault-tolerant tracking control algorithm for smart buildings: a case study. Cybern. Syst. **51**(7), 685–697 (2020). https://doi.org/10.1080/01969722.2020.1798643

16. Leira, A., et al.: One-class-based intelligent classifier for detecting anomalous situations during the anesthetic process. Logic J. IGPL **30**(2), 326–341 (2022). https://doi.org/10.1093/jigpal/jzaa065

17. Fernandez-Serantes, L.A., Casteleiro-Roca, J.L., Calvo-Rolle, J.L.: Hybrid intelligent system for a half-bridge converter control and soft switching ensurement. Revista Iberoamericana de Automática e Informática (2022). https://doi.org/10.4995/riai.2022.16656

18. Gonzalez-Cava, J.M., et al.: Machine learning techniques for computer-based decision systems in the operating theatre: application to analgesia delivery. Logic J. IGPL **29**(2), 236–250 (2020). https://doi.org/10.1093/jigpal/jzaa049

19. Fernandez-Serantes, L.A., Casteleiro-Roca, J.-L., Berger, H., Calvo-Rolle, J.-L.: Hybrid intelligent system for a synchronous rectifier converter control and soft switching ensurement. Eng. Sci. Technol. Int. J. 101189 (2022)

20. Basurto, N., Arroyo, Á., Vega, R., Quintián, H., Calvo-Rolle, J.L., Herrero, Á.: A hybrid intelligent system to forecast solar energy production. Comput. Electr. Eng. **78**, 373–387 (2019)

21. Varol, Y., Koca, A., Oztop, H.F., Avci, E.: Forecasting of thermal energy storage performance of phase change material in a solar collector using soft computing techniques. Expert Syst. Appl. **37**(4), 2724–2732 (2010)

22. Benali, L., Notton, G., Fouilloy, A., Voyant, C., Dizene, R.: Solar radiation forecasting using artificial neural network and random forest methods: application to normal beam, horizontal diffuse and global components. Renew. Energy **132**, 871–884 (2019)

23. Multiple linear regression (2017). http://www.stat.yale.edu/Courses/1997-98/101/linmult.htm. Accessed 01 June 2022

24. Huang, G.-B., Zhu, Q.-Y., Siew, C.-K.: Extreme learning machine: theory and applications. Neurocomputing **70**(1–3), 489–501 (2006)

25. Huang, G.-B., Zhu, Q.-Y., Siew, C.-K.: Extreme learning machine: a new learning scheme of feedforward neural networks. In: 2004 IEEE International Joint Confer-

ence on Neural Networks (IEEE Cat. No. 04CH37541), vol. 2, pp. 985–990. IEEE (2004)

26. Bartlett, P.: The sample complexity of pattern classification with neural networks: the size of the weights is more important than the size of the network. IEEE Trans. Inf. Theory **44**(2), 525–536 (1998). https://doi.org/10.1109/18.661502

27. Huang, G.-B., Chen, L., Siew, C.K., et al.: Universal approximation using incremental constructive feedforward networks with random hidden nodes. IEEE Trans. Neural Netw. **17**(4), 879–892 (2006)

28. Smola, A.J., Schölkopf, B.: A tutorial on support vector regression. Stat. Comput. **14**(3), 199–222 (2004)

29. Vapnik, V.: The Nature of Statistical Learning Theory. Springer, New York (1999)

30. Specht, D.F., et al.: A general regression neural network. IEEE Trans. Neural Netw. **2**(6), 568–576 (1991)

31. Mathworks documentation home (2017). https://mathworks.com/help/index.html Accessed 15 May 2022

32. Jain, A.K., Murty, M.N., Flynn, P.J.: Data clustering: a review. ACM Comput. Surv. (CSUR) **31**(3), 264–323 (1999)

33. MacQueen, J., et al.: Some methods for classification and analysis of multivariate observations. In: Proceedings of the Fifth Berkeley Symposium on Mathematical Statistics and Probability, vol. 1, pp. 281–297, Oakland, CA, USA (1967)

34. Arlot, S., Celisse, A.: A survey of cross-validation procedures for model selection. Stat. Surv. **4**, 40–79 (2010)

Deep Learning

Companion Losses for Ordinal Regression

David Díaz-Vico[2]([✉]), Angela Fernández[1]([✉]), and José R. Dorronsoro[1,2]([✉])

[1] Department Computer Engineering, Universidad Autónoma de Madrid,
Madrid, Spain
{a.fernandez,jose.dorronsoro}@uam.es
[2] Inst. Ing. Conocimiento, Universidad Autónoma de Madrid, Madrid, Spain
david.diaz.vico@outlook.com

Abstract. In Ordinal Regression (OR) class labels contain ranking information about the underlying samples and, thus, the goal is not only to minimize classification errors but also the rank distance of misclassified patterns. Thus, while class rankings are not metric values, they add a regression-like character to OR. Within this perspective, we propose here deep OR models built from losses which mix classification and regression components. While conceptually simple, our experiments will show their performance to be comparable to that of models representative of the state of the art in OR.

1 Introduction

Class labels in standard classification usually reflect qualitative information on the underlying patterns which in general does not allow any quantitative comparison among them. Taking for instance digit classifications, a 0 is obviously quite different from a 9 but, talking about digits, it makes no sense to state that the 9 is bigger than the 0 or that this digit is worse than the 9. But, on the other hand, if the labels between 0 and 9 correspond to exam gradings, they not only provide labellings for the underlying exams but also a measure of their relative quality. This suggests that an adequate classifier not only should attend to reduce misclassification errors but also should minimize the deviation of the classifier prediction from the true label. In other words, even if we cannot assign metric information to the labels themselves, in the sense of, for instance, that a grade of 8 does not imply that an exam is twice as good as one with a grade of 4, these grades imply that the 8-graded exam is better than the 4-graded one and also that a 7 misclassification of an 8-graded exam is preferred to a 4 one.

To achieve such a classifier is the goal of ordinal regression (OR), also known as ordinal classification. This problem appears in many application fields and

The authors acknowledge financial support from the European Regional Development Fund and the Spanish State Research Agency of the Ministry of Economy, Industry, and Competitiveness under the project PID2019-106827GB-I00. They also thank the UAM–ADIC Chair for Data Science and Machine Learning and gratefully acknowledge the use of the facilities of Centro de Computación Científica (CCC) at UAM.

P. García Bringas et al. (Eds.): HAIS 2022, LNAI 13469, pp. 211–222, 2022.
https://doi.org/10.1007/978-3-031-15471-3_19

has received a substantial attention in the literature (see for instance [1] for an early study) with a large number of contributions on the past 20 years. A very good survey is presented in [12] by P.A. Gutiérrez *et al.*, which not only gives an overview of the literature but establishes a taxonomy of OR methods, providing reasonably details on a large number of models and performing substantial numerical comparisons among them.

By its own nature, OR lies between multiclass classification and regression but, as discussed in [18], it is different from plain classification as the ranks encode not just labels but ranking information. And it is also different from regression because there is no actual metric distance between the rank targets as they are just labels that, as such, do not carry metric information and, hence, provide more of a qualitative information than of a quantitative goal. Since OR problems are ultimately classification problems, it is natural to look for classification costs specifically suited to the OR problem; in particular, the ordinal target structure suggests to look for closeness between predictions and targets. This is often expressed by the requirement of the cost function being V shaped [18]; more precisely, assuming targets $\{0, 1, \ldots, K-1\}$ for a K-class problem, if $c_y(k)$ is the cost of predicting class k for a y target, the cost should verify

$$c_y(k-1) \geq c_y(k) \text{ for } 1 \leq k \leq y;$$
$$c_y(k+1) \geq c_y(k) \text{ for } y \leq k \leq K-2.$$

For instance, the standard 0–1 classification cost $c_y(k) = 1 - \delta_{yk}$ is V shaped, and so it is the absolute cost $c_y(k) = |y - k|$. Notice that the 0–1 cost would place OR inside standard classification, while the absolute cost suggests some kind of L_1 regression. Similarly, the squared error $c_y(k) = (y - k)^2$, another V shaped cost, would link it with L_2 regression. The problem, however, is that in any regression approach the cost would be hardly differentiable, as model outputs must have discrete $\{0, 1, \ldots, K-1\}$ values.

Nevertheless, this cost discussion suggests that a possible OR approach could be to build models that combine both a (discrete) classification loss and a (continuous) regression one. This is what we propose here, taking advantage of the ability of modern neural network frameworks to combine distinct losses into a common cost function whose gradient can then be automatically computed. We showed in [10] how this approach can be followed to combine several classification losses into an enhanced classifier and here we will do so by combining the categorical cross entropy loss, standard in deep multiclass classification, with the mean squared loss typical of deep regression. While we will just consider networks with one, three or five hidden layers and, hence, not all too deep, we will retain throughout the paper the term deep network to refer to neural networks using some of the characteristic features of modern deep networks, such as ReLU activations, Adam optimizers [14], Glorot-Bengio initializations [11] or automatic backpropagation. To these two losses, and after the results in [10], we will add a third, the mean squared Fisher loss that, while not aiming to yield directly a classifier or regressor, may be able to induce on the representations learned by the network in its last hidden layer a structure that may enhance the performance of an accompanying model. Thus, our contributions here are

- The proposal of deep models for ordinal regression that combine classification and regression losses.
- A substantial experimental work to first compare the proposed methods against themselves.
- The comparison of the proposed models against the strongest performances obtained with four representative OR models of the current state of the art, showing that their results are clearly comparable.

The rest of the paper is structured as follows. In Sect. 2 we will briefly review the main OR framework with a focus on the four methods against which we will compare ours. Then, the proposed models will be detailed in Sect. 3 and numerical experiments will be presented in Sect. 4. A final section will offer some conclusions as well as pointers to further work.

2 OR Overview

As mentioned, OR has received a long and sustained attention in the literature since, at least, the early 1980s [1,19], with a large number of models being proposed. The concrete approaches vary largely, but in the very good and quite recent survey in [12] many of them are classified into a taxonomy that goes from simple regression or classification models (either standard or cost sensitive) to decompositions of the ordinal targets into a number of binary ones whose corresponding model outputs are then recombined into ordinal ones, going through a third group where fall threshold methods in which latent variable models are sought and whose predictions are transformed into ordinal labels using a set of thresholds that are also learned. To that first taxonomy one can also add an overlapping taxonomy of underlying ML models such as perceptrons [7], neural networks [6], SVMs [4,5,18], boosting [17] and other ensemble models [13,16], or Gaussian processes [3].

The performance of such approaches may greatly vary and of course will largely depend on what problems the different models are applied to. An interesting contribution of [12] is an extensive experimental analysis of 16 representative OR models over an also large number of datasets with respect to both the Mean Zero-One Error (MZE) or Mean Absolute Error (MAE) losses. We will work in our experiments with 16 of the datasets considered in [12]. Unfortunately, the MZE and MAE values reported in [12] for them do not allow for a clear cut selection of top models. As a simple expedient for such a choosing, we have computed and sorted the average MZE and MAE ranks of each model over the 16 datasets we will consider here and, in order to achieve a relatively compact exposition, have retained four of them with the smallest average ranks sum. The first three positions go to the SVOREX, SVORIM and REDSVM models while GPOR and SVMOP, which have very similar scores and rankings, tie for the fourth place. Given that SVMOP, SVOREX, SVORIM and REDSVM have the same underlying SVM models, we have opted to drop SVMOP, retaining instead GPOR (notice that, although not SVM based, this is also a kernel based method). We briefly review them next.

In the linear version of SVOREX, a common vector w is sought but individual thresholds b_j are determined by imposing constraints only among adjacent classes. More precisely, if x_p^j is the p-th pattern in class j with n_j patterns in total, the threshold b_j should be such that

$$w \cdot x_p^j - b_j \leq -1 + \xi_p^j, \qquad \xi_p^j \geq 0, \qquad 1 \leq p \leq n_j$$
$$w \cdot x_q^{j+1} - b_j \geq 1 - \widetilde{\xi}_q^{j+1}, \qquad \widetilde{\xi}_q^{j+1} \geq 0, \qquad 1 \leq q \leq n_{j+1}, \qquad (1)$$

and the loss to be minimized is

$$\frac{1}{2}\|w\|^2 + C \sum_0^{K-2} \left(\sum_1^{n_j} \xi_p^j + \sum_1^{n_{j+1}} \widetilde{\xi}_q^{j+1} \right).$$

This basic SVM ordinal regression formulation was proposed in [20] and in [4] the extra threshold constraints $b_0 \leq \ldots \leq b_{K-2}$ were added. In contrast with (1), in SVORIM [5] patterns in all classes are allowed to enter the constraints, that now are

$$w \cdot x_p^k - b_j \leq -1 + \xi_p^k, \quad 0 \leq \xi_p^k, \quad 0 \leq k \leq j, \qquad 1 \leq p \leq n_k,$$
$$w \cdot x_q^k - b_j \geq 1 - \widetilde{\xi}_q^k, \quad 0 \leq \widetilde{\xi}_q^k, \quad j+1 \leq k \leq K-1, \quad 1 \leq q \leq n_k; \quad (2)$$

now in (2) each sample appears in $K-1$ constraints, while it did so only in two constraints in (1). While the ordering $b_0 \leq \ldots \leq b_{K-2}$ is not initially required, it is a consequence of the problem's solution.

We turn our attention to the SVM reduction method REDSVM. In its simplest form, the reduction scheme proposed in [15,18] defines from an initial sample S with patterns (x, y), a new extended sample S_E with patterns (x^k, y^k), $0 \leq k \leq K - 2$, where $x^k = (x, k)$ and $y^k = 2\chi[k < y] - 1$ (see also [2]); here $\chi[B]$ is 1 when B is true, and 0 otherwise. Notice that the new sample S_E corresponds to a binary classification problem with targets ± 1 (different coding matrices M can be used to work with extensions (x, M_k) instead of the simple (x, k) here). Now for a binary classifier $g(x, k)$ that correctly classifies all the (x, k), the ranker

$$r_g(x) = \sum_0^{K-2} \chi[g(x, k) > 0]$$

would verify $r_g(x) = y$. Moreover, for a V shaped cost c, it is shown in [15,18] that the cost $c_y(r(x))$ is bounded by an appropriately weighted 0–1 loss of g on x. This general reduction scheme offers a great flexibility for the choice of the binary classifier g, the coding matrix M and the cost function c, and it turns out that a large number of OR proposals (SVOREX and SVORIM among them) can be expressed as particular instances of the reduction scheme. The REDSVM results reported in [12] correspond to a binary soft margin SVM with identity coding and absolute costs.

Finally, in Gaussian Processes (GP) one assumes the existence of a latent function f associated to a sample $S = \{x_p, y_p\}$ by a zero mean Gaussian process specified by a kernel covariance matrix

$$\Sigma_{pq} = \text{cov}(f(x_p), f(x_q)) = K(x_p, x_q) = e^{-\frac{\gamma}{2}\|x_p - x_q\|^2}. \tag{3}$$

Following [3], for GP Ordinal Regression a set of thresholds $b_0 < \ldots < b_{K-2}$ determine the ideal likelihood of target y_i given $f(x_i)$ by $\widetilde{p}(y_p|f(x_p)) = 1$ if $b_{y_p-1} < f(x_p) \le b_{y_p}$ and 0 otherwise. However, it is more realistic to assume that normal $\mathcal{N}(\delta; 0, \sigma^2)$ noise is added to the $f(x_p)$, which leads to the posteriors

$$p(y_p|f(x_p)) = \int \widetilde{p}(y_p|f(x_p) + \delta)\mathcal{N}(\delta; 0, \sigma^2)d\delta.$$

The sample's posterior probability is then $p(f|S) = \frac{1}{p(S)}\prod_i p(y_p|f(x_p)p(f)$ where $p(f) = c_\Sigma e^{-\frac{1}{2}f^t \Sigma^{-1}f}$. Here the thresholds b_i, the kernel constant γ and the noise variance σ^2 are the model's hyperparameters that have to be estimated. In [3] two Bayesian techniques for model hyperparametrization are considered, a maximum a posteriori estimate (MAP) using the Laplace approximation, and expectation propagation with variational methods, respectively. The publicly available implementation of GPOR used in [12] applies MAP estimation with the Laplace approximation, plus automatic relevance determination, where instead of a common kernel parameter γ, individual γ_j values weight the coordinate differences $(x_p^j - x_q^j)^2$ in (3).

3   Companion Losses for OR

We will denote as $F(x, \mathcal{W})$ the outputs of a DNN acting on x; here $\mathcal{W}$ denotes the entire set of the network's weight matrices and bias vectors. The targets y will be either categorical for classification losses or numerical for the regression ones. If $z = \Phi(x, \widetilde{\mathcal{W}})$ are the network outputs at the last hidden layer, with $\widetilde{\mathcal{W}}$ the weights and biases of all layers up to the last one, the network activations at the output layer will be $Wz + B$, where W, B are either a matrix and a vector for multidimensional targets or just a vector, scalar pair for one dimensional ones.

For classification, the network output function is the softmax

$$F_j(x; \mathcal{W}) = \frac{e^{w_j \cdot z + b_j}}{\sum_{k=0}^{K-1} e^{w_k \cdot z + b_k}}.$$

Obviously then $\sum_j F_j(x; \mathcal{W}) = 1$ and we assume $P(j|x) \simeq F_j(x; \mathcal{W})$. Since the targets are now one-hot class encodings, given an i.i.d. sample $S = (X, Y)$, the probability of getting targets in class k_p, $1 \le p \le n$, for patterns x^p (i.e., to have $y_k^p = 1$) is

$$P(Y|X; \mathcal{W}) = \prod_{p=1}^{n} P(k_p|x^p; \mathcal{W}) = \prod_{p=1}^{n} \prod_{m=0}^{K-1} P(m|x^p; \mathcal{W})^{y_m^p} \simeq \prod_{p=1}^{n} \prod_{m=0}^{K-1} F_m(x; \mathcal{W})^{y_m^p}.$$

and we estimate the DNN's weights $\mathcal{W}$ by minimizing the categorical cross–entropy loss, i.e., the minus log of the approximate sample's likelihood

$$\ell_{ce}(\mathcal{W}) = -\log \widetilde{P}(Y|X; \mathcal{W}) = -\sum_{p=1}^{n} \sum_{m=0}^{K-1} y_m^p \log F_m(x^p; \mathcal{W}).$$

Once we obtain an optimal weight set $\mathcal{W}^*$, a new x is assigned to the class with the maximum posterior probability.

For regression models the network output is just the scalar $w \cdot z + b$, and we will use the standard mean squared error loss

$$\ell_{rr}(\mathcal{W}) = \frac{1}{2n} \sum_{p=1}^{n} (y^p - w \cdot z^p - b)^2 = \frac{1}{2n} \sum_{p=1}^{n} (y^p - w \cdot \Phi(x^p; \widetilde{\mathcal{W}}) - b)^2.$$

However, at prediction time, the continuous network outputs are mapped to the nearest class label; because of this we may call such a model a rounded regressor.

We will finally consider what we will call Fisher DNNs. Let's denote by S_B and S_T the between-class and total covariance matrices of the sample patterns and s_B and s_T are their counterparts for the projections $z = Ax$. As discussed in [9,21], a projection equivalent to the one given by a matrix A that maximizes the trace criterion, $g(A) = \text{trace}(s_T^{-1} s_B) = \text{trace}\left((A^t S_T A)^{-1}(A^t S_B A)\right)$, can be obtained by solving the least squares problem

$$\min \frac{1}{2} \|Y^f - XW - \mathbf{1}_n B\|^2, \tag{4}$$

where W is a $d \times K$ matrix, B a $1 \times K$ vector, $\mathbf{1}_n$ is the all-ones vector and, in the p-th row of the target matrix Y^f associated to a pattern x^p in class m, we have $Y_{pm}^f = \frac{n - n_m}{n\sqrt{n_m}}$ and $Y_{pj}^f = -\frac{\sqrt{n_j}}{n}$ for $j \neq m$, with n_j the number of patterns in class j; see [9,21] for more details. As proposed in [9], this can be extended to a DNN setting by minimizing

$$\ell_f(y, \widehat{y}_f) = \frac{1}{2} \|Y^f - F(X, W)\|^2 = \frac{1}{2} \|Y^f - \Phi(X; \widetilde{\mathcal{W}})W - \mathbf{1}_n B\|^2, \tag{5}$$

where $\widehat{y}_f$ denotes the network's prediction. Such a loss may enforce that the last hidden layer projections z have to be concentrated around their class means while keeping these apart, as it is the goal in Fisher's classical analysis, and this may enhance the performance of accompanying classifiers and regressors.

Based on the above, a natural idea in an OR setting is to define a companion loss network with multiple outputs and targets and try to minimize a combination of these losses such as, say,

$$\ell(y, \widehat{y}_{ce}, \widehat{y}_{rr}) = \ell_{ce}(y, \widehat{y}_{ce}) + \lambda \ell_{rr}(y, \widehat{y}_{rr}), \tag{6}$$

with λ an appropriately chosen mixing parameter. The rationale is clear: such a network could in principle minimize simultaneously a purely classification loss ℓ_{ce} and a regression one ℓ_{rr} and, hence, return a useful OR model. However, such a network could also yield conflicting predictions $\widehat{y}_{ce}(x), \widehat{y}_{rr}(x)$ for the same input x. Several ways of solving this are possible but here we will opt by relying in just one of them, taking one of the models as preferential. In this vein, we will denote by ce-rr a model built with the loss (6) but whose predictions are $\widehat{y}_{ce}$, while an rr-ce model will minimize the same loss but use the $\widehat{y}_{rr}$ predictions. We will also consider mixed models ce-fisher and rr-fisher, which would minimize the losses $\ell_{ce} + \lambda \ell_f$ and $\ell_{rr} + \lambda \ell_f$, respectively; in this case, the network predictions would be those of the ce or rr outputs, but never the fisher ones.

4   Experimental Results

In this section we will give our experimental results. We are going to consider two sets of problems taken from the review [12]. The first one is made of binned regression problems, that is, standard regression problems whose targets have been grouped in several range-based classes and relabelled to turn them into classification problems with ordered labels that reflect the initial target order. The problems in this set are abalone, housing, machine, pyrim and stock. The second set corresponds to standard classification problems with ordered labels and, hence, amenable of an ordinal regression (OR) treatment. The problems here are automobile, balance-scale, car, eucalyptus, newthyroid, ERA, ESL, LEV, SWD, toy and winequality-red. This subset correspond to the medium sized samples among all the datasets in [12], i.e. neither having a too small nor a rather large number of patterns; see [12] for a more detailed description as well as data availability. In all cases the class labels are taken as consecutive integers between 0 and the number of classes minus 1. As mentioned, these problems can be in principle dealt with by either OR-specific models or by plain classifiers or regressors, where the real outputs of the latter are rounded to the nearest label rank.

4.1   Companion Loss Models

Four companion models are considered: ce-rr, ce-fisher, rr-ce and rr-fisher, to which we add a basic crossentropy classifier ce and a rounded regressor rr. We have applied them to the 5 binned regression and 11 ordinal regression problems just mentioned. To do so, we have used the 20 binned and 30 OR stratified splits used in [12]; and we select the 5 bin version of the binned regression splits.

We have considered L_2 regularized models with 1, 3 or 5 hidden layers with 100 units on each, and have selected the optimal regularization parameter α and DNN architecture plus the mixing parameter λ for the companion losses by 5-fold CV on the train partition; after this is done, we refit the best model on the entire train fold and apply it to the test fold. The explored α and λ values are of the form 10^k, $-7 \leq k \leq 2$ for α and $-2 \leq k \leq 2$ for λ. Given the nature of the OR problem and the methods we use, two different scores, accuracy (that is, $1 - \text{MZE}$) and MAE can be used as the CV scores. As done in [12], we shall use both and report MZE results for the models selected using accuracy as the CV score, and MAE values for those selected using it as the CV score.

Table 1 shows the MZE scores (left) and the MAE ones (right) of the selected companion models. In order to check for differences in the performance of the companion loss models, we have applied a Friedman test [8] on the MZE and MAE scores separately. The test cannot reject the null hypothesis of the MAE score ranks being equal but it does so for the MZE scores. A post hoc Nemeny [8] test shows that the MZE ranks of the ce-fisher model (the one with the best average rank) are statistically different from those of rr and rr-fisher. Because of this and even if any such a difference does not appear for the MAE scores, we will drop these two models from our subsequent analysis for a more compact presentation. Table 2 contains the scores and ranks of the selected models.

Table 1. MZE (left) and MAE (right) values of the companion loss models.

	MZE						MAE					
	ce	rr	ce-rr	rr-ce	ce-f	rr-f	ce	rr	ce-rr	rr-ce	ce-f	rr-f
Abalone	**0.499**	0.554	0.501	0.553	0.501	0.551	0.694	0.672	0.691	**0.668**	0.690	0.677
Housing	**0.341**	0.356	0.347	0.358	0.344	0.362	0.390	**0.383**	0.392	0.386	0.394	**0.383**
Machine	0.417	0.394	0.423	0.406	0.414	**0.381**	0.468	**0.400**	0.461	0.426	0.474	0.406
Pyrim	0.519	0.548	0.544	0.527	**0.517**	0.544	0.727	0.690	0.783	**0.685**	0.746	0.700
Stock	0.110	0.108	0.108	0.113	**0.106**	**0.106**	0.111	0.109	0.108	0.113	0.109	**0.104**
Balance-scale	0.021	0.023	0.028	**0.018**	0.021	0.021	**0.020**	0.022	0.027	**0.020**	0.023	**0.020**
ERA	0.726	0.767	**0.723**	0.767	**0.723**	0.768	**1.209**	1.221	1.211	1.216	1.211	1.218
Eucalyptus	0.360	0.362	**0.342**	0.363	0.363	0.362	0.395	0.377	**0.376**	0.385	0.393	0.390
LEV	0.374	0.377	**0.370**	0.378	0.373	0.375	0.411	0.413	**0.408**	0.415	**0.408**	0.411
SWD	**0.420**	0.430	0.422	0.431	**0.420**	0.431	**0.441**	0.444	0.442	0.446	0.444	0.445
Winequality-red	0.366	0.373	0.359	0.368	**0.349**	0.374	0.412	0.420	0.394	0.415	**0.390**	0.406
ESL	0.288	0.294	0.300	**0.287**	0.290	0.292	0.305	0.306	0.308	**0.299**	0.305	0.303
Automobile	0.263	0.322	0.267	0.298	**0.265**	0.285	0.381	0.412	**0.363**	0.371	0.394	0.374
Car	0.006	0.007	**0.005**	0.006	0.006	0.009	0.008	0.007	**0.004**	0.005	0.005	0.009
Toy	0.068	0.097	0.062	0.113	**0.055**	0.063	0.068	0.095	**0.056**	0.087	0.057	**0.056**
Newthyroid	0.039	0.039	0.040	**0.036**	0.038	0.041	0.039	0.040	0.040	**0.036**	0.039	0.041

Table 2. MZE (left) and MAE (right) values and ranks of the selected companion models.

	MZE				MAE			
	ce	ce-rr	ce-f	rr-ce	ce	ce-rr	ce-f	rr-ce
Abalone	**0.499 (1.0)**	0.501 (2.0)	0.501 (3.0)	0.553 (4.0)	0.694 (4.0)	0.691 (3.0)	0.69 (2.0)	**0.668 (1.0)**
Housing	**0.341 (1.0)**	0.347 (3.0)	0.344 (2.0)	0.358 (4.0)	0.39 (2.0)	0.392 (3.0)	0.394 (4.0)	**0.386 (1.0)**
Machine	0.417 (3.0)	0.423 (4.0)	0.414 (2.0)	**0.406 (1.0)**	0.468 (3.0)	0.461 (2.0)	0.474 (4.0)	**0.426 (1.0)**
Pyrim	0.519 (2.0)	0.544 (4.0)	**0.517 (1.0)**	0.527 (3.0)	0.727 (2.0)	0.783 (4.0)	0.746 (3.0)	**0.685 (1.0)**
Stock	0.11 (3.0)	0.108 (2.0)	**0.106 (1.0)**	0.113 (4.0)	0.111 (3.0)	**0.108 (1.0)**	0.109 (2.0)	0.113 (4.0)
Balance-scale	0.021 (3.0)	0.028 (4.0)	0.021 (2.0)	**0.018 (1.0**	0.02 (2.0)	0.027 (4.0)	0.023 (3.0)	**0.02 (1.0)**
ERA	0.726 (3.0)	0.723 (2.0)	**0.723 (1.0)**	0.767 (4.0)	**1.209 (1.0)**	1.211 (2.0)	1.211 (3.0)	1.216 (4.0)
Eucalyptus	0.36 (2.0)	**0.342 (1.0)**	0.363 (4.0)	0.363 (3.0)	0.395 (4.0)	**0.376 (1.0)**	0.393 (3.0)	0.385 (2.0)
LEV	0.374 (3.0)	**0.37 (1.0)**	0.373 (2.0)	0.378 (4.0)	0.411 (3.0)	0.408 (2.0)	**0.408 (1.0)**	0.415 (4.0)
SWD	**0.42 (1.5)**	0.422 (3.0)	**0.42 (1.5)**	0.431 (4.0)	**0.441 (1.0)**	0.442 (2.0)	0.444 (3.0)	0.446 (4.0)
Winequality-red	0.366 (3.0)	0.359 (2.0)	**0.349 (1.0)**	0.368 (4.0)	0.412 (3.0)	0.394 (2.0)	**0.39 (1.0)**	0.415 (4.0)
ESL	0.288 (2.0)	0.3 (4.0)	0.29 (3.0)	**0.287 (1.0)**	0.305 (2.0)	0.308 (4.0)	0.305 (3.0)	**0.299 (1.0)**
Automobile	**0.263 (1.0)**	0.267 (3.0)	0.265 (2.0)	0.298 (4.0)	0.381 (3.0)	**0.363 (1.0)**	0.394 (4.0)	0.371 (2.0)
Car	0.006 (3.0)	**0.005 (1.0)**	0.006 (4.0)	0.006 (2.0)	0.008 (4.0)	**0.004 (1.0)**	0.005 (3.0)	0.005 (2.0)
Toy	0.068 (3.0)	0.062 (2.0)	**0.055 (1.0)**	0.113 (4.0)	0.068 (3.0)	**0.056 (1.0)**	0.057 (2.0)	0.087 (4.0)
Newthyroid	0.039 (3.0)	0.04 (4.0)	0.038 (2.0)	**0.036 (1.0)**	0.039 (2.5)	0.04 (4.0)	0.039 (2.5)	**0.036 (1.0)**

4.2   Comparison with Classical or Models

We next compare the performance of the companion loss models with the four OR models reviewed in Sect. 2 that can be considered representative of the state of the art in OR; these are REDSVM, SVORIM, SVOREX and GPOR. We have selected them according to the results reported in [12]; more precisely, and as mentioned in Sect. 2, we have computed the average MZE and MAE ranks of each model over the 16 datasets that we will consider here, have sorted the models according to the sum of these MZE and MAE average ranks and have retained the four models with the smallest sum. While this is admittedly an ad hoc procedure, we point out that a Friedman test over the entire 16 models set detects significant differences in their MZE and MAE ranks but a post hoc Nemenyi test shows these differences to be significative for just a few model pairings well outside

Table 3. MZE (left) and MAE (right) values and ranks of the redsvm, svorex, svorim and gpor models for all problems.

	MZE				MAE			
	REDSVM	SVORIM	SVOREX	GPOR	REDSVM	SVORIM	SVOREX	GPOR
Abalone	0.524 (3.5)	0.524 (3.5)	0.512 (2.0)	**0.509 (1.0)**	**0.65 (1.5)**	**0.65 (1.5)**	0.66 (3.0)	0.68 (4.0)
Housing	0.324 (2.0)	0.325 (3.0)	0.329 (4.0)	**0.31 (1.0)**	0.36 (3.0)	0.36 (3.0)	0.36 (3.0)	**0.34 (1.0)**
Machine	**0.4 (1.0)**	0.416 (3.0)	0.432 (4.0)	0.403 (2.0)	0.46 (3.0)	**0.44 (1.5)**	0.47 (4.0)	**0.44 (1.5)**
Pyrim	0.506 (2.0)	**0.498 (1.0)**	0.508 (3.0)	0.515 (4.0)	0.64 (2.0)	**0.63 (1.0)**	0.66 (4.0)	0.65 (3.0)
Stock	0.111 (2.0)	0.113 (3.5)	0.113 (3.5)	**0.109 (1.0)**	**0.11 (2.5)**	**0.11 (2.5)**	**0.11 (2.5)**	**0.11 (2.5)**
Balance-scale	**0.001 (1.0)**	0.002 (2.5)	0.002 (2.5)	0.034 (4.0)	**0.0 (2.0)**	**0.0 (2.0)**	**0.0 (2.0)**	0.03 (4.0)
ERA	0.751 (3.5)	0.751 (3.5)	0.714 (2.0)	**0.712 (1.0)**	1.22 (3.0)	**1.21 (1.5)**	**1.21 (1.5)**	1.24 (4.0)
Eucalyptus	0.362 (3.0)	0.361 (2.0)	0.364 (4.0)	**0.314 (1.0)**	0.4 (3.5)	0.39 (2.0)	0.4 (3.5)	**0.33 (1.0)**
LEV	**0.373 (1.0)**	0.38 (3.0)	0.375 (2.0)	0.388 (4.0)	**0.41 (2.0)**	**0.41 (2.0)**	**0.41 (2.0)**	0.42 (4.0)
SWD	0.429 (2.0)	0.431 (3.0)	0.432 (4.0)	**0.422 (1.0)**	0.45 (3.0)	0.45 (3.0)	0.45 (3.0)	**0.44 (1.0)**
Winequality-red	**0.373 (2.0)**	**0.373 (2.0)**	**0.373 (2.0)**	0.394 (4.0)	**0.42 (2.5)**	**0.42 (2.5)**	**0.42 (2.5)**	**0.42 (2.5)**
ESL	0.287 (2.5)	**0.284 (1.0)**	0.29 (4.0)	0.287 (2.5)	0.31 (4.0)	**0.3 (2.0)**	**0.3 (2.0)**	**0.3 (2.0)**
Automobile	0.317 (2.0)	0.323 (3.0)	**0.316 (1.0)**	0.389 (4.0)	0.4 (2.0)	**0.39 (1.0)**	0.42 (3.0)	0.59 (4.0)
Car	**0.012 (2.0)**	**0.012 (2.0)**	**0.012 (2.0)**	0.037 (4.0)	**0.01 (2.0)**	**0.01 (2.0)**	**0.01 (2.0)**	0.04 (4.0)
Toy	0.023 (3.0)	**0.02 (1.5)**	**0.02 (1.5)**	0.046 (4.0)	**0.02 (2.0)**	**0.02 (2.0)**	**0.02 (2.0)**	0.05 (4.0)
Newthyroid	**0.032 (1.0)**	0.034 (3.0)	0.034 (3.0)	0.034 (3.0)	**0.03 (2.5)**	**0.03 (2.5)**	**0.03 (2.5)**	**0.03 (2.5)**

Table 4. Mean and standard deviation of MZE (top) and MAE (bottom) score ranks for all models sorted by increasing means.

MZE	REDSVM	ce-f	ce	GPOR	ce-rr	SVORIM	SVOREX	rr-ce
mean	**3.84**	**3.84**	4.19	4.34	4.47	4.59	5.03	5.69
std	2.10	1.97	1.97	2.74	2.70	2.27	2.19	1.95
MAE	SVORIM	ce-rr	rr-ce	SVOREX	REDSVM	GPOR	ce-f	ce
mean	**3.50**	4.38	4.44	4.47	4.59	4.62	4.91	5.09
std	1.78	2.71	2.18	2.03	2.26	2.77	2.19	1.98

our selection. Thus, while our selected models cannot rigorously called the top performers, they are certainly among the strongest ones. Table 3 contains the scores and ranks of these four models.

We have thus eight models in total and we have put their MZE and MAE scores together and applied again a Friedman test separately on the MZE and MAE scores of the eight models. Now the test fails to reject the null hypothesis of non equal ranks and, in fact, the p-values are relatively large, 0.392 for the MZE scores and 0.734 for the MAE ones. This can be indirectly seen in Table 4, where the sorted rank averages of the MZE and MAE scores of all models are quite close and well within one standard deviation from each other.

This would preclude the application of post hoc tests, but just for illustration purposes, we have applied a Wilcoxon rank sum test of two control companion loss models, ce-fisher and ce-rr, against the other models. These appear in Table 4 as the companion models with best MZE and MAE ranks, respectively. The tests' results are shown in the leftmost four columns of Table 5 for the MZE score and on the rightmost four for the MAE one. In turn, columns 2 and 6 contain the p values of the Wilcoxon MZE and MAE comparisons of the ce-fisher model against all others, and columns 4 and 8 those of the ce-rr model. All the model columns are sorted by increasing p values.

Table 5. p values when comparing the MZE (half left) and MAE (half right) scores of the ce-fisher (columns 1, 2, 5 and 6) and ce-rr (columns 3, 4, 7, 8) models against the other models considered.

MZE				MAE			
model-1	p-val	model-2	p-val	model-1	p-val	model-2	p-val
rr-ce	0.016	ce-f	0.038	SVORIM	0.044	ce	0.193
ce-rr	0.038	rr-ce	0.144	rr-ce	0.313	SVORIM	0.252
ce	0.286	ce	0.597	SVOREX	0.323	SVOREX	0.562
SVORIM	0.495	GPOR	0.816	REDSVM	0.323	REDSVM	0.706
SVOREX	0.623	REDSVM	0.900	ce-rr	0.736	ce-f	0.736
GPOR	0.782	SVORIM	0.900	GPOR	0.744	GPOR	0.782
REDSVM	0.816	SVOREX	0.980	ce	0.856	rr-ce	0.940

As this involves a multiple model comparison, some correction should be applied. There are several options for that (see [8]) which usually end up in lowering the significance threshold, which can be done by either dividing the desired significance by the number of model comparisons (as in the Bonferroni-Dunn test [8]) or the significance value is divided by values decreasing inversely as the test's significance increases (other approaches work with inverse orderings). For instance, the Holm correction sorts the different models by increasing p values and corrects the α significance for model i by dividing it by $k - i$, $0 \leq i \leq k - 1$, with k the total number of models considered; whenever a null hypothesis cannot be rejected, all the following ones are also retained.

In our case, we have 8 models in total and, for a 10% significance level, the Holm corrected p values would range from $\alpha = 0.0125$ for the first model to $\alpha = 0.1$ for the last one. It is clear from the p values in Table 5 that in no case the null hypothesis could be rejected, and we should conclude that MZE and MAE performance differences between the considered models are not statistically significant. On the other hand, from the first columns in Table 5, it would appear that the MZE differences between ce-fisher against rr-ce and ce-rr are significant at the 0.1 level on their favor as they have a smaller rank mean. Similarly, SVORIM MAE is significantly different from that of ce-fisher. We also point out that if we repeat this experiment but taking now as control models REDSVM and SVORIM (the best MZE and MAE classical models, respectively), REDSVM MZE would outperform that of rr-ce, and SVORIM MAE those of SVOREX, REDSVM, ce-f and ce.

5   Discussion and Conclusions

In this work we have proposed new ordinal regression (OR) models based on deep networks with companion losses that mix properly balanced classification and regression penalties. The losses considered are the cross entropy loss for classification (denoted by ce), the mean squared loss for regression (denoted by rr) and the deep Fisher loss (denoted by fisher), proposed in [9] which is based on a squared loss but whose targets seek to achieve in a network's last hidden layer the class separation effects of Fisher's classical discriminant analysis. In between them, we consider the ce and rr losses as the dominant ones, which reinforce, respectively, the classification and regression aspects of OR; on the other hand, the `fisher` loss is used only in an auxiliary role. While in principle six pure and companion loss combinations are possible, we have dropped after a preliminary analysis the `rr` and `rr-fisher` losses for a more compact presentation, as they appear to underperform the others. We have added to the remaining four models another four OR models, REDSVM, SVORIM, SVOREX and GPOR, that can be considered representative of the state of the art in OR.

We have compared the eight final models over 16 datasets considered in [12] using two scores, the MZE that should favor classification oriented models, and the MAE, likely to favor regression oriented ones. A Friedman test on both the MZE and MAE scores fails to reject the null hypothesis of all models having similar ranks. This allows us to conclude that the `ce-fisher`, `ce-rr` and `rr-ce` plus the base classification model `ce`, have a performance similar to that of the state of the art models. On the other hand, a pairwise Wilcoxon rank test would indicate the `ce-fisher` model has better MZE scores than `ce-rr` and `rr-ce`, although its MAE scores would be worse than those of SVORIM; no significant difference is observed for the other models. It may also be worth noting that model performance depends on the score considered: for MZE, the best three models seem to be REDSVM, `ce-fisher` and `ce` in that order, while for the MAE score the best three would be SVORIM, `ce-rr` and `rr-ce`.

In summary, the mixed classification-regression companion losses we propose appear to be a sensible addition to models solving OR problems. This has to be substantiated working with a larger number of datasets but besides their good behavior, the deep companion loss approach present three advantages. The first one is its simplicity, in contrast with the more involved approaches of their OR counterparts presented here. The second one is their computational costs, likely to be smaller than those of the state of the art models discussed here. Notice that three of them rely on a kernel SVM core while GPOR is also kernel based; moreover, REDSVM enlarges an initial sample size N to $(K - 1) \times N$, with the subsequent impact on the training of the underlying SVM (with a cost at least quadratic on sample size). It is likely that they may be too costly to apply to large size samples, upon which DNN costs may still be reasonable. The final one derives from the flexibility deep networks offer, both in terms of the underlying architecture and by the possibility of using other losses perhaps better suited to OR; an example could be the replacement of the MSE loss used here by a MAE loss. We are currently pursuing these and related venues.

References

1. Anderson, J.A.: Regression and ordered categorical variables. J. Roy. Stat. Soc. Ser. B (Methodol.) **46**(1), 1–22 (1984)
2. Cardoso, J.S., da Costa, J.F.P.: Learning to classify ordinal data: the data replication method. J. Mach. Learn. Res. **8**, 1393–1429 (2007)
3. Chu, W., Ghahramani, Z.: Gaussian processes for ordinal regression. J. Mach. Learn. Res. **6**, 1019–1041 (2005)
4. Chu, W., Keerthi, S.S.: New approaches to support vector ordinal regression. In: Proceedings of the Twenty-Second International Conference on Machine Learning, vol. 119, pp. 145–152. ACM (2005)

5. Chu, W., Keerthi, S.S.: Support vector ordinal regression. Neural Comput. **19**(3), 792–815 (2007)
6. da Costa, J.P., Cardoso, J.S.: Classification of ordinal data using neural networks. In: Gama, J., Camacho, R., Brazdil, P.B., Jorge, A.M., Torgo, L. (eds.) ECML 2005. LNCS (LNAI), vol. 3720, pp. 690–697. Springer, Heidelberg (2005). https://doi.org/10.1007/11564096_70
7. Crammer, K., Singer, Y.: Online ranking by projecting. Neural Comput. **17**(1), 145–175 (2005)
8. Demsar, J.: Statistical comparisons of classifiers over multiple data sets. J. Mach. Learn. Res. **7**, 1–30 (2006)
9. Díaz-Vico, D., Dorronsoro, J.R.: Deep least squares fisher discriminant analysis. IEEE Trans. Neural Netw. Learn. Syst. **31**(8), 2752–2763 (2020)
10. Díaz-Vico, D., Fernández, A., Dorronsoro, J.R.: Companion losses for deep neural networks. In: Sanjurjo González, H., Pastor López, I., García Bringas, P., Quintián, H., Corchado, E. (eds.) HAIS 2021. LNCS (LNAI), vol. 12886, pp. 538–549. Springer, Cham (2021). https://doi.org/10.1007/978-3-030-86271-8_45
11. Glorot, X., Bengio, Y.: Understanding the difficulty of training deep feedforward neural networks. In: Proceedings of the Thirteenth International Conference on Artificial Intelligence and Statistics, AISTATS 2010, vol. 9, pp. 249–256 (2010)
12. Gutiérrez, P.A., Pérez-Ortiz, M., Sánchez-Monedero, J., Fernández-Navarro, F., Hervás-Martínez, C.: Ordinal regression methods: survey and experimental study. IEEE Trans. Knowl. Data Eng. **28**(1), 127–146 (2016)
13. Gutiérrez, P.A., Pérez-Ortiz, M., Suárez, A.: Class switching ensembles for ordinal regression. In: Rojas, I., Joya, G., Catala, A. (eds.) IWANN 2017. LNCS, vol. 10305, pp. 408–419. Springer, Cham (2017). https://doi.org/10.1007/978-3-319-59153-7_36
14. Kingma, D.P., Ba, J.: Adam: a method for stochastic optimization. In: 3rd International Conference on Learning Representations, ICLR 2015 (2015)
15. Li, L., Lin, H.: Ordinal regression by extended binary classification. In: Proceedings of the Twentieth Annual Conference on Neural Information Processing Systems, pp. 865–872. MIT Press (2006)
16. Lin, H.-T., Li, L.: Large-margin thresholded ensembles for ordinal regression: theory and practice. In: Balcázar, J.L., Long, P.M., Stephan, F. (eds.) ALT 2006. LNCS (LNAI), vol. 4264, pp. 319–333. Springer, Heidelberg (2006). https://doi.org/10.1007/11894841_26
17. Lin, H.T., Li, L.: Combining ordinal preferences by boosting. In: Proceedings ECML/PKDD 2009 Workshop on Preference Learning, pp. 69–83 (2009)
18. Lin, H., Li, L.: Reduction from cost-sensitive ordinal ranking to weighted binary classification. Neural Comput. **24**(5), 1329–1367 (2012)
19. McCullagh, P.: Regression models for ordinal data. J. Roy. Stat. Soc. Ser. B (Methodol.) **42**(2), 109–127 (1980)
20. Shashua, A., Levin, A.: Ranking with large margin principle: two approaches. In: Advances in Neural Information Processing Systems 15, pp. 937–944. MIT Press (2002)
21. Zhang, Z., Dai, G., Xu, C., Jordan, M.I.: Regularized discriminant analysis, ridge regression and beyond. J. Mach. Learn. Res. **11**, 2199–2228 (2010)

Convex Multi-Task Learning with Neural Networks

Carlos Ruiz[1]([✉]), Carlos M. Alaíz[1], and José R. Dorronsoro[1,2]

[1] Dept. Computer Engineering, Universidad Autónoma de Madrid,
Madrid, Spain
`carlos.ruizp@uam.es`
[2] Inst. Ing. Conocimiento, Universidad Autónoma de Madrid, Madrid, Spain

Abstract. Multi-Task Learning aims at improving the learning process by solving different tasks simultaneously. The approaches to Multi-Task Learning can be categorized as feature-learning, regularization-based and combination strategies. Feature-learning approximations are more natural for deep models while regularization-based ones are usually designed for shallow ones, but we can see examples of both for shallow and deep models. However, the combination approach has been tested on shallow models exclusively. Here we propose a Multi-Task combination approach for Neural Networks, describe the training procedure, test it in four different multi-task image datasets and show improvements in the performance over other strategies.

Keywords: Multi-task learning · Deep learning · Convex combination

1 Introduction

In Machine Learning (ML) it is often assumed that the data is independently identically distributed, and the empirical risk minimization principle [19], typically used in supervised learning, bases its generalization abilities in this claim. However, we often find problems with different but possibly related data distributions. Multi-Task Learning (MTL) [2] solves jointly those similar problems, each of which is considered a task.

Extending the taxonomy of [24], the MTL approaches can be divided in three main blocks: feature-learning models, regularization-based methods and combination approaches. Feature-learning models try to learn a space of features useful for all tasks at the same time. The regularization-based methods impose some soft constraints on the task-models so that there exists a connection across them. Finally, the combination approach combines task-specific models with a

The authors acknowledge financial support from the European Regional Development Fund and the Spanish State Research Agency of the Ministry of Economy, Industry, and Competitiveness under the project PID2019-106827GB-I00. They also thank the UAM–ADIC Chair for Data Science and Machine Learning and gratefully acknowledge the use of the facilities of Centro de Computación Científica (CCC) at UAM.

P. García Bringas et al. (Eds.): HAIS 2022, LNAI 13469, pp. 223–235, 2022.
https://doi.org/10.1007/978-3-031-15471-3_20

common one shared for all tasks. Recently a convex combination formulation was proposed in [16].

In ML we call deep models those who have a feature learning process that construct new features along the training process. The standard example are deep neural networks where the nonlinear process up to the last hidden layer builds new extended features presumably better than the original ones. The shallow models by which we essentially mean models different from deep NNs, in contrast, use directly the original features or a fixed transformation of them. Shallow models are not limited to linear approaches. Although kernel models are very expressive due to the implicit transformation of the original features in some Reproducing Kernel Hilbert Space (RKHS), this transformation is non-learnable and they can be considered shallow. In feature-learning-based MTL we can find examples of both deep models [5,13,15] and shallow ones [12]. Also in regularization-based approaches we have examples with deep [23] and shallow approaches [1,7,17]. However, the combination-based approach has only been applied to shallow models [8,18,22].

In this work we propose a convex formulation for a combination-based MTL approach based on deep models. To the best of our knowledge this is the first combination-based approach to MTL using deep models. The convex formulation we use enables an interpretable parametrization. The goal of this work is to define this approach and test its properties. More precisely, our main contributions are:

- Review the taxonomy for MTL, where we include a third category, the combination-based approaches, different from the original feature-learning and regularization-based approaches.
- Show a general formulation for convex combination-based MTL. Where the hypotheses that are combined can be taken from a wide range of models, not limited to kernel models.
- Propose a combination-based MTL with deep models and use a convex formulation for better interpretability.
- Implement this approach and test it with four image datasets.

This rest of the paper is organized as follows. In Sect. 2 we revise the Multi-Task Learning paradigm, reviewing different approaches and proposing a taxonomy. In Sect. 3 we present the general formulation for convex combination-based approaches and propose its application using Neural Networks. In Sect. 4 we show the experiments carried out to test our proposal and analyze their results. The paper ends with some conclusions and pointers to further work.

2   Multi-Task Learning Approaches

Multi-Task Learning (MTL) tries to learn multiple tasks simultaneously with the goal of improving the learning process of each task. Given T tasks, a Multi-Task (MT) sample is $z = \{(\boldsymbol{x}_i^r, y_i^r) \in \mathbb{R}^d \times \mathcal{Y}; i = 1, \ldots, m_r; r = 1, \ldots, T\}$, where $\mathcal{Y}$ can be $\mathbb{R}$ in the case of regression or $\{0, 1\}$ in the case of classification; the superindex $r \in \{1, \ldots, T\}$ indicates the task and m_r the number of examples in each task. The pair $(\boldsymbol{x}_i^r, y_i^r)$ can be also expressed as the triplet $(\boldsymbol{x}_i, y_i, r_i)$. The MT regularized risk for hypotheses h_r, that will be minimized, is defined as:

$$\sum_{r=1}^{T} \sum_{i=1}^{m_r} \ell(h_r(\boldsymbol{x}_i^r), y_i^r) + R(h_1, \dots, h_T), \tag{1}$$

where ℓ is some loss function and R some regularizer. One strategy to minimize this risk, denoted Common Task Learning (CTL), consists in using a common model for all the tasks, $h_1, \dots, h_T = h$. On the other side, in the Independent Task Learning (ITL) approach we minimize the risk independently for each task, without any transfer of information between them. Between these two extreme approaches lies MTL. The coupling between tasks can be enforced using different strategies. The choice of a strategy is influenced by the properties of the underlying models performing the learning process. In this paper we will focus on deep models, but in this section we will also discuss shallow models.

2.1   Multi-Task Learning with a Feature-Learning Approach

The feature-based approaches implement transfer learning by sharing a representation among tasks; that is, $h_r(\boldsymbol{x}) = g_r(f(\boldsymbol{x}))$, where f is some common feature transformation that can be learned and g_r is a task-specific function over these features. The first approach, *Hard Sharing*, is introduced in [5], where a Neural Network with shared layers and multiple outputs is used. The hidden layers are common to all tasks and, using the representation from the last hidden layer, a linear model is learned for each task; see Fig. 1 for an illustration. In the figure, a sample belonging to task 1 is used, the updated shared weights are represented in red, and in blue the updated specific weights. The input neurons are shown in yellow, the hidden ones in cyan and the output ones in magenta. The regularized risk corresponding to feature-learning MTL can be expressed as

$$\sum_{r=1}^{T} \sum_{i=1}^{m_r} \ell(g_r(f(\boldsymbol{x}_i^r)), y_i^r) + \mu_1 \sum_{r=1}^{T} \Omega_r(g_r) + \mu_2 \Omega(f), \tag{2}$$

where Ω_r and Ω are regularizers that penalize the complexity of the functions g_r and the function f, respectively; μ_1 and μ_2 are hyperparameters. The regularization over the predictive functions g_r can be done independently because the coupling is enforced by sharing the feature-learning function f.

A relaxation of the *Hard Sharing* approach consists in using the hypotheses $h_r(\boldsymbol{x}) = g_r(f_r(\boldsymbol{x}))$ where a coupling is enforced between the feature functions f_r. This is known as *Soft Sharing*, where specific networks are used for each task and some feature sharing mechanism is implemented at each level of the networks; examples are *cross-stitch networks* [13] or *sluice networks* [15]. In deep models, where a good representation is learned in the training process, Feature-Learning MTL is the most natural approach; however some Feature-Learning MTL approaches for shallow models can be found [12].

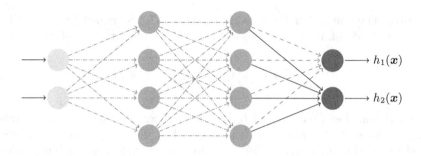

Fig. 1. *Hard sharing* neural network for two tasks and a two-dimensional input.

2.2 Multi-Task Learning with a Regularization-Based Approach

The regularization-based approaches are used when the hypothesis for each task can be expressed as $h_r(\boldsymbol{x}) = \boldsymbol{w}_r^\mathsf{T}\boldsymbol{\phi}(\boldsymbol{x})$, where $\boldsymbol{\phi}(\boldsymbol{x})$ is a non-learnable transformation which is the same for all tasks, and $\boldsymbol{w}_r$ are the parameters of interest to establish a relation between tasks. The transformation $\boldsymbol{\phi}(\boldsymbol{x})$ can be just the identity, using then the original features $\boldsymbol{x}$ in linear models, or some non-linear transformation of $\boldsymbol{x}$, explicit in deep models and implicit in kernel models. Here, the coupling is enforced by imposing some penalty over the matrix W whose columns are the vectors $\boldsymbol{w}_r$. The Multi-Task regularized risk is

$$\sum_{r=1}^{T}\sum_{i=1}^{m_r}\ell(\boldsymbol{w}_r^\mathsf{T}\boldsymbol{\phi}(\boldsymbol{x}_i^r), y_i^r) + \mu\Omega(W),\tag{3}$$

where $\Omega(W)$ is some regularizer of the matrix W and μ is a hyperparameter. For example, in [1,6] a low-rank constraint $\Omega(W) = \operatorname{rank} W$ is imposed over W, while in [7,17] a graph connecting the tasks is defined and a Laplacian regularization is used to penalize the distances between parameters, i.e., $\Omega(W) = \sum_{r,s=1}^{T} A_{rs} \|\boldsymbol{w}_r - \boldsymbol{w}_s\|^2$, where A is the adjacency matrix of the graph that encodes the pairwise task relations. These strategies can be more suitable for MTL with shallow models, but they are also applicable for deep ones [23]. In any case, in this work we use the standard L2 regularization common in deep networks; in particular, coupling is not necessarily enforced by the regularizer.

2.3 Multi-Task Learning with a Combination Approach

Another strategy, different to both the feature-learning and regularization-based approaches, is a combination $h_r(\boldsymbol{x}) = g(\boldsymbol{x}) + g_r(\boldsymbol{x})$ of a shared common model and task-specific ones. This approach was introduced in [8], where a combination of models $h_r(\boldsymbol{x}) = (\boldsymbol{w} + \boldsymbol{v}_r)^\mathsf{T}\boldsymbol{\phi}(\boldsymbol{x}) + b + b_r$ is defined; here $\boldsymbol{w}$ and $\boldsymbol{v}_r$ are the common and task-specific weights, respectively, whereas b and b_r are the corresponding biases. The regularized risk is here

$$\sum_{r=1}^{T}\sum_{i=1}^{m_r}\ell((\boldsymbol{w} + \boldsymbol{v}_r)^\mathsf{T}\boldsymbol{\phi}(\boldsymbol{x}_i^r) + b + b_r, y_i^r) + \mu_1 \|\boldsymbol{w}\|^2 + \mu_2 \sum_{r=1}^{T} \|\boldsymbol{v}_r\|^2.\tag{4}$$

If $\boldsymbol{w}_r = \boldsymbol{w} + \boldsymbol{v}_r$, the risk in (4) is equivalent to that in (3) with the regularizer

$$\Omega(W) = \rho_1 \sum_{r=1}^{T} \left\| \boldsymbol{w}_r - \left(\sum_{r=1}^{T} \boldsymbol{w}_r \right) \right\|^2 + \rho_2 \sum_{r=1}^{T} \|\boldsymbol{w}_r\|^2$$

for some values of the hyperparameters $\rho_1(\mu_1, \mu_2)$ and $\rho_2(\mu_1, \mu_2)$. That is, it imposes a regularization that penalizes the complexity of the parameters w_r and the variance between these parameters. Observe that both the common and specific parts belong to the same RKHS defined by the transformation ϕ.

An extension proposed in [4] uses $h_r(\boldsymbol{x}) = \boldsymbol{w}^\mathsf{T}\phi(\boldsymbol{x}) + b + \boldsymbol{v}_r^\mathsf{T}\phi_r(\boldsymbol{x}) + b_r$, where different transformations are used: ϕ for the common and ϕ_r for each of the specific parts. That is, the common part and each of the specific parts can belong to different spaces and, hence, capture distinct properties of the data. In [4] the connection of this MT approach with the *Learning Under Privileged Information* paradigm [20] is also outlined. A convex formulation for this approach, named Convex MTL, is presented in [16], where we have $h_r(\boldsymbol{x}) = \lambda\{\boldsymbol{w}^\mathsf{T}\phi(\boldsymbol{x}) + b\} + (1 - \lambda)\{\boldsymbol{v}_r^\mathsf{T}\phi_r(\boldsymbol{x}) + b_r\}$ and λ is a hyperparameter in the $[0, 1]$ interval. This parameter controls how much to share among the tasks. When $\lambda = 1$, the model is equivalent to the CTL approach, whereas $\lambda = 0$ represents the ITL approach. The regularized risk corresponding to this convex formulation is

$$\sum_{r=1}^{T} \sum_{i=1}^{m_r} \ell(\lambda\{\boldsymbol{w}^\mathsf{T}\phi(\boldsymbol{x}) + b\} + (1 - \lambda)\{\boldsymbol{v}_r^\mathsf{T}\phi_r(\boldsymbol{x}) + b_r\}, y_i^r) + \mu \left(\|\boldsymbol{w}\|^2 + \sum_{r=1}^{T} \|\boldsymbol{v}_r\|^2 \right),$$
(5)

where the hyperparameters μ_1 and μ_2 from (4) have been changed for λ and μ for a better interpretability: μ is the single regularization parameter and λ determines the specificity of our models. We can find the combination approach in the context of shallow models in [18, 22].

3   Convex MTL Neural Networks

3.1   Definition

The Convex MTL formulation described above in terms of linear models in some RKHS, can be generalized as the problem of minimizing the regularized risk

$$\sum_{r=1}^{T} \sum_{i=1}^{m} \ell(\lambda g(\boldsymbol{x}_i^r) + (1 - \lambda)g_r(\boldsymbol{x}_i^r), y_i^r) + \mu \left(\Omega(g) + \sum_{r=1}^{T} \Omega_r(g_r) \right), \qquad (6)$$

where Ω and Ω_r are regularizers and g and g_r are functions. Observe that (6) is not an *a posteriori* combination of common and specific models, but the objective function is minimized jointly on g and the specific models $g_1, \ldots, g_T$. In (5) each model acts in a different space determined by the implicit transformations ϕ and ϕ_r, that is $g(\boldsymbol{x}_i^r; \boldsymbol{w}) = \boldsymbol{w}^\mathsf{T}\phi(\boldsymbol{x}_i^r) + b$ and $g_r(\boldsymbol{x}_i^r; \boldsymbol{w}_r) = \boldsymbol{w}_r^\mathsf{T}\phi_r(\boldsymbol{x}_i^r) + b_r$. This permits a great flexibility but also imposes, for instance, the challenge of finding the optimal kernel width that implicitly defines the space for each model if kernels are used to define the underlying RKHS.

The Convex MTL neural network can be defined using a convex combination of common and task-specific models. The output of the overall model can be expressed as

$$h_r(\boldsymbol{x}_i^r) = \lambda\{\boldsymbol{w}^\mathsf{T} f(\boldsymbol{x}_i^r; \Theta) + b\} + (1 - \lambda)\{\boldsymbol{w}_r^\mathsf{T} f_r(\boldsymbol{x}_i^r; \Theta_r) + b_r\}. \tag{7}$$

That is, we use neural networks as the models $g(\boldsymbol{x}_i^r; \boldsymbol{w}, \Theta) = \boldsymbol{w}^\mathsf{T} f(\boldsymbol{x}_i^r; \Theta) + b$ and $g_r(\boldsymbol{x}_i^r; \boldsymbol{w}_r, \Theta_r) = \boldsymbol{w}_r^\mathsf{T} f_r(\boldsymbol{x}_i^r; \Theta_r) + b_r$. where Θ and Θ_r are the sets of hidden weights, and w, w_r are the output weights of the common and specific networks, respectively, and b and b_r the output biases. In this formulation, the common and specific feature transformations $f(\boldsymbol{x}_i^r; \Theta)$ and $f_r(\boldsymbol{x}_i^r; \Theta_r)$, the feature-building functions of the hidden layers, are automatically learned in the training process.

This formulation offers multiple combinations since we can model each common or independent function using different architectures. For example, we can use a larger network for the common part, since it will be fed with more data, and simpler networks for the specific parts. Even different types of neural networks, such as fully connected and convolutional, can be combined depending on the characteristics of each task. This combination of neural networks can also be interpreted as an implementation of the LUPI paradigm [20], i.e., the common network captures the privileged information for each of the tasks, since it can learn from more sources. To the best of our knowledge, this is the first joint-learning MTL approach for deep models, in contrast with previous feature-based or parameter-based approaches.

3.2 Training Procedure

The goal of the Convex MTL NN is to minimize the regularized risk

$$\sum_{r=1}^{T}\sum_{i=1}^{m} \ell(h_r(\boldsymbol{x}_i^r), y_i^r) + \mu\left(\|\boldsymbol{w}\|^2 + \sum_{r=1}^{T}\|\boldsymbol{w}_r\|^2 + \Omega(\Theta) + \Omega(\Theta_r)\right). \tag{8}$$

Here, h_r is defined as in equation (7), and $\Omega(\Theta)$ and $\Omega(\Theta_r)$ represents the L_2 regularization of the set of hidden weights of the common and specific networks, respectively. Given a loss $\ell(\hat{y}, y)$ and a pair $(\boldsymbol{x}_i^t, y_i^t)$ from task t, the gradient with respect to some parameters $\mathcal{P}$ is

$$\nabla_\mathcal{P} \ell(h_t(\boldsymbol{x}_i^t), y_i^t) = \frac{\partial}{\partial \hat{y}_i^t} \ell(\hat{y}_i^t, y_i^t)|_{\hat{y}_i^t = h_t(\boldsymbol{x}_i^t)} \nabla_\mathcal{P} h_t(\boldsymbol{x}_i^t). \tag{9}$$

Recall that we are using the formulation $h_t(\boldsymbol{x}_i^t) = \lambda\{\boldsymbol{w}^\mathsf{T} f(\boldsymbol{x}_i^t; \Theta) + b\} + (1 - \lambda)\{\boldsymbol{w}_t^\mathsf{T} f_t(\boldsymbol{x}_i^t; \Theta_t) + b_t\}$, where we make a distinction between output weights $\boldsymbol{w}, \boldsymbol{w}_t$ and hidden parameters Θ, Θ_t. The corresponding gradients are

$$\begin{aligned}
&\nabla_{\boldsymbol{w}} h_t(\boldsymbol{x}_i^t) = \lambda\{f(\boldsymbol{x}_i^t, \Theta)\}, && \nabla_\Theta h_t(\boldsymbol{x}_i^t) = \lambda\{\boldsymbol{w}^\mathsf{T} \nabla_\Theta f(\boldsymbol{x}_i^t, \Theta)\} : \\
&\nabla_{\boldsymbol{w}_t} h_t(\boldsymbol{x}_i^t) = (1 - \lambda)\{f_t(\boldsymbol{x}_i^t, \Theta)\}, && \nabla_{\Theta_t} h_t(\boldsymbol{x}_i^t) = (1 - \lambda)\{\boldsymbol{w}^\mathsf{T} \nabla_{\Theta_t} f_t(\boldsymbol{x}_i^t, \Theta_t)\}; \\
&\nabla_{\boldsymbol{w}_r} h_t(\boldsymbol{x}_i^t) = 0, && \nabla_{\Theta_r} h_t(\boldsymbol{x}_i^t) = 0, \text{ for } r \neq t.
\end{aligned}$$
$$\tag{10}$$

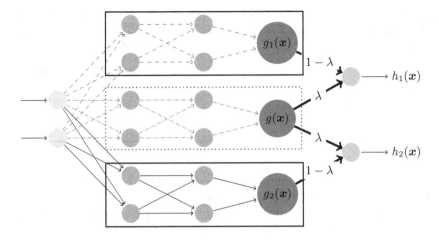

Fig. 2. Convex MTL neural network for two tasks and a two-dimensional input.

The convex combination information is transferred in the back propagation step as follows: the gradients of the loss function with respect to the common network parameters are scaled by λ, the gradients with respect to the t-th specific network parameters are scaled by $1 - \lambda$, and the rest of the task-specialized networks parameters have null gradients, so they are not updated. The regularization is independent in each network, so the gradients of the regularizers are also computed independently. That is, no specific training algorithm has to be developed for the Convex MTL NN, so (8) can be minimized with any stochastic gradient descent strategy using back propagation. In Fig. 2, a Convex MTL NN is shown. In particular, the updated shared weights are represented in red, and in blue the updated specific weights. Specific networks are framed in black boxes and the common one in a blue box. The input neurons are shown in yellow, the hidden ones in cyan (except those in grey), and the output ones in magenta. We use the grey color for hidden neurons containing the intermediate functions that will be combined for the final output: $g_1(x)$, $g_2(x)$ and $g(x)$. The thick lines are the hyperparameters λ and $1 - \lambda$ of the convex combination.

3.3 Implementation Details

Our implementation of the Convex MTL neural network is based on PyTorch [14]. Although we include the gradients expressions in equation (10), the PyTorch package implements automatic differentiation, so no explicit gradient formulation is necessary. The Convex MTL is implemented using (possibly different) PyTorch modules for the common model and each of the specific modules. In the forward pass of the network, the output for an example x from task r is computed using a forward pass of the common module and the specific module corresponding to task r, and the final output is simply the convex combination of both outputs. In the training phase, in which minibatches are used,

---

Algorithm 1: Forward pass for Convex MTL neural network.

Input: $X_{\text{mb}}, t_{\text{mb}}$ `// Minibatch data and task labels`
Output: f `// Forward pass for the minibatch`
Data: λ `// Parameter of convex combination`
Data: $g; g_1, \ldots, g_T$ `// Modules of the common and specific networks`
for $x_i, t_i \in (X_{mb}, t_{mb})$ **do**
    $f_i \leftarrow \lambda g(x_i) + (1 - \lambda) g_{t_i}(x_i)$ `// Convex combination`
end

---

the full minibatch is passed through the common model, but the minibatch is partitioned using only the corresponding examples for each task-specific modules. As mentioned above, with the adequate forward pass, the `PyTorch` package automatically computes the scaled gradients in the training phase.

In Algorithm 1 we show the pseudo-code of this Convex MTL forward pass, where g and $g_1, \ldots, g_T$ are the common and task-specific modules whose predictions are combined. As mentioned above, for the backward pass we rely on PyTorch automatic differentiation, so we do not need an explicit algorithm.

4   Experimental Results

4.1   Problems Description

To test the performance of the Convex MTL deep neural network approach we use four different image datasets: `var-MNIST`, `rot-MNIST`, `var-FMNIST` and `rot-FMNIST`. Datasets `var-MNIST` and `rot-MNIST` are the result of applying two different procedures, which will be detailed below, to the MNIST dataset [11], while for `var-FMNIST` and `rot-FMNIST` we apply the same procedure over the fashion-MNIST dataset [21]. Both MNIST and fashion-MNIST datasets are composed of 28×28 grey-scale images, with 10 balanced classes; also both problems have 70 000 examples. The procedures considered divide the original datasets and apply a different transformation to each resulting subset. To do this, we shuffle the original data and divide it equally among the tasks considered.

For datasets `var-MNIST` and `var-FMNIST` we consider two transformations described for the MNIST Variations datasets in [3]: *background random*, adding random noise to the original image, and *background image*, adding random patches of natural images. Using these transformations we define three tasks: `standard`, `random` and `images`, where either no transformation or one of the *background random* and the *background image* transformations is applied, respectively, to define each task. That is, two tasks have 23 333 examples each, and there are 23 334 in the third one.

The datasets `var-MNIST` and `var-FMNIST` are generated using the procedure specified in [9]. We define six different tasks, each one corresponding to a rotation of 0, 15, 30, 45, 60 and 75°, respectively; therefore, there are four tasks with 11 667 examples and two with 11 666. In Fig. 3, examples of the tasks for the four problems considered are shown.

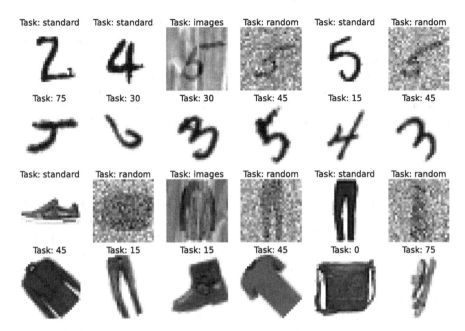

Fig. 3. Images of the four classification problems used. Each image has a title indicating the corresponding task. The rows correspond to var-MNIST, rot-MNIST, var-FMNIST and rot-FMNIST (from top to bottom).

4.2   Experimental Procedure

We compare four different models, all based on deep neural networks: a Common-Task Learning approach ctlNN, an Independent-Task learning approach itlNN, a Convex Multi-Task Learning approach cvxmtlNN and a *hard sharing* Multi-Task Learning approach hsNN. The base architecture of all models is a convolutional NN that we will name *convNet*. The architecture is based on the Spatial Transformer Network (STN) [10] architecture proposed in Pytorch[1], for further work using STN's. This *convNet* has 2 convolutional layers of kernel size 5, the first one with 10 output channels and the second one with 20; then we add a dropout layer, a max pooling layer and two hidden linear layers with 320 and 50 neurons each. In the ctlNN approach, a single *convNet* with 10 outputs, one for each class, is used. For the itlNN approach, an independent *convNet* with 10 outputs is used for each task. In cvxmtlNN both the common and task-specific networks are modelled using a *convNet* with 10 outputs; hsNN uses a *convNet* and a group of 10 outputs for each task.

All the models considered are trained using the *AdamW* algorithm and the optimal weight decay parameter μ is selected using a cross-validation grid search over the values $\{10^{-4}, 10^{-3}, 10^{-2}, 10^{-1}, 10^{0}\}$. The rest of the parameters corresponding to the algorithm are set to the default values: the dropout rate is 0.5

---

[1] www.pytorch.org/tutorials/intermediate/spatial_transformer_tutorial.html.

and a stride of 2 for the 2 × 2 max pooling layer. Additionally, in the cvxmtlNN model the mixing parameter λ is also included in the grid search using the values $\{0, 0.2, 0.4, 0.6, 0.8, 1\}$. The training and test sets are generated using a task-stratified 70 % and 30 % random split of the complete datasets. Grid-search is done by cross-validation using 5 folds over the training subset. We use task-stratified folds, that is, we divide the training set in 5 differents subsets where the task proportions are kept constant. Notice that the problems are class-balanced, so no further stratification is needed.

4.3   Analysis of the Results

To obtain results less sensitive to the randomness in deep networks, once the hyperparameters have been selected by cross-validation, we refit the models with optimal hyperparameters five times using the entire training set and the predictions are combined as described below. The final goal in classification problems is typically to maximize accuracy; however it cannot be used as a loss, so we will minimize the categorical cross entropy. We show both scores in the results.

In Table 1 we compute a single accuracy score for each model using the majority voting prediction of the 5 refitted models. In Table 2 we show the average cross entropy loss of the 5 different models. We also show in the tables the optimal values for hyperparameter λ^* selected in CV for cvxmtlNN. In both tables, cvxmtlNN obtains the best results in all four problems and the itlNN comes second except for the var-MNIST problem using majority voting. That is, training a specific model for each task obtains better results than the more rigid ctlNN or hsNN models. Also, although the ctlNN model obtains the worst results, the difference is not that large, so it suggests that the tasks are not very different, or that there exists information shared across tasks. The hsNN model consistently outperforms the ctlNN model and it seems to capture some shared information; however this hard sharing approach seems too rigid to fully exploit this common knowledge. Our proposal, the cvxmtlNN model, has the adequate flexibility because it trains specific modules for each task, but it also captures the shared information through the common model. Moreover, in cvxmtlNN the training of the common and specific models is made jointly and since this results in better models, we can conclude that, although the common and specific parts do not learn totally overlapping information, they complement each other's learning. This is supported by the fact that the λ values selected in CV are away from the extremes 0 and 1 of the independent and common models.

Table 1. Test accuracy with majority voting.

	var-MNIST	rot-MNIST	var-FMNIST	rot-FMNIST
ctlNN	0.964	0.973	0.784	0.834
itlNN	0.968	0.981	0.795	0.873
hsNN	0.971	0.980	0.770	0.852
cvxmtlNN	**0.974**	**0.984**	**0.812**	**0.880**
	$(\lambda^* = 0.6)$	$(\lambda^* = 0.8)$	$(\lambda^* = 0.6)$	$(\lambda^* = 0.6)$

Table 2. Test mean categorical cross entropy.

	var-MNIST	rot-MNIST	var-FMNIST	rot-FMNIST
ctlNN	1.274 ± 0.143	1.145 ± 0.039	2.369 ± 0.183	1.757 ± 0.075
itlNN	1.072 ± 0.029	0.873 ± 0.058	2.356 ± 0.130	1.598 ± 0.042
hsNN	1.087 ± 0.253	0.898 ± 0.073	3.067 ± 0.888	1.888 ± 0.075
cvxmtlNN	$\mathbf{0.924 \pm 0.024}$	$\mathbf{0.831 \pm 0.029}$	$\mathbf{2.147 \pm 0.090}$	$\mathbf{1.482 \pm 0.063}$
	$(\lambda^* = 0.6)$	$(\lambda^* = 0.8)$	$(\lambda^* = 0.6)$	$(\lambda^* = 0.6)$

5   Conclusions and Further Work

Here we have proposed a combination-based MTL approach using deep networks to define common and task-specific models that work together through a convex formulation. We have revised a taxonomy of previous MTL proposals adding the combination-based models as a distinct category; to the best of our knowledge, ours is the first proposal in this new category which uses neural networks.

The most popular NN based approach to MTL has been *hard sharing*, where tasks share the hidden parameters and different outputs are used for each task. In our experiments we have observed that our model outperforms the *hard sharing* approach in the four image problems considered. Moreover, our proposal also obtains better results than the baseline neural models for common- or independent-task learning. From this fact, we can infer that our MTL approach is able to extract and jointly exploit the information learned by the common and task-specific parts. We point out that the convex combination approach to MTL can also be applied to other models, such as SVMs. However, their potentially very high computational cost is well known and, in fact, we have not been able to apply them to our image classification problems.

As lines of further work, we point out that the mixing λ coefficient selected here as a hyperparameter by CV, can be alternatively seen as another network weight to be learned. It is also interesting to fully exploit the flexibility of our approach by using different architectures for each one of the common and task-specific modules. We are currently pursuing these and other ideas.

References

1. Ando, R.K., Zhang, T.: A framework for learning predictive structures from multiple tasks and unlabeled data. J. Mach. Learn. Res. **6**, 1817–1853 (2005)
2. Baxter, J.: A model of inductive bias learning. J. Artif. Intell. Res. **12**, 149–198 (2000)
3. Bergstra, J., Bengio, Y.: Random search for hyper-parameter optimization. J. Mach. Learn. Res. **13**, 281–305 (2012)
4. Cai, F., Cherkassky, V.: SVM+ regression and multi-task learning. In: International Joint Conference on Neural Networks, IJCNN 2009, pp. 418–424. IEEE Computer Society (2009)
5. Caruana, R.: Multitask learning. Mach. Learn. **28**(1), 41–75 (1997)
6. Chen, J., Tang, L., Liu, J., Ye, J.: A convex formulation for learning shared structures from multiple tasks. In: ACM International Conference Proceeding Series, ICML 2009, vol. 382, pp. 137–144 (2009)
7. Evgeniou, T., Micchelli, C.A., Pontil, M.: Learning multiple tasks with kernel methods. J. Mach. Learn. Res. **6**, 615–637 (2005)
8. Evgeniou, T., Pontil, M.: Regularized multi-task learning. In: Proceedings of the Tenth ACM SIGKDD International Conference on Knowledge Discovery and Data Mining, pp. 109–117. ACM (2004)
9. Ghifary, M., Kleijn, W.B., Zhang, M., Balduzzi, D.: Domain generalization for object recognition with multi-task autoencoders. In: IEEE International Conference on Computer Vision, ICCV, pp. 2551–2559. IEEE Computer Society (2015)
10. Jaderberg, M., Simonyan, K., Zisserman, A., Kavukcuoglu, K.: Spatial transformer networks (2015). https://doi.org/10.48550/ARXIV.1506.02025. https://arxiv.org/abs/1506.02025
11. LeCun, Y., Bottou, L., Bengio, Y., Haffner, P.: Gradient-based learning applied to document recognition. Proc. IEEE **86**(11), 2278–2324 (1998)
12. Maurer, A., Pontil, M., Romera-Paredes, B.: Sparse coding for multitask and transfer learning. In: Proceedings of the 30th International Conference on Machine Learning, ICML 2013, vol. 28, pp. 343–351. JMLR.org (2013)
13. Misra, I., Shrivastava, A., Gupta, A., Hebert, M.: Cross-stitch networks for multitask learning. In: 2016 IEEE Conference on Computer Vision and Pattern Recognition, CVPR 2016, pp. 3994–4003. IEEE Computer Society (2016)
14. Paszke, A., et al.: Pytorch: an imperative style, high-performance deep learning library. In: Advances in Neural Information Processing Systems 32, pp. 8024–8035 (2019)
15. Ruder, S.: An overview of multi-task learning in deep neural networks. CoRR abs/1706.05098 (2017)
16. Ruiz, C., Alaíz, C.M., Dorronsoro, J.R.: A convex formulation of SVM-based multitask learning. In: Pérez García, H., Sánchez González, L., Castejón Limas, M., Quintián Pardo, H., Corchado Rodríguez, E. (eds.) HAIS 2019. LNCS (LNAI), vol. 11734, pp. 404–415. Springer, Cham (2019). https://doi.org/10.1007/978-3-030-29859-3_35
17. Ruiz, C., Alaíz, C.M., Dorronsoro, J.R.: Convex graph Laplacian multi-task learning SVM. In: Farkaš, I., Masulli, P., Wermter, S. (eds.) ICANN 2020. LNCS, vol. 12397, pp. 142–154. Springer, Cham (2020). https://doi.org/10.1007/978-3-030-61616-8_12
18. Ruiz, C., Alaíz, C.M., Dorronsoro, J.R.: Convex formulation for multi-task L1-, L2-, and LS-SVMs. Neurocomputing **456**, 599–608 (2021)

19. Vapnik, V.: Estimation of Dependences Based on Empirical Data. Springer, New York (1982)
20. Vapnik, V., Izmailov, R.: Learning using privileged information: similarity control and knowledge transfer. J. Mach. Learn. Res. **16**, 2023–2049 (2015)
21. Xiao, H., Rasul, K., Vollgraf, R.: Fashion-MNIST: a novel image dataset for benchmarking machine learning algorithms (2017)
22. Xu, S., An, X., Qiao, X., Zhu, L.: Multi-task least-squares support vector machines. Multimedia Tools Appl. **71**(2), 699–715 (2013). https://doi.org/10.1007/s11042-013-1526-5
23. Yang, Y., Hospedales, T.M.: Trace norm regularised deep multi-task learning. In: 5th International Conference on Learning Representations, ICLR 2017. OpenReview.net (2017)
24. Zhang, Y., Yang, Q.: An overview of multi-task learning. Natl. Sci. Rev. **5**(1), 30–43 (2017)

Smash: A Compression Benchmark with AI Datasets from Remote GPU Virtualization Systems

Cristian Peñaranda[1]([✉])[ID], Carlos Reaño[2][ID], and Federico Silla[1][ID]

[1] Universitat Politècnica de València, Valencia, Spain
{cripeace,fsilla}@gap.upv.es
[2] Universitat de València, Valencia, Spain
carlos.reano@uv.es

Abstract. Remote GPU virtualization is a mechanism that allows GPU-accelerated applications to be executed in computers without GPUs. Instead, GPUs from remote computers are used. Applications are not aware of using a remote GPU. However, overall performance depends on the throughput of the underlying network connecting the application to the remote GPUs. One way to increase this bandwidth is to compress transmissions made within the remote GPU virtualization middleware between the application side and the GPU side.

In this paper we make a two-fold contribution. On the one hand, we present a new compression benchmark with more than 40 compression libraries. On the other hand, we have gathered the internal transmissions of a remote GPU virtualization middleware while executing 4 popular artificial intelligence applications. With these data we have created a new dataset for testing compression libraries in this specific domain. Both, the new compression benchmark and the new dataset are publicly available.

Keywords: Compression · Benchmark · Artificial Intelligence · GPU

1 Introduction

GPU-based (Graphics Processing Unit) computing has revealed to be an enabling technology for many areas. Among them, Artificial Intelligence (AI) largely benefits from these powerful accelerators [2]. Frameworks such as TensorFlow [1] or PyTorch [12] would have no real impact without GPUs, which are typically programmed by leveraging the CUDA technology created by NVIDIA [11].

In a similar way to other devices, GPUs have finally been reliably virtualized. In this regard, NVIDIA created the vGPU (virtual GPU) and MIG (Multi-Instance GPU) technologies, which allow a given GPU to be concurrently shared

This work was supported by the project "AI in Secure Privacy-Preserving Computing Continuum (AI-SPRINT)" through the European Union's Horizon 2020 Research and Innovation Programme under Grant 101016577.

among several applications [10,13]. The MIG technology is a generalization of the previous vGPU mechanism. It provides more flexibility for partitioning a GPU among several processes being executed in the computer hosting the GPU.

Virtualizing GPUs can be further enhanced by concurrently sharing a GPU located in a computer among processes being executed in other computers. This mechanism is known as remote GPU virtualization. By using this approach, a CUDA application being executed in a computer without a GPU is provided a GPU (or a part of it) located in another computer. The application is not aware of using a remote GPU because the CUDA calls performed by the application are transparently forwarded, across the network, to the remote GPU.

Using remote virtual GPUs can be very useful. For example, in the context of a cluster, instead of installing one or more GPUs in all the nodes of the cluster, GPUs could only be installed in some of the nodes and shared among the rest of the nodes. In a different scenario, remote GPUs could be used by an AI application to detect whether people wear face masks or not. In this case, a small embedded hardware with a camera could take images and process them using TensorFlow, for instance, while using a remote GPU located in the cloud to increase the computing power of the embedded system.

The performance of applications using the remote GPU virtualization mechanism greatly depends on the network connecting the computer where the application is being executed and the remote server with the GPU. To prevent loss of performance in comparison to using local GPUs, the throughput of the network used to transfer CUDA calls and data to/from the remote GPU should be similar to the throughput of the bus (e.g. PCIe) connecting the GPU to the rest of the system in the traditional (local) scenario. In case the network fabric does not present the required bandwidth, data might be compressed on the fly before being transmitted to/from the remote GPU. This compression would artificially increase the available bandwidth. Nevertheless, given that compressing and decompressing data requires some computations, the communication layer responsible for transferring data between the CUDA application and the remote GPU should be carefully designed. A pipeline approach, for instance, would reduce the impact of compression and decompression times while still reducing overall data size. In practice, this would increase the available bandwidth.

Designing a compression-capable communication layer within a remote GPU virtualization middleware requires to consider different choices for the compression library. There are many of them, although not all of them provide the same features (e.g. compression ratio and compression speed). Moreover, when analyzing the compression/decompression capabilities of the different libraries, it is important to use the right dataset [16]. In this regard, many datasets are available. However, not all of them are suitable for every study. Actually, this is one of the reasons why there are so many different datasets, as every specific domain requires a representative dataset.

In the case of remote GPU virtualization solutions using on-the-fly compressed transmissions, the dataset to be used in the design stage should be representative of the traffic between the CUDA application and the remote GPU.

This traffic is basically composed of the data copied to/from the remote GPU (from/to main memory) during the execution of the application using the remote GPU virtualization middleware. Furthermore, different CUDA applications are expected to move different data to/from the GPU. For instance, in the AI application domain, the traffic generated by TensorFlow should be considered. However, as far as we know, currently such dataset does not exist. This is precisely one of the problems that this work aims to address.

This paper presents the following main contributions:

- We present a new dataset to be used by compression libraries. This new dataset is composed of the data copied to/from the GPU when TensorFlow is executed using representative samples.
- A new compression benchmark, referred to as Smash, is introduced. This compression benchmark includes more than 40 different compression libraries popular in current literature.
- We show that current compression libraries behave very differently when dealing with standard datasets than when dealing with data moved between main memory and GPU memory during the execution of TensorFlow tests.

The work presented in this paper fills two gaps: (1) a novel dataset is presented, demonstrating its significance, and (2) a new benchmark for comparing compression libraries is introduced. Both of them are publicly available.

The rest of the paper is organized as follows. Section 2 presents the required background to better understand this paper. Then, Sect. 3 introduces the new compression benchmark as well as the new AI dataset. Later, in Sect. 4 we show the influence on the consequences of this study when the new AI dataset is used. Finally, in Sect. 5 we summarize the main conclusions of the paper and also present some directions for future work.

2   Related Work

In this section we provide the necessary background to put our work into context. In particular, we describe the state of the art of (i) remote GPU virtualization solutions, (ii) compression libraries and, (iii) datasets commonly used to evaluate those libraries.

2.1   Remote GPU Virtualization

In the context of the CUDA technology developed by NVIDIA, remote GPU virtualization allows a CUDA application being executed in a computer which does not have a physical GPU to access a GPU installed in a remote computer. Figure 1 depicts the architecture usually deployed by remote GPU virtualization solutions, which follows a distributed client-server approach. The client part of the middleware is a library providing the same API as CUDA. It is installed in the computer executing the application requesting GPU services. The server

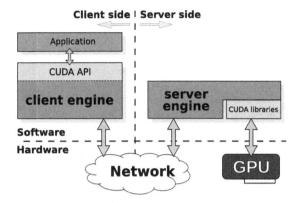

Fig. 1. General architecture of remote GPU virtualization solutions.

side of the middleware is a daemon that runs in the computer owning the actual GPU. Both client and server make use of the network to communicate.

The architecture depicted in Fig. 1 is used in the following way: every time an application makes a CUDA request, the client middleware intercepts that request, appropriately processes it, and forwards it to the server middleware. In the server side, the daemon receives the request, interprets and forwards it to the GPU, which executes the request and returns the results to the server middleware. Finally, the server sends back the results to the client middleware, which forwards them to the initial application. Notice that GPU virtualization solutions provide GPU services in a transparent way and, therefore, applications are not aware that their requests are actually serviced by a remote virtual GPU instead of a local one. In this way, the source code of applications using remote GPU virtualization solutions does not need to be modified.

Currently, the most prominent remote GPU virtualization solutions supporting CUDA are rCUDA [14] and GVirtuS [4]. Both of them present the same architecture, although rCUDA supports a wider set of CUDA functions. For this reason, in this work we focus on rCUDA.

2.2   Compression Libraries

Data compression is a well-defined area [3,9] with some widely used libraries, such as the LZ4 or brotli ones. Table 1 shows the 42 compression libraries used in our novel benchmark. Note that all these libraries already existed and we have just put them together in a single benchmark. Due to space constraints, we do not detail the libraries. More information is available in the links provided.

2.3   Datasets Used with Compression Libraries

In addition to use different compression libraries, different application domains require different datasets to evaluate those libraries. Actually, compression libraries can behave differently depending on the data to be compressed [5].

Table 1. Compression libraries considered in our novel benchmark.

Library	Domain	Link
BriefLZ	generic	github.com/jibsen/brieflz
brotli	generic	github.com/google/brotli
bzip2	generic	www.sourceware.org/bzip2
c-blosc2	binary data	github.com/Blosc/c-blosc2
csc	English text	github.com/fusiyuan2010/CSC
density	generic	github.com/k0dai/density
FastARI	generic	github.com/davidcatt/FastARI
FastLZ	data with lots of repetitions	ariya.github.io/FastLZ
flzma2	generic	github.com/conor42/fast-lzma2
fse	generic	github.com/Cyan4973/FiniteStateEntropy
gipfeli	generic	github.com/google/gipfeli
heatshrink	generic	github.com/atomicobject/heatshrink
libbsc	generic	libbsc.com
libdeflate	generic	github.com/ebiggers/libdeflate
liblzg	generic	liblzg.bitsnbites.eu
lizard	generic	github.com/inikep/lizard
lodepng	PNG images	lodev.org/lodepng
LZ4	generic	lz4.github.io/lz4
LibLZF	data with repeated blocks	oldhome.schmorp.de/marc/liblzf.html
LZFSE	generic	github.com/lzfse/lzfse
LZFX	scientific data	code.google.com/archive/p/lzfx
LZHAM	generic	github.com/richgel999/lzham_codec
LZJB	crash dumps	github.com/nemequ/lzjb.git
lzma	generic	tukaani.org/lzma
lzmat	generic	www.matcode.com/lzmat.htm
lzo	generic	www.oberhumer.com/opensource/lzo
LZSSE	generic	github.com/ConorStokes/LZSSE
miniz	generic	github.com/richgel999/miniz
ms	generic	github.com/coderforlife/ms-compress
pithy	text	github.com/johnezang/pithy
QuickLZ	generic	www.quicklz.com
snappy	generic	google.github.io/snappy
ucl	generic	www.oberhumer.com/opensource/ucl
wfLZ	game engines	github.com/ShaneYCG/wflz
xpack	generic	github.com/ebiggers/xpack
yalz77	generic	bitbucket.org/tkatchev/yalz77/src
z3lib	generic	scara.com/ schirmer/o/z3lib
zlib	generic	www.zlib.net
zlib-ng	generic	github.com/Dead2/zlib-ng
Libzling	generic	github.com/richox/libzling
zpaq	generic	mattmahoney.net/dc/zpaq.html
zstd	generic	github.com/facebook/zstd

For instance, some libraries are very good when compressing images but perform bad with text; some libraries perform better with larger datasets, while others are better with smaller ones. As a result, many standard datasets have been created, each of them intended for a particular domain.

Table 2. Common datasets used to test compression libraries.

Name	Source	Description	Size
alice29.txt	Canterbury Corpus	English text	148.52 KB
asyoulik.txt	Canterbury Corpus	Shakespeare	122.25 KB
cp.html	Canterbury Corpus	HTML source	24.03 KB
fields.c	Canterbury Corpus	C source	10.89 KB
grammar.lsp	Canterbury Corpus	LISP source	3.63 KB
kennedy.xls	Canterbury Corpus	Excel Spreadsheet	1005.61 KB
lcet10.txt	Canterbury Corpus	Technical writing	416.75 KB
plrabn12.txt	Canterbury Corpus	Poetry	470.57 KB
ptt5	Canterbury Corpus	CCITT test set	501.19 KB
sum	Canterbury Corpus	SPARC Executable	37.34 KB
xargs.1	Canterbury Corpus	GNU manual page	4.13 KB
dickens	Silesia Corpus	Collected works of Charles Dickens	9.72 MB
mozilla	Silesia Corpus	Tarred executables of Mozilla 1.0	48.85 MB
mr	Silesia Corpus	Medical magnetic resonanse image	9.51 MB
nci	Silesia Corpus	Chemical database of structures	32 MB
ooffice	Silesia Corpus	A dll from Open Office.org 1.01	5.87 MB
osdb	Silesia Corpus	Sample database in MySQL format	9.62 MB
reymont	Silesia Corpus	Text from Chopi by W. Reymont	6.32 MB
samba	Silesia Corpus	Tarred source code of Samba 2-2.3	20.61 MB
sao	Silesia Corpus	The SAO star catalog	6.92 MB
webster	Silesia Corpus	1913 Webster Unabridged Dictionary	39.54 MB
x-ray	Silesia Corpus	X-ray medical picture	8.08 MB
xml	Silesia Corpus	Collected XML files	5.1 MB
enwik8	Text Comp. Benchmark	First 10^8 bytes of English Wikipedia	95.37 MB
fireworks.jpeg	Snappy	A JPEG image	120.21 KB
geo.protodata	Snappy	A set of Protocol Buffer data	115.81 KB
paper-100k.pdf	Snappy	A PDF	100 KB
urls.10K	Snappy	List of 10000 URLs	685.63 KB

In order to properly assess the features of our benchmark, in this paper we have used several standard datasets (in addition to the novel AI dataset we detail later in the paper). These datasets have data with different contents and sizes, covering a wide range of scenarios. Table 2 gives a summary of the standard datasets used in the paper.

3   The Smash Compression Benchmark for AI

In order to carry out a thorough performance analysis of compression libraries for the data moved to/from the remote GPU when AI frameworks are executed using a remote GPU virtualization solution such as rCUDA, we need two different elements: (1) an appropriate dataset to evaluate the compression libraries, and (2) a tool for easily comparing the behavior of different compression libraries with that dataset.

3.1   A New Dataset for AI Applications

As commented in previous sections, in addition to use different compression libraries, different application domains require different datasets to evaluate those libraries. This is also the case of remote GPU virtualization solutions with on-the-fly compression. Data is compressed in the client side before being sent to the network and is later decompressed once it is received at the remote GPU server. In the opposite direction, from the remote GPU server to the client application, compression is also used in a similar way.

Notice that in this scenario what is compressed is the data that is moved to/from the remote GPU. For instance, every time the CUDA application performs a call to a cudaMemcpy function, some data is moved across the network to/from the remote GPU. It is important to remark that this data moved to/from the GPU is not necessarily the same data used as input to the application. For instance, the input data to the Cifar10 TensorFlow sample used in this paper consists of 60,000 32×32 colour images classified in 10 classes. Those images are stored in the disk and read during the execution of the Cifar10 sample. Every time one of these images is read from disk, it is appropriately transformed and the result of such transformation is inserted into the Convolutional Neural Network (CNN) used within the Cifar10 sample. During the insertion into the CNN, several copies will be made to the GPU and data will be later copied back to main memory after processing. Data moved in those copies to the GPU memory and back to main memory are what will be compressed in a remote GPU virtualization middleware with on-the-fly compression. Thus, the aggregation of those data copies will compose the novel dataset. Notice that users do not have access to that data moved to/from the GPU because that data is part of the internals of the execution of AI applications. Notice also that data in those copies is only available during part of the execution of the AI application. Before the program begins its execution, that data does not exist. In a similar way, once the AI application completes its execution, that data is not available anymore.

In order to gather the data moved to/from the remote GPU so that the novel dataset for AI applications is created, the rCUDA middleware can be slightly modified. The modification to the rCUDA middleware consists of storing on disk the data sent to the network. In this way, every time the CUDA application performs a data copy to/from the remote GPU, a new file is created on disk containing the data sent to the network before it is compressed. At the end of the execution of a TensorFlow sample, we will have a copy on disk of all the data exchanged with the remote GPU. That data can be later used to test different compression libraries.

In this work, we have gathered the data exchanged with the remote GPU during the execution of four TensorFlow samples, namely: Alexnet, Cifar10, Mnist and Inception, all available at https://git.dst.etit.tu-chemnitz.de/external/tf-models. Each sample uses different CNNs. Table 3 shows more details about them. The new AI dataset has been created with representative parts of these data. Table 4 shows detailed information about the dataset, which is available at https://github.com/cpenaranda/AI-dataset.

Table 3. Samples used to create the AI dataset.

Name	Description	# transmissions	Total size
Alexnet	Alexnet [7] CNN time inference test	36073	800 MB
Cifar10	Simple CNN using Cifar10 [6] dataset	74698	2.25 GB
Mnist	LeNet-5-like CNN using Mnist dataset [8]	262068	556 MB
Inception	Inception-V3 [15] CNN using flowers dataset	204194	4.71 GB

Table 4. New AI dataset created to evaluate compression libraries.

Name	Source	Description	Total size
AI-5K	Artificial Intelligence	Data exchanged with	5 KB
AI-10K	Tests	the remote GPU	10 KB
AI-25K		during the execution	25 KB
AI-50K		of four TensorFlow	50 KB
AI-100K		samples: Alexnet,	100 KB
AI-250K		Cifar10, Mnist and	250 KB
AI-500K		Inception	500 KB
AI-1000K			1000 KB
AI-2500K			2500 KB
AI-5M			5 MB
AI-10M			10 MB
AI-25M			25 MB
AI-50M			50 MB
AI-100M			100 MB

3.2   The Smash Compression Benchmark

Once a suitable dataset is available to test compression in remote GPU virtualization solutions, a tool is required to automate the analysis of the compression features of different compression libraries. In this regard, an existing compression benchmark could be used just by feeding the benchmark with the new dataset. At this point we initially considered the Squash compression benchmark (https://quixdb.github.io/squash-benchmark/). Despite being able to run 33 compression libraries, this benchmark is not maintained since 2018, being its last release in 2015. Also, we have not been able to make it work.

For these reasons, we decided to create a new compression benchmark named Smash, which is also available at https://github.com/cpenaranda/smash. This new benchmark comprises the 42 compression libraries previously mentioned and allows to specify different parameters for each one. Thus, by using Smash it is possible to select among others: (i) different levels of compression ratio and compression speed, (ii) the size used to store the dictionary to compress data, or (iii) how the compression library behaves with repetitive data. Furthermore,

```
/smash/build/bin:./smash_benchmark --compression_library all --input_file dickens
```

Library	Level	Window	Mode	WF	Flags	Threads	BR	Ratio	Total
zstd	1						---	2.39	152.63 MB/s
zpaq	0						---	2.47	17.64 MB/s
zling	0						---	3.00	33.97 MB/s
zlib-ng	0						---	1.00	2090.30 MB/s
zlib	0						---	1.00	816.18 MB/s
z3lib					None		---	2.58	6.88 MB/s
yalz77	1	10					---	1.44	43.42 MB/s
xpack	1						---	2.41	63.95 MB/s
wflz	0						---	1.71	0.06 MB/s
ucl	1		NRV2B				---	2.10	25.65 MB/s
snappy							---	1.61	173.17 MB/s
quicklz							---	1.75	151.77 MB/s
pithy	0						---	1.52	178.55 MB/s
ms			Lznt1				---	1.61	32.32 MB/s
miniz	1	10	None				---	1.90	50.20 MB/s
lzsse	1		LZSSE2				---	1.94	12.77 MB/s
lzo	0		LZO1				---	1.61	110.73 MB/s
lzmat							---	2.26	14.60 MB/s
lzma			Default			1	---	3.60	1.53 MB/s
lzjb							---	1.32	73.36 MB/s
lzham	0	15			None		---	2.34	2.21 MB/s
lzfx							---	1.61	106.37 MB/s
lzfse							---	2.64	36.90 MB/s
lzf	0						---	1.74	133.01 MB/s
lz4	0		Fast				---	1.59	268.46 MB/s
lodepng	1	10		1	Fast		1	1.45	20.99 MB/s
lizard	0		FastLZ4				---	1.58	191.78 MB/s
liblzg	1						---	1.47	34.71 MB/s
libdeflate	0		Deflate				---	1.00	2818.21 MB/s
libbsc	1	10	Bwt		None		4	1.00	14.81 MB/s
heatshrink		4					3	1.76	13.83 MB/s
gipfeli							---	2.06	101.03 MB/s
fse			FSE				---	1.76	194.92 MB/s
flzma2	1					1	---	2.83	12.13 MB/s
flz	1						---	1.68	121.20 MB/s
fari							---	2.64	11.34 MB/s
density			Chameleon				---	1.75	577.07 MB/s
csc	1	15			None		---	2.56	10.65 MB/s
c-blosc2	0				None	1	---	1.00	2812.31 MB/s
bzip2	1		Faster	0			---	3.09	7.85 MB/s
brotli	0	10	Generic				---	1.58	40.65 MB/s
brieflz	1						---	2.22	50.25 MB/s

Fig. 2. Smash benchmark execution using all the available compression libraries.

Smash allows to: (i) run multiple times the same libraries with different datasets showing the standard deviation to obtain more statistically significant results, (ii) run the compression libraries varying their parameters, and (iii) sort the results either by compression ratio, compression/decompression speed or total speed. As an example, Fig. 2 shows an execution of the Smash benchmark where all the compression libraries are compressing and decompressing the dickens dataset using default parameters. The ratio value displayed is defined in Eq. 1 as the ratio between the size of the original data and the size of the compressed data. Moreover, the total speed represents the speed it takes for the compression library to compress and decompress the original data. It is formally defined in Equation 2 as the ratio between (i) the size of the original data and (ii) the time taken to compress the original data plus decompress the compressed data.

$$Ratio = \frac{Size_{original}}{Size_{compressed}} \tag{1}$$

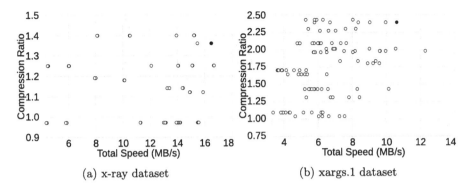

(a) x-ray dataset   (b) xargs.1 dataset

Fig. 3. LodePNG library using all possible combinations of parameters.

$$Speed = \frac{Size_{original}}{Time_{compression} + Time_{decompression}} \tag{2}$$

4   Experiments

The Smash benchmark is composed of multiple compression libraries and it allows to specify different parameters for each one. Due to space limitations, it is not possible to show all the results for all the libraries and datasets using all possible combinations of parameters. As an example, Fig. 3 shows the results for two of the datasets. As it can be seen, there is a lot of information. For clarity, it will be sensible to synthesize the information in some way.

To summarize the results, we have first run all the compression libraries with all datasets using all possible combinations of parameters. This results into figures similar to Fig. 3. Next, we have obtained the maximum compression ratio and the maximum total speed achieved by each compression library and dataset. Then, we have used these maximum values to get normalized numbers (i.e. dividing each value by the maximum value). Finally, we have aggregated the normalized compression ratio and the normalized total speed by using α and β coefficients. The sum of the α and β weights must be 1 to get a final result between 0 and 1. Note that the closer the sum is to 1, the better the result (i.e. the better the compression library behaves with that dataset). In the experiments shown in this paper, we have assumed that the compression ratio is as important as the total speed. Thus, we have set the α and β weights to 0.5. Equation 3 summarizes the above methodology.

$$f(result) = (\frac{Ratio_{result}}{Ratio_{max}} * \alpha) + (\frac{Speed_{result}}{Speed_{max}} * \beta) \tag{3}$$

As an example, in Fig. 3a and Fig. 3b we can see the results obtained with *LodePNG* compression library when it is run with *x-ray* and *xargs.1* datasets,

● fari	◆ gipfeli	■ lzfse	▲ lzfx	▼ lzjb
● quicklz	◆ snappy	■ lzmat	▮ lzf	▼ zpaq
● brieflz	✦ flz	▧ xpack	▲ zlib	▼ zlib-ng
● zstd	◆ bzip2	▨ brotli	▲ c-blosc2	▼ density
◉ libdeflate	◆ lzham	▦ lzma	▲ ms	▽ yalz77
◎ z3lib	◇ lodepng	▣ flzma2	▲ lizard	▽ lzsse
○ pithy	◇ zling	▢ csc	△ lz4	▽ miniz
○ liblzg	◇ fse	▢ ucl	△ wflz	▽ heatshrink
○ libbsc	◇ lzo			

Fig. 4. Legend used in Fig. 5 and Fig. 6.

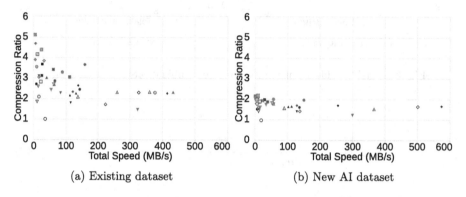

(a) Existing dataset   (b) New AI dataset

Fig. 5. Average of all datasets for each compression library.

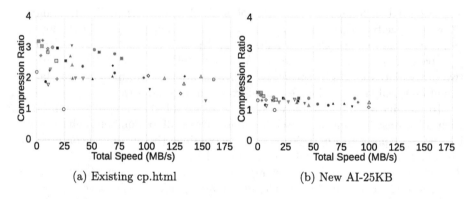

(a) Existing cp.html   (b) New AI-25KB

Fig. 6. Comparison between a standard 25 KB data and the AI 25 KB data.

respectively. Tests are executed using all possible combinations of parameters. The black dot represents the best result calculated using Eq. 3, while white dots are the rest of results obtained using all possible combinations of parameters. Following this methodology, we have calculated the best results for all the libraries and datasets using all possible combinations of parameters. This allows to have a single best result for each pair {compression library, dataset}.

Figure 4 shows the legend of all subsequent figures. Figure 5 shows the average of the best results from all datasets for every compression library: Fig. 5a shows results with the standard dataset shown in Table 2, whereas Fig. 5b shows results for the AI dataset introduced in Sect. 3.1. Most compression libraries obtain a total speed between 0 and 160 MB/s in both datasets, but we can observe significant differences in the compression ratio. Despite most compression libraries get a compression ratio between 2 and 5 with standard datasets, they barely achieve a compression ratio higher than 2 with the AI dataset. That makes sense, because the compression libraries focus on text, image or binaries datasets instead of data transferred to/from the remote GPU when running AI applications.

To better observe the differences, we show specific examples with *cp.html* and *AI-25KB* datasets in Fig. 6. Both datasets are around 25 KB in size. We can see that some compression libraries achieve a total speed higher than 100 MB/s with *cp.html*, while with *AI-25KB* they do not exceed that speed. Compression libraries behave slower due to the complexity of the data. Furthermore, no compression library gets a compression ratio higher than 1.8 with the *AI-25KB* dataset, while most of them get compression ratios higher than 2 with *cp.html*. These results again support the need for additional compression libraries that perform well with the particular characteristics of the AI dataset.

5   Conclusion

In this paper we have introduced a new open source benchmark, called Smash, comprising 42 different compression libraries. Its main focus is to evaluate the performance of compression libraries when running AI applications in remote GPU virtualization environments. Initial results showed that existing datasets were not appropriate for that purpose. Thus, a new dataset was created with data exchanged with the remote GPU during the execution of popular TensorFlow codes. Both the Smash benchmark and the AI dataset are publicly available.

As future work, we plan to improve the Smash benchmark by (i) adding GPU-accelerated compression libraries (in this work only CPU-accelerated libraries were considered), (ii) developing new ad hoc compression libraries for remote GPU virtualization environments, (iii) studying the influence of the network when data is transferred to a remote GPU located in a different computer.

References

1. Abadi, M., et al.: TensorFlow: large-scale machine learning on heterogeneous systems (2015). https://www.tensorflow.org/
2. Talib, M.A., Majzoub, S., Nasir, Q., Jamal, D.: A systematic literature review on hardware implementation of artificial intelligence algorithms. J. Supercomput. **77**(2), 1897–1938 (2020). https://doi.org/10.1007/s11227-020-03325-8
3. Alakuijala, J., et al.: Brotli: a general-purpose data compressor. ACM Trans. Inf. Syst. (TOIS) **37**(1), 1–30 (2018)
4. Giunta, G., Montella, R., Agrillo, G., Coviello, G.: A GPGPU transparent virtualization component for high performance computing clouds. In: D'Ambra, P., Guarracino, M., Talia, D. (eds.) Euro-Par 2010. LNCS, vol. 6271, pp. 379–391. Springer, Heidelberg (2010). https://doi.org/10.1007/978-3-642-15277-1_37

5. Gupta, A., et al.: Modern lossless compression techniques: review, comparison and analysis. In: ICECCT, pp. 1–8 (2017)
6. Krizhevsky, A., et al.: Learning multiple layers of features from tiny images (2009)
7. Krizhevsky, A., et al.: ImageNet classification with deep convolutional neural networks. In: Advances in Neural Information Processing Systems 25, pp. 1097–1105 (2012)
8. LeCun, Y., Cortes, C., Burges, C.: MNIST handwritten digit database. ATT Labs 2 (2010). http://yann.lecun.com/exdb/mnist
9. Liu, W., et al.: Data compression device based on modified LZ4 algorithm. IEEE Trans. Consum. Electron. **64**(1), 110–117 (2018)
10. NVIDIA Corporation: NVIDIA multi-instance GPU and NVIDIA virtual compute Server. Technical brief (2020)
11. NVIDIA Corporation: CUDA (Compute Unified Device Architecture) (2022). https://developer.nvidia.com/cuda-toolkit
12. Paszke, A., et al.: Pytorch: an imperative style, high-performance deep learning library. In: Advances in Neural Information Processing Systems, pp. 8024–8035 (2019)
13. Prades, J., Silla, F.: GPU-job migration: the rCUDA case. IEEE Trans. Parallel Distrib. Syst. **30**(12), 2718–2729 (2019). https://doi.org/10.1109/TPDS.2019.2924433
14. Silla, F., et al.: On the benefits of the remote GPU virtualization mechanism: the rCUDA case. Concurr. Comput. Pract. Exp. **29**(13), e4072 (2017)
15. Szegedy, C., et al.: Rethinking the inception architecture for computer vision. In: IEEE Conference on Computer Vision and Pattern Recognition, pp. 2818–2826 (2016)
16. Venu, D., et al.: An efficient low complexity compression based optimal homomorphic encryption for secure fiber optic communication. Optik **252**, 168545 (2022)

Time Series Forecasting Using Artificial Neural Networks
A Model for the IBEX 35 Index

Daniel González-Cortés[1]([⊠]) [iD], Enrique Onieva[2] [iD], Iker Pastor[2] [iD], and Jian Wu[1] [iD]

[1] NEOMA Business School, rue du Maréchal Juin,
76825 Mont Saint Aignan Cedex, France
{daniel-alejandro.gonzalez-cortes.20,jian.wu}@neoma-bs.com
[2] Faculty of Engineering, University of Deusto, 48007 Bilbao, Spain
{enrique.onieva,iker.pastor}@deusto.es

Abstract. The amount of data generated daily in the financial markets is diverse and extensive; hence, creating systems that facilitate decision-making is crucial. In this paper, different intelligent systems are proposed and tested to predict the closing price of the IBEX 35 using ten years of historical data with four different neural networks architectures. The first was a multi-layer perceptron (MLP) with two different activation functions (AF) to continue with a simple recurrent neural network (RNN), a long-short-term memory (LSTM) network and a gated recurrent unit (GRU) network. The analytical results of these models have shown a strong, predictable power. Furthermore, by comparing the errors of predicted outcomes between the models, the LSTM presents the lowest error with the highest computational time in the training phase. Finally, the empirical results revealed that these models could efficiently predict financial data for trading purposes.

Keywords: Machine learning · IBEX35 · Stock market prediction · Artificial neural networks · Recurrent neural network · Gated recurrent unit · Long-short-term memory

1 Introduction

The main concern of many economic agents is to forecast the future trends of the financial markets to make better decisions. The methods used and the time frames to predict are diverse. The stock markets represent a fundamental piece of any modern economy by letting investors exchange financial instruments at an agreed price with many fluctuations over time. These variations are considered chaotic and non-stationary; however, there is some empirical evidence [18] suggesting that stock returns can have some predictable components rejecting the hypothesis of the random walks.

All the economic agents need to be aware of the stock market's implications at different economic levels. As seen in the global financial crisis of 2007–2009,

© Springer Nature Switzerland AG 2022
P. García Bringas et al. (Eds.): HAIS 2022, LNAI 13469, pp. 249–260, 2022.
https://doi.org/10.1007/978-3-031-15471-3_22

the financial contagion affected different sectors of the real economy, such as consumer goods, industrials, telecommunications and technology [4]. Therefore, forecasting future stock prices and trends can be crucial for better financial decisions. However, this is not an easy task because the nature of the stock market is intrinsically nonlinear, non-parametric and chaotic, where many variables interact, making prices move in one direction or another.

The prediction process in the stock market has been approached by two different methodologies, fundamental and technical analysis. The first one is based on the valuation of the intrinsic value of stocks by using the current and future earnings of the company to evaluate the fair value and then contrast this information with the market value indexed in the stock exchange. The second methodology does not count on the company's financial statements as the primary source of information. However, it merely relies on data using historical stock prices to make predictions trying to identify statistical trends.

Many investors use both methodologies to make buying or selling decisions, and the 87 % of fund managers use some technical analysis [19]. However, the increasing expansion and evolution of datification and automation prompted the financial markets to find new processes to remain competitive and reinvent their services. Then artificial intelligence and machine learning became a powerful tool for institutions, financial advisers, banks and wealth managers and disruptively transformed their business model [17].

This research focuses on studying Artificial Neural Networks (ANN) and attempts to clarify further the use of its different variations in predicting the stock market. This work starts with a bibliographic review in Sect. 2, about the use of ANN to solve some financial problems and continues with the Materials and methods in Sect. 3 to explain the initial settings of the experimentation. Then, to continue in Sect. 4 with the description of ANNs, starting with the description and creation of an Multi-Layer Perceptron (MLP) network in Sect. 4.1. Subsequently, the structure of a Recurrent Neural Network (RNN) is presented, introducing the Simple RNN, Long-Short-Term Memory (LSTM) and Gated Recurrent Unit (GRU) architectures in Sect. 4.2. Finally, to finish with the results and discussion and conclusions about the performance of these models predicting financial times series in section in Sect. 5 and 6 respectively.

2 Background

Over the years, ANNs have played a vital role in the decision-making process of banks and financial institutions due to the adaptation of new and automated systems to their operations. These new technologies have been quickly adopted because they are consistent and objective, eliminating human bias or wrong assumptions. In addition, ANNs empower investors and institutions to create new powerful models by extracting information from past observations and improving preconceived models.

The ANN are a bio-inspired computing system based on many connected processors called neurons activated by different types of activation functions

(AF) triggered according to specific weights and bias. This system aims to minimize the prediction error by using a feed-forward optimization that will change the weights of the different interconnected neurons. This model can capture nonlinearities and has been used in different business applications to predict financial distress, bankruptcy analysis, stock price predictions, and credit scoring Tkac et al. [26].

Different ANN applications to predict the stock market can be found in the literature. For example, Qui et al. [23] used an ANN to predict the return of the Japanese Nikkei 225 index by using a hybrid approach based on a genetic algorithm and simulated annealing. Additionally, Pyo et al. [22] analyzed the prediction of a stock exchange index, building three hypotheses to forecast the daily closing prices of the Korea Stock Price Index 2000 by using an ANN and two SVMs models. Also, Kara et al. [14] using a three-layered feed-forward ANN and an SVM model, predicted the direction of the Istanbul stock exchange index, concluding that the 75.74% performing prediction for the ANN was significantly better than for the SVM. Moreover, to predict the stock market index Moghaddam et al. [20] created an ANN using the four and nine previous days as inputs, concluding that there is no distinct difference between using different days.

Sagir et al. [24] presented a contrast between ANNs and classical statistical techniques to predict the Malaysian stock market index using three variables, showing that the ANN was more accurate than the multiple linear regression model in this research, with a coefficient of determination of 0.9256. Also, Ariyo Adebiyi et al. [2] compared an autoregressive integrated moving average with an ANN to predict the price of a single stock using 5680 observations. The authors concluded that the ANN model is better, having a higher forecasting accuracy; however, there is not statistically significant.

An example of the integration of metaheuristics to predict financial data with neural networks was performed by Gocken et al. [11], where an ANN is hybridized with a genetic algorithm and harmony search to make a feature selection to reduce the complexity of variable selection. In their model, the inputs were technical indicators concluding that hybrid ANN can be successfully used to forecast the stock market price movement. Additionally, Enke et al. [10] introduced an information gain technique to evaluate the prediction of stock market returns using data mining and ANN for level estimation and classification with macroeconomic variables as inputs. As a result, the ANN model was more accurate, showing more consistency than a linear regression forecast and generating higher profits than other strategies with the same risk exposure.

Another example of a hybrid model using ANN to predict financial time series is the work made by Kim et al. [15], combining LSTM, GARCH models and moving averages. These authors concluded that LSTM single models could effectively learn temporal patterns of time-series data with fewer prediction errors than deep feed-forward network-based integrated models.

3   Materials and Methods

This study coded and tested four different models using Python programming language. The data set used to create the inputs for this model was the Spanish Exchange Index, known as IBEX 35. The entire data set used in this research contains 2454 observations, covering the closing prices and volume values from June 1, 2010, to December 31, 2019, covering almost ten years of daily prices in which different trends took place that may represent a normal market cycle. The Anderson-Darling test was performed on the sample data and rejected the null hypothesis; therefore, the assumption of normal distribution in the data sample cannot be allowed. Because the values of the data-set did not follow a normal distribution, the closing prices and the volume values were rescaled between $(-1 < x < 1)$ to be used with the hyperbolic tangent functions and between $(0 < x < 1)$ with the sigmoid activation function. The performance metrics to measure the predictive ability of the different models used in this research were the mean square error (MSE), mean absolute error (MAE), mean squared log error (MLE) metrics, and the determination coefficient (R_2).

The common input layer for the ANN architectures use in this study is presented in Eq. 1:

$$X_{input}(t) = f[v(t-4), v(t-3), v(t-2), v(t-1),$$
$$c(t-4), c(t-3), c(t-2), c(t-1)] \tag{1}$$

where $c(t)$ is the function for closing price and $v(t)$ for volume value at a given time t.

For the training phase, the data-set was split into two parts. The first portion contains the initial 80% of the data selected for the training set, while the remaining 20% for the test sets.

4   ANN Architectures

This section first introduces a model that consists of an ANN with a multi-layer network structure coded by the authors, using NumPy library for matrix multiplication, and continues with another three models constructed using Keras and TensorFlow [1] to build a simple RNN, LTSM and GRU commonly used in different artificial intelligence projects.

4.1   Multi-layer Neural Network

The most widely implemented neural network topologies [12] is the MLP, a multi-layer network structure where the neurons are displayed as input, output, and hidden layers. The other components in the models are weights, connecting coefficients between layers, and activation functions that trigger a signal given a weighted sum of its input. In the first part of this research, two AFs used will be used, the sigmoid (SF) and hyperbolic tangent functions (HTF), while for the simple RNN, LSTM and GRU the rectified linear units (RELU) activation function.

The training phase of an ANN model consists of four steps: Initialization of weights to small random values, forward pass, backward pass and updating of the weights and biases [24]. In this study, for the MLP architecture, two hidden layers will be used and one input and output layer; therefore, three weights matrices (SW_i) were created and initialized with random values for the initialization of weights. Where n_i is the numbers of nodes of the input layer, n_j for the first hidden layer and n_k for the second hidden layer.

For the forward pass, the first step was the multiplication between the input values of the vector $X_{input}(t)$ and the SW_1 matrix (2) were evaluated in the AF.

$$AF_1 = X_{input}(t) \cdot SW_1 \tag{2}$$

in the following steps, when i is greater than one:

$$AF_i = AF_{i-1} \cdot SW_i \tag{3}$$

Once the final layer is reached, the output value must be compared with the target value (4).

$$\delta = y_{output} - y_{target} \tag{4}$$

In the back pass, the error derivate goes back to the input layer updating the weights (SW_1, SW_2, SW_3). The first vector in the back pass is the derivative of the activation function of the δ multiplied by a learning rate (α); the remaining steps will be the multiplication of the first back pass vector by the transpose matrix of SW_3 and by the derivative of the second activation AF_2 and the same procedure for a third error vector (5)

$$e_1 = e_2 \cdot SW_2^T \cdot AF'(X_{input}) \tag{5}$$

Then e_1 multiply by the transpose matrix of X_{input} is updating SW_1 and then the following steps (6) until the last weight.

$$\Delta SW_i = AF_{i-1}^T \cdot e_i \tag{6}$$

4.2   Recurrent Neural Networks

RNN are one type of ANN that deals with data that has sequential inputs. This architecture has been used to process speech, language and sentiment data [27], specially predicting the next character and word in a given text [5], and for more complex tasks [16]. As mentioned previously, RNNs can process sequential inputs using an internal memory to process these incoming inputs and as Le Cun et al. [16] pointed, this model is able to keep in their hidden units a sort of state vector, which can enclose details of previous parts of the sequence. Therefore RNN can process at the same time the previous and recent flow of inputs data by using this hidden unit or layer to keep a historical record. However, the RNN only takes one sequence at any given time. In this research, three commonly used types of RNN will be tested. The following section will describe the LSTM and the GRU architecture.

LSTM. The advantage of an RNN is that it learns long term dependencies over time; however, there are some error back-flow problems [13], showing difficulties to achieve a proper learning process. Therefore the information can not be stored for a long time. A novel solution was introduced by hochreiter and schmidhuber [13] proposing the LSTM model in order to correct this problem, augmenting the network with explicit memory, using hidden units to remember short and long term values. The total number of units of the LSTM is displayed to create a network with an input node, an input, an output, and a forget gates, where the gates will regulate the flow of the information. In the following equations (7–12) is possible to see the forward pass of the LSTM unit [13].

$$g_t = \tilde{C} = \varphi(W_g x_t + U_g h_{t-1} + b_g) \tag{7}$$

$$i_t = \sigma(W_i x_t + U_i h_{t-1} + b_i) \tag{8}$$

$$f_t = \sigma(W_f x_t + U_f h_{t-1} + b_f) \tag{9}$$

$$o_t = \sigma(W_o x_t + U_o h_{t-1} + b_o) \tag{10}$$

$$C_t = g_t * i_t + f_t * C_{t-1} \tag{11}$$

$$h_t = o_t * \varphi(C_t) \tag{12}$$

where i_t represents the input gate activation vector, f_t is the forget gate activation vector, o_t the output gate activation vector, C_t represents the cell state vector, h_t is the output vector of the LSTM unit, σ is SF function, b: biases for the respective gates, W, U are weight matrices for the respective gates, φ represents a HTF and $*$ is an element-wise product.

Several studies used LSTM to predict the stock market in the last five years, where this technique and its modifications dominate the financial time series forecasting [25]. However, despite their popularity, there are some variations on this model. One modification of LSTM is the GRU model that aims to resolve the vanishing and exploding gradient problem presented in the previous model [6].

GRU. This model was introduced by Cho et al. [8] in 2014, proposing a novel ANN called RNN Encoder-Decoder consisting of two RNNs. This model has fewer parameters and has proven a competitive performance to others models like LSTM. Furthermore, it is possible to observe that GRU and LSTM have gating units that modulate the flow of information inside the unit cell. However, it does not have a separate memory cell [9]. According to Alom et al. [3] this model is now popular among people who work with RNN because the computational cost and the simplicity of GRU are better compared to others. Furthermore, the decrease in the computational cost is due to the absence of an output gate, accelerating the speed of the model [21]. GRUs have been successfully applied to many applications for pattern analysis where sequential data is used as input and multivariate time series with missing values [7]. Among the classical GRU's disadvantages is that it is very easy to fall into local minimum with small time series and complex time order. Apart from the sensitivity to the time order, Pei et al. [21] describe some disadvantages that this model has with the data, quoting that for GRUs is hard to detect the implication information of time series and that an imbalanced data can affect the performance of the model by influencing the convergence.

5   Results and Discussion

Diverse ANN structures were tested with the SF and HTF functions to determine the best topology for the MLP net, and in this investigation we have tested different values for the different layers, ranging from one to fifty nodes. As shown in Table 1 different results were obtained by changing the configuration of the ANN. The learning process was made with ten thousand iterations and an $\alpha = 0.01$, in all the MLP architectures tested.

Table 1. R_2 of the ANN model using the SF and the HTF.

n°	Structure	R_2	
		SF	HTF
1	5-5	0.4789	0.6294
2	5-40	<0	0.5775
3	10-20	0.8646	0.7428
4	10-30	<0	0.7819
5	20-20	0.7172	0.7694
6	20-30	0.9242	0.7522
7	30 15	0.8213	0.8485
8	30-30	0.9213	0.8950
9	39-21	0.8166	0.9241
10	40-40	<0	0.9065
11	40-25	**0.9275**	0.9159
12	50-50	<0	0.8633

The prediction accuracy of the ANN model that uses the SF function is statistically different from the others that used HTF. The Wilcoxon signed-rank test was performed to compare the accuracy with the null hypothesis as follows:

$$H_0 : \mu R_2/SF = \mu R_2/HTF$$
$$H_1 : \mu R_2/SF \neq \mu R_2/HTF$$

(13)

Even though it is possible to accept differences because the null hypothesis was rejected is not possible to affirm which activation function works better with this model, but the best coefficient of determination found was 0.9275, obtained by using the SF.

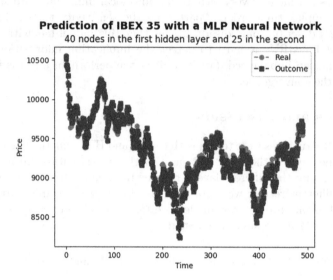

Fig. 1. Output and predicted values of the IBEX 35 index using SF (40-25), performed in test set.

When the number of nodes in the hidden layer is low, the model performs worse than with more numbers of nodes, this phenomenon was clearer using SF than HTF, but this might not always necessarily true. For example when using SF the R_2 is not constantly growing at a given rate. At first, a low number of nodes returned a low R_2, and as the number of nodes starts to increment also, the R_2 increases. However, when there are 40 nodes in each layer, the R_2 is negative; therefore, there is no optimal number of nodes for each layer. Also, negative R_2 values can be found in some specific ANN structures. The highest R_2 found was 0.9275 by using SF with 40 nodes in the first hidden layer and 25 in the second as shown in Fig. 1. For the ANN with HTF, the best R_2 found was 0.9241, as shown in Table 1, with an ANN with 39 and 21 nodes.

It is possible to observe that the ANN with SF has lower MSE, MAE, and MLE levels than those using HFT when the number of nodes is relatively low. However, this trend changed when the number of nodes was more extensive than 30 in the first hidden layer.

5.1   Recurrent Neural Network Performance

This section will present the implementation of the simple RNN, LSTM and GRU models. These models were implemented using the Keras library for Python backup by the Tensorflow library. The data were pre-processed in the same way that was done for the previous models, but in the following cases, it was normalized only between $(0 < x < 1)$. The simple RNN sequential model was implemented using one Keras Simple RNN layer of 8 nodes and a hidden dense layer with 32 nodes and a RELU activation function. The MSE loss function was applied for compilation with an RMSprop optimizer, while epochs were 50 with a batch size of 1. The MSE for this model was 17520.86 while the $R_2 = 0.8814$. The MAE was 116.30, and MLE was equal to 0.00020.

For the LSTM, a sequential model was applied using one Keras LSTM layer of 32 nodes and one dense layer with eight nodes as an input to match the shape of the matrix that holds the arranged data. The loss function applied was the MSE with an ADAM optimizer and a RELU activation function. The number of epochs was equal to 50 with a batch size of 1. The results were, MAE = 69.33 with a MSE = 7192.27, while the $r_2 = 0.9513$ with a MLE = 0.000083. The GRU model was implemented using one Keras GRU layer of 8 nodes, one hidden dense layer with 32 nodes with a RELU activation function and an MSE loss function for compilation with an RMSprop optimizer. The number of epochs was 50 with a batch size of 1. The MSE for this model was 8608.07 while the $R_2 = 0.9417$. The MAE was 71.40, and MLE was equal to 0.000092.

5.2   Results Comparison

The best result in terms of R_2 is obtained by LSTM, with a $R_2 = 0.9513$ and GRU with a $R_2 = 0.9417$. Also, these two models presented the lowest errors compared to the others; however, it is also important to consider the computational time to analyse the model's performance fully. It is possible to observe in Table 2 the results of Simple RNN, LSTM and GRU models.

Table 2. Performance of MLP (SF 20-30), Simple RNN, LSTM and GRU models, using the test set.

	R_2	MSE	MAE	MLE	Time (seconds)
MLP	0.9275	10701.29	80.18	0.00012	**136.3**
RNN	0.8814	17520.86	116.30	0.00020	141.9
LSTM	**0.9513**	**7192.27**	**69.33**	**0.000083**	273
GRU	0.9417	8608.07	71.40	0.000092	228.6

The simple RNN showed the worst performance in terms of R_2, MAE, MSE and MLE errors; however, this is true compared to the best MLP results because some MLP configurations showed worst performance than RNN as shown in Table 2.

In this study, in order to have an analysis of the computational time, the platform Google Colaboratory was used as a tool for accelerating the learning applications to have similar performance levels to those acquired with a dedicated hardware. The computational time results show that the model with the lowest computational time was the MLP made by the authors, with 40 and 25 nodes using an SF activation function. This model took 136.3 s to complete, while the Simple RNN was the second-fastest, taking 141.9 s. The model that took more time to be completed was the LSTM model taking 273 s to complete all the routines, followed by the GRU, which took 228.6 s. These performances took almost twice the time that MLP and Simple RNN.

6 Conclusions

Implementing different neural network architectures to forecast financial time series has shown a predictive capacity with low errors. Although all forms of ANN have successfully predicted the IBEX-35, LSTM has the best results. It is essential to consider that the ANN structure and the number of iterations in the training phase will determine the model's predictive capacity. However, there are no clear procedures to define a proper structure because the error does not decrease linearly. Therefore is necessary to include in further studies different types of heuristics to optimize the computational time of the training phase and the search for the optimal ANN architecture. Even though these ANNs have a high R2 at predicting the closing price of a stock market index, using these models as a tool for financial trading can be challenging because investors need to consider price predictions and the risk and the size of any given trading position. Hence, any potential strategy needs to consider rigorous risk management with a robust backtest.

In addition, because financial markets are interconnected, there might be other variables impacting the prices and bias of the market so that an extensive feature selection could improve the prediction of the model. Consequently, further research can be done to increment the number of significant variables used by the model and search for different ANN architectures to predict closing values and make a trading system that can be a reliable tool to predict financial markets.

In order to have proper management and performance of deep learning methods applied to financial time series, it is crucial to consider a robust data set with a proper and reliable testing phase for the model. Especially during times of high uncertainty, when the complexity and size of the financial markets grow. Due to the nature of the markets, the predicted outcomes will not necessarily represent future outcomes because there are rapid changes in the market dynamics. Thus, when picking the training data, it is necessary to keep in mind that one asset or instrument can only have one historical price record; therefore, any model trained to forecast financial data is vulnerable or prone to overfitting.

Not choosing a correct data set is dangerous because the model can predict outcomes similar or exact to the underlying data set but will fail to predict future values. Therefore, it is important to consider this vulnerability for future works, where models can be tested with historical data from multiple assets and adding some synthetic data. Furthermore, using other deep learning models, such as generative adversarial models, could create endless data-sets for multiple scenarios that are not real but close enough to a potentially real scenario.

The forecast of future financial values is essential to investors and private companies and for government policymakers who need to make an appropriate asset allocation of scarce resources. In this way, better financial predictive models will not only help financial agents but could potentially affect everyone.

References

1. Abadi, M., Agarwal, A., Barham, P., Brevdo, E., Chen, Z., et al.: TensorFlow: large-scale machine learning on heterogeneous systems (2015). https://www.tensorflow.org/. Software available from tensorflow.org
2. Adebiyi, A.A., Adewumi, A.O., Ayo, C.K.: Comparison of ARIMA and artificial neural networks models for stock price prediction. J. Appl. Math. **2014**, 1–7 (2014). https://doi.org/10.1155/2014/614342
3. Alom, M.Z., et al.: A state-of-the-art survey on deep learning theory and architectures. Electronics **8**(3), 292 (2019). https://doi.org/10.3390/electronics8030292
4. Baur, D.: Financial contagion and the real economy. J. Banking Financ. **36**(10), 2680–2692 (2012)
5. Bengio, Y.: Probabilistic neural network models for sequential data. In: Proceedings of the IEEE-INNS-ENNS International Joint Conference on Neural Networks. IJCNN 2000. Neural Computing: New Challenges and Perspectives for the New Millennium (2000). https://doi.org/10.1109/ijcnn.2000.861438
6. Bengio, Y., Simard, P., Frasconi, P.: Learning long-term dependencies with gradient descent is difficult. IEEE Trans. Neural Netw. **5**(2), 157–166 (1994). https://doi.org/10.1109/72.279181
7. Che, Z., Purushotham, S., Cho, K., Sontag, D., Liu, Y.: Recurrent neural networks for multivariate time series with missing values. Sci. Rep. **8**(1), 1–12 (2018). https://doi.org/10.1038/s41598-018-24271-9
8. Cho, K., et al.: Learning phrase representations using RNN encoder-decoder for statistical machine translation. In: Proceedings of the 2014 Conference on Empirical Methods in Natural Language Processing (EMNLP) (2014). https://doi.org/10.3115/v1/d14-1179
9. Chung, J., Gulcehre, C., Cho, K., Bengio, Y.: Empirical evaluation of gated recurrent neural networks on sequence modeling (2014)
10. Enke, D., Thawornwong, S.: The use of data mining and neural networks for forecasting stock market returns. Expert Syst. Appl. **29**(4), 927–940 (2005). https://doi.org/10.1016/j.eswa.2005.06.024
11. Gocken, M., Ozcalici, M., Boru, A., Dosdogru, A.T.: Integrating metaheuristics and artificial neural networks for improved stock price prediction. Expert Syst. Appl. **44**, 320–331 (2016). https://doi.org/10.1016/j.eswa.2015.09.029
12. Guresen, E., Kayakutlu, G., Daim, T.U.: Using artificial neural network models in stock market index prediction. Expert Syst. Appl. **38**(8), 10389–10397 (2011). https://doi.org/10.1016/j.eswa.2011.02.068
13. Hochreiter, S., Schmidhuber, J.: Long short-term memory. Neural Comput. **9**(8), 1735–1780 (1997). https://doi.org/10.1162/neco.1997.9.8.1735

14. Kara, Y., Boyacioglu, M.A., Baykan, Ö.K.: Predicting direction of stock price index movement using artificial neural networks and support vector machines: the sample of the Istanbul stock exchange. Expert Syst. Appl. **38**(5), 5311–5319 (2011). https://doi.org/10.1016/j.eswa.2010.10.027
15. Kim, H.Y., Won, C.H.: Forecasting the volatility of stock price index: a hybrid model integrating LSTM with multiple GARCH-type models. Expert Syst. Appl. **103**, 25–37 (2018). https://doi.org/10.1016/j.eswa.2018.03.002
16. Lecun, Y., Bengio, Y., Hinton, G.: Deep learning. Nature **521**(7553), 436–444 (2015). https://doi.org/10.1038/nature14539
17. Lee, J., Suh, T., Roy, D., Baucus, M.: Emerging technology and business model innovation: the case of artificial intelligence. MDPI (2019). https://www.mdpi.com/2199-8531/5/3/44
18. Lo, A.W., MacKinlay, A.C.: Stock market prices do not follow random walks: evidence from a simple specification test. Rev. Financ. Stud. **1**(1), 41–66 (1988)
19. Menkhoff, L.: The use of technical analysis by fund managers international evidence. J. Banking Financ. **34**(11), 2573–2586 (2010). https://doi.org/10.1016/j.jbankfin.2010.04.014
20. Moghaddam, A.H., Moghaddam, M.H., Esfandyari, M.: Stock market index prediction using artificial neural network. J. Econ. Financ. Adm. Sci. **21**(41), 89–93 (2016). https://doi.org/10.1016/j.jefas.2016.07.002
21. Pei, S., Shen, T., Wang, X., Gu, C., Ning, Z., Ye, X., Xiong, N.: 3DACN: 3D augmented convolutional network for time series data. Inf. Sci. **513**, 17–29 (2020). https://doi.org/10.1016/j.ins.2019.11.040
22. Pyo, S., Lee, J., Cha, M., Jang, H.: Predictability of machine learning techniques to forecast the trends of market index prices: hypothesis testing for the Korean stock markets. PLoS One **12**, e0188107 (2017). https://doi.org/10.1371/journal.pone.0188107
23. Qiu, M., Song, Y., Akagi, F.: Application of artificial neural network for the prediction of stock market returns: the case of the Japanese stock market. Chaos Solitons Fractals **85**, 1–7 (2016). https://doi.org/10.1016/j.chaos.2016.01.004
24. Sagir, A., Sathasivan, S.: The use of artificial neural network and multiple linear regressions for stock market forecasting. Matematika **33**, 1–10 (2017)
25. Sezer, O.B., Gudelek, M.U., Ozbayoglu, A.M.: Financial time series forecasting with deep learning?: a systematic literature review: 2005–2019. Appl. Soft Comput. **90**, 106181 (2020). https://doi.org/10.1016/j.asoc.2020.106181
26. Tkáč, M., Verner, R.: Artificial neural networks in business: two decades of research. Appl. Soft Computi. **38**, 788–804 (2016). https://doi.org/10.1016/j.asoc.2015.09.040
27. Wang, J., Zhang, Y., Yu, L.C., Zhang, X.: Contextual sentiment embeddings via bidirectional GRU language model. Knowl.-Based Syst. **235**, 107663 (2022). https://doi.org/10.1016/j.knosys.2021.107663

A Fine-Grained Study of Interpretability of Convolutional Neural Networks for Text Classification

Maite Giménez$^{(\boxtimes)}$(iD), Ares Fabregat-Hernández, Raül Fabra-Boluda,
Javier Palanca(iD), and Vicent Botti(iD)

Valencian Research Institute for Artificial Intelligence (VRAIN), Universitat
Politècnica de València, Camino de Vera s/n, 46022 Valencia, Spain
{mgimenez,rafabbo,jpalanca,vbotti}@dsic.upv.es,
arfabher@etsii.upv.es

Abstract. In this work, we proposed a new interpretability framework for convolutional neural networks trained for text classification. The objective is to discover the interpretability of the convolutional layers that composes the architecture. The methodology introduced explores the most relevant words for the classification and more generally look for the most relevant concepts learned in the internal representation of the CNN. Here, the concepts studied were the POS tags.

Furthermore, we have proposed an iterative algorithm to determine the most relevant filters or neurons for the task. The outcome of this algorithm is a threshold used to mask the least active neurons and focus the interpretability study only on the most relevant parts of the network.

The introduced framework has been validated for explaining the internal representation of a well-known sentiment analysis task. As a result of this study, we found evidence that certain POS tags, such as nouns and adjectives, are more relevant for the classification. Moreover, we found evidence of the redundancy among the filters from a convolutional layer.

Keywords: Interpretability · Convolutional neural networks · Natural Language Processing · Sentiment analysis

1 Introduction

Deep learning (DL) networks allowed to improve the performance of a wide variety of tasks radically. Notably, in Natural Language Processing, deep learning approaches are demoting other approximations, and the improvements had led to implementing commercial applications that made the technology available for a broad public [6]. However, network interpretability is one of the main concerns that arise. Practitioner bias, ethics and the possibility of explaining the performance of these systems raised the interest of the community, which is reflected in a growth of literature on the topic and workshops [7,11]. Therefore, in this work, we investigated how to help to interpret Natural Language Processing tasks.

© Springer Nature Switzerland AG 2022
P. García Bringas et al. (Eds.): HAIS 2022, LNAI 13469, pp. 261–273, 2022.
https://doi.org/10.1007/978-3-031-15471-3_23

Hereafter, we introduce a fine-grained definition of the interpretability of each filter in a Convolutional Neural Network (CNN) trained for predicting the class of a text using the saliency maps. Concretely, we studied the interpretability of a CNN trained for Sentiment Analysis.

The purpose of this work is to discover the interpretability of each filter, namely, if a filter is learning some concepts for its internal representation. In order to do so, we studied the n-grams that get more activated in a filter and investigated what characteristics share these top activated n-grams in a filter. To this aim, we propose an iterative algorithm that will select the most influential n-grams for classifying a sentence in a given class. In addition, we generalise the concept of purity presented in [9], which separates n-grams into informative and uninformative for the classification.

The main contributions presented in this paper are:

- An algorithm for selecting the interpretability threshold of filters iteratively in a CNN is presented
- We propose a novel fine-grained interpretability method to explain the most relevant n-grams learned by a CNN.
- Finally, we also propose a novel methodology for retrieving the ratio of relevant n-grams required for classifying a sentence correctly.
- We have validated these proposed methods analysing a sentiment analysis classification task and discussed the relevant findings.

The rest of the paper is organised as follows. The following section gives a brief overview of the literature related to our work. Section 3 is devoted to describing the methodology proposed. In Sect. 4, the task and the details of the experimental phase selected for evaluating the methodology proposed are presented, as well as the results achieved. Finally, in Sect. 5, we discussed our results, and future work is proposed.

2   Related Work

Nowadays, the application of DL models to NLP tasks has grown massively, becoming the state-of-the-art algorithms for most NLP tasks. However, these trailblazing algorithms present a common flaw that prevents them from being applied in sensitive domains where interpretability and explainability are required. This has lead to the development of a new field that aims to shed light on all DL models. The goal is to lead the advancement of the field towards Responsible Artificial Intelligence as described in the work of [2] which is to develop models focused on fairness, model explainability and accountability.

The related work is divided into two subsections. Firstly, a review of the literature on interpretability is presented. Secondly, a definition of the CNNs that we are studying in this research is shown.

2.1 Network Interpretability

Initially, we must present a definition of interpretability. The definition proposed by [7] characterises interpretability as a methodology that describes the internals of a system in a way that is understandable to humans considering the cognition, knowledge, and biases of the user.

The approaches proposed to achieve network interpretability can be divided into two categories, on the one hand, frameworks developed for interpreting existing models and on the other hand, deep learning models designed with interpretability systems in place. In this work, we are focus on the former approach.

Particularly relevant for the development of the approach proposed in this paper is the work of [9] where they introduced a method for understanding how CNNs classify text. They proposed a system to investigate if the filters of CNNs were learning different semantic classes of n-grams studying the saliency maps of the filters learned by the model. Moreover, they proposed a formula to discard uninformative n-grams that reduce the complexity of analysing the model pruning the problem. We extended this approach with ideas from [3] and proposed a fine-grained model that, in addition, computes metrics to evaluate the influence of each filter in the inferred class. That is, we make a case by case analysis of the relevance of each filter for each POS tag and each class which give us a more precise metric rather than an aggregated one as in [9].

In [8] the authors defined Convolutional Networks as neural networks that use convolutions instead of matrix multiplication in at least one of its layers. A convolution is a linear operation that takes two multidimensional arrays: an input $\vec{x} \in \mathbb{R}^{m \times n}$ and a kernel $\vec{w} \in \mathbb{R}^{h \times k}$ where $h < m$ and $k < n$. Notice that the dimension of the word embedding affects m and n and consequently the sizes of filters allowed, that is h and k. This has been taken into account in the Sect. 4. After applying a convolution to these two arrays, it will produce multidimensional arrays called feature maps. Therefore, each element of a feature map is obtained as result of applying the convolution across a window of words $\{\vec{x}_{0:h-1}, \vec{x}_{1:h}, \dots \vec{x}_{n-h+1:n}\}$, and it is defined as:

$$c_i = f(\vec{w} \cdot \vec{x}_{i:i+h-1} + b_i) \tag{1}$$

where $\vec{c} \in \mathbb{R}^{n-h+1}$ is the feature map obtained, $b_i \in \mathbb{R}$ is a bias term, and f a non linear function. Then, in the backward phase, CNN learns the values of its kernels.

3 Methodology

In this section, we describe the methodology we proposed for shedding light on what a network is learning. The key objectives of the methodology proposed for interpretability are:

– The interpretability methodology should not impact the performance of the model. A filter is discarded when it does not add relevant information. Hence,

the performance of the classification task must be preserved, up to a certain point, even when we are studying the interpretability of the model.
- Do not modify the architecture of the model. The objective that we seek is to be able to interpret a particular model without modifications.
- Following the same philosophy described in the previous point, this methodology will not modify the learned weights of the model.

These objectives need further clarification. First of all, the interpretability framework consists of two phases. In the first phase we probe the model: this is where we learn about the interpretability. In the second phase we use the model and we interpret the results according to the results of the first phase. In order to probe the model, and learn about its inner workings, we need to modify it but, when we reach the second phase, we reestablish every parameter to its original state. This is what is meant by the second and third objectives. Analogously, a significant drop in performance, caused by masking a small percentle of filters, is necessary to determine the interpretability threshold (see Fig. 2) in phase one. Those changes are reverted back in the second phase to make a comparison (see Sect. 4).

A Convolutional Neural Network is composed of convolutions with different kernel sizes. In order to take into consideration the different behaviours of each filter with the same kernel configuration, we computed the top percentile for each layer $l \in L$, where L is the set of convolutional layers of the network. This process relies on selecting the top quantile level as [3] proposed. In this paper, the authors suggested to filter the more relevant units considering the saliency maps $A_k(x)$ of each convolutional unit k obtained for each image in the dataset. Afterwards, for each unit k the top quantile level T_k is defined as $P(a_k > T_k) = 0.005$ and the units below this threshold are set to $\vec{O}$.

Conversely, we propose to use the weights and biases directly from each convolutional neural layer instead of the activation maps. In this way, we simplified the computation since the top percentile is computed only once for each layer and computing the activation maps for selecting the top percentile is not required. The metric considered to evaluate the performance of the model, while the least relevant neurons are discarded, is the F_1 score computed on the development set ($F_1(dev)$). The variable $F_1'(dev)$ stores the F_1 score from the current iteration whilst $F_1(dev)$ stores the F_1 score from the previous iteration. The development dataset $dev = (x_j, y_j)$ contains pairs of labeled examples. The algorithm has been designed and implemented to select the threshold after training the model using the development split of the dataset.

A Metric for Fine-Grained Interpretability of Convolutional Neural Networks. At this point, we can define the proposed method for evaluating the interpretability of each one of the feature maps learned at a level of the CNN.

Given a labelled dataset, which we call $dev = X \times Y$, with M tuples (x_i, y_i), where $x_i \in X$ is a text in a known language and $y_i \in Y$ the class from the ground truth assigned to this text, we try to validate or discard the hypothesis

Data:
- A development dataset $dev = X \times Y$ as above.
- A CNN model trained for the task at hand.
- A parameter ϵ as described above.

Result: The threshold to discard n-grams in every layer of the CNN model without compromising the performance of the model.

top_percentile = 0.99;
We initialize the F_1-scores to be equal: $F_1'(dev), F_1(dev) = F_1(dev), F_1(dev)$;
while $F_1'(dev) \geq F_1(dev) + \epsilon$ **do**
    forall the $l \in L$: *layer in the CNN model* **do**
        threshold_weights = get_top_percentile_layer($\vec{W_l}, top\_percentile$);
        threshold_bias = get_top_percentile_layer($\vec{b_l}, top\_percentile$);
    end
    $F_{1,dev}'$ = Compute $F_1(dev)$ masking the positions where the weights of the convolution are below the threshold_weights with $\vec{0}$; similarly set to $\vec{0}$ the positions where the biases are below the threshold_bias ;
    if $F_1'(dev) \geq F_1(dev) + \epsilon$ **then**
        top_percentile -= 0.01;
    end
end

Algorithm 1: Iterative algorithm for finding the interpretability threshold.

that a CNN learns semantic concepts. To that end, we automatically annotated each word with semantic concepts such as part of speech tags, obtaining a vector $An(\vec{x_i})$ with the annotation labels[1] found in a sentence. The set $An_t(\vec{x_i})$ are the n-grams of $\vec{x_i}$ that have annotation tag t. Moreover, $R_k(\vec{x_i})$ is the set of relevant n-grams of unit k from sentence i that was labeled as class y_i. The relevance vector is obtained by selecting the n-grams with the value over the threshold computed applying the iterative algorithm described in 1. It is important to stress that this process will select the most relevant units at the convolutional layer level, considering both the weights ($\vec{W_l}$) and the biases ($\vec{b_l}$) of the convolution operation. The ϵ parameter in the algorithm controls how much decay we allow for the percentage of neurons disconnected during an iteration. Therefore, Eq. 2 defines the interpretability, for a specific POS tag t, of each filter in each layer of a convolutional network considering only those examples where the CNN predicted the class correctly. That is, $x \in R_k(\vec{x_i})$ is a particular word within the string $\vec{x_i}$ that lies above the threshold for the convolutional unit k (relevant word for unit k) such that $\vec{x_i}$ gets classify correctly. The intersection on the nominator part means that we only look at the relevant words x that have POS tag t. The union on the denominator signifies that we also look at the words x with POS

---

[1] The annotation classes considered are discussed in Sect. 4.

tag t that are not relevant for unit k, that is words x in $\vec{x}_i$ with POS tag t such that $x \notin R_k(\vec{x}_i)$.

$$interpretability_{k,c,t} = \frac{\sum_{i=0}^{M-1} |\{x \in R_k(\vec{x}_i)|y_i = c\} \bigcap An_t(\vec{x}_i)|}{\sum_{i=0}^{M-1} |\{x \in R_k(\vec{x}_i)|y_i = c\} \bigcup An_t(\vec{x}_i)|} \qquad (2)$$

Meanwhile, Eq. 3 considers what the filter learns in each sentence independently of the correctness of the prediction. This interpretability measure gives insights into global concepts learned in a feature map from a convolutional layer, regardless of how the following layers combine this information to predict the label of the sentence.

$$interpretability_{k,t} = \frac{\sum_{i=0}^{M-1} |R_k(\vec{x}_i) \bigcap An_t(\vec{x}_i)|}{\sum_{i=0}^{M-1} |R_k(\vec{x}_i) \bigcup An_t(\vec{x}_i)|} \qquad (3)$$

Similarly, Eq. 4 measures the ability of each convolutional unit to predict the class of the sentence correctly. In this case, we aim to uncover units specialised in identifying a class disregarding any semantic concepts.

$$interpretability_{k,c} = \frac{\sum_{i=0}^{M-1} |\{x \in R_k(\vec{x}_i)| \quad y_i = c\}|}{\sum_{i=0}^{M-1} |R_k(\vec{x}_i)|} \qquad (4)$$

From the previous definitions, we can derive general metrics such as accuracy, precision, recall and the F_1 score for each unit in a layer.

$$TP_k = \sum_{i=0}^{M-1} |\{x \in R_k(\vec{x}_i)| \, y_i = c\}| \quad TN_k = \sum_{i=0}^{M-1} |\{x \in NR_k(\vec{x}_i)| \, y_i \neq c\}|$$

$$FP_k = \sum_{i=0}^{M-1} |\{x \in R_k(\vec{x}_i)| \, y_i \neq c\}| \quad FN_k = \sum_{i=0}^{M-1} |\{x \in NR_k(\vec{x}_i)| \, y_i = c\}|$$

Being $NR_k(\vec{x}_i)$ the set of not relevant n-grams, this is the complementary of $R_k(\vec{x}_i)$ previously presented.

$$accuracy_k = \frac{TP_k + TN_k}{TP_k + TN_k + FP_k + FN_k} \qquad (6a)$$

$$precision_k = \frac{TP_k}{TP_k + FP_k} \qquad (6b)$$

$$recall_k = \frac{TP_k}{TP_k + FN_k} \qquad (6c)$$

$$F_{1,k} = 2\frac{precision_k \cdot recall_k}{precision_k + recall_k} \qquad (6d)$$

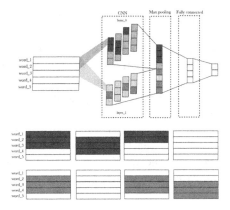

Fig. 1. Schema of the interpretability framework proposed. The top of the Figure depicts a CNN model to study. The shadowed areas represent the neurons above the interpretability threshold selected using the interpretability algorithm. The lower part of the Figure illustrates which words from the input were attending each of the most relevant neurons of each filter of the CNN.

Figure 1 shows an schema of the framework proposed. The top of the Figure depicts a CNN model with two layers of convolutional filters, the first layer is attending to bigrams of words, whilst the second layer is attending to trigrams of words. Each layer of the CNN has four filters and a fully connected layer. The shadowed areas represent the neurons above the interpretability threshold. Considering the receptive field of these neurons, the lower part of the Figure illustrates the words where the model was paying attention to classify a sentence with a polarity label. The first row represents the words activated by the first layer and the second row to the second layer of the CNN.

4   Evaluation

Hereafter, we presented the evaluation of the interpretability framework proposed in this paper. Firstly, we describe the task and the details of the model that we want to interpret. Subsequently, we present the results of the experimental phase.

4.1   Corpora

In order to validate the proposed interpretability methodology, we have selected a Sentiment Analysis (SA) classification task. Therefore, we studied if there are convolutional filters specialised in the sentiment carried out by a text. SA is a well-studied task that allowed us to focus on the analysis of the interpretability of the model. Provided that in the literature, there is evidence that suggests that longer texts are challenging for CNN models [14], we have selected a task that aims to predict the sentiment of a movie review. Sentences with different sizes were studied, allowing us to analyse the effect of this parameter in the learned inner representation.

The dataset used is commonly known as IMDB because it is composed of movie reviews extracted from the popular website[2]. [12] curated this dataset, establishing a popular benchmark for evaluating SA models. This corpus is composed of 50.000 reviews, divided into two balanced splits, one for training and one for testing. Each split has 25000 reviews, and in each split, there are 12.500 positive reviews and 12.500 negative reviews. A review is considered negative, and therefore, labelled with the tag 0, if the score is less than four. Whereas positive reviews, labelled with the tag 1, are those with a score greater than seven. Therefore, neutral rated reviews are not included in this corpus.

4.2 Model Studied

The model that we have proposed to visualise in this work to validate the framework introduced replicates the architecture of a well-known model in text classification proposed in [10]. This CNN architecture is a manageable deep learning model which allows us to interpret and visualise the learned.

This CNN architecture has a convolutional layer with 100 filters of size $\mathbb{R}^{3 \times e}$, $\mathbb{R}^{4 \times e}$, and $\mathbb{R}^{5 \times e}$ respectively, being e the size of the embeddings studied; we used rectified linear (RELU) units after applying the convolution operation; the dropout rate was 0.5, and the fully connected layer consisted on 150 neurons. The model was also trained using mini-batches of size 100 through Adam gradient descent. Custom convolutional and the max-pooling layers were implemented using Keras [5] and TensorFlow [1] to allow saving and analysing the activation maps. The word embeddings used were pre-trained embeddings based on the skipgram version of *word2vec* proposed by [13].

4.3 Experimental Phase

Interpretability Threshold. The objective of the first experiment we carried out was to select the interpretability threshold. Initially, we train the previously described architecture until we achieved a satisfactory F_1 of 0.864. Using this trained model, we applied the iterative algorithm described in 1.

Figure 2 shows the decay in the achieved F_1 as we decreased the top percentile of neurons that are kept active. Interestingly, the F_1 plummeted when we masked a small percentile of filters. However, the F_1 plateau until only less than 20% of the filters is still active. From that point, going forward, a rapid decay is appreciated. This behaviour could be explained because the CNN architecture used has a high capacity for modelling the problem and, therefore, high tolerance for missing connections.

---

[2] https://www.imdb.com/.

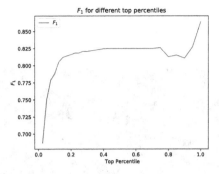

Fig. 2. Evolution of the F_1 selecting different thresholds.

We set the top percentile to 0.3. Masking the positions below the threshold of this top percentile, the model achieved an 82.08%. Therefore, the thresholded model is performing only a 5% worse than the original model. The weights of one of the convolutional layers before and after thresholding is applied can be seen in Fig. 3. The scale is the same as in Figs. 4 and 5. Unlike in computer vision, interpreting these filters is not straightforward. Hence, the analytical framework proposed will allow investigating what the model learned systematically.

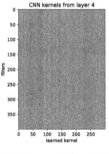

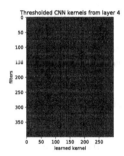

(a) Learned weighths.   (b) Thresholded weighths.

Fig. 3. Weigths from the convolutional layer with a kernel of size four before and after thresholding the top percentile.

To illustrate an example of what the convolutional neural layers are learning, Fig. 4 shows a truncated example of the activation map for one of the sentences in the corpora. The rows correspond to the n-gram used by the filter, and the columns are each of the 100 filters for a layer; brighter positions in the heatmap indicate the relevance of the n-gram in the filter. Although there are other active areas, in the highlighted ones, where the filters seemed to be slightly more active than in most of other areas, sentiment loaded words can be found.

Fig. 4. Activation map for one of the sentences. Brighter positions in the heatmap indicate the relevance of the n-gram in the layer with filter of size $\mathbb{R}^{3 \times e}$.

Notably, shorter sentences imply having most of the input matrix padded, and therefore the activation map presents more deactivated positions. As discussed previously, using that region with semantic-based padding could boost the performance of the model.

4.4   Study of the Interpretability of the Convolutional Layers

Once an appropriate threshold has been selected, the analytical study of the internal representation learned by the convolutional layers is performed. In this section, we computed the interpretability metrics presented in 3. The annotation tags considered in this experiment were part of speech tags (POS). The library NLTK was used to label the words in the corpus [4].

Firstly, we explored the interpretability of the filters where the class of the examples was predicted correctly, as defined in Eq. 2. Forty POS tags were observed in the developmental corpus. Table 1 presents the mean and the standard deviation of the ten most relevant POS tags aggregated by the class predicted correctly. The abbreviations of the POS tags are: nouns in singular (NN), adjectives (JJ), prepositions (IN), determiners (DT), adverbs (RB), nouns in plural (NNS), conjunctions (CC), verbs in present tense and in third person singular (VBZ), verbs in past tense (VBD), and verbs in base form (VB).

Figure 5 shows the interpretability of the top five POS tags is shown. The first column showed the interpretability when the sentence was correctly classified as negative, whilst the second column shows the interpretability when the sentence

Table 1. Interpretability of the top ten POS tags considered examples correctly predicted.

POS tag	Class	$Interpretability_{k,c,t}$
NN	Negative	23.88% (3.99)
	Positive	24.81% (4.23)
JJ	Negative	10.85% (1.40)
	Positive	11.34% (1.49)
IN	Negative	10.39% (2.19)
	Positive	10.66% (2.11)
DT	Negative	8.97% (2.17)
	Positive	8.49% 2.19
RB	Negative	5.69% (1.52)
	Positive	5.44% (1.39)
NNS	Negative	5.16% (0.93)
	Positive	5.33% (1.39)
VBZ	Negative	4.43% (1.06)
	Positive	4.96% (1.12)
CC	Negative	4.66% (3.45)
	Positive	4.89% (3.49)
VBD	Negative	3.79% (0.92)
	Positive	3.30% (0.77)
VB	Negative	2.97% (1.56)
	Positive	2.51% (1.33)

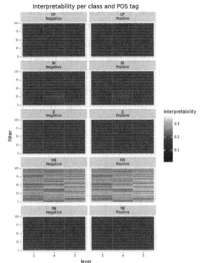

Fig. 5. Interpretability of each filter in the CNN aggregated per tag and class.

was correctly classified as positive. Noteworthy, the filters were more activated for words that identify something and sentiment charged words like adjectives and adverbs.

5 Conclusions and Future Work

In summary, in this paper, we have proposed a new interpretability framework that seeks to shed light on what the internal layers of a convolutional neural network are learning in an NLP task. The framework has been validated using a sentiment analysis task.

Among the most relevant findings is the resilience of the network, which was evidenced in the iterative algorithm that selected the interpretability threshold –most of the weights needed to be disabled in order to see the performance plummeted. Another relevant finding was the validation that certain POS tags such as nouns, adjectives and prepositions, that serve as connectors between different words, were relevant in the internal representation of the convolutional layers. Intuitively, this finding could have been foreseen, but we have validated it methodologically. Further work could be done by applying similar methods to explain the dense part of the network and different problems and architectures.

To conclude, stress that the development of interpretability frameworks will impulse the expansion of reliable, robust and trustworthy systems. Considering the ubiquity of the deep learning models in different real-world NLP tasks, this effort is crucial.

Acknowledgments. We gratefully acknowledge the support of NVIDIA Corporation with the donation of the Titan Xp GPU used for this research. This work is partially supported by the TAILOR project, a project funded by the EU Horizon 2020 research and innovation programme under GA No 952215.

References

1. Abadi, M., et al.: TensorFlow: large-scale machine learning on heterogeneous systems (2015). https://www.tensorflow.org/. Software available from tensorflow.org
2. Arrieta, A.B., et al.: Explainable artificial intelligence (XAI): concepts, taxonomies, opportunities and challenges toward responsible AI. Inf. Fusion **58**, 82–115 (2020)
3. Bau, D., Zhou, B., Khosla, A., Oliva, A., Torralba, A.: Network dissection: quantifying interpretability of deep visual representations. In: Proceedings of the IEEE Conference on Computer Vision and Pattern Recognition, pp. 6541–6549 (2017)
4. Bird, S., Klein, E., Loper, E.: Natural Language Processing with Python: Analyzing Text with the Natural Language Toolkit. O'Reilly Media, Inc. (2009)
5. Chollet, F., et al.: Keras (2015). https://github.com/fchollet/keras
6. Deng, L., Liu, Y.: Deep Learning in Natural Language Processing. Springer, Singapore (2018). https://doi.org/10.1007/978-981-10-5209-5
7. Gilpin, L.H., Bau, D., Yuan, B.Z., Bajwa, A., Specter, M., Kagal, L.: Explaining explanations: an overview of interpretability of machine learning. In: 2018 IEEE 5th International Conference on Data Science and Advanced Analytics (DSAA), pp. 80–89. IEEE (2018)
8. Goodfellow, I., Bengio, Y., Courville, A.: Deep Learning. MIT Press (2016). http://www.deeplearningbook.org
9. Jacovi, A., Shalom, O.S., Goldberg, Y.: Understanding convolutional neural networks for text classification. In: Proceedings of the 2018 EMNLP Workshop BlackboxNLP: Analyzing and Interpreting Neural Networks for NLP, pp. 56–65 (2018)
10. Kim, Y.: Convolutional neural networks for sentence classification. In: Proceedings of the 2014 Conference on Empirical Methods in Natural Language Processing (EMNLP), pp. 1746–1751. Association for Computational Linguistics, Doha (2014). https://doi.org/10.3115/v1/D14-1181, https://aclanthology.org/D14-1181
11. Linzen, T.T., Chrupała, G., Alishahi, A.: Proceedings of the 2018 EMNLP workshop BlackboxNLP: analyzing and interpreting neural networks for NLP. In: Proceedings of the 2018 EMNLP Workshop BlackboxNLP: Analyzing and Interpreting Neural Networks for NLP (2018)
12. Maas, A., Daly, R.E., Pham, P.T., Huang, D., Ng, A.Y., Potts, C.: Learning word vectors for sentiment analysis. In: Proceedings of the 49th Annual Meeting of the Association for Computational Linguistics: Human Language Technologies, pp. 142–150 (2011)

13. Mikolov, T., Chen, K., Corrado, G.S., Dean, J.: Efficient estimation of word representations in vector space. In: ICLR (2013)
14. Yin, W., Kann, K., Yu, M., Schütze, H.: Comparative study of CNN and RNN for natural language processing. CoRR abs/1702.01923 (2017). http://dblp.uni-trier.de/db/journals/corr/corr1702.html#0001KYS17

Olive Phenology Forecasting Using Information Fusion-Based Imbalanced Preprocessing and Automated Deep Learning

Andrés Manuel Chacón-Maldonado[1], Miguel Angel Molina-Cabanillas[2],
Alicia Troncoso[1], Francisco Martínez-Álvarez[1],
and Gualberto Asencio-Cortés[1(✉)]

[1] Data Science and Big Data Lab, Pablo de Olavide University, 41013 Seville, Spain
{amchamal,atrolor,fmaralv,guaasecor}@upo.es
[2] easytosee AgTech S.L., Diego Martínez Barrio 10 (3rd floor), 41013 Seville, Spain
miguelangel.molina@ec2ce.com

Abstract. A new methodology has been applied to improve the prediction accuracy on the olive phenology forecasting problem, applying deep learning with hyperparameter optimization to handle with imbalanced data. The application of hyperparameter optimization to optimize the architecture of the deep neural network along with both class balancing preprocessing and the introduction of new variables allowed to improve the phenological forecast classification problem in 16 different plots from 4 different areas in Spain is introduced in this work. The results obtained have been shown to be promising and encourage further research in this field, where the potential for improvement is very high. The improvements, in terms of prediction accuracy, achieved are around 4% on average and, in some cases, exceeding 20%.

Keywords: Deep learning · Classification · Phenology forecasting · Imbalanced learning

1 Introduction

The agricultural sector or farming sector is the set of productive initiatives in society that are dedicated to obtain agricultural products. Historically, it has positioned itself as the most important economic sector in the world. However, in recent years several problems have arisen that endanger the proper livelihood of this sector in order to feed the entire population and livestock. For example, climate change is causing difficulties in the maintenance of crops that increasingly need to be controlled in order to feed a growing population. In addition, food safety is increasingly important and is subject to strict controls. For all these reasons, it is becoming increasingly common to find applications of new technologies that help farmers to make decisions and manage all their resources properly.

For the application of these technologies we need to collect data. The implementation of sensors in cultivation fields is becoming more and more widespread,

P. García Bringas et al. (Eds.): HAIS 2022, LNAI 13469, pp. 274–285, 2022.
https://doi.org/10.1007/978-3-031-15471-3_24

which, together with the state of cultivation sampling and the help of data from public sources, we have a large set of data on which to apply artificial intelligence techniques. Even so, there is still little data belonging to this sector (except meteorological data) so it is necessary to apply new algorithmic techniques to optimize the use available information and, if possible, extend this knowledge to other geographies without any information.

The availability of the phenology data is weekly, so the predictions must be at least weekly. On the other hand, having a forecast of the ripening of the olive a week in advance allows the farmer to detect possible outbreaks of pests in time and be able to intervene with phytosanitary products. Due to the evolution of crop phenology, different states of maturation are not equally represented in datasets, generally producing inaccurate classification results. For such reason, the proposed methodology includes the application of imbalanced preprocessing techniques (both undersampling and oversampling).

In this work, an information fusion of phenological stages in different olive grove plots taken from the field, along with satellite index data and vegetative indices (obtained from images), in addition to climatological data have been used. This dataset is formulated as a multi-class imbalanced classification problem with four phenological states. Different techniques are also evaluated to compare the accuracy, precision, recall and F1, achieving quite promising results.

This paper shows an advance in the prediction of olive phenology by the use of automated deep learning, including an internal validation to perform a hyperparameter optimization. To validate the proposal, a fair comparison with other algorithms (support vector machines, decision trees and k-nearest neighbors) has been performed. The rest of the paper is structured as follows. Section 2 overviews recent and relevant papers in the field of deep learning and its application to classification datasets. Section 3 describes the proposed methodology. Section 4 reports and discusses the results achieved. Finally, Sect. 5 summarizes the conclusions.

2   Related Works

Deep learning is becoming one of the research fields in which much effort is being put into [1]. In fact, many applications can be found in the literature currently. However, these techniques are not yet widespread in agriculture sector. Some works have been published introducing these techniques. For example, in [2] where basic algorithms are used to yield predictions, disease detection or crop quality. It should be noted that this work is a continuation of the experiments started in [3], where the same techniques were applied but without adding parameter optimization or deep data pre-processing to improve the results.

In [4], deep learning techniques are applied to detect phenological stages of the rice crop through images taken by aerial vehicles to make estimates of production and harvest dates. Additionally, in [5,6], some works can be seen using deep learning for the recognition of phenological patterns or for the prediction of the phenology and incidence of pests and diseases in crops. Especially in reference [5] the data taken for the analysis comes from half-hourly captures from cameras mounted on agronomic stations.

In this kind of study, satellite images play a key role. In recent years, satellite observations have been widely used in research work requiring difficult to obtain field data, due to their public accessibility and low cost, covering a large area with increasingly accurate resolutions.

Hence, time series of the MODIS NDVI index were used to distinguish different crop types according to their phenological evolution in [7]. Even in [8], these images were used to detect the optimal conditions for a given crop type and sowing areas, being able to distinguish each one before harvest.

Another work that also used the MODIS satellite (and the corresponding NDVI index) to perform a spatiotemporal analysis of crop phenology in a given area can be found in reference [9] which aimed to classify crop types. A similar study can be found in [10] and in [11].

In other sectors, another deep neural network model was proposed in [12] for person re-identification. In particular, deep feature representation learning, deep metric learning and ranking optimization are included to analyses different perspectives of a person. In other cases, deep transfer learning is used under small training set conditions, with a very limited number of instances for training [13].

3   Methodology

This section describes the proposed methodology, which is based on the use of imbalanced preprocessing techniques and a feed-forward fully-connected neural network [14], whose architecture is optimized by a random search algorithm, producing a classification model to forecast the olive phenology at seven days ahead (see Fig. 1). This work is based on the same olive crop places and satellite images than in [3], but including new attributes: vegetative indices (NDVI) and meteorological data.

Section 3.1 describes the data preparation. Section 3.2 describes the different imbalanced techniques applied in the studies. Section 3.3 introduces the proposed neural network model generation (ADL). Finally, the different benchmark algorithms to perform the comparison are introduced in Sect. 3.4.

3.1   Data Preparation

The first step in the preparation of the data for further use is to build the same database that relates the information obtained from the bands from the satellite images with the data of the phenological states in each of the plots and climatological data. From now on, phenology will be understood as the study of the changes in the calendar of seasonal events such as sprouting, flowering, fruit formation, and so forth, characteristic for each crop and, as a plot, an extension or piece of land.

Once all the information available for all the plots has been mapped and unified, it is necessary to relate all the information that has occurred during the season with the evolution of the phenological states. The objective of these studies is to predict the phenology of the crop one week in advance, so all the

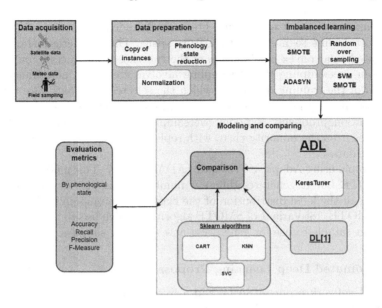

Fig. 1. General procedure carried out in the proposed methodology.

relationships between the input data and the target variable are oriented in this direction.

The second step consists of encoding the data labels in order to obtain as many neural network outputs as data labels are defined. In this way, it is possible to obtain a predicted probability for each of the defined labels. At this point, a normalization of the data is also carried out. Due to the minimum threshold of number of instances required by the imbalanced learning algorithms, three copies of the unique instance of class 1 (the first phenological state) have been made specifically in the plot 11.

3.2 Imbalanced Techniques

In order to balance the 4 classes corresponding to the 4 phenological states defined for the olive grove in this work, different imbalanced learning algorithms are applied. In Sect. 4.1, the unbalancing of the classes is shown, being very remarkable the unbalancing of the states 1 and 4, which makes this technique a very relevant improvement in this work. The different algorithms used for this are:

1. Synthetic Minority Oversampling Technique (SMOTE) [15]. SMOTE works by selecting examples that are close in the feature space, drawing a line between the examples in the feature space and drawing a new sample at a point along that line.
   SMOTE first selects a minority class instance a at random and finds its k nearest minority class neighbors. The synthetic instance is then created by

choosing one of the k nearest neighbors b at random and connecting a and b to form a line segment in the feature space. The synthetic instances are generated as a convex combination of the two chosen instances a and b.

2. Random oversampling [16] Resampling involves creating a new transformed version of the training dataset in which the selected examples have a different class distribution. This is a simple and effective strategy for imbalanced classification problems. Random oversampling involves randomly selecting examples from the minority class, with replacement, and adding them to the training dataset.

3. Oversample using Adaptive Synthetic (ADASYN) [17] This method is similar to SMOTE but it generates different number of samples depending on an estimate of the local distribution of the class to be oversampled.

4. SVMSMOTE [18] Variant of SMOTE algorithm which use an SVM algorithm to detect sample to use for generating new synthetic samples.

3.3   Automated Deep Learning Proposal

The neural network proposed in [3] had a fixed configuration. The aim of this improvement is to optimize the configuration of these parameters.

KerasTuner has been used for the optimization of these parameters. Keras-Tuner is an easy-to-use, scalable hyperparameter optimization framework that solves the pain points of hyperparameter search. The base Tuner class is the class that manages the hyperparameter search process, including model creation, training, and evaluation. RandomSearch Tuner has been used in this case, whose configuration will be indicated in Sect. 4.3.

In general, a grid has been created that allows a random search for the best grid configuration for those data rather than setting an exhaustive parametrization. In the end, the best individual corresponding to a particular network architecture is found. The number of layers as well as the number of neurons per layer and the learning rate have been modified. For this purpose, the dataset (which has 2043 instances and 422 attributes, including the class to predict, the ID of the plots and the feature corresponding to the time) is divided into training and test, being the test the last campaign of each plot (from January 1, 2019 to December 31, 2019 except in plots 11 and 13 which is from January 1, 2018 to December 31, 2018) and the training the remaining instances. We have used 80% for train and 20% for test.

3.4   Benchmark Algorithms

In order to assess the model's fitness, it will be compared to the network implemented at [3] from now on DL [3], a fixed neural network with 6 dense layers. In addition, it will be compared with:

1. CART (Classification And Regression Tree). The goal is to create a model that predicts the value of a target variable by learning simple decision rules inferred from the data features. Default values are used (gini as the function to measure the quality of a split 1 as the minimum number of samples required at a leaf node)

2. KNN (K-nearest neighbors). The principle behind nearest neighbor methods is to find a predefined number of training samples closest in distance to the new point, and predict the label from these. The number of neighbors to use by default is 5.
3. SVC (Support Vector Classifier). The goal is to match the data you provide, returning the hyperplane that divides or categorizes your data (C was set to 1).

4 Experimentation and Results

This section presents and discusses the results achieved. Thus, Sect. 4.1 describes how the data set has been generated and interpreted. The metric used to evaluate the performance of the proposal is introduced in Sect. 4.2. The experimental settings are listed and discussed in 4.3. Finally, the results themselves are reported and commented in Sect. 4.4.

4.1 Dataset

The data set used to implement the studies detailed in the methodology has been taken from two different open access data sources.

Table 1. Domains characteristics.

Code	Region	Coordinates	Altitude (m)	Surface (Ha)	Slope (%)	Soil	Density (plant/Ha)	Main variety
P01	Jaén	37.71, −2.96	700	6.02	2	Irrigated	178	Picual
P02	Jaén	37.68, −2.94	700	2.86	5	Irrigated	200	Picual, Marteño
P03	Jaén	37.66, 2.93	700	1.00	1	Irrigated	140	Picual, Marteño
P04	Cádiz	36.86, −5.43	360	8.54	19	Dry	58	Lechín, Zorzaleño, Ecijano
P05	Cádiz	36.87, −5.45	770	2.26	19	Dry	138	Lechín, Zorzaleño, Ecijano
P06	Cádiz	36.94, −5.25	350	1.04	5	Dry	134	Picual, Marteño
P07	Cádiz	36.94, −5.30	425	3.10	11	Dry	74	Lechín
P08	Córdoba	37.44, −4.69	280	6.30	1	Dry	194	Picual, Marteño
P09	Córdoba	37.67, −4.33	300	42.53	3	Irrigated	154	Picual, Marteño
P10	Córdoba	37.60, −4.42	460	9.58	9	Dry	76	Picual, Picudo
P11	Córdoba	37.44, −4.71	320	19.52	1	Irrigated	208	Manzanillo
P12	Jaén	37.67, −2.91	700	0.96	3	Irrigated	92	Picual
P13	Sevilla	37.05, −5.01	600	1.93	15	Dry	119	Picual, Marteño
P14	Sevilla	37.10, −5.04	460	31.59	15	Irrigated	156	Manzanillo
P15	Sevilla	37.05, −5.08	520	1.78	20	Dry	120	Hojiblanco
P16	Sevilla	37.07, −5.06	510	2.87	25	Dry	150	Hojiblanco

First, from the images taken by the SENTINEL satellite, information is obtained from the different spectral bands, which in this case are composed of 12 of the visible, near-infrared and short-wave infrared part of the spectrum.

On the other hand, from the open data set available at *Red de Alerta e Información Fitosanitaria* [19], property of *Junta de Andalucía*, the samples of the different phenological stages of the olive tree are obtained.

In the olive crop, 14 phenological states are detected. In order to facilitate the evolution of the crop and to ensure that each stage is minimally represented in the data set, these 14 states are summarised in 4. State 1 includes all the states prior to flowering. These are the winter bud and moved bud stages, which comprise 2.25% of the samples studied. State 2 comprises the states related to flowering. These are inflorescence, corolla bloom and petal fall. These states comprise 29% of the samples studied. State 3 comprises the fruit formation stages, which are fruit set and stone hardening. These states comprise 65.13% of the data. State 4 includes the last phenological states of the olive grove, which are veraison and ripe fruit, accounting for 3.62% of the remaining samples.

Finally, the last dataset is the meteorological data. Different meteorological stations measure different variables, such as temperature, humidity, solar radiation and precipitation. These stations have a daily sampling periodicity. All this information has been extracted from 16 plots taken from all over Andalusia, grouped in 4 areas with 4 plots each. The geographical distribution of the plots can be seen in Fig. 2. Each area and plot has different characteristics (see Table 1).

Fig. 2. Localization of the parcels used for the study (Andalusia, Spain).

4.2   Evaluation Metrics

To show the effectiveness of the studies conducted, accuracy, recall, precision and F-measure were calculated. These metrics were obtained from the 4 × 4 confusion matrix obtained from the phenological classification problem with four phenological stages. Specifically, for each phenological state, true positives (TP), true negatives (TN), false positives (FP) and false negatives (FN) were considered. Specifically, TP describes the correctly predicted cases for each phenological state and FP are positive outcomes that the model predicted incorrectly. A similar reasoning was done for the negative cases. The first metrics were computed for all four phenology classes. We started from the 4 × 4 confusion matrix mentioned before and the accuracy was calculated as defined in Eq. (1).

$$Acc = \frac{TP + TN}{TP + FP + FN + TN} \qquad (1)$$

Subsequently, the metrics of precision, recall and F-1 were calculated for each phenological stage, and the weighted mean was also calculated.

Specifically, precision was the ratio of correctly predicted positive observations to the total predicted positive values, as defined in Eq. (2).

$$Prec = \frac{TP}{TP + FP} \tag{2}$$

Recall was the ratio of correctly predicted positive observations with respect to all actual positive instances, as defined in Eq. (3).

$$Recall = \frac{TP}{TP + FN} \tag{3}$$

The F1 score was the weighted average of precision and recall (Eq. (4)).

$$F1 = 2 * \frac{Recall * Precision}{Recall + Precision} \tag{4}$$

Furthermore, all those metrics were also computed for each class (phenological state) independently. Each TP, TN, FN, FP referred to each phenological stage versus the rest.

4.3   Experimental Settings

In order to set up the neural network architecture for the experimentation carried out, the experimental settings established for the random search-based optimization algorithm were the presented in Table 2.

Table 2. Settings for the random search-based optimization algorithm.

Parameter	Description
Output shape (NN)	Parametrization of the output space of the neural network on a grid of 10 to 100 with steps of 10
Optimizer (NN)	The optimizer used was SG (Gradient descent (with momentum) optimizer)
Learning rate (NN)	Hyperparameter that controls how much to change the model in response to the estimated error each time the model weights are updated. Grid [1e−4, 1e−1]
Loss function (NN)	Sparse categorical crossentropy
Optimization metric (NN)	Accuracy
Epochs (model)	One epoch is when an entire dataset is passed forward and backward through the neural network only once. The number of epochs used was 100
Activation function	The activation function used in all layers has been relu, except in the output layer which has been softmax

Each base algorithm combination (ADL, DL, CART, KNN, SVC), together with each imbalanced learning algorithm produces a different optimal network architecture. The best results were obtained with ADL together with ADASYN and the optimized neural network architecture was composed of 4 dense-type layers with 10, 100, 10 and 4 neurons respectively, with a learning rate of 0.01967. The initializer used has been GlorotNormal. The internal validation used in the adjustment of hyperparameters has been a hold-out of 70–30%.

4.4   Results and Discussion

The results obtained after applying the methodology proposed in the previous section are shown in this section. Table 3 shows a breakdown of the results, in terms of the metrics proposed in Sect. 4.2, for each one of the proposed algorithms and the different imbalanced learning techniques previously introduced.

Table 3. Effectiveness metrics for each algorithm and imbalanced learning techniques averaging all parcels (all metrics are in %).

Algorithm	Imbalanced learning	Accuracy	Precision	Recall	F1-score
ADL	Original	83.3	86.7	83.3	82.1
	SMOTE	83.3	88.5	83.3	83.8
	RandomOverSampling	83.7	86.8	83.7	83.1
	ADASYN	83.7	88.2	83.7	84.3
	SVMSMOTE	79.1	87.9	79.1	81.3
DL [3]	Original	81.7	81.7	81.7	80.6
	SMOTE	77.9	84.9	77.9	78.8
	RandomOverSampling	78.3	84.0	78.3	78.8
	ADASYN	80.5	88.6	80.5	82.4
	SVMSMOTE	74.5	79.6	74.5	75.5
CART	Original	80.5	86.3	80.5	82.0
	SMOTE	79.7	84.0	79.7	80.4
	RandomOverSampling	80.9	84.7	80.9	80.8
	ADASYN	78.7	85.0	78.7	80.2
	SVMSMOTE	79.0	84.6	79.0	80.2
KNN	Original	81.6	86.8	81.6	82.0
	SMOTE	76.9	87.4	76.9	79.7
	RandomOverSampling	81.3	89.5	81.3	83.0
	ADASYN	76.8	87.8	76.8	79.2
	SVMSMOTE	79.7	88.3	79.7	82.1
SVC	Original	59.0	35.8	59.0	44.3
	SMOTE	54.3	73.9	54.3	56.7
	RandomOverSampling	53.2	73.2	53.2	55.6
	ADASYN	29.0	25.0	29.0	25.0
	SVMSMOTE	62.1	72.7	62.1	63.4

The best results were obtained by applying the methodology proposed in this paper (ADL) using the ADASYN imbalanced preprocessing algorithm. With such combination, an accuracy of 83.7% and a F1 of 84.3% were obtained. In particular, looking at the F1 metric, ADL improves by 1.9% on the best result of the algorithm cited in [3], by 2.3% on the best result of the $CART$ algorithm, by 1.3% on the best result applying KNN and by 21.1% applying SVC.

Figure 3 shows the percentage of success for each of the proposed metrics for the ADL algorithm with $ADASYN$. According to such figure, we can conclude that the plot with the best results was the parcel $P2$ with about 97% of success in all its metrics. On the other hand, the worst plot was the $P13$, with a F1 metric of 60%.

Analyzing the areas of the studied olive crops, the two areas with the highest percentage of success were Jaén (P01, P02, P03, P12) and Seville (P13, P14, P15, P16). These areas have an high percentage of sampling during the whole campaign and are much more represented in the historical data than the other areas, except for the parcel $P13$, that has less samplings compared to the rest.

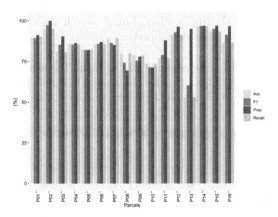

Fig. 3. Effectiveness metrics achieved by ADL for each olive crop parcel.

If we compare the results of the original set with those obtained in paper [3], ADL has improved the results by 1.6%, which also means that the introduction of the new input features of the information fusion proposed in this work in itself improves the results of both the deep learning models (ADL and DL [3]).

These new attributes introduced come from 3 different datasets: the first dataset has satellite data (since 2015, every 5 days and up to 12 bands with different wavelengths), the second contains meteorological data (variables of temperature, humidity, solar radiation and precipitation) and the third is field sampling data (consists of phenological sampling obtained by crop experts).

5   Conclusions

In this work, the benefits of automated deep learning have been empirically demonstrated using a classification dataset of phenological states of olive crops, with vegetative indices (GCI, NDVI, EVI, NDMI and NDWI obtained from satellite images) and meteorological data, improving the accuracy, precision, recall and F1 with respect to other proposals. The methodology proposed, ADL, is based on a preprocessing stage with imbalanced techniques in order to balance the representation of each of the phenological states in the dataset. Moreover, the proposed training process includes a random-search based optimization of the architecture of a feed-forward fully-connected neural network. The proposed model has been compared to three well-known classifiers along with the deep neural network proposed in [3]. The results achieved by ADL were promising, obtaining improvements of around 2% for most cases when comparing the proposed algorithm with other well-known classifiers and the deep neural network without parameter optimization. These works are a starting point to continue exploring the benefits of automated deep learning on agriculture problems. As future works, the prediction of new longer horizons, the application of feature selection techniques and the application of more search algorithms for the hyperparameter optimization will be developed.

Acknowledgements. The authors would like to thank the Spanish Ministry of Science and Innovation for the support under the project PID2020-117954RB, the European Regional Development Fund and Junta de Andalucía for projects PY20-00870 and UPO-138516.

References

1. Dong, S., Wang, P., Abbas, K.: A survey on deep learning and its applications. Comput. Sci. Rev. **40**, 100379 (2021)
2. Liakos, K.G., Busato, P., Moshou, D., Pearson, S., Bochtis, D.: Machine learning in agriculture: a review. Sensors **18**(8), 2674 (2018)
3. Molina, M.Á., Jiménez-Navarro, M.J., Martínez-Álvarez, F., Asencio-Cortés, G.: A model-based deep transfer learning algorithm for phenology forecasting using satellite imagery. In: Sanjurjo González, H., Pastor López, I., García Bringas, P., Quintián, H., Corchado, E. (eds.) HAIS 2021. LNCS (LNAI), vol. 12886, pp. 511–523. Springer, Cham (2021). https://doi.org/10.1007/978-3-030-86271-8_43
4. Yang, Q., Shi, L., Han, J., Yu, J., Huang, K.: A near real-time deep learning approach for detecting rice phenology based on UAV images. Agric. For. Meteorol. **287**, 107938 (2020)
5. Yalcin, H.: Phenology recognition using deep learning. In: Proceedings of the Electric Electronics, Computer Science, Biomedical Engineerings' Meeting, pp. 1–5 (2018)
6. Grünig, M., Razavi, E., Calanca, P., Mazzi, D., Wegner, J.D., Pellissier, L.: Applying deep neural networks to predict incidence and phenology of plant pests and diseases. Emerg. Technol. **12**, e03791 (2021)

7. Skakun, S., et al.: Early season large-area winter crop mapping using MODIS NDVI data, growing degree days information and a Gaussian mixture model. Remote Sens. Environ. **195**, 244–258 (2017)

8. Hao, P., Zhan, Y., Wang, L., Niu, Z., Shakir, M.: Feature selection of time series MODIS data for early crop classification using random forest: a case study in Kansas, USA. Remote Sens. **7**(5), 5347–5369 (2015)

9. Wang, Y., Xue, Z., Chen, J., Chen, G.: Spatio-temporal analysis of phenology in Yangtze River Delta based on MODIS NDVI time series from 2001 to 2015. Front. Earth Sci. **13**(1), 92–110 (2019). https://doi.org/10.1007/s11707-018-0713-0

10. Xue, Z., Du, P., Feng, L.: Phenology-driven land cover classification and trend analysis based on long-term remote sensing image series. IEEE J. Sel. Top. Appl. Earth Observ. Remote Sens. **7**(4), 1142–1156 (2014)

11. Melgar, L., Gutiérrez-Avilés, D., Godinho, M.T., et al.: A new big data triclustering approach for extracting three-dimensional patterns in precision agriculture. Neurocomputing **500**, 268–278 (2022)

12. Ye, M., Shen, J., Lin, G., Xiang, T., Shao, L., Hoi, S.C.H.: Deep learning for person re-identification: a survey and outlook. CoRR, abs/2001.04193 (2020)

13. Feng, S., Zhou, H., Dong, H.: Using deep neural network with small dataset to predict material defects. Mater. Des. **162**, 300–310 (2019)

14. Torres, J.F., Hadjout, D., Sebaa, A., Martínez-Álvarez, F., Troncoso, A.: Deep learning for time series forecasting: a survey. Big Data **9**, 3–21 (2021)

15. Chawla, N.V., Bowyer, K.W., Hall, L.O., Kegelmeyer, W.P.: SMOTE: synthetic minority over-sampling technique. J. Artif. Intell. Res. **16**, 321–357 (2002)

16. Branco, P., Torgo, L., Ribeiro, R.: A survey of predictive modelling under imbalanced domains. ACM Comput. Surv. **49**(a30), 1–50 (2017)

17. He, H., Bai, Y., Garcia, E. A., Li, S.: ADASYN: adaptive synthetic sampling approach for imbalanced learning. In: Proceedings of the IEEE International Joint Conference on Neural Networks, pp. 1322–1328 (2008)

18. Nguyen, H.M., Cooper, E.W., Kamei, K.: Borderline over-sampling for imbalanced data classification. Int. J. Knowl. Eng. Soft Data Paradig. **3**(1), 4–21 (2011)

19. de Andalucia, J.: RAIF website of the Consejeria de Agricultura, pesca y desarrollo rural (2020). https://www.juntadeandalucia.es/agriculturapescaydesarrollorural/raif. Accessed 26 Mar 2020

Architecture for Fault Detection in Sandwich Panel Production Using Visual Analytics

Sebastian Lopez Florez[1,2](✉) [iD], Marcos Severt Silva[1],
Alfonso González-Briones[1], and Pablo Chamoso[1] [iD]

[1] Grupo de Investigacin BISITE, Departamento de Informática y Automática, Facultad de Ciencias, University of Salamanca, Instituto de Investigación Biomédica de Salamanca, Calle Espejo 2, 24.2, 37007 Salamanca, Spain
{sebastianlopezflorez,marcos_ss,alfonsogb,chamoso}@usal.es
[2] Universidad Tecnológica de Pereira, Cra. 27 N 10-02, Pereira, Risaralda, Colombia
sebastianlopezflorez@utp.edu.co

Abstract. Technology can give industries the ability to create products/materials/services that meet customer needs and comply with applicable regulatory obligations. In this context, an automatic damage detection system is proposed for sandwich panels. Instead of relying on manual inspection, the system is based on artificial vision and operates with high accuracy in an industrial environment, ensuring traceability in product quality, reducing the percentage of returns caused by imperfections. The adaptive thresholding method seeks to identify the pixel intensities found on the surface of the sandwich panel. Unlike existing methods, the proposed algorithm is based on an adaptive threshold that uses the local characteristics of an image to segment and classify damage on the surfaces of sandwich panels, seeking to reject or accept a product according to the quality levels defined by the standard. The experimental results propose to generate a comparison with a sandwich panel damage detection method based on a convolutional neural network. The results of the experiment show that the proposed thresholding-based method has better accuracy and F1Score than deep learning methods. Moreover, this system is able to improve the industrial standards of sandwich panel manufacturing according to the standard, which limits the allowable imperfections, pointing out only the maximum admissible value of manufacturing imperfections to obtain a quality product.

Keywords: Damage identification · Deep learning · Feature extraction · Computer vision

1 Introduction

The ISO 9000 standard has been created in 1947 with the aim of making the different international producers and services safe, reliable and compliant with

© Springer Nature Switzerland AG 2022
P. García Bringas et al. (Eds.): HAIS 2022, LNAI 13469, pp. 286–297, 2022.
https://doi.org/10.1007/978-3-031-15471-3_25

all manufacturing quality requirements, by reducing the level of waste, producing sandwich panels [3] with fewer defects, avoiding future returns, minimizing customer complaints and increasing product quality. Moreover, ISO 9001:2000 establishes a quality benchmark in the manufacture of industrial products [5]. This is because damage is inevitable during the manufacturing process of sandwich panels, such as the fracture and buckling of the reinforcement, burning of the front sheets or nodes of the reinforcement that are not attached to the front sheet [12]. In some cases, sandwich panel flaps create a V-shaped gap due to their manufacture where insulation is not always placed, causing thermal bridging and air ingress. This is one of the many examples where manual inspection does not meet regulatory quality standards, the integrity of the structure can lead to poor product quality which results in a loss of reputation in the competitive landscape. Therefore, it is important to realise that quality is not only achieved through good and motivated workers; more competitive standards can be achieved through emerging technologies which automatically speed up processes by relying less on a subjective approach that takes a long time to be identified by the expert eye. This paper presents the Deep-Panel project, which aims to contribute to the value chain by providing a solution to three aspects of production, which are included in the main business quality requirements a general outline is shown in Fig. 1. The aspects being worked on are (1) the improvement of the produced product, (2) the optimisation of the production process and (3) the increase in production quality according to the standard offered to the client. Firstly, automation through the detection of non-conformities, as well as process optimisation and cost savings associated with the improvement in quality achieved in compliance with the standard. The plant is production areas will be equipped with technology for process monitoring and collection of the data required for automatic analysis based on the technique of unmbralisation and deep networks (DL) providing models both in the production area and in the storage area.

This document is organised as follows: Sect. 1 gives the background and the introduction. Section 2 is an overview of the fundamentals of CNN. Section 3 presents the proposed method. Section 4 describes the experimental study and analyzes the results. Section 6 draws conclusions from the conducted research

2   Related Works

This section describes the techniques used for damage detection in the manufacture of sandwich panels. The techniques are based on deep neural networks, as well as some classical techniques that use handmade features for their classification, in order to justify the use of these two techniques to provide a model capable of complying with regulatory quality standards in a real environment, similar implementations were sought in order to achieve reliability in the project.

Research works that have used deep learning as a method for detecting damage at the level of structure, finishes and manufacturing in different processes, have made a series of contributions to the field. Namely, in [17] authors considered a non-destructive evaluation method for the identification of disbonding

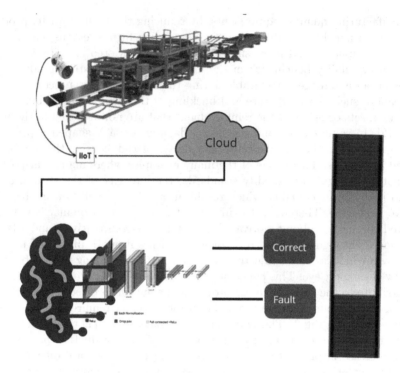

Fig. 1. Diagramatic representation of the system.

in a honeycomb sandwich beam using the frequency response function (FRF). However, when detecting damage in large structures, the method cannot effectively identify small detachments, which have little influence on the low frequency range of a FRF. In [1], the authors proposed an approach based on the DFNN model to predict the location and severity of damage in civil engineering structures. With the capability of massive processing of incomplete modal data with noise as well as fast training through the deep architecture of neural networks, machine learning provides an objective solution with quantitative accuracy. In the paper, they used 1D-CNN to automatically extract damage-sensitive features from raw acceleration signals and performed preliminary experiments to verify the proposed method. In [9] the authors proposed a method based concrete spalling damage detection based on a faster region-based convolutional neural network (R-CNN) with an inexpensive depth sensor to quantify separate, multiple instances of simultaneous spalling on the same surface and consider multiple surfaces. In [6], the authors presented a DL-based method that extracts damage features from mode shapes without using any hand crafted features or prior knowledge. Data sets based on numerical simulations were used, along with two data sets based on laboratory measurements.

Several methods have been proposed in the literature for the detection and localisation of damage in sandwich structures. A strain energy method for identifying the location of surface cracks in composite laminates was proposed in [8]. Another approach by et al. [15], combined thermographic and vibration measurements and trained artificial neural networks with the test data to detect damage in sandwich structures. Most of these studies were able to detect and locate the damage, however, to more accurately detect the extent of damage, some intelligent techniques, such as genetic algorithm (GA) or convolutional neural networks (CNN), have been combined with traditional methods because they have performed better in industrial failure detection [9,16].

3   The CNN Fundamentals

Thanks to the unique feature of its convolutional layers, which contain learnable kernel filters as parameters, the CNN can obtain spatially correlated local features from a small region of previous layers where its activation functions, such as rectified linear units (ReLU) and sigmoid functions provide these using the output of the nodes from one layer to the next layer, allowing for the modelling of non-linear input data to be incorporated in the network [7].

To learn the optimal parameters through weight updates formulas (2), (3) define the prediction value and loss function, respectively. Minimising the loss function brings the predicted value as close as possible to the actual value.

$$\hat{y} = f(W^T X + b) \tag{1}$$

$$L_i = -\sum_j y_{ij} \log(\sigma(f_j)) + (1 - y_{ij}) \log(1 - \sigma(f_j)) \tag{2}$$

where y_i is the true label with 1 if the class is a cart and 0 when it is not, 0 when it is not, it computes a difference between the actual and predicted probability distributions to predict the 1 class. The score is minimised and a perfect value is 0. $\sigma(.)$ is the sigmoid function, f is the score vector, which is positive when the class is present and negative otherwise. Thus, the sigmoid function is derived.

4   Proposed Surface Damage Detection Using Thresholding

It is the most important part of image processing. The goal of segmentation is to make it easier to analyze images by changing how they are represented. Segmentation divides an image into regions with similar features. Image processing is aimed at identifying regions of interest (ROIs) [14]. This method is described below.

1. Symmetrically distributed camera which ensures an image capture over the process (Fig. 2).
2. Images are resized (to 30% of their size) to be able to work with 4K images.
3. The sections with the highest illumination are cropped, to eliminate high variability in lighting and reflections.

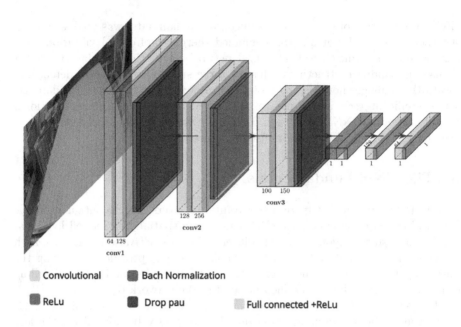

Fig. 2. Architecture of the applied convolutional neural network consisting of 3 convolutional layers and three fully connected layers.

4. The four points that define our region of interest are selected; the ones that define the section of the panel.
5. We apply a perspective transformation on those 4 points to get a perspective perpendicular to the panel (as if the image had been taken from above) and eliminate the effects of the tilt.
6. We transform to greyscale, to be able to apply a threshold that binarises the light intensities (black and white).
7. We apply a Gaussian adaptive thresholding, which detects the defects due to the differences in intensity with respect to their closest surroundings. The value of the parameter that best captures the defects is specified.
8. We apply a function that finds contours on the basis of the above thresholding.
9. We add a counter of the contours, which is the counter of the defects in case they exist. We select the child contours contained in the parent contour of the panel and set a minimum distance between contours to consider whether or not two points belong to the same contour.
10. In this case, if more than 20 defects are detected, the image is considered to contain rollers and is discarded, since we have observed that in valid images no more than 10 defects are detected.

In the global threshold method, a specific threshold Fixed are pixels whose intensity is above the threshold assumed to be black or foreground, the rest assumed to be white or background. This approach will not yield favorable results for all species picture. For images with variable background and variety Across the whole object, the performance of the global thresholding technique is not satisfactory. Using this type of image is a suitable result Shows in some regions and results in others Not acceptable. To overcome this shortcoming of the global threshold, the local Threshold method to use. The original of this sleeper technique Divide the image into different sub-images and calculate the threshold value for each subimage. Calculate each threshold These partial images can be analyzed using various statistical tools such as mean and/or or standard deviation, etc. In the reported work, adaptive thresholding method. These are Gaussian adaptive thresholding [10].

Minimum distance between contours is a widely used soft clustering algorithm, where data points belong to two or more clusters at the same time. In this algorithm, each data point is assigned a corresponding membership Each cluster center is based on the distance between cluster centers and data points [11].

The conventional methods are implemented by the OpenCV [2], where After clustering and thresholding the images, the next step is to apply some morphological methods The process of obtaining the final segmented image. Morphological operations Mainly used to remove distortion Threshold the image. The binary image has several internal holes is filled. Next, erase the edges of the image by removing the incomplete parts object. There are also some unwanted small objects in the image rejected in the final image. After applying these morphological operations.

5   Experimental

To evaluate the effectiveness of the proposed CNN-based approach and thresholding model, it is proposed to verify the algorithms on sample images of burn damage and imperfections in honeycomb structures. The algorithm was trained to identify the presence of damage and discard the product that does not meet the required quality standards. To perform this process, a series of images, which are described below, were collected and the different techniques are compared.

5.1   Data Sets

The data acquisition system is based on a camera positioned on the production lines where data capture and transmission tests are performed to devices in the cloud where they will be processed with the proposed model. For the failure cases we have the panel wear in production line and the burn-in damage, 10 failure cases are created with different positions according to the most common defects exposed by the technicians along the panel. For imperfection damage, we create 6 damage cases with different number of folds. To collect more abnormal samples along the supply chain, we monitored the process online, so for each defect, 30 images were taken. Each time the operator generated an alarm about a change in the panel, the camera assigned a label of the damage of the receptive register, which makes the captured images different. We collected 1,200 images in total: 720 images are of failures, 480 images are of wear, burns, imperfections. The resolution of the images is 3840×2160 pixels. To avoid the data imbalance problem that may occur during the training process, we randomly selected 300 images from each category to establish the training data set and uniformly scaled the images to 500×500 pixels.

Looking for a better generalisation of the model, the "ImageDataGenerator" function in the Keras deep learning framework has been used to generate batches of augmented image data making the database as diverse as possible, avoiding network overfitting problems. Data augmentation. In particular, for training and testing as appropriate, it is intended that the original image generate a series of randomly rotated figures within $30°$, or a shift to the left or right within 10% of the original image, a sample of this from the Fig. 3 database.

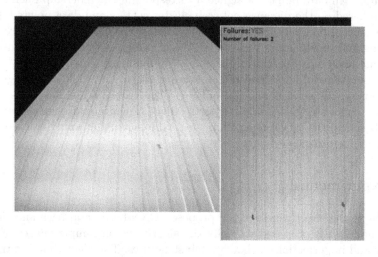

Fig. 3. Examples of the identified damage.

5.2   Evaluation MetricS

This section describes the metrics used to evaluate the proposed model because the task consists of a supervised classification problem commonly used to evaluate system performance [4].

These metrics are based on a confusion matrix that incorporates information about the prediction outcome of each test sample. TP stands for the number of true positive predictions, TN - True negative predictions, FP - False positive predictions, and FN - False negative predictions.

$$A = \frac{TP + TN}{TP + TN + FP + FN} \tag{3}$$

$$Precision = \frac{TP}{TP + FP} \tag{4}$$

$$Recall = \frac{TP}{TP + FN} \tag{5}$$

$$F1Score = \frac{2 \times Precision \times Recall}{Precision + Recall} \tag{6}$$

5.3 Implementation Details

In this paper, damage detection in sandwich panels is a binary classification problem. Given the relative simplicity of the binary classification task, compared to proposals presented in the state of the art, a complex CNN is not required in the network design. In the proposed CNN, convolutional layers and clustering layers are mainly used for feature extraction, which is automatic and needs no prior expert knowledge. Figure 6 shows the structure of the proposed CNN, where conv, pool and dense represent the convolutional layer, the pooling layer and the fully connected layer, such as drop-pau, Bach nominalisation operations respectively. The proposed CNN consists of three pairs of convolutional and pooling layers and three fully connected layers. The lattice structure of this paper is based on [13]. The damage on the surface of sandwich panels is caused by faults in the operation which causes imperfections in the structure leading to local features, therefore the used convolution kernel is smaller and similarly to the step size it is more useful for extracting local damage information and improving the classification performance. Therefore, the size of the convolution kernel in the proposed CNN is 3×3, and the step is 1. The 'Dropout' function is used later as a regularisation system in some capabilities. Hinton et al. showed that the overfitting problem can be reduced by using the 'Dropout' function to avoid complex coadaptations in the training data. At each presentation of each training set, each hidden unit is randomly deactivated from the network with a probability of 0.5 so that a hidden unit cannot depend on the presence of other hidden units. In this way, we achieve maximum randomness.

5.4 Framework Evaluation

We use different sizes of the training data to explore whether the model is robust. we choose 300 images for training and 600 images for testing from each category; we randomly select in another set 300 images for training and 600 images for testing from each category, and so on. We repeat the above data selection processes ten times, and calculate the average of the ten accuracy tests. According to Fig. 4, the scores basically stabilize around 0.67.

One of these training sets is applied according to the noise conditions that an image may suffer from including illumination, occlusion, scale, etc. Are avoided by positioning the camera at a fixed angle, which ensures that there is little variation in the lighting in that area, since the camera is fixed there are no scaling problems and the operators do not generate occlusion in the process. For the training and validation protocol, the database was divided into 0.75 for training and 0.25 for validation. This step was repeated 5 times, reporting a confidence interval that ruled out possible over-training or bias in the results.

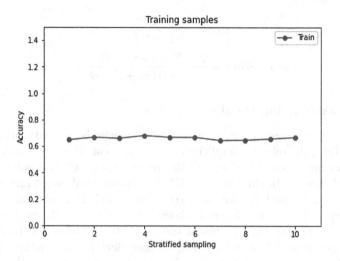

Fig. 4. Under different numbers of training sample.

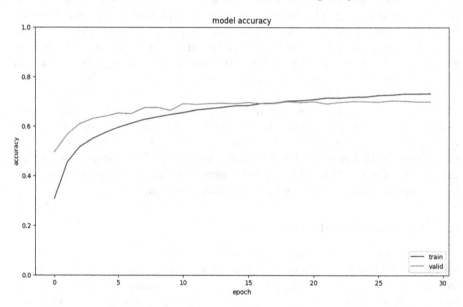

Fig. 5. Average accuracy and standard deviation on the validation data set, based on 30 epochs.

We show the results in the graphs Fig. 5, whit the performance obtained during the training and validation process. The system begins to converge from epoch 25 with a loss close to zero and an accuracy close to 67%.

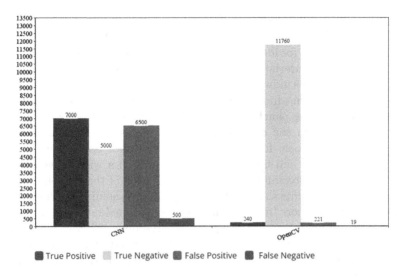

Fig. 6. Comparative metrics of the CNN model and thresholding (OpenCV) for a series of text images.

5.5   Comparison of Detection Performance

In this section, we compare the performance between the proposed method and the conventional machine learning method, which requires manual feature extraction. A series of steps are involved in the conventional method for the detection of features on the basis of the adaptive thresholding of pixel intensities (Table 1).

Table 1. Conjuntos de pruebas mAP para el modulo de deteccion SSD300

Model	Accuracy	Precision	Recall	F1Score
CNN	0.63	0.434	0.90	0.585
OpenCV	0.98	0.9719	0.98	0.975

The current graph is the result of testing both solutions with the total set of photos, i.e., 12000 photos of both faulty and non-faulty panels is presented in Fig 6.

As we can see, the second of the proposed options has an accuracy of 98% with hardly any false negatives, making it the solution of choice.

6   Conclusion and Future Work

In this paper, a method for identifying damage in sandwich panels by integrating a technique based on thresholding and CNN is proposed. When acquiring data from a set of images of sandwich panels with normal and manufacturing imperfections, the proposed machine learning methods could not reliably identify the damaged features on the panel surfaces due to the complex task of discriminant feature extraction from the convolutional. In contrast, the thresholding algorithm uses local statistical features of the panel surface image improving the accuracy of the system and finding the acceptability of a wide range of images of various qualities, overcoming the intrinsic limitation of the convolutional features. Experimental results show that the adaptive thresholding method achieves 98% Accuracy and F1score 97.5%, while the convolutional method is at $67 \pm 3\%$ and F1score 58.5%. Furthermore, the proposal presents an alternative for ensuring that quality reaches the standard established for industrial sandwich panel manufacturing processes. This standard restricts the permitted defects in order to obtain a quality product.

In future research, it will be necessary to address the initial hypothesis in which the deep learning techniques applied in this engineering problem are related to the effectiveness of classical thresholding models which performed better in preventing panel damage. To obtain a model that is robust to changes and adverse conditions, using convolutional features, it will be necessary to identify why the hand-crafted features performed better. This will make it possible to construct a deep network that can address the panel damage detection problem, enabling dynamic damage tracking and obtaining lifetime inference metrics will be of great use as a complement to the present research.

Acknowledgements. This research has been supported by the project "Intelligent and sustainable mobility supported by multi-agent systems and edge computing (InEDGEMobility): Towards Sustainable Intelligent Mobility: Blockchain-based framework for IoT Security", Reference: RTI2018-095390-B-C32, financed by the Spanish Ministry of Science, Innovation and Universities (MCIU), the State Research Agency (AEI) and the European Regional Development Fund (FEDER).

References

1. Ali, R., Zeng, J., Cha, Y.J.: Deep learning-based crack detection in a concrete tunnel structure using multispectral dynamic imaging. In: Smart Structures and NDE for Industry 4.0, Smart Cities, and Energy Systems, vol. 11382, pp. 12–19. SPIE (2020)
2. Bradski, G.: The OpenCV library. Dr. Dobb's J. Softw.Tools **25**, 120–123 (2000)
3. Cabal, J.V.Á., Pérez, F.R., Gutiérrez, N.R., Cuiñas, M.C.: Revisión del proceso de adaptación al marcado ce de los paneles sandwich autoportantes. In: XI International Congress on Project Engineering: [celebrado en] Lugo, do 26 September 2007 ao 28 September 2007, pp. 493–502. Departamento de Ingeniería Agroforestal (2007)

4. Géron, A.: Aprende machine learning con scikit-learn, keras y tensorflow. Anaya (2020)
5. Guasco García, V.J.: Propuesta para reducir el número de no conformidades en la sección corrugadora en la Empresa Industria Cartonera Ecuatoriana SA en base a la norma ISO 9001: 2000. B.S. thesis, Universidad de Guayaquil. Facultad de Ingeniería Industrial. Carrera de ... (2014)
6. Guo, T., Wu, L., Wang, C., Xu, Z.: Damage detection in a novel deep-learning framework: a robust method for feature extraction. Struct. Health Monit. **19**(2), 424–442 (2020)
7. Hinton, G.E., Srivastava, N., Krizhevsky, A., Sutskever, I., Salakhutdinov, R.R.: Improving neural networks by preventing co-adaptation of feature detectors. arXiv preprint arXiv:1207.0580 (2012)
8. Hu, H.W., Wang, B.T., Lee, C.H.: Damage detection of surface crack in composite quasi-isotropic laminate using modal analysis and strain energy method. In: Key Engineering Materials, vol. 306, pp. 757–762. Trans Tech Publ (2006)
9. Huang, X., Liu, Z., Zhang, X., Kang, J., Zhang, M., Guo, Y.: Surface damage detection for steel wire ropes using deep learning and computer vision techniques. Measurement **161**, 107843 (2020)
10. Lee, C., Li, C.: Adaptive thresholding via gaussian pyramid. In: China, 1991 International Conference on Circuits and Systems, vol. 1, pp. 313–316 (1991). https://doi.org/10.1109/CICCAS.1991.184348
11. Lei, T., Jia, X., Zhang, Y., He, L., Meng, H., Nandi, A.K.: Significantly fast and robust fuzzy c-means clustering algorithm based on morphological reconstruction and membership filtering. IEEE Trans. Fuzzy Syst. **26**(5), 3027–3041 (2018)
12. Lu, L., Wang, Y., Bi, J., Liu, C., Song, H., Huang, C.: Internal damage identification of sandwich panels with truss core through dynamic properties and deep learning. Front. Mater. **7**, 301 (2020)
13. Modarres, C., Astorga, N., Droguett, E.L., Meruane, V.: Convolutional neural networks for automated damage recognition and damage type identification. Struct. Control. Health Monit. **25**(10), e2230 (2018)
14. Sharma, N., Aggarwal, L.M.: Automated medical image segmentation techniques. J. Med. Phys./Assoc. Med. Phys. India **35**(1), 3 (2010)
15. Shrestha, R., Choi, M., Kim, W.: Thermographic inspection of water ingress in composite honeycomb sandwich structure: a quantitative comparison among lock-in thermography algorithms. Quantit. InfraRed Thermogr. J. **18**(2), 92–107 (2021)
16. Zhang, Y., Sun, X., Loh, K.J., Su, W., Xue, Z., Zhao, X.: Autonomous bolt loosening detection using deep learning. Struct. Health Monit. **19**(1), 105–122 (2020)
17. Zhu, K., Chen, M., Lu, Q., Wang, B., Fang, D.: Debonding detection of honeycomb sandwich structures using frequency response functions. J. Sound Vib. **333**(21), 5299–5311 (2014)

Deep Reinforcement Learning-Based Resource Allocation for mm-Wave Dense 5G Networks

Jerzy Martyna[✉][iD]

Institute of Computer Science, Faculty of Mathematics and Computer Science,
Jagiellonian University, ul. Prof. S. Lojasiewicza 6, 30-348 Cracow, Poland
jerzy.martyna@uj.edu.pl

Abstract. In microwave technology, directional beams are used for the propagation of radio waves. Nevertheless, significant errors occur in localizing the receiver. The paper presents the method for radio resource allocation and beam management based on the double deep Q-learning algorithm. Simulation studies confirm that the proposed method significantly improves the efficiency of the millimeter 5G network.

1   Introduction

The fifth generation (5G) of mobile networks is characterized by ultra-high data throughput, mobile service capabilities, the use of radio computing clouds, etc. [1]. With a growing number of band combinations used in these networks, space constraints appear that limit the number of antennas, especially for low frequencies. Moreover, the existing spectrum for new mobile services becomes overcrowded. To solve the problem, millimeter waves are used (30–300 GHz) that offer the possibility of collecting large continuous-spectrum fragments. However, millimeter waves in a dense 5G millimeter network have a downside, namely significant propagation losses and the resultant limited range. Therefore, directional beams are used that provide high antenna gains with reduced interference between beams. The typical method of management in bundles is the division of all users into numerous clusters, each of them being served by a separate bundle [2].

In dense millimeter 5G networks, appropriate beam management is used to compensate for interference between the beams. An effective method for localizing the user equipment (UE) is useful in the management process. Suitable solutions are provided e.g. by the positioning methods known as AOA [3] or TOA [4]. However, all these methods give an error that reduces the network performance. Hence, there are reasons to use an artificial intelligence method - that produces accurate results at relatively low computational costs.

Artificial intelligence methods are widely used for localization purposes in 5G networks. For example, Prasad *et al.* [5] propose machine learning based on a nonlinear regression process with parameter estimation using Gaussian methods

© Springer Nature Switzerland AG 2022
P. García Bringas et al. (Eds.): HAIS 2022, LNAI 13469, pp. 298–307, 2022.
https://doi.org/10.1007/978-3-031-15471-3_26

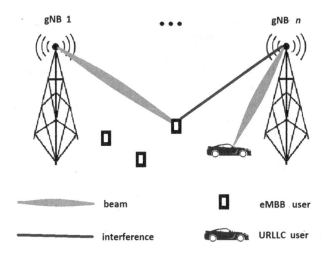

Fig. 1. System model.

for positioning users in a system with distributed MIMO antennas. AI methods are suitable for positioning in 5G using neural networks, as demonstrated in [6], where data from the GNSS system and an analysis of the beam signal lead to a very precise location obtained in an urban environment. In another study, by Gante *et al.* [7], recursive neural network methods are proposed for precise positioning in 5G networks.

In recent years, the possibility of using *deep learning* (DL) in wireless communication has been explored. For example, the paper by Luo *et al.* [8] proposed a beamforming method designed to improve downstream transmission efficiency without line of sight, based on a convolutional neural network. The article by Wang *et al.* [9] proposes transmission maximization in dynamic, correlated multi-channel access using the *Q*-learning algorithm combined with deep learning. In another paper, Wang *et al.* [10] used a deep learning method for automatic modulation recognition in cognitive radio networks.

The main aim of this article is to present a new method of deep learning application for the purposes of resource allocation in the 5G millimeter network. The original solution consists in the proposal of a new deep learning algorithm that uses distance estimation and distance parameters from the UE base station. Simulation studies confirmed the effectiveness of the proposed method.

2 System Model

It is assumed that the system model includes n 5G base stations (see Fig. 1), referred to as nodes, gNodeB (gNB), and m user equipments (UEs) with different QoS parameters. Each base station is assumed to be a gNB that has information about the location of user devices and can perform clustering to form user groups. Based on the observed location, each cluster is served by a separate beam marked *b.* Two types of user devices are considered here: URLLC (ultra-reliable low latency communication) and eMBB (enhanced mobile broadband). The first type of users require high data transmission reliability and allow only low latency. On the other hand, eMBB users require a very high transmission speed. In addition, the system uses OFDM modulation, preventing interference inside the beams. It is assumed that the available bandwidth of each beam b is divided into Resource Blocks (RB), including 12 subcarriers each. A Resource Block Group (RBG) is formed by connecting adjacent RBs. It is assumed to be the smallest radio resource that can be assigned to a user [11].

Spectral efficiency b_{nm} beams in UE can be represented as:

$$U(b_{nm}) = \log_2(1 + \frac{P_r(n,m)}{I(m) + N(m)})$$ (1)

where $P_r(n,m)$ is the received signal power in the time slot for m-th user device, $I(m)$ are respectively interference power and $N(m)$ noise power. The received signal strength by m-th user device is as follows:

$$P_r(n,m)(dB) = P_n + G_n(m) - PL(d)$$ (2)

where P_n is the transmission power of n-th gNB base station, $G_n(m)$ is the beamforming gain at the n-th gNB, $PL(d)$ is the path loss component as a function of the distance d from the gNB in n.

The energy gain $G_n(m)$ of antenna is given by:

$$G_n(m) =| \mathbf{w}_{nm}^H \mathbf{h}_{nm} |^2$$ (3)

where $\mathbf{h}_{nm}$ is the communication channel between n-th gNB base station and m-th user device, $\mathbf{w}_{nm}$ is a beamforming vector for n-th gNB base station in m-th user device. The $\mathbf{h}_{nm}$ communication channel can be represented as [12]:

$$\mathbf{h}_{nm} = \sqrt{\frac{N}{L}} \sum_{l=1}^{L} \alpha_l \mathbf{a}_{UE}(\phi_l^{UE}) \mathbf{a}_{gNB}^*(\phi_l^{gNB})$$ (4)

where α_l is the complex gain of the l-th path ($1 \leq l \leq L$), L is the number of scatterers. $\phi_l^{UE} \in [0, 2\pi]$, $\phi_l^{gNB} \in [0, 2\pi]$ are uniformly distributed variables that denote angles of l-th path for arrival and departure, respectively. (For the sake of simplicity, only one-dimensional is assumed here beamforming). The vectors $\mathbf{a}_{UE}$ and $\mathbf{a}_{gNB}^*$ denote antenna arrays of UE and gNB base station, respectively. $\mathbf{a}_{gNB}$ can be expressed as

$$\mathbf{a}_{gNB} = \frac{1}{\sqrt{N}} \left[1, \ldots, e^{j(N_{gNB}-1)\frac{1\pi}{\lambda} d\sin(\phi_l^{gNB})}\right]^T$$ (5)

where λ is the radio wavelength. In the case of a single antenna UE, is obtained

$$\mathbf{a}_{ml} = 1$$ (6)

3 Double Deep Reinforcement Learning for Resource Allocation in mm-Wave Dense 5G Networks

This section presents the concept of deep reinforcement learning (DRL), being an enhanced technique of machine learning. It enables the agent to discover what action should be taken to maximize the expected future reward in an interactive environment. The DRL method uses the ability of convolutional neural networks to learn representations of the individual with a growing efficiency, surpassing mere function approximations. The disadvantage of the DRL method is its high accuracy in approximating the Q function, as well as its poor state discrimination in the case of a continuous system. Hence, the double deep Q-learning (DDQL) method is used here, as proposed e.g. in the study by van Hasselt *et al.* [13]. The method consists in generating a neural network to determine the value of Q function for the problem considered, and then creating a target neural network that finds the correct parameters of the Q-function.

In the model, it was assumed that each beam is directed to the EU at a specific angle. Resource allocation requires QoS parameters for both eMBB and URLLC flows. The Markov decision-making process is defined here by specifying the following elements.

1) *Agents.* Each gNB is assumed to be a single agent that can independently perform resource allocation in the form of a certain number of RBG assigned to each beam b.

2) *States.* The states are defined as the CQI (channel quality indicator), sent as feedback via m-th UE device. The state of bth beam in m-th UE can be represented as:

$$s_i^b := q_m^b \qquad (7)$$

where q_m^b is an indicator of CQI for both types of data transmissions, namely: eMBB and URLCC.

3) *Actions.* For beam b, the i RBGs allocated to m-th UE are defined as an action, namely:

$$a_i^b := u_m^b \qquad (8)$$

4) *Reward.* The reward functions take into account the QoS parameters for both adopted flows: URLLC and eMBB. It was assumed that for ith RBG in beam b there is a reward function for URLLC devices expressed as follows:

$$r_i^b := SignRelu(\frac{SINR_i}{SINR^{QoS}} \frac{T^{QoS}}{T_m^{que}}) \qquad (9)$$

where $SINR_i$ is the signal-to-interference-plus-noise ratio in the link to m-th of the user receiving beam b, $SINR_{QoS}$ with the required SINR value is the flow with the specified QoS, I_{QoS} represents the delay required for a user with URLLC flow, T_m^{que} is the queue delay.

The reward feature for the eMBB device comes in the form of

$$r_i^b := SignReLu(\frac{SINR}{SINR^{QoS}}) \qquad (10)$$

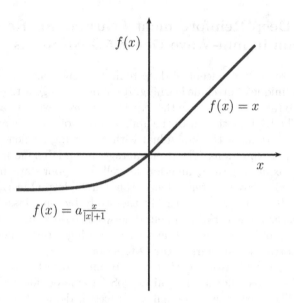

Fig. 2. *SignReLu* function image.

where *SignReLu* is the neuron segment activation function introduced by Lin and Shen [14] as an improvement to the *sigmoid* function. It is defined as follows (see Fig. 2):

$$f(x) = \begin{cases} x. & \text{for } x \geq 0 \\ a\frac{x}{|x|+1} & \text{for } \quad x < 0 \end{cases} \tag{11}$$

where x represents the nonlinear neuron segment activation function, a is the superparameter variable. If $a = 0$, then the $f(x)$ function is the *ReLu* function.

Figure 3 shows a schematic of the DDQL-based system used here. Each i-th agent, which is a single gNB base station, sends its beams (b_m^i, k_m^i) at the specified angles k_m^i. For each of them, the rewards are calculated - of being then transferred to the environmental system. On their basis, the state of the system is computed. The convolutional neural network (CNN), given the Q-function, calculates the predicted values of the Q-function in the next state, taking into account the penalties. At the same time, it makes use of temporary memory. It then selects new action values and sends them to the agents.

To train the parameters of the double deep Q-learning system, have been choosen Max_{sample} transition from the set of training samples. For each action a was observed reward r and state s'. From s' have been choosen action a' using the greedy policy π. In double deep Q-learning, the update is decoupled by use the two tables, A^A and Q^B, namely:

$$Q_{t+1}^A(s,a) := Q_t^A(s,a) + \alpha \left[r + \gamma Q_t^B(s', \text{argmax}_a Q_t^A(s',a')) - Q_t^A(s,a) \right] \tag{12}$$

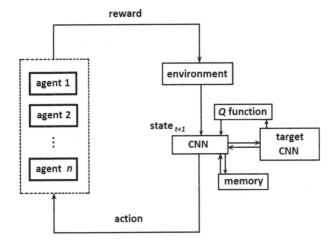

Fig. 3. Double deep Q-learning system for multi-user beam-selection in 5G system.

where α is the learning rate, γ is the discounting factor, $\gamma \in [0, 1]$.

The probability of taking action a from state s is given by:

$$\pi(a \mid s) = \begin{cases} 1 - \epsilon, & \text{if} \quad a = \text{argmax}_a(Q^A(s, a') + Q^B(s, a')) \\ \frac{\epsilon}{N_a - 1}, & \text{otherwise} \end{cases} \tag{13}$$

where ϵ is the exploration probability, N_a is the number of actions that can be taken from state s.

The presented learning scheme for training the parameters in used DDQL system is illustrated in Algorithm 1.

Algorithm 1 Double Deep Q-learning Algorithm

1: **procedure** DDQL ALGORITHM
2: **Require:** $Max_{episode}$, Max_{sample}
3: **Initialisation:**
4:     Q^A, Q^B, s
5:     Choose a from s using arbitrary policy
6:     **for** $episode \leftarrow 1, Max_{episode}$ **do**
7:         **for** $sample \leftarrow 1, Max_{sample}$ **do**
8:             Take action a, observe reward r and s'
9:             Choose a' from s' using policy π
10:            $Q^A(s, a) := Q^A(s, a) + \alpha\left[r + \gamma Q^B(s', \text{argmax}_a Q^A(s', a')) - Q^A(s, a)\right];$
11:            $s \leftarrow s', a \leftarrow a'$
12:            Swap Q^A and Q^B with probability equal to 0.5;
13:        **end for**
14:    **end for**
15: **end procedure**

Table 1. Simulation parameters.

Parameter	Value
5G network	
Number of URLLC per cell	7
Number of eMBB per cell	3
Beam angle	20°
Radius of cell	120 m
DDQL algorithm	
Learning rate (α)	0.6
Explorating probability (ϵ)	0.2
Discount factor (γ)	0.9
Max. number of samples	720
Simulation parameters	
Simulation time	0.5 s
Number of TTIs in every run	1200
Number of runs	12

4 Simulation Results

In this section, results of simulation tests will be presented, showing the effectiveness of the double deep Q-learning method in the allocation and management of beams in the 5G cellular network. It is assumed that there are only two gNB stations with a power of 10 W, and a range of 150 m. It is also assumed that they can generate four beams characterized by an angle of inclination equal 30°. Users keep track of them with a delay allowed for URLLC flow, being here 1 ms; $SINR_{QoS}$ for eMBB is assumed to be 15 dB. The remaining parameters are contained in Table 1.

The realistic dataset Raymobtime s009 presented in the paper by Klautau [15] were used for the simulation of the 5G network. The model achieved efficiency on the test set within 90%. On the training set, 94% of the predictions turned out to be correct. The results of the learning episodes are shown in Fig. 4.

The first scenario compares the operation of the DDQL method with that of the DCA (Dynamical Channel Allocation) method. It is assumed that each channel has only one bundle and that each of the two cells is associated with 10 users (3 eMBB and 7 URLLC devices per cell). Figure 5 shows the system spectral efficiency as a function of the number of users. It is clear that the use of the DDQL method results in an increase in the spectral efficiency by approximately 25%.

The second simulation scenario was designed to investigate the interdependencies between both types of users. Figure 6 shows the average number of eMBB user admissions depending on the average number of URLLC users for

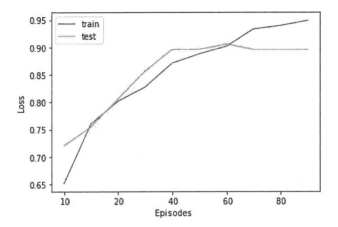

Fig. 4. Learning efficiency for individual episodes.

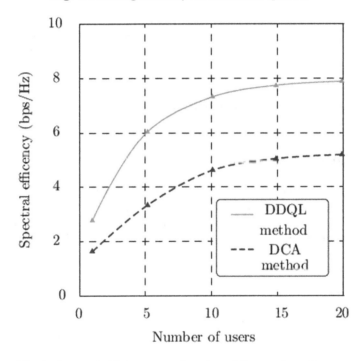

Fig. 5. Spectral efficiency as a function of the number of users.

the adopted bandwidth values of 10 and 20 MHz. The chart shows that a doubled system capacity increases the average number of eMBB users admitted by approximately 30% at the same number of URLLC users.

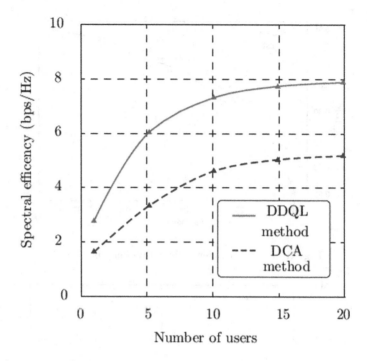

Fig. 6. Average number of eMBB users depending on the average number of URLLC users.

5   Conclusion

The study presents the use of double deep reinforcement learning (DDQL) for beamforming and radio resource management in millimeter 5G networks. A calculation scheme is proposed, designed to practically implement this method for allocating beams with specific QoS parameters of eMBB and URLLC flows. The simulation results confirm the effectiveness of this method.

References

1. Andreev, S., Petrov, V., Dohler, M., Yanikomeroglu, H.: Future of ultra-dense networks beyond 5G: harnessing heterogeneous moving cells. IEEE Commun. Mag. **57**(16), 86–92 (2019). https://doi.org/10.1109/MCOM.2019.1800056
2. Filippini, I., Sciancalepore, V., Capone, A.: Context information for fast cell discovery in mm-wave 5G networks. In: 21th European Wireless Conference, pp. 1–6 (2016). https://doi.org/10.48550/arXiv1501.02223
3. Menta, E.Y., Malm, N., Jantti, R., Ruttik, K., Costa, M.J., Leppanen, K.: On the performance of AoA-based localization in 5G ultra-dense networks. IEEE Access **7**, 33870–33880 (2019). https://doi.org/10.1109/ACCESS.2019.2903633
4. Huang, J., Liang, J., Luo, S.: Method and analysis of TOA based localization in 5G ultra-dense networks with randomly distributed nodes. IEEE Access **7**, 174986–175002 (2019). https://doi.org/10.1109/ACCESS.2019.2957380

5. Prasad, K.N.R.S.V., Hossain, E., Bhargava, V.K.: Machine earning methods for RSS-Based user positioning in distributed massive MIMO. IEEE Trans. on Wireless Comm. **17**(12), 8402–8417 (2018). https://doi.org/10.1109/TWC.2018.2876832

6. Klus, R., Talvitie, J., Valkama, M.: Neural network fingerprinting and GNSS data fusion for improved localization in 5G. In: 2021 IEEE International Conference on Localization and GNSS (ICL-GNSS), pp. 1–6 (2021). https://doi.org/10.1109/ICL-GNSS51451.2021.9452245

7. Gante, I., Falcao, G., Sousa, L.: Deep learning architectures for accurate millimeter wave positioning in 5G. Neural Process. Lett. **51**, 487–514 (2020). https://doi.org/10.1007/s11063-019-10073-1

8. Luo, C., Ji, J., Wang, Q., Yu, L., Li, P.: Online power control for 5G wireless communications: a deep Q-network approach. In: Proceedings of IEEE International Conference on Communication, pp. 1–6 (2018). https://doi.org/10.1109/ICC.2018.8422442

9. Wang, S., Liu, H., Gomes, P.H., Krishnamachari, B.: Deep reinforcement learning for dynamic multichannel access in wireless networks. IEEE Trans. Cogn. Commun. Netw. **4**(2), 257–265 (2018). https://doi.org/10.1109/TCCN.2018.2809722

10. Wang, Y., Liu, M., Yang, J., Gui, G.: Data-driven deep learning for automatic modulation recognition in cognitive radios. IEEE Trans. Veh. Techn. **68**(4), 4074–4077 (2019). https://doi.org/10.1109/TVT.2019.2900460

11. Elsayed, M., Erol-Kantarci, M.: Radio resource and beam management in 5G mmWave using clustering and deep reinforcement learning. In: GLOBECOM 2020, pp. 1–6 (2020). https://doi.org/10.1109/GLOBECOM42002.2020.9322401

12. Alkhateeb, A., El Ayach, O., Leus, G., Heath, R.W.: Hybrid precoding for millimeter wave cellular systems with partial channel knowledge. In: Information Theory and Applications Workshop (ITA), pp. 1–5 (2013). https://doi.org/10.1109/ITA.2013.6522603

13. van Hasselt, H., Guez, A., Silver, D.: Deep reinforcement learning with double Q-learning. CoRR abs/1509.06461 (2015). http://arxiv.org/1509.06461, https://doi.org/10.48550/ARXIV.1509.06461

14. Lin, G., Shen, W.: Research on convolutional neural network based on improved Relu piecewise activation function. Proc. Comput. Sci. **131**, 977–984 (2018). https://doi.org/10.1016/j.procs.2018.04.239

15. Klautau, A., Batista, P., González-Prelcic, N., Wang, Y., Heath, R.W.: 5G MIMO data for machine learning: application to beam-selection using deep learning. In: Information Theory and Applications Workshop (ITA), pp. 1–9 (2018). https://doi.org/10.1109/ITA.2018.8503086

Evolutionary Computation

Evolutionary Triplet Network of Learning Disentangled Malware Space for Malware Classification

Kyoung-Won Park[1(✉)], Seok-Jun Bu[2], and Sung-Bae Cho[1,2]

[1] Deptartment of Artificial Intelligence, Yonsei University, Seoul 03722, Korea
{pkw408,sbcho}@yonsei.ac.kr
[2] Department of Computer Science, Yonsei University, Seoul 03722, Korea
sjbuhan@yonsei.ac.kr

Abstract. With the advent of sophisticated deep learning models, various methods for classifying malware from structural features of source codes have been devised. Nevertheless, recent advanced detection-avoidance techniques actively imitate structural features of benign programs and share vulnerable subroutines, making it difficult to distinguish malicious attacks. Therefore, a method to distinguish and classify similar malicious attacks is urgent and significant. In this paper, we propose a method based on a triplet network of learning the disentangled malware space from assembly-level features beyond the structural characteristics of malware. The method comprises two major components, which are 1) triplet loss-trained network to disentangle deep representation between malware being close in the latent vector space, and 2) genetic optimization of assembly-level features to resolve collisions between thousands of assembly-level features. Experiments with the assembly and binary code dataset released from Microsoft show that the proposed method outperforms existing methods based on structural features, achieving the highest performance in 10-fold cross-validation. Moreover, we demonstrate the superiority of disentangled representation for malware classification by visualizing the latent space and ROC curves.

Keywords: Cybersecurity · Deep learning · Triplet network · Genetic optimization

1 Introduction

Malicious code is any software designed to damage a single computer, server, or computer network. The damage caused by malicious code has increased over the past few decades, along with the development of many technologies. Starting with assembly-based features, many security and AI experts have developed the detection technology of malicious behavior from the codes, which is nonexistent, using traditional deep learning techniques through static analysis and dynamic analysis. Despite advances in malicious code detection technologies, several methods to avoid and neutralize malicious code have also developed to a high level [1, 2].

© Springer Nature Switzerland AG 2022
P. García Bringas et al. (Eds.): HAIS 2022, LNAI 13469, pp. 311–322, 2022.
https://doi.org/10.1007/978-3-031-15471-3_27

New types of data are essential besides the known features. Large amounts of computation through deep learning have led to superior performance with large amounts of data over the past few years. Thus, approaches based on structural features have been promising in new malware detection approaches by converting code into images [3, 4].

Nevertheless, conventional convolutional neural networks (CNNs) performance has limited accuracy and recall performance compared to the traditional methods with code features. Figure 1 shows the reason for the degradation in performance. First, the vector representations of Trojan-style malicious codes (e.g., Vundo, Tracur, Obfuscator.ACY, and Gatak) are entangled closely. Simda malware has a problem that cannot be effectively represented in vector space through a significantly smaller amount of data compared to other classes. These two significant reasons affect performance degradation in the traditional approaches.

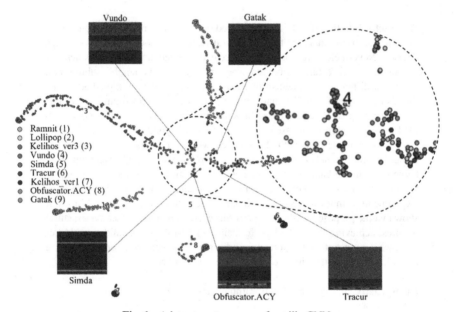

Fig. 1. A latent vector space of vanilla CNN.

In this paper, we propose a triplet network to map the latent vectors distinctly into the space between entangled structural features with Trojan characteristics 1,008 code-level features are collected from assembly files to support image-based approaches. However, some attributes do not correlate with malware families with zero, as shown in Fig. 2. Therefore, the proposed method extracts the optimal code-level features from the genetic optimization algorithm and uses them as additional information to the image-based approach [5–7]. Finally, we fuse the two different networks through the ensemble network, considering which networks are weighted to make the final decision [8]. To demonstrate the feasibility of the proposed method, we conduct thorough experiments with the malware Kaggle challenge dataset provided by Microsoft in 2015. Furthermore, we show the effectiveness of disentangled representation by visualizing the spatial latent embedding space of the proposed model and its usefulness through analyzing ROC curves.

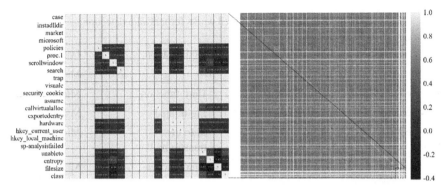

Fig. 2. Correlation map between 1,008 code-level attributes from assembly and byte code.

2   Related Works

Many security and AI experts have detected malicious code by static analysis such as control flow graph-based analysis [9, 10], byte-level analysis [11], similarity analysis [12], and dynamic analysis such as log information and API call sequence analysis. However, advanced attack techniques with avoidance have recently emerged, resulting in hidden malicious behavior patterns that existing detection methods can no longer detect. Thus, additional information is necessary for malware detection. With the advancement of methods capable of handling large resource computations, an approach to applying images transformed from assembly or binary code to high-performance deep learning models is promising [2].

Table 1 summarizes the flow concerning image-based deep learning methods over the last four years. For the existing transformation to images, binary codes are converted to a decimal number ranging from 0 to 255. Gradually, several methods of image generation are advanced, such as images based on the hash function [2], a Markov image using stochastic frequency [13], or a pre-trained model using ImageNet weights are utilized for performance improvement [14].

Nonetheless, attempts have been made to use possible multiple levels of information together to solve the ultimate problem of requiring additional information for malware detection [13, 15]. Furthermore, data augmentation through generative adversarial learning methods has been studied to obtain more data samples [16, 17].

To detect zero-day attacks, distance metric learning methods have been recently studied actively to detect attacks that have only been seen once or never before. The reason is that Siamese neural networks learn whether a pair of images are identical or not, and performance degradation is observed for multiclass classification problems [18, 19]. This paper proposes a method to separate latent embedding vectors between different classes in the latent space using a similarity-triplet network, learning similarity between the images, and fusing additional code-level features through a genetic optimization process.

Table 1. The recent image-based works for malware detection.

Data representation	Model	Optimization	Dataset
Malware image	CNN	–	Microsoft
Malware image [2]	CNN	Hash function	Microsoft
Malware image [14]	CNN	Pre-trained	Malimg
Malware image & 184 Opcodes [13]	CNN	Markov transfer prob	Microsoft
Malware image & Word frequency [15]	CNN	Autoencoder	Microsoft
Malware image [16, 17]	CNN	GAN augmentation	Microsoft
Malware image	Siamese Net	–	Own data
Malware image	Siamese Net	Batch norm	Own data

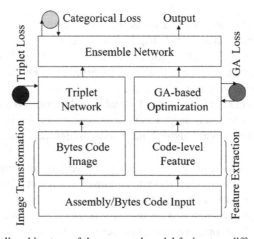

Fig. 3. The overall architecture of the proposed model fusing two different-level features.

3   Proposed Method

Figure 3 shows the overall architecture consisting of a triplet loss-trained neural network and the genetic optimization process of the features. At first, the model receives assembly code as input, transforms it into the image, and extracts optimized code features through the genetic algorithm. The converted images are performed a neural network optimization through the triplet loss, and the optimized code features are trained by the categorical loss-based multi-layer perceptron (MLP). To combine the two different levels of features extracted from two neural networks, we perform the final learning process of an optional combination of two-level features via an ensemble network. An ensemble function combines vector y_1 extracted from a triplet network with code-level vector y_2 derived by learning attributes extracted from genetic operations corresponding to the different weight values. The sum of the total weights is 1, and each weight of the neural network is defined as g_1 and g_2 as follows.

$$\sum_{u=1}^{N} g_u = 1 \qquad (1)$$

This weight not only determines the contribution of the neural network to the final result by multiplying the output value of each neural network, but also gives more learning opportunity to the network that contributes more to the final output as follows.

$$\widehat{y_k} = \sum_{i=1}^{N} y_i \cdot g_i \qquad (2)$$

3.1 Triplet Network for Disentangled Malware Representation

As a distance-based learning method, triplet network projects similar data in the near space, while it projects dissimilar data far away from each point. Therefore, we construct a triplet loss-based neural network in which similar structural features with anchor image are represented close to each other in the latent space while dissimilar images are represented far apart. To proceed with the learning as shown in Fig. 4, separate datum anchor x^a, positive x^p, which is judged to be the same, and negative x^n data that are not determined to be the same, and then go through a triplet sampling process to form a similar (x^a, x^p) pair and dissimilar (x^a, x^n) pairs. We perform distance-based learning from mapped latent vectors after calculating neural network $f(\cdot)$ that extracts feature vectors to encode latent vectors in the next latent space. The learning process is conducted in a direction to optimize the triplet loss. The model minimizes $\left\| f\left(x_i^a\right) - f\left(x_i^p\right) \right\|_2^2$ distance to increase the similarity of (x^a, x^p) pairs, while maximizing $\left\| f\left(x_i^a\right) - f\left(x_i^n\right) \right\|_2^2$ distance to decrease the (x^a, x^n) pairs. The learning for going through this process is summarized as follows.

$$Triplet\ Loss = \sum_{i=1}^{N} \left(\left\| f\left(x_i^a\right) - f\left(x_i^p\right) \right\|_2^2 - \left\| f\left(x_i^a\right) - f\left(x_i^n\right) \right\|_2^2 + \alpha \right) \qquad (3)$$

3.2 Genetic Optimization of Assembly-Level Features

The feature collision in the given attribute causes performance degradation and interferes with exploring the optimal solution. Thus, various studies have been conducted to find the optimal combination of attributes using genetic optimization approaches [5–7]. Genetic optimization can optimize code-level features specialized for malware detection in the current security field. The number of combinations of 1,008 properties collected from assemblies and binaries is 2^{1008}. Finding the optimal combination from many cases requires huge computational quantities and is cost-ineffective. We schematize the evolutionary process of finding the optimal combination of the collected properties to optimize the code-level features with low computational cost, as shown in Fig. 5.

The first step in the evolutionary process is an initialization that produces as many as a specified population of all possible combinations from 1,008 properties.

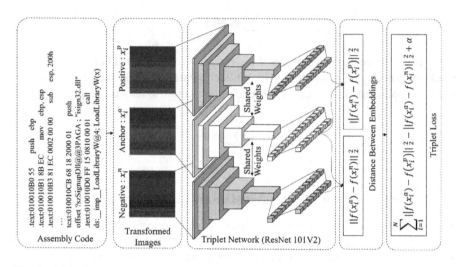

Fig. 4. Triplet loss-based network to learn distance-based similarity among different classes.

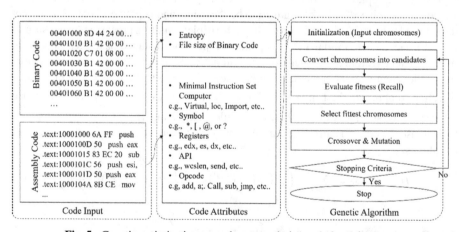

Fig. 5. Genetic optimization procedure to optimize code-level features.

The genes of the generated phenotype chromosomes are represented as 0 or 1 and represent the use of each attribute. The process of converting phenotypic chromosomes into genotypic chromosomes is necessary to calculate the actual fitness score. Genotypic chromosomes represent values of the actual properties of the gene, denoted "1" by the phenotypic chromosome. For example, if the gene 'edx' of a phenotype chromosome is "1", the corresponding "edx" genetic information value in the genotype chromosome is used. A population generated in this way defines a recall-based function to calculate the fitness, using chromosomes with high recall value during the selection process for the next generation. Afterward, selected chromosomes go through crossover and mutation to maintain chromosome diversity. Chromosome or gene diversity leads to obtaining the optimal chromosome without being trapped in local minima. Eventually, we can remove possible collisions in the neural network and zero correlation between code-level features through the genetic optimization process.

4   Experiments

4.1   Data Specification and Experiment Settings

We utilize assembly malware code for the 2015 Kaggle competition provided by Microsoft [20]. The dataset has 10,869 samples with nine different classes. We transform the given assembly code into two different data levels: grayscale image and assembly-code features. We explained how grayscale image is converted from assembly code in the related works [21]. Regarding code-level features, we extract operation symbols, opcode type, register type, API, and assembly modules collected from assembly codes. We also merge 1,006 assembly-level features with entropy index and file size features from the binary code. The total number of code-level features is 1,008. Regarding the structural data, we resize the transformed images to 224 × 224 images with one channel and normalize to scale pixels to the range of 0 to 1. We normalize the collected code-level features as well. Data for the experiment is divided by 10 fold cross-validation. The comparable experiments between the proposed model and other comparative methods use the data achieving the highest accuracy out of the 10-fold.

We perform 1000 generations on 1000 populations for evolutionary optimization. Generally, we confirm that learning ended by satisfying the stopping criterion before the set number of generations. As generations pass, we set 90% crossover between chromosomes, and 10% mutation.

4.2   10-Fold Cross-Validation

Table 2 summarizes 10-fold cross-validation results to compare the recent image-based and code feature-based approaches with the proposed method. We implement the state-of-the-art models among byte image-based approaches [2]. Image-based approaches improve detection performance with more profound architecture and distance-based learning. Especially, the proposed model shows 0.0422 and 0.0256 performance improvement compared to the state-of-the-art methods in accuracy and recall, respectively. The following experiments analyze the cause of performance enhancement.

In terms of the features of assembly and byte malware code with deep learning, we conduct the experiments before and after optimization. A combination of 518 code attributes extracted from evolutionary optimization achieves 0.0071 (0.0024) accuracy (recall) improvement compared to when using all 1,008 attributes.

Table 2. 10-fold cross-validation with different methods

Metrics	Accuracy ± std	Recall
Structural features		
Vanilla CNN [2]	0.9458 ± 0.0186	0.9155
ResNet101V2	0.9503 ± 0.0137	0.9203
DCGAN-based Augmentation [16]	0.9601 ± 4.60e−05	0.9298
tDCGAN-based Augmentation [16]	0.9674 ± 3.03e−05	0.9328
Triplet Network	0.9783 ± 0.0268	0.9381
Assembly-level features		
MLP: Non-optimization	0.9360 ± 1.58e−04	0.9189
MLP: GA-based Optimization	0.9431 ± 0.0109	0.9213
Ensemble method		
Concatenation (Triplet Network and Non-optimization)	0.9773 ± 3.96e−05	0.9367
Concatenation (Triplet Network and Optimization)	0.9798 ± 0.0093	0.9413
Ensemble with different weights (Triplet Network and Non-optimization)	0.9819 ± 0.0102	0.9431
Ours: Ensemble with different weights (Triplet Network and Optimization)	**0.9880 ± 0.0208**	**0.9411**

Ultimately, we split four different cases considering both two ensemble approaches and before/after evolutionary optimization. It shows a performance degradation compared to the code-level method regardless of whether or not it is optimized when combining through general concatenation. The reason is that the meaning of each feature vector is lost by simply combining two different feature vectors without considering any weights. On the other hand, the ensemble network considering the weights of each neural network shows approximately 0.0107 (0.0044) performance improvement in the accuracy (recall) performance compared to the simple combination method since it extracts the final output vector considering the contribution between the two neural networks.

4.3 ROC Curves and AUC

A receiver operating characteristics curve (ROC) is a graph that determines classification performance based on accuracy, sensitivity, and specificity. Suppose that there is a high possibility of predicting positive for the actual positive cases and negative for the negative cases. In that case, the ROC curve converges on the upper left corner of the graph. Figure 6-(b) shows that the proposed method quickly converges on the left top compared to Fig. 6-(a), which is the triplet-based method.

The area under curves (AUC) measures a model's performance through the area below the ROC curve. As shown in Fig. 6-(a), the triplet-based approach performs well,

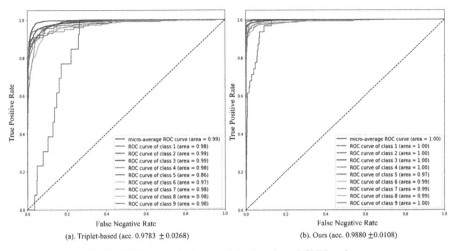

Fig. 6. ROC curves between triplet loss-based CNN and ours.

except for Simda class. On the other hand, Fig. 6-(b) shows that the proposed model achieves an improved performance over Simda cases, 0.97, compared to the triplet-based method, 0.86, and the proposed model gets better AUC on the rest of the malware families as well. We have proven that integration of the triplet network with evolutionary optimization of code features increases classification performance despite the lack of the number of Simda cases.

4.4 Visualization of the Latent Space

The triplet network can disentangle Trojan-based representation, such as Gatak, Tracur, Vundo, and Obfuscator.ACY, compared to the vanilla CNN. Furthermore, our model improves to capture more features in Simda cases. When visualizing the latent vector space of the model proposed as shown in Fig. 7-(a), the Trojan-based Obfuscator.ACY and Gatak images, which are not classified in triplet-based CNN, are set apart from each other. However, latent features for Vundo(4), Tracur(6), and Simda are not fell apart from each other in the space. On the other hand, as shown in Fig. 7-(b), the proposed method disentangles all existing Trojan-based malware images and maps Simda(5)'s extracted vectors on the separated latent space.

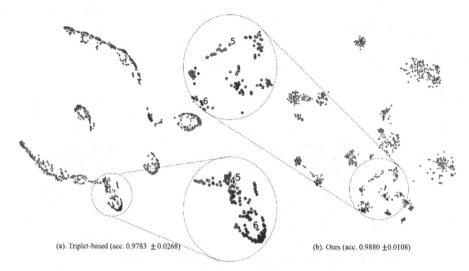

(a). Triplet-based (acc. 0.9783 ±0.0268)   (b). Ours (acc. 0.9880 ±0.0108)

Fig. 7. Visualization of the latent vector spaces between two modeling methods.

5 Conclusions

This paper proposes a triplet neural network to learn distance-based similarities between vectors in latent space via the separation of entangled representation. In addition, an evolutionary optimization process in code-level attributes discovers the optimal combination by excluding attributes that cause performance degradation. At last, we utilize the ensemble network with different weight parameters to return the outcome by reflecting each neural network's contributions (i.e., triplet network and evolutionary optimization) in the learning process. In addition to accuracy and performance improvements, the proposed method proves the validity of separating entangled representations between Trojan-based structural features, and confirms that Simda-typed data, significantly less than other malware families, is segregated in the latent space.

With effective mapping results for such a small amount of data, this paper provides the opportunity to classify various unknown malicious codes beyond zero-day malicious codes. For the future works, we will conduct experiments on other malicious data to prove generalization performance and remove Simda from the collected data to support the validity of the proposed method through a scenario verification method to see if actual classification is possible.

Acknowledgements. This work was supported by an IITP grant funded by the Korean government (MSIT) (No.2020-0-01361, Artificial Intelligence Graduate School Program (Yonsei University)) and an ETRI grant funded by the Korean government (22ZS1100, Core Technology Research for Self-Improving Integrated Artificial Intelligence System).

References

1. Han, K., Lim, J.H., Im, E.G.: Malware analysis method using visualization of binary files. In: Proceedings of the Conference on Research in Adaptive and Convergent Systems, pp. 317–321 (2013)
2. Jung, B., Kim, T., Im, E.G.: Malware classification using byte sequence information. In: Proceedings of the Conference on Research in Adaptive and Convergent Systems, pp. 143–148 (2018)
3. Bu, S.J., Park, N., Nam, G.-H., Seo, J.-Y., Cho, S.-B.: A Monte Carlo search-based triplet sampling method for learning disentangled representation of impulsive noise on steering gear. In: IEEE International Conference on Acoustics, Speech and Signal Processing, pp. 3057–3061 (2020)
4. Qin, C., Zhang, Y., Liu, Y., Coleman, S., Kerr, D., Lv, G.: Appearance-invariant place recognition by adversarially learning disentangled representation. Robot. Auton. Syst. **131**, 103561 (2020)
5. Afan, H.A., et al.: Input attributes optimization using the feasibility of genetic nature inspired algorithm: application of river flow forecasting. Sci. Rep. **10**, 1–15 (2020)
6. Cho, S.-B., Shimohara, K.: Evolutionary learning of modular neural networks with genetic programming. Appl. Intell. **9**, 191–200 (1998)
7. Zhang, Q., Deng, D., Dai, W., Li, J., Jin, X.: Optimization of culture conditions for differentiation of melon based on artificial neural network and genetic algorithm. Sci. Rep. **10**, 1–8 (2020)
8. Eigen, D., Ranzato, M.A., Sutskever, I.: Learning factored representations in a deep mixture of experts. arXiv preprint arXiv:1312.4314 (2013)
9. Cesare, S., Xiang, Y.: A fast flowgraph based classification system for packed and polymorphic malware on the endhost. In: IEEE International Conference on Advanced Information Networking and Applications, pp. 721–728 (2010)
10. Kinable, J., Kostakis, O.: Malware classification based on call graph clustering. J. Comput. Virol. **7**(4), 233–245 (2011)
11. Tabish, S.M., Shafiq, M.Z., Farooq, M.: Malware detection using statistical analysis of byte-level file content. In: Proceedings of the ACM SIGKDD Workshop on CyberSecurity and Intelligence Informatics, pp. 23–31 (2009)
12. Sung, A.H., Xu, J., Chavez, P., Mukkamala, S.: Static analyzer of vicious executables (SAVE). In: Annual Computer Security Applications Conference, pp. 326–334 (2004)
13. Yuan, B., Wang, J., Liu, D., Guo, W., Wu, P., Bao, X.: Byte-level malware classification based on Markov images and deep learning. Comput. Secur. **92**, 101740 (2020)
14. Vasan, D., Alazab, M., Wassan, S., Safaei, B., Zheng, Q.: Image-based malware classification using ensemble of CNN architectures (IMCEC). Comput. Secur. **92**, 101748 (2020)
15. Li, L., Ding, Y., Li, B., Qiao, M., Ye, B.: Malware classification based on double byte feature encoding. Alex. Eng. J. **61**, 91–99 (2022)
16. Kim, J.Y., Bu, S.J., Cho, S.B.: Malware detection using deep transferred generative adversarial networks. In: Liu, D., Xie, S., Li, Y., Zhao, D., El-Alfy, E.S. (eds.) ICONIP 2017. LNTCS, vol. 10634, pp. 556–564. Springer, Cham (2017). https://doi.org/10.1007/978-3-319-70087-8_58
17. Kim, J.-Y., Cho, S.-B.: Detecting intrusive malware with a hybrid generative deep learning model. In: Yin, H., Camacho, D., Novais, P., Tallón-Ballesteros, A.J. (eds.) IDEAL 2018. LNCS, vol. 11314, pp. 499–507. Springer, Cham (2018). https://doi.org/10.1007/978-3-030-03493-1_52
18. Hsiao, S.C., Kao, D.Y., Liu, Z.Y., Tso, R.: Malware image classification using one-shot learning with Siamese networks. Proc. Comput. Sci. **159**, 1863–1871 (2019)

19. Zhu, J., Jang-Jaccard, J., Watters, P.A.: Multi-loss Siamese neural network with batch normalization layer for malware detection. IEEE Access **8**, 171542–171550 (2020)
20. Ronen, R., Radu, M., Feuerstein, C., Yom-Tov, E., Ahmadi, M.: Microsoft malware classification challenge. arXiv preprint arXiv:1802.10135 (2018)
21. Singh, A., Handa, A., Kumar, N., Shukla, S.K.: Malware classification using image representation. In: Dolev, S., Hendler, D., Lodha, S., Yung, M. (eds.) CSCML 2019. LNCS, vol. 11527, pp. 75–92. Springer, Cham (2019). https://doi.org/10.1007/978-3-030-20951-3_6

A Two-Level Hybrid Based Genetic Algorithm to Solve the Clustered Shortest-Path Tree Problem Using the Prüfer Code

Adrian Petrovan[1,2], Petrică C. Pop[1,2(✉)], Cosmin Sabo[1,2],
and Ioana Zelina[1,2]

[1] Technical University of Cluj-Napoca, Cluj-Napoca, Romania
adrian.petrovan@ieec.utcluj.ro,
{petrica.pop,cosmin.sabo,ioana.zelina}@mi.utcluj.ro
[2] North University Center of Baia Mare, Dr. V. Babes 62A,
430083 Baia Mare, Romania

Abstract. The clustered shortest-path tree (CluSPT) problem is a generalization of the popular shortest path problem, in which, given a graph with the set of vertices divided into a given set of clusters, we look for a shortest-path spanning tree from a predefined source vertex to all the other vertices of the graph, with the property that every cluster should generate a connected subgraph. In this paper, we describe a two-level hybrid algorithm to solve the CluSPT problem, in which a framework with a macro-level genetic algorithm (GA) used to generate trees connecting the clusters using Prüfer code, is combined with a micro-level algorithm based on dynamic programming (DP) that determines the best corresponding feasible solution of the CluSPT problem associated to a given tree spanning the clusters. Finally, some preliminary computational results are stated on a set of 40 standard benchmark non-euclidean instances from the literature to illustrate the performance of our developed two-level hybrid algorithm. The obtained computational results demonstrate that our novel hybrid algorithm is highly competitive against the existing algorithms.

Keywords: Hybrid algorithms · Genetic algorithms · Clustered shortest-path tree problem

1 Introduction

The clustered shortest-path tree (CluSPT) problem extends the classical shortest path problem. CluSPT is defined on a graph whose entire set of vertices is split up into a given number of vertex sets, called clusters, and it searches for a spanning tree with the following features: all the sub-graphs induced by each of the clusters are connected and it minimizes the total cost of the paths from a given source vertex to all the other vertices of the graph.

P. García Bringas et al. (Eds.): HAIS 2022, LNAI 13469, pp. 323–334, 2022.
https://doi.org/10.1007/978-3-031-15471-3_28

D'Emidio et al. [5] investigated for the first time the CluSPT problem and its unweighted version called clustered breadth-first search tree (CluBFS) problem, justified by practical applications in field of communication and social networks. The same authors presented some results concerning the computational hardness and approximation results for CluSPT and CluBFS problems. Thanh et al. [20] developed a random heuristic search algorithm that couples a randomized greedy algorithm with a shortest path tree algorithm. Binh et al. [1] developed an approach which reduces the solution space by decomposing the CluSPT problem into two sub-problems which then are solved separately. Cosma et al. [4] described four cases of the CluSPT in which the problem is solvable in polynomial time, described an exact algorithm that solves the CluSPT problem in polynomial time when the triangle inequality is satisfied, and presented a genetic algorithm for solving the general case. Binh and Thanh [2] proposed a multifactorial evolutionary algorithm with two main tasks: the first one is solving the CluSPT problem, while the second one consists on solving a new optimization problem obtained from CluSPT problem by considering the sub-graph of the largest cluster of the input graph of the CluSPT as its input graph and with the aim to find the minimum spanning tree of that sub-graph. Recently, Hanh et al. [12] described two novel solution approaches: the first one reduces the initial underlying graph to a multi-graph which than is solved by means of an evolutionary algorithm and the second one decomposes the multi-graph into a set of simple graphs and the CluSPT problem is solved using a Multi-tasking Local Search Evolutionary Algorithm, and Cosma et al. [3] proposed a new genetic algorithm that fits the challenges of CluSPT problem, reported the optimal solutions for all the existing euclidean instances by making use of the exact algorithm proposed by Cosma et al. [4] and provided a new collection of databases containing 248 non-euclidean instances split up into six categories and reported the obtained results on this new set of datasets.

We should emphasize on the fact that all the existing methods for solving the CluSPT problem, except the GAs developed by Cosma et al. [3,4] and the two levels solution approach based on multifactorial optimization proposed by Binh and Thanh [2], even though they have presented distinctive ways to explore and exploit the solution space of the CluSPT, they have been only tested in the case of Euclidean instances, which actually can be optimally solved making use of the exact algorithm developed by Cosma et al. [4].

The clustered shortest-path tree problem is part of the group of generalized network design problems. This group of optimization problems extends in a natural way the classical network design problems. The main characteristics of these generalized network design problems are: the vertices of the original graph are partitioned into a given number of clusters, and the feasibility constraints are delivered in relation to the clusters instead of individual vertices of the graph. A closely connected problem to CluSPT problem is the the generalized minimum spanning tree problem (GMSTP), which was first considered by Myung et al. [14], and whose objective is to look for a tree with minimum cost with the property that it spans a subset of vertices that contains exactly one vertex from every cluster. For more information regarding the GMSTP and its extensions, we refer to Pop et al. [15,17]. Some other intensively investigated generalized network design problems are: the generalized traveling salesman problem and its extensions [10], the generalized vehicle routing problem and its extensions [11,16], the selective graph coloring problem [7,9], etc. For more information concerning the group of generalized network design problems, we recommend the following papers: Feremans et al. [8] and Pop [18].

The aim of this paper is to describe a novel solution approach in which a framework with a macro-level genetic algorithm (GA) used to generate trees connecting the clusters using the Prüfer code is combined with a micro-level algorithm based on dynamic programming (DP) that determines the best corresponding feasible solution of the CluSPT problem. One of the principal advantages of our proposed approach against the existing two-level solution approaches for solving the CluSPT problem is that the micro-level sub-problem is solved optimally using dynamic programming.

The present paper is structured as follows: the second section defines the investigated problem and presents the particular cases when the CluSPT problem is optimally solvable in polynomial time, Sect. 3 describes in detail our novel solution approach. In Sect. 4, we make a comparative analysis of the efficiency of our novel proposed solution approach in comparison to the existing methods from the literature, and finally in Sect. 5, we formulate some conclusions and present future research directions.

2   Definition of the Clustered Shortest-Path Tree Problem

Let us consider $G = (V, E)$ an undirected complete graph with the set of vertices $V = \{v_1, v_2, ..., v_n\}$ and the set of edges E defined as follows:

$$E = \{\{v_i, v_j\} | \ v_i, v_j \in V, \ i \neq j \in \{1, 2, ..., n\}\}.$$

In addition, we define a cost function $c : E \rightarrow R_+$, which associates a positive value denoted by $c_e = c(u, v)$ to every edge $e = (u, v) \in E$, value which is called the cost of the edge.

The entire set of vertices V is divided into $k + 1$ nonempty subsets of vertices denoted by $V_0, V_1, ..., V_k$ and which will be called *clusters*. Therefore the following conditions must be fulfilled:

1. $V = V_0 \cup V_1 \cup V_2 \cup ... \cup V_k$
2. $V_l \cap V_p = \emptyset$ for all $l, p \in \{0, 1, ..., k\}$ and $l \neq p$.

We distinguish two kinds of edges in our graph: edges which link vertices from the same cluster, which are called intra-cluster edges and edges which link vertices from different clusters, which are called inter-cluster edges.

If S is a subset of vertices, $S \subseteq V$, then we denote by $G[S]$ the subgraph induced by S, which is a graph with the set of vertices S and the set of edges whose endpoints are both in S. Given a tree T which spans the vertices of the graph G and two vertices $v_i, v_j \in V$, we know that there exists a single path between v_i and v_j whose cost will be denoted by $d_T(v_i, v_j)$. Given a vertex $s \in V$, called the source vertex, we will denote by $TC = \sum_{v \in V} d_T(s, v)$ the total cost of the paths from the given source vertex s to all the other vertices of the graph. In our case, we consider that the cluster V_0 contains the source vertex.

According to D'Emidio et al. [5], the *clustered shortest-path tree (CluSPT) problem* looks for a tree T satisfying the next properties:

1. T spans G, i.e. includes every vertex of G;
2. All the vertex-induced subgraphs $T[V_i]$, for all $i \in \{0, 1, ..., k\}$ must be connected;

and such that the total TC cost of the paths from the given source vertex to all the other vertices of G is minimized, i.e.

$$TC = \sum_{v \in V} d_T(s, v) \to \min.$$

3 The Proposed Hybrid Algorithm

Splitting the complex optimization problems into smaller and easier sub-problems and solving them separately is a well-known technique applied successfully to solve hard problems, see for example Pop et al. [15,16]. Obviously, the achieved solution using such a decomposition method is not as good as the one obtained by solving the entire initial problem, but this disadvantage can be surmounted by making use of an iterative solution process, which improves iteratively the solution of the initial problem.

In this section, we describe our new two-level hybrid algorithm for solving the CluSPT problem obtained by decomposing the problem into two logical and natural smaller sub-problems:

1. a macro-level (global) sub-problem, whose scope is to determine the global tree spanning the clusters (inter-cluster connections) by means of a GA using the Prüfer code;
2. a micro-level (local) sub-problem, whose aim is to determine the links between the vertices from the same cluster and the links between vertices belonging to different clusters, corresponding to the global spanning tree. For each tree spanning the clusters, the micro-level sub-problem is optimally solved using dynamic programming.

Our novel two-level hybrid based genetic algorithm is design to fit the hierarchical structure of the CluSPT problem, i.e. a graph whose vertices are divided into a predefined number of clusters, and provides computational advantages by using effective algorithms for solving both sub-problems and by coupling the obtained results, in order to get a feasible solution of the CluSPT problem without using any post-processing procedure.

It is worth mentioning that one of the principal advantages of our proposed approach against the existing two-level solution approaches for solving the

CluSPT problem is related to the micro-level sub-problem, namely our method determines the optimal solution associated to a given global tree spanning the clusters using dynamic programming.

3.1   The Macro-level (Global) Subproblem

In order to define the macro-level subproblem associated to CluSPT problem, we used a similar approach as the one presented by Pop et al. [18] in the case of the GMSTP. We consider the graph G_{global} resulted from the graph G after substituting all the vertices belonging to a cluster V_i with a vertex representing the cluster V_i, $\forall i \in \{0, 1, ..., k\}$.

We will call this graph the *global graph*. The global graph $G_{global} = (V_{global}, E_{global})$ is defined as follows: every cluster of the graph G is a vertex of V_{global}, $V_{global} = \{V_0, V_1, ..., V_k\}$ with $|V_{global}| = k+1$, and the edges are defined between each pair of the global graph vertices

$$E_{global} = \{(V_i, V_j) \mid 0 \le i < j \le k\}.$$

In the case of the CluSPT problem, we make use of a GA applied to the associated global graph G_{global} to supply a set of trees which span the clusters of the graph. In this way, we are able to reduce considerable the size of the solution space of the original problem. A tree spanning the clusters of the graph will be called a *global spanning tree*.

As we will see in Sect. 3.3, given a global spanning tree, it is quite easy to determine the clustered path tree with minimum cost. There are several clustered path trees associated to a global spanning tree and among these there exists one which is called the clustered shortest-path tree and that can be found using dynamic programming.

Given a graph G with its vertices partitioned into six clusters and its corresponding global graph G_{global}, in Fig. 1, we illustrate an example of a global spanning tree in G_{global} and the clustered shortest-path tree (i.e. a feasible solution of the CluSPT problem) associated to the given global spanning tree.

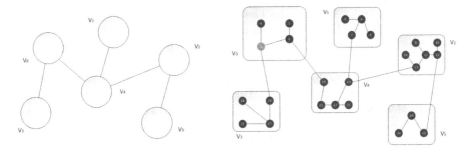

Fig. 1. The one-to-one correspondence between the global spanning tree in G_{global} and the corresponding clustered shortest-path tree

3.2   The Genetic Algorithm

In the case of the CluSPT problem, we used an effective representation at the level of the global graph G_{global}. In this representation the chromosome associated to each candidate solution (i.e. global spanning tree) is represented as a unique Prüfer sequence of length $k - 1$, where $k + 1$ is the number of vertices of the G_{global}. This way of encoding the chromosomes is suitable for representing all the possible global trees, because even after performing any genetic operators, we are sure that the resulted chromosomes are still global spanning trees.

The Prüfer string is determined by discarding iteratively vertices from the tree till just two vertices are left. At step i, the leaf with the smallest label is removed and the i-th element of the Prüfer sequence is given by the label of this leaf's neighbor. Given a labeled tree with the vertices labeled by $0, 1, 2, ..., k$ corresponding to the clusters $V_0, V_1, V_2, ..., V_k$, where V_0 is the cluster that contains the source vertex.

The Prüfer code corresponding to the global spanning tree with six vertices (representing the clusters), illustrated in Fig. 1, is a sequence of length 4 whose entries are: $C = (4, 0, 4, 2)$.

The inverse algorithm which finds the unique labeled tree with $k + 1$ vertices for a given Prüfer string of length $k - 1$, is named the PRÜFER DECODING algorithm.

The tree assigned to the code $C = (4, 0, 4, 2)$ is precisely the global tree illustrated in Fig. 1.

The fitness function associated to each individual chromosome in the population is given by the cost of the best clustered shortest-path tree associated to the global spanning tree specified by the chromosome, which is determined using the dynamic programming described in Subsect. 3.3.

Our aim is to minimize this total cost of the paths from the given source vertex to all the other vertices of the graph.

In order to preserve the diversity of the population, we generated the initial population with completely random chromosomes.

In our novel approach, we used the elite tournament selection technique that combines the principle of competitive selection with the elitist principle of keeping the most befitting individuals. The tournament selection is one of the most popular selection methods with large applicability in the case of many combinatorial problems due to its efficiency in exploitation of the solutions space. By this technique a number of individuals are chosen from the whole population and compete against each other during a number of rounds to determine the winner, the fittest individual who will serve in the matting process to produce offspring. Most notable advantages of the tournament selection are the effective time complexity especially in parallel implementations and the absence of fitness scaling or sorting requirements. The elitism strategy always keeps the most fitted individuals and is helpful for exploitation process. Practically we used in our approach the tournament selection with 5 rounds and an elitist value of 0.05, that is, 5% of the next population will be composed of elite individuals of the previous population.

Our GA uses a standard version of the one point crossover of two strings representing Prüfer strings.

To maintain and introduce diversity in the population, in our approach we use the swap mutation operator by selecting two random positions on the chromosome and interchange the values.

Based on preliminary computational experiments, for solving the CluSPT problem, we selected the following parameters of the GA: the size of the population is proportional to the number of clusters of each instance, by multiplying the number of clusters with a multiplication factor equal to 300. The evolution of populations is maintained as long as there is a favorable evolution of the fitness value of the best chromosome. If there is no improvement or change in the value of fitness over 5 epochs then the evolutionary process is interrupted. The probability of mutation is kept throughout the evolutionary process at 0.05.

3.3   The Micro-level (Local) Subproblem

In this subsection, given a tree spanning the clusters, we present an effective approach, based on dynamic programming, in order to determine the corresponding best clustered shortest-path tree. The tree spanning the clusters is a rooted tree with the root cluster V_0. For every node $V_i, 1 \leq i \leq k$ there is a unique path from the root V_0 to V_i. For each node $V_i, 1 \leq i \leq k$ the node V_j adjacent to V_i on the path from V_0 to V_i is called the parent of V_i. We define the parent array for the rooted tree, an array P with $k + 1$ elements in which $P[0] = -1$ and $P[i] = j, 1 \leq i \leq k$, where V_j is the parent of V_i. The parent array completely defines the rooted tree.

The following are the steps that our proposed dynamic programming algorithm follows:

1) For every cluster V_i of the graph G, with $i \in \{0, 1, ..., k\}$, we compute the values $d_{u,v}$ representing the distance within the cluster V_i between the vertices u and v, with $u, v \in V_i$ using the Dijkstra's algorithm. We compute the value D_v which is the value of the shortest-path tree within the cluster V_i rooted at each vertex belonging to that cluster, $v \in V_i$. In the case of the V_0, the shortest-path tree rooted at the given source vertex $s \in V_0 \subset V$ is determined.

2) We set the cluster that contains the source vertex as the root of the tree spanning the clusters, in our case V_0. We direct all the edges away from vertices of V_0 based on the parent array associated to the global spanning tree. A directed edge $\langle V_i, V_j \rangle$ of G_{global} that results from the orientation of edges of the global spanning tree, will define in a natural way an orientation $\langle l, p \rangle$ of an edge $(l, p) \in E$, where $l \in V_i$ and $p \in V_j$. The root cluster is considered to be at level 0, and we compute for each cluster V_i its level in the rooted tree at V_0. The level of the cluster V_i is the distance from the root cluster to cluster $V_i, i \in \{0, 1, ..., k\}$.

3) The subtree rooted at a vertex $v, v \in V_l$ with $0 \leq l \leq k$, denoted by $T(v)$ contains all the vertices that can be reached from v under the given orientation of the edges. We denote by $W(v)$ the value of the shortest-path tree for

the subtree rooted in the vertex $v \in V$. We want to determine $W(s)$, where $s \in V$ is the given source vertex.

4) We provide next the dynamic programming recursion aiming to solve the subproblem $W(z)$. The initialization is:

$$W(z) = 0$$

for each vertex $z \in V$.

For every vertex $z \in V_l$ of the tree spanning the clusters, the recursion is given by the following formula:

$$W(z) = D_z + \sum_{C \in L(V_l)} \min_{u \in V_l, v \in C} [(d(z, u) + c(u, v))ns(C) + W(v)],$$

where $L(V_l)$ is the set (list) of clusters with the cluster parent V_l, $ns(C)$ is the number of vertices in all the clusters in the subtree rooted at C, $d(z, u)$ is the distance between z and u within the cluster V_l and $c(u, v)$ is the cost of the edge connecting the vertices $u \in V_l$ and $v \in C$.

The computation of the values $W(z)$ is made by traversing the set of vertices in the decreasing order of the cluster level, starting with the vertices belonging to the leaves of the rooted tree. When the computation is made for a node $z \in V_i$ at level l_i, the values $W(v)$ for all the nodes v in the clusters at levels $j > l_i$ are already computed.

4   Computational Results

This section is dedicated to the preliminary experimental results obtaineded by our novel two-level hybrid based GA. In order to assess the efficiency of our developed hybrid algorithm, we used the non-euclidean instances introduced by Cosma et al. [3], obtained by transforming the euclidean instances from the MOM-lib [13] and the non-Euclidean instances introduced by Binh and Thanh [2], which were created from MOM-lib small instances of type 1, 5 and 6, preserving the information regarding the source vertex, the number of clusters and the number of vertices, and adding a random weight matrix with values in $[1, 1000]$. All the non-euclidean instances used in our computational experiments can be found at https://sites.google.com/view/tstp-instances, respectively [19].

For evaluating the performance of our developed two-level hybrid approach, we compared it to the existing algorithms for solving the CluSPT problem, the genetic algorithm developed by Cosma et al. [3], respectively the two level solution approach based on multifactorial optimization proposed by Binh and Thanh [2]. As we have already mentioned, the other existing methods were tested only on euclidean instances, which are actually solvable in polynomial time by the exact algorithm developed by Cosma et al. [4].

Our novel two-level hybrid based GA was implemented in Python and have been evaluated on a PC with AMD Ryzen 9, 12-Core 3.8 GHz, 64 GB RAM,

Windows 11 Education operating system. In our proposed algorithm for every instance we performed 30 independent trials.

In Table 1, we report the solutions obtained by our novel two-level hybrid algorithm for solving 20 non-euclidean instances of Type 1 of the CluSPT problem and compare them against the existing results obtained by Cosma et al. [3].

The structure of the Table 1 is as follows: the first two columns provide information concerning the considered non-euclidean instances: the number and its name, the following two columns provide the best found solutions (BF) and average solutions (Avg) achieved by the GA developed by Cosma et al. [3], together with the average CPU processing times (T_{avg}) in seconds, necessary to obtain the reported solutions, and the gap as a percentage computed by the formula:

$$gap = 100 \times \frac{Avg - BF}{BF},$$

where Avg is the average of the solutions computed in the 30 independent trials for each instance. The following two columns display the best found solutions (BF) and the average solutions (Avg) achieved by our novel two-level hybrid algorithm and the last two columns contain the corresponding running times and the percentage gaps. The symbol "=" indicates that the algorithm reached the same solution in each of the 30 independent trials, i.e. $BF = Avg$ and obviously the gap in these cases is 0. The best solutions are highlighted in bold.

Summarizing the results presented in Table 1, we can notice that our developed two level solution approach is competitive against the GA described by Cosma et al. [3]. In 17 out of 20 instances, our algorithm achieved the same best solution of the CluSPT problem and in 12 out of 20 instances the average solutions coincide to the best found solutions, and for the rest of the instances the percentage gap ranges between 0.13% and 3.64%.

In Table 2, we report the solutions obtained by our novel two-level hybrid algorithm for solving 20 non-euclidean instances of the CluSPT problem introduced by Binh and Thanh [2] and compare them against their results obtained using the two-level approach based on multifactorial optimization.

The structure of the Table 2 is as follows: the first two columns provide information concerning the non-euclidean instances: the number and its name, the following six columns display the best found solutions (BF) and the average (Avg) solutions together with the average CPU processing times (T_{avg}) necessary to obtain the corresponding solutions, reported in seconds, obtained by the two versions of the two-level multifactorial evolutionary algorithm: Single task and Multi-tasks, developed by Binh and Thanh [2]. The average solutions were calculated based on 30 independent trials of each instance and initially the CPU times were reported by the authors in minutes and we transformed them in seconds. The following three columns display the best found solutions (BF) and the average solutions (Avg) achieved by our novel two-level hybrid algorithm and average CPU processing times necessary to achieve the corresponding solutions.

Table 1. Experimental results in the case of small non-euclidean instances of Type 1 introduced by Cosma et al. [3]

Instance		GA [3]				Our two-level hybrid algorithm			
No.	Name	BF	Avg.	T_{avg} [s]	gap [%]	BF	Avg.	T_{avg} [s]	gap [%]
1	nec-5eil51	**986**	=	0.23	0	**986**	=	0.39	0
2	nec-5berlin52	**13456**	=	0.23	0	**13456**	=	0.51	0
3	nec-5st70	**2398**	=	0.27	0	**2398**	=	0.83	0
4	nec-5eil76	**1491**	=	0.25	0	**1491**	=	0.80	0
5	nec-5pr76	**345489**	=	0.28	0	**345489**	=	0.78	0
6	nec-10eil51	**1009**	=	0.78	0	**1009**	=	1.11	0
7	nec-10berlin52	**28027**	=	1.27	0	**28027**	=	1.62	0
8	nec-10st70	**1628**	=	0.97	0	**1628**	=	1.71	0
9	nec-10eil76	**1258**	=	1.25	0	**1258**	=	3.77	0
10	nec-10pr76	**297535**	=	1.03	0	**297535**	=	2.45	0
11	nec-10rat99	**4111**	=	1.15	0	**4111**	=	3.46	0
12	nec-10kroB100	**76274**	=	1.40	0	**76274**	=	4.95	0
13	nec-15eil51	**867**	=	1.88	0	**867**	871.86	7.07	0.56
14	nec-15berlin52	**16836**	=	2.33	0	**16836**	16858.26	8.06	0.13
15	nec-15st70	**2204**	=	2.29	0	**2204**	2209.2	9.60	0.23
16	nec-15eil76	**1578**	=	2.54	0	**1578**	1588.33	10.56	0.65
17	nec-15pr76	**404626**	=	3.98	0	**404626**	408087.68	11.46	0.85
18	nec-25rat99	**3742**	=	8.77	0	3754	3811.5	44.69	1.53
19	nec-25kroA100	**86971**	=	9.78	0	87535	89292.1	52.60	2.00
20	nec-25eil101	**2514**	2516.40	12.26	0.10	2523	2614.86	42.65	3.64

The last column displays the improvement gap (%) to the best found solution by Binh and Thanh [2]. The best solutions are highlighted in bold.

Summarizing the results presented in Table 2, we can notice that our described solution approach outperforms the two versions of the two-level multi-factorial evolutionary algorithm: Single task and Multi-tasks, developed by Binh and Thanh [2]. We were able to improve the best found solutions of the CluSPT problem in all the reported 20 instances and the improvement gap ranges from −27.68% up to −62.66%.

Concerning the CPU processing times, it is hard to do a proper comparison between our developed two-level hybrid algorithm and the existing algorithms from the literature, because they have been assessed on different computers and they have been implemented using distinctive programming languages. The CPU processing time of our proposed two-level hybrid algorithm is proportional with the number of generations.

Table 2. Experimental results in the case of non-euclidean instances introduced by Binh and Thanh [2]

Instance		Single task [2]			Multi-tasks [2]			Our hybrid approach			Gap (%)
No.	Name	BF	Avg.	T_{avg} [s]	BF	Avg.	T_{avg} [s]	BF	Avg.	T_{avg} [s]	
1	N-10berlin52	26892	30236.2	34.8	22906	27082.6	31.8	**13951**	13998.3	2.75	−39.09
2	N-10berlin52-2x5	22870	28002.8	–	20026	25323.2	–	**13154**	13154	4.17	−34.31
3	N-10eil51	33148	36379.1	34.2	24229	30936.4	28.2	**17522**	17614.85	2.47	−27.68
4	N-10eil76	51780	59611.8	69	35900	45377.2	66	**17281**	17281.0	4.16	−51.86
5	N-10KROB100	97807	107727.9	124.2	69214	80291.1	114	**25842**	25924.5	7.68	−62.66
6	N-10pr76	54440	60122.6	73.8	40105	51548.3	57	**21909**	21970.05	5.24	−45.37
7	N-10rat99	79167	84352.8	115.8	51112	66885.9	94.8	**19710**	19757.1	6.64	−61.44
8	N-10st70	60294	63893.1	57	36098	50138.8	55.2	**23976**	24055.55	4.06	−33.58
9	N-12eil51-3x4	29304	34691.9	34.8	28020	32120	30	**15018**	15322.4	4.14	−46.40
10	N-12eil76-3x4	50686	55117.2	72	38587	50748	37.2	**23652**	23914	5.86	−38.70
11	N-12pr76-3x4	44708	51842	67.8	32679	42425.8	45	**16836**	16918	8.27	−48.28
12	N-12st70-3x4	42145	48270.1	60	33741	44716.5	31.2	**19938**	20072.2	5.72	−40.91
13	N-15berlin52	26849	30156.2	34.2	26061	32212.7	31.8	**16095**	16343.15	9.57	−38.24
14	N-15eil51	29256	32213.9	34.8	26816	32807.6	28.8	**15402**	15684.05	7.89	−42.56
15	N-15eil76	54456	60010.4	64.2	47104	58862.4	60	**25326**	25936.65	10.72	−46.23
16	N-15pr76	60103	68192.3	67.8	53677	65326.1	52.8	**28914**	29252.7	11.18	−46.13
17	N-15pr76-3x5	44085	52755.9	67.8	42838	52297.9	43.2	**20069**	20367.6	13.44	−53.15
18	N-15st70	47081	55091.8	54	42945	53459.6	51	**22367**	22646.5	11.96	−47.92
19	N-16eil51-4x4	26252	28109.9	34.8	24218	31763.2	28.2	**13225**	13502.3	7.37	−45.39
20	N-16eil76-4x4	47325	52876.1	58.8	41490	50834.6	36	**21139**	21341.2	13.40	−49.05

5   Conclusions and Further Research Directions

The aim of this paper was to solve the clustered shortest-path tree (CluSPT) problem using a hybrid two-level based genetic algorithm (GA) with a Prüfer encoding strategy. We have shown how to use a GA to generate trees spanning the clusters and how to determine the optimal solution associated to a given global spanning tree using dynamic programming, resulting a powerful and effective novel solution approach for solving the CluSPT problem.

The preliminary experimental results on a set consisting of 40 available non-euclidean instances from the literature prove that our developed two-level hybrid algorithm is highly competitive against the genetic algorithm proposed by Cosma et al. [3] and outperforms in terms of both the quality of the obtained solutions and the associated CPU processing times the two level approach based on multifactorial evolutionary algorithm presented by Binh and Thanh [2].

It is our objective to carry on our research and to investigate some other Cayley-type encoding of spanning trees, such as Blob and Dandelion codes. We also intend to evaluate the generality and scalability of our developed two-level hybrid algorithm by evaluating it on larger non-euclidean instances.

References

1. Binh, H.T.T., Thanh, P.D., Thang, T.B.: New approach to solving the clustered shortest-path tree problem based on reducing the search space of evolutionary algorithm. Knowl.-Based Syst. **180**, 12–25 (2019)

2. Huynh Thi Thanh, B., Pham Dinh, T.: Two levels approach based on multifactorial optimization to solve the clustered shortest path tree problem. Evol. Intell. **15**(1), 185–213 (2020). https://doi.org/10.1007/s12065-020-00501-w

3. Cosma, O., Pop, P.C., Zelina, I.: An effective genetic algorithm for solving the clustered shortest-path tree problem. IEEE Access **9**, 15570–15591 (2021)

4. Cosma, O., Pop, P.C., Zelina, I.: A novel genetic algorithm for solving the clustered shortest-path tree problem. Carpathian J. Math. **36**(3), 401–414 (2020)

5. D'Emidio, M., Forlizzi, L., Frigioni, D., Leucci, S., Proietti, G.: On the clustered shortest-path tree problem. In: Proceedings of Italian Conference on Theoretical Computer Science, pp. 263–268 (2016)

6. D'Emidio, M., Forlizzi, L., Frigioni, D., Leucci, S., Proietti, G.: Hardness, approximability and fixed-parameter tractability of the clustered shortest-path tree problem. J. Comb. Optim. **38**, 165–184 (2019). https://doi.org/10.1007/s10878-018-00374-x

7. Demange, M., Monnot, J., Pop, P.C., Ries, B.: On the complexity of the selective graph coloring problem in some special classes of graphs. Theor. Comput. Sci. **540–541**, 82–102 (2014)

8. Feremans, C., Labbe, M., Laporte, G.: Generalized network design problems. Eur. J. Oper. Res. **148**(1), 1–13 (2003)

9. Fidanova, S., Pop, P.C.: An improved hybrid ant-local search for the partition graph coloring problem. J. Comput. Appl. Math. **293**, 55–61 (2016)

10. Fischetti, M., Salazar-Gonzales, J.J., Toth, P.: A branch-and-cut algorithm for the symmetric generalized traveling salesman problem. Oper. Res. **45**(3), 378–394 (1997)

11. Ghiani, G., Improta, G.: An efficient transformation of the generalized vehicle routing problem. Eur. J. Oper. Res. **122**, 11–17 (2000)

12. Hanh, P.T.H., Thanh, P.D., Binh, H.T.T.: Evolutionary algorithm and multifactorial evolutionary algorithm on clustered shortest-path tree problem. Inf. Sci. **553**, 280–304 (2021)

13. Mestria, M., Ochi, L.S., de Lima Martins, S.: GRASP with path relinking for the symmetric Euclidean clustered traveling salesman problem. Comput. Oper. Res. **40**(12), 3218–3229 (2013)

14. Myung, Y.S., Lee, C.H., Tcha, D.W.: On the generalized minimum spanning tree problem. Networks **26**, 231–241 (1995)

15. Pop, P.C., Matei, O., Sabo, C., Petrovan, A.: A two-level solution approach for solving the generalized minimum spanning tree problem. Eur. J. Oper. Res. **265**(2), 478–487 (2018)

16. Pop, P.C., Fuksz, L., Horvat Marc, A., Sabo, C.: A novel two-level optimization approach for clustered vehicle routing problem. Comput. Ind. Eng. **115**, 304–318 (2018)

17. Pop, P.C.: The generalized minimum spanning tree problem: an overview of formulations, solution procedures and latest advances. Eur. J. Oper. Res. **283**(1), 1–15 (2020)

18. Pop, P.C.: Generalized Network Design Problems. Modeling and Optimization. De Gruyter Series in Discrete Mathematics and Applications, Germany (2012)

19. Thanh, P.D.: CluSPT instances, Mendeley Data v2. https://doi.org/10.17632/b4gcgybvt6.2

20. Thanh, P.D., Binh, H.T.T., Dac, D.D., Binh Long N., Hai Phong, L.M.: A Heuristic Based on Randomized Greedy Algorithms for the Clustered Shortest-Path Tree Problem. In Proceedings of IEEE Congress on Evolutionary Computation (CEC), pp. 2915–2922 (2019)

New Hybrid Methodology Based on Particle Swarm Optimization with Genetic Algorithms to Improve the Search of Parsimonious Models in High-Dimensional Databases

Jose Divasón[ID], Alpha Pernia-Espinoza[ID],
and Francisco Javier Martinez-de-Pison[(✉)][ID]

University of La Rioja, Logroño, Spain
{jose.divason,alpha.pernia,fjmartin}@unirioja.es

Abstract. Our previous PSO-PARSIMONY methodology (a heuristic to search for accurate and low-complexity models with particle swarm optimization) shows a good balance between accuracy and complexity with small databases, but gets stuck in local minima in high-dimensional databases. This work presents a new hybrid methodology to solve this problem. First, we incorporated to PSO-PARSIMONY an aggressive mutation strategy to encourage parsimony. Second, a hybrid method between PSO and genetic algorithms was also implemented. With these changes, particularly with the second one, improvements were observed in the search for more accurate and low-complexity models in high-dimensional databases.

Keywords: PSO-PARSIMONY · Hybrid method · Parsimonious modeling · Auto machine learning · GA-PARSIMONY

1 Introduction

The success of machine learning techniques in practically all fields of science and industry has led to an increasing demand for optimization heuristics and tools to facilitate some typical tasks such as hyperparameter optimization (HO) and feature selection (FS). GA-PARSIMONY [13] is a well-established methodology to search for parsimonious solutions with genetic algorithms by performing HO and FS. It has been successfully applied in many fields, such as steel industrial processes and hospital energy demand. Moreover, previous comparisons with other existing AutoML methodologies (such as Auto-sklearn, H2O and MLJAR) demonstrated its effectiveness [13]. However, certain limitations of GA-PARSIMONY are also well-known: it requires a high number of individuals and also several repetitions of the method to guarantee an optimal solution (because it is unstable). It often converges to local minima instead of global minima. In addition, it sometimes generates twin individuals that do not contribute new information to the search process.

© Springer Nature Switzerland AG 2022
P. García Bringas et al. (Eds.): HAIS 2022, LNAI 13469, pp. 335–347, 2022.
https://doi.org/10.1007/978-3-031-15471-3_29

A first attempt to improve the results of GA-PARSIMONY was to develop a new algorithm that combined the particle swarm optimization technique (PSO) and the parsimony criteria to obtain high-accuracy and low complexity models. The algorithm was named PSO-PARSIMONY [1] and improved GA-PARSIMONY on small databases, but was stuck in local minima on large databases. To solve this problem, this paper describes new proposals to improve PSO-PARSIMONY. On the one hand, PSO is combined with an aggressive mutation strategy to foster parsimonious models. On the other hand, a hybrid model between PSO and genetic algorithms is proposed, in which the particles with worse fitness are replaced in each iteration by new ones generated from typical genetic algorithm operations: selection, crossover and mutation. The accuracy and complexity of the new proposals are tested with public databases of different sizes and compared with GA-PARSIMONY.

2   Related Work

The importance of HO and FS when solving a machine learning problem is well-known: they permit improving the predictive accuracy of algorithms. However, the right choice of hyperparameters and a subset of features is a difficult combinational problem for which efficient heuristic methods are usually required. Currently approaches are usually inspired from nature, mainly from biological systems such as animal herding, bacterial growth and so on. They usually consist of a population of simple individuals interacting both locally and globally with each other following some simple rules. For example, Mirjalili et al. [11] proposed a new meta-heuristic called Grey Wolf Optimizer (GWO) inspired by grey wolves and was successfully applied to several classical engineering design problems. Mirjalili et al. also proposed the Salp Swarm Algorithm [10], being inspired by the swarming behavior of salps when navigating and foraging in oceans. Other techniques related to animals include bat [15], glowworm [7] and bee colonies [5]. One of the most commonly used optimization techniques is the Particle Swarm Optimization (PSO), originally proposed by Kennedy and Eberhart [6]. There has been much research on this technique and numerous improvements have been proposed [14,16], for instance, in terms of topology, parameter selection, and other technical modifications, including quantum-behaved and chaotic PSO, extensions to multiobjective optimization, cooperation and multi-swarm techniques. Indeed, the study of PSO modifications is an active area of research due to the success of the algorithm. Some hybridizations of PSO with other meta-heuristic methods have been also proposed. For instance, for feature selection Chuang et al. [2] proposed an improved binary particle swarm optimization using the catfish effect, that is, new particles are introduced into the search space if the best solution does not improve in some number of consecutive iterations. This is done by replacing the 10% of original particles with the worst fitness values by new ones at extreme positions. Some PSO hybridizations include operations from genetic algorithms, for example, there are several modifications that include the crossover [3] operator. In [4] the crossover is taken between each particle's individual best position. After the crossover, the fitness of the individual best position is compared with two offspring produced after crossing. Then, the best one is chosen as the new individual best position. In [12], the standard crossover and mutation operations from GA are applied to PSO.

3   Previous PSO

In a previous work by the authors [1], the particle swarm optimization (PSO) was combined with a parsimony criterion to find parsimonious and, at the same time, accurate machine learning models. The PSO algorithm works by having a population (called a swarm) of possible solutions (called particles). The position of a particle is simply a vector $X = (H, F)$ where H corresponds to the values of model's hyperparameters and F is a vector with values between 0 and 1 for selecting the input features. These particles are moved around in the search-space of the combinational problem according to simple formulas:

$$V_i^{t+1} = \omega V_i^t + r_1\varphi_1 \times \left(pbest_i^t - X_i^t\right) + r_2\varphi_2 \times \left(lbest_i^t - X_i^t\right) \tag{1}$$

$$X_i^{t+1} = X_i^t + V_i^{t+1} \tag{2}$$

where V_i^t and X_i^t denote the velocity and position of the i-th particle in iteration t, respectively. Such formulas just state that the movement of a particle is influenced by three components: its previous velocity, its own experience (its best position achieved so far, $pbest_i$) and also by the experience of other particles (the best position within a neighborhood, $lbest_i$). This permits particles to explore the search space based on their current momentum, each individual particle thinking (cognitive component) and the collaborative effect (cooperation component). More concretely, ω is the inertia weight used to control the displacement of the current velocity. φ_1 and φ_2 are positive constant parameters that balance the global exploration and local exploitation. r_1 and r_2 are uniformly distributed random variables used to maintain the diversity of the swarm.

Our modified PSO includes a strategy where the best position of each particle (thus, also the best position of each neighborhood) is computed considering not only the goodness-of-fit, but also the principle of parsimony, see [1] for further details. Algorithm 1 presents a pseudo-code version of it.

Experiments were conducted with various databases. A subset of them is shown in Table 1. PSO always improved accuracy with respect to GA. In databases with a low number of features, the difference with respect to parsimony was usually small (GA found solutions about 10% simpler). However, datasets with a larger number of features caused trouble for PSO, which found better solutions than GA but with twice as many features. Moreover, GA required much less computational effort: approximately, GA was three times faster and needed halving the iterations. This showed that PSO-PARSIMONY is a good alternative to GA-PARSIMONY if the number of features is relatively small, because it finds better solutions with a good balance between accuracy and parsimony.

Algorithm 1. Pseudo-code of the modified PSO algorithm

1: Initialization of positions $\mathbf{X}^0$ using a random and uniformly distributed Latin hypercube within the ranges of feasible values for each input parameter
2: Initialization of velocities according to $\mathbf{V}^0 = \frac{random_{LHS}(s, \ D)-\mathbf{X}^0}{2}$
3: **for** $t = 1$ to T **do**
4:     Train each particle $\mathbf{X}_i$ and validate with CV
5:     Fitness evaluation and complexity evaluation of each particle
6:     Update $pbest_i$, $pbest_{p,i}$ and the $gbest$
7:     **if** early stopping is satisfied **then**
8:         **return** $gbest$
9:     **end if**
10:     Generation of new neighborhoods if $gbest$ did not improve
11:     Update each $lbest_i$
12:     Update positions and velocities according the formulas
13:     Mutation of features
14:     Limitation of velocities and out-of-range positions
15: **end for**
16: **return** gbest

---

4   New Proposals

The previous section shows that PSO-PARSIMONY works well, but has room for improvement in terms of parsimony (especially for datasets with a large number of features) and computational time. The experiments also provided some insight on the accuracy and the evolution of parsimony in both GA and PSO; see also Fig. 1 for further results with the crime database. Specifically, Fig. 1a shows an exponential-like decrease in the number of features selected by GA in the first iterations, whereas PSO performs a more linear decrease. Figure 1b shows how good PSO is in terms of precision.

Table 1. PSO-PARSIMONY vs GA-PARSIMONY with 10 databases (results are the average of 5 runs with each methodology and $tol = 10^{-3}$).

Database	#rows	#feats	$\overline{PSO_J}$	$\overline{GA_J}$	$\overline{PSO_{N_{FS}}}$	$\overline{GA_{N_{FS}}}$	$\overline{PSO_{time}}$	$\overline{GA_{time}}$	$\overline{PSO_{iters}}$	$\overline{GA_{iters}}$
strike	625	7	**0.83856**	0.86479	**1.8**	3.0	105.1	**16.3**	88.0	**39.6**
no2	500	8	**0.65608**	0.66007	6.0	6.0	59.8	**17.5**	102.8	**46.4**
concrete	1030	9	**0.28943**	0.29526	**7.4**	7.8	468.0	**160.9**	107.2	**41.6**
housing	506	14	**0.31261**	0.32559	11.0	**10.0**	215.8	**74.6**	104.0	**61.0**
bodyfat	252	15	**0.10709**	0.10806	3.4	**2.0**	97.3	**40.2**	128.2	**69.2**
cpu_act	8192	22	**0.12405**	0.12473	14.8	**13.4**	1241.7	**788.3**	84.6	**50.0**
bank	8192	33	**0.63420**	0.63792	23.6	**19.8**	1298.8	**629.1**	177.8	**101.6**
puma	8192	33	**0.18049**	0.18097	4.6	**4.2**	2188.6	**1028.4**	96.0	**51.2**
ailerons	13750	41	**0.38228**	0.38258	17.8	**10.4**	2663.2	**1144.7**	115.2	**65.6**
crime	2215	128	**0.59336**	0.59565	49.8	**19.8**	797.8	**410.5**	185.6	**143.6**

To increase parsimony in PSO, two new variants of the algorithm are proposed in this section. The main goal is to promote the parsimonious behavior of GA in PSO. The first is a straightforward variant in which the mutation phase in PSO is performed exactly as in GA. The second is a hybrid approach between GA and PSO. They have been implemented in Python and are available at https://github.com/jodivaso/Hybrid-PSOGAParsimony.

4.1 PSO with a New Mutation

PSO-PARSIMONY already includes a mutation operator, for which the mutation rate was set to $1/D$ by default, where D is the dimensionality of the problem. In contrast, GA has a much more aggressive strategy; it also performs uniform random mutation, but three parameters are involved: `pmutation` represents the percentage of parameters to be muted, `feat_mut_thres` represents the probability of select a feature (include it in the selected features of the individual) when muting it, and `not_muted` is the number of best individuals that will not be muted. Note that `not_muted` prevents losing the best individuals in the mutation step. The default values are set to 0.1, 0.1 and 3, respectively. With these default values, GA approximately excludes 9% of the features in each iteration. In contrast, only $1/2D$ of the features are excluded in each step with PSO, which are very few if the dataset has many variables. This explains why the PSO algorithm performs worse in terms of parsimony with high-dimensional databases. To solve this problem, Algorithm 1 was modified in Line 13 to include the mutation step as is done in GA.

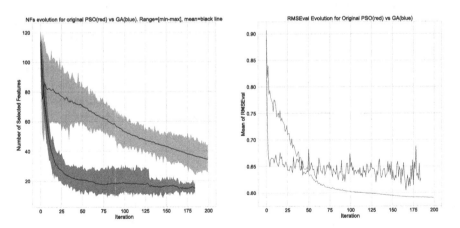

(a) Features evolution in GA and PSO   (b) Evolution of RMSE in GA and PSO

Fig. 1. PSO-PARSIMONY (red) vs. GA-PARSIMONY (blue) in terms of number of selected features and accuracy with the crime dataset. (Color figure online)

4.2 Hybrid Method: PSO with Crossover and Mutation

To further encourage parsimony in PSO and make its behavior closer to GA, especially in the first iterations, a hybrid model was developed. For this purpose, Algorithm 1 was modified in several parts.

A crossover phase is added just after calculating the local bests of the neighborhoods (i.e., after Line 11). The purpose of the crossover is to distribute good parts of the genome among individuals. To perform this crossover, a selection phase is also added at that point, which is based on a nonlinear-rank selection following Michalewicz [8]. In this way, the selection of other individuals in addition to the best ones maintains the diversity of the population and prevents premature convergence. Furthermore, the best individuals are more likely to be selected for crossover. Thus, they are selected for breeding more times to foster good offspring. The crossover function was implemented by using heuristic blending [9] for hyperparameters and random swapping for features. It was also adapted to work properly with PSO: the positions are crossed with each other, as well as the velocities according to the crossover performed at the positions.

In addition, the way of replacing the particles differs from the typical GA crossover: In this case, the new particles created from the crossover replace the worst particles (those with the worst fitness value) that appeared in the population. For this purpose, a parameter `pcrossover` is incorporated, which fixes the percentage of worst individuals to be substituted from crossover. This parameter can be either a constant (such a percentage of particles is substituted in all iterations) or an array to indicate a different percentage for each iteration. In this way, one can vary and encourage the crossover process in the first iterations by setting high values of `pcrossover` (to obtain a behavior similar to GA) and in further iterations decrease the percentage or even make it equal to 0 to obtain a pure PSO algorithm. Once the crossover step is done, PSO algorithm requires updating the positions and velocities according the formulas. In this case, this step is only applied to the particles that have not been substituted by the crossover. The mutation phase is also modified to include the changes proposed in the previous subsection, i.e., a more aggressive mutation strategy to encourage parsimony. The rest of steps of the algorithm are preserved.

5 Experiments

In order to test the capacity of the proposed methodologies to find accurate and parsimonious models, databases with a high number of features were selected. In particular, experiments compared the PSO-PARSIMONY method with the new mutation (New-PSO) and the hybrid HYB-PARSIMONY (HYB) against GA-PARSIMONY (GA) and the previous PSO method (Old-PSO).

All experiments were similar to previous works with a population size of $P = 40$, $tol = 0.001$, a maximum number of generations of $G = 200$, and an early stopping of 35. Experiments were implemented in 9 separately 24-core servers from the Beronia Cluster at the University of La Rioja.

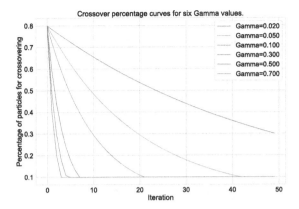

Fig. 2. Example of six curves created with different Γ values to establish the percentage of individuals to be replaced by crossover in each iteration.

For the hybrid method, the following equation was defined to calculate the percentage of particles to be substituted by crossover in each iteration *iter*:

$$\%particles = max(0.80 \cdot e^{(-\Gamma \cdot iter)}, 0.10) \tag{3}$$

Figure 2 shows six curves obtained with different Γ values. In the first iterations the hybrid method performs the substitution by crossing a high percentage of particles. As the optimization process progresses, the number of substituted particles is reduced exponentially until it ends up fixed at a percentage of 10%. Thus, the hybrid method begins by facilitating the search for parsimonious models using GA-based mechanisms and ends up using more PSO optimization.

Table 2 presents the results with the *crime* database with 128 features. It shows results for the GA, the Old-PSO, the New-PSO and 26 Γ values of the hybrid method. The second and third columns indicate respectively the validation error (J) and the number of features (N_{FS}) of the best model obtained. The last four columns correspond to the mean of J, N_{FS}, *time* and the number of iterations (*iters*) of five runs for each algorithm. The hybrid method with $\Gamma = 0.10$ obtained the best model reducing J to 0.57844 versus the previous best model achieved with GA $(J = 0.58070)$. However, the improvement in J involved the selection of 24 features (5 more) versus 19 in GA. On the other hand, the hybrid method with $\Gamma = 0.04$ obtained the most parsimonious model with only 17 features and an error of $J = 0.58142$, slightly higher than the J of GA. With respect to the mean values obtained from the five runs of each algorithm, it is observed that the hybrid method with $\Gamma = 0.32$ obtained the best mean values of J and N_{FS}.

Figure 3 shows in blue the range (minimum and maximum) and in a solid black line the mean value of N_{FS} (left) and J (right) for five runs of the algorithm with different values of Γ for the database *crime*. It also includes the range and mean value for GA (red) and for New-PSO (green). Regarding N_{FS}, the hybrid method with Γ between 0.32 and 0.36 was more stable, since it obtained lower

Table 2. Hybrid with different Γ vs previous methods for *crime* database.

Method	Γ	J_{best}	N_{FSbest}	$\overline{J}$	$\overline{N_{FS}}$	iters	time
GA	0.00	0.58070	19	0.58503	20.4	146.8	86.2
OLD PSO	0.00	0.58333	29	0.58773	33.0	200.0	121.0
NEW PSO	0.00	0.58155	26	0.58419	25.2	362.8	211.9
HYB	0.02	0.57981	22	0.58650	24.4	229.6	132.0
HYB	0.04	0.58142	**17**	0.58571	22.6	200.8	118.8
HYB	0.06	0.58397	23	0.58747	26.8	206.6	119.1
HYB	0.10	**0.57844**	24	0.58402	26.6	200.4	115.6
HYB	0.12	0.58106	24	0.58576	25.8	184.4	97.6
HYB	0.14	0.58143	24	0.58856	28.8	190.0	109.9
HYB	0.16	0.58151	19	0.58304	23.4	228.8	121.5
HYB	0.18	0.58603	26	0.58773	27.4	148.6	85.9
HYB	0.20	0.58251	23	0.58517	25.2	185.6	111.0
HYB	0.22	0.57964	23	0.58434	23.4	241 4	138.0
HYB	0.24	0.58229	25	0.58554	26.4	167.4	98.0
HYB	0.26	0.58368	23	0.58656	25.4	176.0	101.2
HYB	0.28	0.58054	24	0.58340	23.6	235.6	135.0
HYB	0.30	0.58343	29	0.58540	25.6	**143.8**	**82.7**
HYB	0.32	0.58050	22	**0.58193**	**20.2**	242.2	139.1
HYB	0.34	0.58247	**17**	0.58421	23.6	233.2	123.5
HYB	0.36	0.58001	23	0.58378	21.6	221.6	127.4
HYB	0.38	0.58119	20	0.58544	23.2	197.2	117.9
HYB	0.40	0.58117	27	0.58493	24.6	209.8	120.6
HYB	0.45	0.58304	22	0.58494	24.8	176.8	101.9
HYB	0.50	0.57938	24	0.58319	23.6	213.4	123.0
HYB	0.55	0.58314	24	0.58555	25.4	193.8	111.5
HYB	0.60	0.58158	22	0.58378	24.2	218.2	115.8
HYB	0.70	0.58080	24	0.58582	26.6	195.8	116.5
HYB	0.80	0.58065	19	0.58232	23.2	215.0	123.7
HYB	0.90	0.58101	25	0.58437	24.8	187.4	108.5

ranges than the one obtained with GA and with minimum values similar to the latter. On the other hand, it clearly outperformed the New-PSO method. Figure 4 shows that the hybrid method reduced the number of features more drastically and converged earlier than the New-PSO method, similar than GA method. With respect to J, Fig. 3b clearly shows that the new-PSO method was more robust than GA as it had a much smaller range of J in the five runs of the

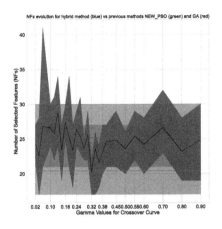
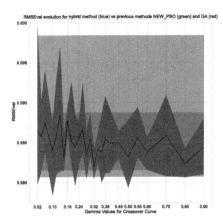

(a) N_{FS} evolution with different Γ values   (b) J evolution with different Γ values

Fig. 3. Comparison between HYBRID-PARSIMONY (blue) vs. PSO-PARSIMONY (green) and GA-PARSIMONY (red) in terms of the number of selected features and accuracy with the *crime* dataset. (Color figure online)

Table 3. NEW PSO-PARSIMONY vs HYBRID-PARSIMONY with a population size of $P = 40$ and $tol = 0.001$ (results are the average of the 5 runs).

Dataset	#rows	#feats	Γ	$\overline{PSO}_J$	$\overline{HYB}_J$	$\overline{PSO}_{N_{FS}}$	$\overline{HYB}_{N_{FS}}$	$\overline{PSO}_{time}$	$\overline{HYB}_{time}$
slice	5000	379	0.34	0.0238	**0.0231**	146.8	**132.2**	819.4	**609.0**
blog	4999	277	0.70	0.4087	**0.3983**	127.6	**113.8**	1117.5	**1051.6**
crime	2215	128	0.32	0.5842	**0.5819**	25.2	**30.2**	211.9	**139.1**
tecator	240	125	0.50	0.0331	0.0331	55.0	**48.6**	11.9	**8.7**
ailerons	5000	41	0.70	0.3947	**0.3934**	10.6	**10.2**	473.4	**466.1**
bank	8192	33	0.50	0.6514	**0.6511**	21.4	21.4	2146.4	**1536.6**
puma	8192	33	0.50	0.1817	0.1817	4.0	4.0	1063.8	**933.2**

algorithm. However, the hybrid method obtained with $\Gamma = 0.32$ better J values than PSO with a significantly lower range.

Finally, Fig. 5 presents the mean values of N_{FS} and J for the four methods and with the *crime* database. In this case, the hybrid model with $\Gamma = 0.32$ (blue) improved the reduction of N_{FS} and J in a balanced way, reducing the convergence time and obtaining accurate but low-complexity solutions.

Similar results can be observed with other high-dimensional databases. Tables 3 and 4 show respectively the average results and the best model obtained with the Hybrid Method and the New-PSO. In almost all databases, the hybrid method obtained more accurate models with less complexity, although it was necessary to find a suitable Γ value.

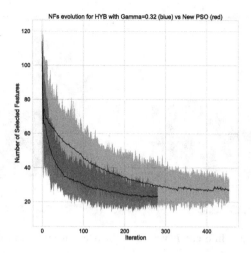

Fig. 4. Comparison of N_{FS} evolution between the Hybrid method (blue) (with $\Gamma = 0.32$) and the new PSO-PARSIMONY (red). (Color figure online)

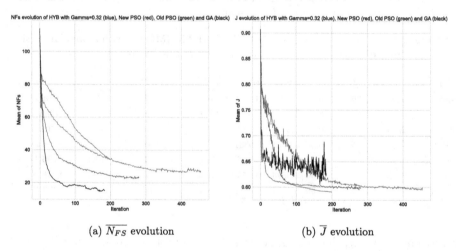

(a) $\overline{N_{FS}}$ evolution   (b) $\overline{J}$ evolution

Fig. 5. Comparison with *crime* dataset of $\overline{N_{FS}}$ and $\overline{J}$ between the four methods: the Hybrid method (blue) (with $\Gamma = 0.32$), the new PSO-PARSIMONY (red), the old PSO-PARSIMONY (green) and GA-PARSIMONY (black). (Color figure online)

Table 4. Best individual obtained with PSO-PARSIMONY vs HYBRID-PARSIMONY using a population size of $P = 40$ and $tol = 0.001$.

Dataset	Γ	PSO_J	HYB_J	$PSO_{J_{TST}}$	$HYB_{J_{TST}}$	$PSO_{N_{FS}}$	$HYB_{N_{FS}}$	PSO_{time}	HYB_{time}
slice	0.70	0.0228	**0.0218**	**0.0012**	0.0017	124	**112**	1050.1	**627.5**
blog	0.38	0.3948	**0.3879**	0.2523	**0.2023**	**115**	129	1304.0	**1277.4**
crime	0.10	0.5815	**0.5784**	0.5021	**0.4780**	26	**24**	263.5	**138.5**
tecator	0.38	0.0328	**0.0327**	0.0207	**0.0206**	**48**	51	16.3	**10.7**
ailerons	0.15	0.3935	**0.3922**	**0.3675**	0.3698	13	**10**	**484.2**	494.3
bank	0.70	0.6510	**0.6507**	**0.5839**	0.5865	22	**21**	2428.5	**1675.0**
puma	0.38	0.1817	0.1817	0.1776	0.1776	4	4	1191.8	**712.9**

6   Conclusions

This paper presents two new proposals to improve our previous PSO-PARSIMONY methodology for the simultaneous search of the best model hyper-parameters and input features, with a balance between accuracy and complexity. Concretely, a PSO with an aggressive mutation strategy as well as a hybrid method have been implemented. The main novelty relies on the hybrid model between GA and PSO, where the optimization is based on the PSO formulas, but the common genetic operations of selection, crossover and mutation are included to replace the worst particles. The percentage of variables to be substituted in each iteration can be customized. In this work, functions that depend on a Γ parameter have been used to promote parsimony in the first iterations (a high percentage of particles is substituted), but in further iterations the percentage is decreased. This differs from other hybrid methods where the crossover is applied between each particle's individual best position or other approaches where the worst particles are also substituted by new ones, but at extreme positions. Experiments show that, in general and once the appropriate gamma is fixed, this HYB-PARSIMONY methodology allows one to obtain better, more parsimonious and more robust models compared to our previous PSO-based methodology and the PSO with mutation. The computational effort is also reduced, since it requires less time.

Although it is a promising method, further research is required to provide an explicit formula that fixes the Γ value for each dataset, for instance, depending on the number of instances and features or by means of adaptive strategies.

Acknowledgements. We are greatly indebted to *Banco Santander* for the REGI2020/41 fellowship. This study used the Beronia cluster (Universidad de La Rioja), which is supported by FEDER-MINECO grant number UNLR-094E-2C-225. The work is also supported by grant PID2020-116641GB-I00 funded by MCIN/ AEI/ 10.13039/501100011033.

References

1. Ceniceros, J.F., Sanz-Garcia, A., Pernia-Espinoza, A., Martinez-de-Pison, F.J.: PSO-PARSIMONY: a new methodology for searching for accurate and parsimonious models with particle swarm optimization. Application for predicting the force-displacement curve in T-stub steel connections. In: Sanjurjo González, H., Pastor López, I., García Bringas, P., Quintián, H., Corchado, E. (eds.) HAIS 2021. LNCS (LNAI), vol. 12886, pp. 15–26. Springer, Cham (2021). https://doi.org/10. 1007/978-3-030-86271-8_2
2. Chuang, L.Y., Tsai, S.W., Yang, C.H.: Improved binary particle swarm optimization using catfish effect for feature selection. Expert Syst. Appl. **38**(10), 12699–12707 (2011). https://doi.org/10.1016/j.eswa.2011.04.057
3. Engelbrecht, A.P.: Particle swarm optimization with crossover: a review and empirical analysis. Artif. Intell. Rev. **45**(2), 131–165 (2015). https://doi.org/10.1007/s10462-015-9445-7
4. Hao, Z.F., Wang, Z.G., Huang, H.: A particle swarm optimization algorithm with crossover operator. In: 2007 International Conference on Machine Learning and Cybernetics, vol. 2, pp. 1036–1040 (2007)
5. Karaboga, D., Basturk, B.: Artificial bee colony (ABC) optimization algorithm for solving constrained optimization problems. In: Melin, P., Castillo, O., Aguilar, L.T., Kacprzyk, J., Pedrycz, W. (eds.) Foundations of Fuzzy Logic and Soft Computing, pp. 789–798. Springer, Heidelberg (2007). https://doi.org/10.1007/978-3-540-72950-1_77
6. Kennedy, J., Eberhart, R.: Particle swarm optimization. In: Proceedings of ICNN 1995 - International Conference on Neural Networks, vol. 4, pp. 1942–1948 (1995). DOI: https://doi.org/10.1109/ICNN.1995.488968
7. Marinaki, M., Marinakis, Y.: A glowworm swarm optimization algorithm for the vehicle routing problem with stochastic demands. Expert Syst. Appl. **46**, 145–163 (2016). https://doi.org/10.1016/j.eswa.2015.10.012
8. Michalewicz, Z.: Genetic Algorithms + Data Structures = Evolution Programs, Third Revised and Extended Edition. Springer, heidelberg (1996)
9. Michalewicz, Z., Janikow, C.Z.: Handling constraints in genetic algorithms. In: Icga, pp. 151–157 (1991)
10. Mirjalili, S., Gandomi, A.H., Mirjalili, S.Z., Saremi, S., Faris, H., Mirjalili, S.M.: Salp swarm algorithm: a bio-inspired optimizer for engineering design problems. Adv. Eng. Softw. **114**, 163–191 (2017). https://doi.org/10.1016/j.advengsoft.2017. 07.002
11. Mirjalili, S., Mirjalili, S.M., Lewis, A.: Grey wolf optimizer. Adv. Eng. Softw. **69**, 46–61 (2014). https://doi.org/10.1016/j.advengsoft.2013.12.007
12. Nazir, M., Majid-Mirza, A., Ali-Khan, S.: PSO-GA based optimized feature selection using facial and clothing information for gender classification. J. Appl. Res. Technol. **12**(1), 145–152 (2014). https://doi.org/10.1016/S1665-6423(14)71614-1
13. Martinez-de Pison, F.J., Ferreiro, J., Fraile, E., Pernia-Espinoza, A.: A comparative study of six model complexity metrics to search for parsimonious models with GAparsimony R Package. Neurocomputing **452**, 317–332 (2021). https://doi.org/10.1016/j.neucom.2020.02.135
14. Shami, T.M., El-Saleh, A.A., Alswaitti, M., Al-Tashi, Q., Summakieh, M.A., Mirjalili, S.: Particle swarm optimization: a comprehensive survey. IEEE Access **10**, 10031–10061 (2022). https://doi.org/10.1109/ACCESS.2022.3142859

15. Yang, X.S.: A new metaheuristic bat-inspired algorithm. In: González, J.R., Pelta, D.A., Cruz, C., Terrazas, G., Krasnogor, N. (eds.) NICSO 2010, vol. 284, pp. 65–74. Springer, Heidelberg (2010). https://doi.org/10.1007/978-3-642-12538-6_6
16. Zhang, Y., Wang, S., Ji, G.: A comprehensive survey on particle swarm optimization algorithm and its applications. Math. Probl. Eng. 1–38 (2015). https://doi.org/10.1155/2015/931256

Evolving Dynamic Route Generators for Open-Ended ARPs

Alejandro González Casal⬭, Pedro José Trueba Martínez⬭,
and Abraham Prieto García(✉)⬭

Universidade da Coruña, A Coruña, Spain
`abraham.prieto@udc.es`

Abstract. This paper addresses the resolution of an open-ended cost-based formulation of arc routing problems (ARPs) by means of a dynamic route generator (DRG) based on a neural network architecture and trained by an specific version of a differential evolution algorithm (DDR-Gen). The propopsed formulation brings together the main features present in realistic ARPs along with a metric based on a time-dependent non-service penalty. The operation of the dynamic route generator uses the current state information of the graph to produce efficient routes in real time. The method is first validated by solving one of the main benchmarks in the field of ARPs (GDB). Then, we present a sensitivity analysis and parameter tuning of the main components of the method when solving the proposed formulation on the same benchmark. Thirdly, we test the ability of DDRGen to solve the mentioned formulation on higher dimension instances (TCARP) and the generalizability of the obtained solutions to instances different from those of their training. Finally, we have studied the use of generalist DRGs as fast evaluators in a design process which optimises service networks to improve their efficiency.

Keywords: Neural networks · Differential evolution · Arc routing problems

1 Introduction

The field of Arc Routing Problems (ARPs) has experienced a remarkable increase in its impact due to the recent number of applications that have emerged and the advantages compared with Node Routing Problems in scenarios with poorly discretised or easily grouped demands.

ARPs deal with scenarios in which the elements requiring service are distributed along the existing connections between nodes. Some of the most common applications are garbage collection [7], street cleaning, snow removal and salt spreading [4]. Some other applications have recently been adapted to ARP problems such as network inspection [8] as well as some novel manufacturing problems, namely: pattern generation for sheet metal cutting or the design of 3D printing nozzle paths [6]. [2] provides a comprehensive review of the most important applications of ARPs nowadays, as well as potential new applications.

© Springer Nature Switzerland AG 2022
P. García Bringas et al. (Eds.): HAIS 2022, LNAI 13469, pp. 348–359, 2022.
https://doi.org/10.1007/978-3-031-15471-3_30

Depending on the application, different formulations of the ARPs can be considered, the first and most basic of these is the Chinese Postman Problem (CPP), where a single agent (the postman) must, starting from an initial node (depot), cover all the arcs that make up the graph and return to the depot. This formulation, although very popular in the field, is not suitable to represent realistic applications so, more complex formulations were soon developed to include additional elements and restrictions. The k-CARP variation combines the use of multiple agents to perform the service with the addition of capacity constraints. The importance of this formulation lies on the fact that it represents a trade-off among a sufficiently realistic and complex model for many applications and a simple and clear definition of the problem which is why it has been the subject of extensive study in the literature.

Many optimization methods can be found in literature for these problems as well as frameworks which combine several existing methods to solve more complex or dynamic instances [12]. In particular, we are interested in metaheuristics, non-exact methods which use search strategies that are independent, to a degree, from the problem to solve at the expenses of an increment of the solving time and which allow for a higher generalizability. From the diverse metaheuristics methods that exist, the three most used in the literature to solve ARPs are tabu search, ant colony optimization and genetic algorithms [5,11,13].

Regarding the encoding of the optimal path, the usual approach is to define a fixed sequence of the network arcs to be covered. This method, although very efficient for simple and static formulations, which are the ones usually solved in the field of ARPs, shows serious weaknesses when dealing with realistic problems where the dynamism and stochasticity of the environment may require a modification of the route in real time. To overcome these weaknesses, this work proposes a less common method, **a dynamic route generator**, which will design the route in real time based on the current state of the graph and in some other relevant information of the environment.

2 Open-ended Cost-Based Dynamic k-CARP Formulation

We will consider the following formalization in order to define metrics, constraints and different variants of the problem:

Let G be an undirected graph which will represent the network to be served, v_i the i-th node of the set of V nodes that compose the graph, e_{ij} the arc between nodes v_i and v_j of the set of arcs A. Each arc will have a series of parameters associated with l_{ij} being the length or cost of the connection, c_{ij} being the failure cost, st_{ij}^n being the time when the arc receives its n-th service and dem_{ij} the demand required to serve the arc.

Let K be the number of agents, Q their capacity to serve arcs without reloading in the depot and $del_{k,ij}$ the delivery served by the k-th agent to arc e_{ij}.

Let N_k be the number of trips realized by the k-th agent, E_k, t be the set of arcs served by the k-th agent during its p-th trip (movements between two visits to the depot) and T_k, p be the set of arcs travelled by the m-th agent to serve E_k, p.

As for the cost, it is defined from the product of the probability of failure times the impact of the failure. The potential cost of for each arc $C_{ij}(t)$ is calculated according to the time without service (t) by adding the cost of each of the possible failures during the time that they would have remained active up to the current instant (Eq. 1). The probability of a failure occurring in the ij-th arc after the s-th time step is defined by equation.

$$C_{ij}(t) = \sum_{s=0}^{t} p_{ij,s} c_{ij}(t - s) \quad with \quad p_{ij,s} = p_{ij}(1 - p_{ij})^s \tag{1}$$

The accumulated cost during the whole simulation for each arc is calculated as a sum of all the accumulated potential costs among services as defined in Eq. 2, where T is the total duration of the service.

$$C_{cum,ij}(T) = C_{ij}(st_{ij}^1) + C_{ij}^n(T - st_{ij}^N) + \sum_{n=1}^{N} C_{ij}^n(st_{ij}^{n+1} - st_{ij}^n) \tag{2}$$

The global objective funcion to minimize is defined by the sum of the total costs for all the arcs during the whole simulation subject to the constrain set by the limited capacity regarding the service provided by each agent on one trip (Eq. 3).

$$Min(\sum_{\forall ij} C_{cum,ij}(T)) \quad with \quad \sum_{E_k,p} del_{ij} \leq Q \tag{3}$$

3   Differential Evolution Dynamic Route Generation (DDRGen)

The dynamic route generator (DRG) proposed consists in an neural network based evaluator which, as part of a deliberative model, allows to determine in real time the next arc to be travelled and whether or not it will be served. The optimization DRG will consist in adjusting the weights of the network so that, when it is used as the evaluator of the route generator, the global penalty-cost is minimized. This optimization process will be carried out using differential evolution (DE) [10]. The combination of DRG and the DE based optimization operator constitutes the resolution method proposed: Differential Evolution Dynamic Route Generation (DDRGen).

3.1   Dynamic Route Generator

Since the proposed formulation include capacity constrains and the optimal solution will probably include travelling some arcs without serving them just to reach some others, each of the arcs connected to a node (N) can be either served or travelled without being served. Therefore, there will be 2·N candidate actions to be evaluated by DRG: N real arcs and another N virtual arcs with the same parameters but with zero required demand.

For a better understanding of the DRG inputs we introduce the following concepts:

- Distances matrix. Consists in a precalculated matrix which contains the minimum distances among each pair of nodes of the graph.
- Favourite paths matrix. Based on the distances matrix, this matrix defines for each node and adjacent arc a set of arcs whose shortest access from the current node is achieved through the adjacent arc considered. This is also precalculated for each graph with an iterative search algorithm developed to that end.
- Downstream arcs. Those arcs stored on the favourite paths matrix and associated to an specific node and adjacent arc.

The following features are used (classified in four groups):

1. **Arc cost:** potential cost (1), cuadratic coefficient (2), linear coefficient (3).
2. **Arc properties:** arc length (4), arc demand (5), time since last service of the arc (9).
3. **Downstream information:** crowdedness in downstream arcs (7), potential cost in downstream arcs (8), time since last service of downstream arcs (10).
4. **Remaining capacity of the agent (6).**

Based on those 10 inputs the NN will produce a single output corresponding to the score obtained by the evaluated arc. The alternative with the highest score is selected, which will determine the route to follow.

The algorithm used to optimise the weights of the network is Differential Evolution. The notable variants of the algorithm are the ones that adjust their operating parameters, the mutation factor and the crossover rate. Some popular versions, JDE [1] and SADE [9], adapt those parameters during the operation of the algorithm, depending on the success of the parameters. Those two variants will be tested in addition to the standard DE.

4   Preliminary Experiments

In a first stage, DDRGen has been validated using a benchmark for one of the ARP formulations (k-CARP) and a sensitivity analysis of its main parameters has been performed.

4.1   Validation of the Solver

We will use the GDB benchmark [3], a repository of 23 k-CARP graph instances that has been widely studied in the literature and whose optimal solutions are known. The repository instances are multi-agent, include capacity constraints, and require the full service of the graph arcs.

First, we trained one single DRG network to solve the 23 instances of the GDB benchmark, later we trained one DRGs for each instance of the benchmark. The single network experiment is designed to test the generalization capability of the solver, so the results shown in this section correspond to this experiment.

In terms of managing the capacity of the agents, several strategies are considered, one standard and two more flexible strategies that allow the agents to perform several trips (multi-trip) and to deliver parcial services (split-service). Six configurations of the algorithms (three algorithms with two selection strategies) will be tested in each capacity configuration, with 10 evolutionary processes for each variant, making 60 evolutionary processes for each configuration.

The network topology for the DRG of this experiment is a two-layer network with two neurons per layer, with a hyperbolic tangent as activation function. The parameters of the algorithms (DE, JDE, SADE) follow recommendations in the literature, with a population of at least ten times the dimensionality of the problem.

Table 1 contains the results for both single and multi-network approach with the rows containing the results for each formulation in terms of the average error to the optimal solution. The multi-network column also includes the number of optimal results achieved (out of 23). The split-service configurations obtains the best results in terms of errors.

Table 1. Single network results

Strategies	Multi network error	Single network error
Standard	4.62% (8 optimal solutions)	32.99%
Multi-trip	3.96% (6 optimal solutions)	12.96%
Split-service	3.17% (11 optimal solutions)	11.79%

In all of the configuration solutions obtain an average relative error below 5%, which is particularly remarkable considering that the static benchmark problem is not a particularly suitable environment for the performance of reactive methods such as the DDRGen.

Although the optimization time is higher than in specific heuristics, once trained the DRG can provide solutions to new graphs by running the DRG network much faster than defining a new route with an heuristic.

4.2   Parameter Tunning

The metric used to evaluate the performance of the method is a relative error calculated as the deviation (error) of the potential cost obtained by the best solution of each of the evolutionary processes with respect to the best costs obtained in previous tests used as reference. The evolution time will be set at 10 min.

Twelve versions of the algorithm will be tested corresponding to combinations of the three variants of the algorithm [DE, JDE and SADE], two standard and target-to-best mutation schemes [est, ttb] and two types of selection, standard and binary tournament [est, bt] [9].

The first process consisted of adjusting the specific parameters of each version of the algorithm, F and CR for the DE and the adaptive parameters for JDE and SADE. The DE parameters have shown a significant effect on performance in contrast to the adaptive variants which have shown little sensitivity to their parameters.

The next step is the population size adjustment, which is set up identically for all variants. Populations from 4 to 1000 individuals are studied. The population sizes affected notably to the results obtained and the performance exhibited an U-shaped response curve for all the cases. The optimal performance of all the variants fits to the interval [60, 140].

Finally, a comparison of each of the variants in their optimal configuration is made in terms of their performance for the resolution of the complete benchmark. Figure 1 shows the error of all the evolutions performed per variant and we see a clear dominance of the *ttb-bt* variants in terms of mean error due to their great ability to quickly exploit promising areas of the search space. The best result is obtained by the SADE-ttb-bt, showing a great robustness for the different instances solved.

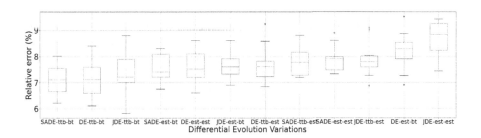

Fig. 1. Performance by differential evolution variant

Secondly, the effect of the inputs used by the DRG was tested. Ten configurations have been tested, the original version containing all the inputs, 8 configurations in which one of the 8 inputs (except for time inputs (9 and 10)) is not included and a last one in which only the inputs corresponding to the costs have been left. The results are shown in Fig. 2 (A: all inputs/C: only cost inputs (1 and 7)/i: without i-th input) and show that the cost inputs are indispensable (the error is multiplied by 10 if they are eliminated) but if the model is reduced to one that only includes these inputs, the results lose performance notably.

Finally, the sensitivity to the network architecture and size has been studied in terms of the number of hidden layers and number of adjustable parameters

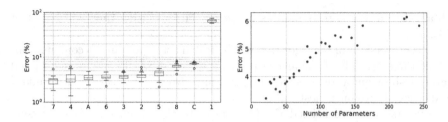

Fig. 2. Effect of inputs and network size on DRG performance

used. The aim is to find a compromise solution between the difficulty of training networks with more parameters and their ability to model more complex effects. The results show in general U-curves that represent a compromise between the number of parameters to be tuned and the ability of the network to capture the necessary knowledge. Figure 2 shows the results of the best architectures, from which we can set a range between 20 and 50 parameters for the most satisfactory solutions.

5   Solving Real Networks (Real Networks Application and Generalization)

Once the different elements and parameters of the proposed method have been analysed and adjusted, its performance will be evaluated by solving real routing problems with the formulation proposed in Sect. 2. This experiment has two main objectives: to validate the ability of the method to solve the proposed dynamic formulation in large instances, which is usually one of the limiting factors for state-of-the-art methods, and to study the generalisation capacity of the DRG to solve instances other than those used for its training.

A set of graphs generated using GIS from a real road network in Ireland will be used. The repository consists of a total of five graphs of considerably larger dimension than those used in the GDB benchmark (about 10 times more arcs).

5.1   Experimental Setup

The methodology followed for this experiment consists of training one DRG with each of the five graphs. The training process for each DRG will involve ten evolutionary processes of 30 min which will allow to evaluate both the performance (the best of them) and the stability of the optimisation process (their variance). The costs (penalty for non-service: the higher, the worse) obtained will be used as a reference, c_i^{ref}, for cross-evaluating the i-th DRG with the other four graphs to test their generalisation capacity and normalize regarding size and costs of the graph.

The associated crossed cost, c_{ij}, obtained solving the graph Ri by a DRG trained on the graph Rj is normalized with the reference solutions for each graph producing the relative cost: $c_{ij}^{rel} = (c_{ij} - c_i^{ref})/c_i^{ref}$.

Regarding the network topology, we have studied, on the one hand, different network sizes, and on the other hand, the use of an adaptive normalisation layer, which consists of a layer prior to the input layer, whose objective is to scale each input value individually in order to improve the capacity of the network to adequately deal with values of very diverse magnitude through its inputs. By performing experiments with various network sizes with and without the adaptive normalisation layer, it was determined that the optimal topology to use would be a network [5,5] with adaptive normalisation. This implies a total of 113 parameters.

Each of the evolutionary processes will use the configuration that has shown more robustness against size changes (SADE-ttb-bt; lr = 20) and with a population size that maintains the optimal parameter/individual ratio found in the previous phase (600 individuals), which has allowed to perform evolutions of at least 60 generations.

5.2   Results

The result is an averaged (among evolutionary runs) relative cost for each graph-DRG combination, $\overline{c_{ij}^{rel}}$, which is presented in the Table 2. Lower values imply either a general DRG or an easy graph or just similarity among the graphs used for training and testing. The diagonal of the table represents the averaged relative cost of the instances trained with each network excluding the reference one (the one with the lowest cost), i.e. it provides an indication of the stability of the results. In order to characterize each of the graphs as good for training or for testing DRGs, Table 2 also shows the average in rows and colums (excluding the diagonal). The row average represents the performance of the DRGs obtained by training on one of the instances to solve the rest of the instances in the set. The column average represents the average error obtained by solving the current graph using the DRGs trained in other graphs. At a conceptual level, two attributes have been determined for each instance based on these values. Richness, defined as the percentage complementary to the first of the averages, which reflects the generality that can be obtained in a DRG by training it on that instance, and difficulty, defined as the second of the averages, which represents the difficulty of generalising DRGs trained on other graphs to solve the current graph.

The results show a high generalisation capacity of the dRGs obtained by training in the different instances showing average errors below 10% with the cross testing. This type of cross-testing also allows us to determine which graphs would be most suitable for training a DRG, which in this case would be R1, R3 and R4, as they are the ones that produce the greatest richness.

Table 2. Relative cost cross-evaluation

	R1	R2	R3	R4	R5	Richness
R1	2.49%	6.11%	3.97%	8.06%	9.28%	94.02%
R2	20.97%	4.97%	4.06%	9.54%	8.40%	90.41%
R3	9.50%	5.96%	1.79%	7.73%	4.24%	94.16%
R4	12.70%	3.01%	3.36%	3.49%	7.64%	93.96%
R5	20.80%	7.41%	2.88%	7.24%	6.35%	91.06%
Difficulty	13.29%	5.49%	3.21%	7.21%	7.18%	

6   DDRgen as a Design Tool

Taking advantage of the capacity of the DDRGen method to provide generalist solvers and therefore to generate satisfactory solutions in a short time (the time to run the DRG on the graph) by skipping the optimisation process for each new graph, DDRGen will be used to carry out a meta-optimisation which will configure the graph to be solved to improve the efficiency of the network regarding the inspection task. The parameters considered to be optimized will be the number and position of depots, the number of agents and their service capacity.

6.1   Experimental Setup

First, an objective function is defined including, on the one hand, the cost associated with the lack of service of the arcs of the graph (coming from the previous proposed formulation) and, on the other hand, the cost associated with the means used to perform the service (number and capacity of the available agents and the number of depots included in the graph). This new cost function implies the search for a trade-off among both cost sources. The cost of each configuration is based in a cost per agent and cost per unit of capacity per agent and cost per depot estimated from real cases of a salt spreading scenario.

The experiment starts from the generation of a DRG that has the required generalisability to evaluate the range of different scenarios considered. Using the results of the previous section, this DRG has been trained with those 3 instances that showed the highest richness [R1, R3 and R4]. Additionally, as the number of depots will also be a parameter to optimise, the DRG has been trained with variations of the selected graphs with 1, 3 and 5 depots. The training of the DRG will be carried out using the previously selected SADE-ttb-bt variant and a population size of 600 individuals.

The meta-optimisation process will be carried out on one of the 5 graphs of the studied repository [R5] and the ranges of the studied parameters will be: agents (1–10), capacity per agent (1100–6600 l), number of depots (1–5).

The design process has been carried out in two stages. In the first stage, the position of the depots for each number of depots (1–5) will be optimised by means of an evolutionary process based on a specific recombination operator for depot configurations and a mutation operator that moves the nodes to contiguous nodes. This process will be carried out independently for each number of depots (from 1 to 5), obtaining the most suitable configuration for each case. In a second stage, the optimal size and capacity of the fleet and the optimal number of depots has been studied through an exhaustive search process.

6.2   Results

Figure 3 shows the best configurations obtained for each number of depots. In it we can see the tendency of the algorithm to select depot locations in areas with a high level of interconnection of the graph, although in many cases this means that they are less homogeneously distributed. On the other hand, there is hardly any recurrent selection of nodes for different numbers of depots, so that the suitability of a node to act as a depot is more dependent on the overall configuration of the graph than on its own characteristics.

Fig. 3. Optimized depot layouts TCAPR-R5

As for the second stage of the meta-optimisation, the results are displayed in five heat maps (Fig. 4) showing the final cost of each configuration.

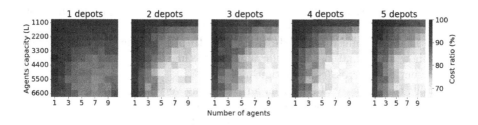

Fig. 4. Cost distribution by scenario TCARP-R5

The results obtained show a considerable sensitivity to the number of depots. This sensitivity is especially clear after the first additional depot is included and becomes less and less relevant as the number of depots increases. Regarding the number of agents and their capacity, we observe, independently of the number of depots, a clear increase in cost for lower values of both parameters (due to the inability to serve the graph arcs conveniently). On the other hand, from above one value, the cost increase associated with these two parameters is not compensated by the service improvement, showing a trade-off solution located, in all configurations, close to the maximum capacity and number of agents.

As an overall result, the meta-optimisation process generates a 19.9 % reduction in the cost generated with respect to the average of the different configurations evaluated and a 30.8 % reduction compared to not carrying out any service process.

7   Conclusion

This paper introduces a new method (DDRGen) for the generation of solutions for open-ended versions of arc routing problems based on dynamic costs. The method uses a neural network which, as part of a deliberative-reactive model determines in real time the next arc to be served. The method is optimized by a differential evolution.

The method is first validated by solving the GDB benchmark to verify that the method is a competitive alternative to the state of the art approaches. The comparison shows that for different variations of the problem DDRGen obtains an average relative error below 5 %, which is particularly remarkable considering that the static benchmark problem is not a particularly suitable environment for the performance of reactive methods.

A sensitivity analysis of the parameters of the has been performed. For the optimiser, a strong influence of the population size and the variant of the DE algorithm used is observed. As for the neural decision maker, the error obtained is very sensitive to the dimensionality of the network and the optimal number of parameters falls in the range of [20–50], also the relevance of the information provided to the network has been evaluated by testing different combinations of inputs. The results have shown the huge relevance of the information of the cost of each arc (up to 20 times more error).

DDRGen has been trained and tested with the proposed open formulation in larger instances, and subsequently cross-evaluated to solve graphs other than those used for training to validate the ability to produce general solvers. The results show a minimal loss in performance, less than 10% percent deviation in 95% percent of the cross combinations.

Finally, the capacity of DDRGen for producing very fast solutions to similar graphs once it is trained is used as part of a design tool to optimize service scenario of one of the studied networks (TCARP-R5), reaching a compromise solution between the cost associated with the lack of service of the arcs and the cost of the infrastructure necessary for such service, achieving a reduction of 19.9% of the cost, compared to the average of the evaluated scenarios and 30.8% compared to the cost of not performing any service.

Acknowledgements. The authors wish to acknowledge the support received from the CITIC research center, funded by Xunta de Galicia and European Regional Development Fund by grant ED431G 2019/01.

References

1. Brest, J., Greiner, S., Bošković, B., Mernik, M., Zumer, V.: Self-adapting control parameters in differential evolution: a comparative study on numerical benchmark problems. IEEE Trans. Evol. Comput. **10**, 646–657 (2006). https://doi.org/10.1109/TEVC.2006.872133
2. Corberán, Á., Eglese, R., Hasle, G., Plana, I., Sanchis, J.M.: Arc routing problems: a review of the past, present, and future. Networks **77**, 88–115 (2021). https://doi.org/10.1002/net.21965
3. Golden, B.L., Dearmon, J.S., Baker, E.K.: Computational experiments with algorithms for a class of routing problems. Comput. Oper. Res. **10**, 47–59 (1983). https://doi.org/10.1016/0305-0548(83)90026-6
4. Gáspár, L., Bencze, Z.: Salting route optimization in Hungary. Trans. Res. Procedia **14**, 2421–2430 (2016)
5. Harris, N., Liu, S., Louis, S.J., La, J.H.: A genetic algorithm for multi-robot routing in automated bridge inspection. In: GECCO 2019 Companion - Proceedings of the 2019 Genetic and Evolutionary Computation Conference Companion, pp. 369–370 (2019). https://doi.org/10.1145/3319619.3321917
6. Iori, M., Novellani, S.: Optimizing the nozzle path in the 3D printing process. In: Rizzi, C., Andrisano, A.O., Leali, F., Gherardini, F., Pini, F., Vergnano, A. (eds.) ADM 2019. LNME, pp. 912–924. Springer, Cham (2020). https://doi.org/10.1007/978-3-030-31154-4_78
7. Jin, X., Qin, H., Zhang, Z., Zhou, M., Wang, J.: Planning of garbage collection service: an arc-routing problem with time-dependent penalty cost. IEEE Trans. Intell. Transp. Syst. 1–14 (2020). https://doi.org/10.1109/tits.2020.2973806
8. Liu, Y., Shi, J., Liu, Z., Huang, J., Zhou, T.: Two-layer routing for high-voltage powerline inspection by cooperated ground vehicle and drone. Energies **12**, 1385 (2019). https://doi.org/10.3390/en12071385
9. Qin, A.K., Huang, V.L., Suganthan, P.N.: Differential evolution algorithm with strategy adaptation for global numerical optimization. IEEE Trans. Evol. Comput. **13**, 398–417 (2009). https://doi.org/10.1109/TEVC.2008.927706
10. Storn, R., Price, K.: Differential evolution - a simple and efficient heuristic for global optimization over continuous spaces. J. Glob. Optim. **11**, 341–359 (1997). https://doi.org/10.1023/a:1008202821328
11. Tirkolaee, E.B., Mahdavi, I., Esfahani, M.M.S., Weber, G.W.: A hybrid augmented ant colony optimization for the multi-trip capacitated arc routing problem under fuzzy demands for urban solid waste management. Waste Manage. Res. **38**, 156–172 (2020). https://doi.org/10.1177/0734242X19865782
12. Tong, H., Minku, L.L., Menzel, S., Sendhoff, B., Yao, X.: A novel generalised meta-heuristic framework for dynamic capacitated arc routing problems. arXiv preprint arXiv:2104.06585 (2021)
13. Yu, M., Jin, X., Zhang, Z., Qin, H., Lai, Q.: The split-delivery mixed capacitated arc-routing problem: applications and a forest-based tabu search approach. Transp. Res. Part E: Logistics Transp. Rev. **132**, 141–162 (2019)

HAIs Applications

SHAP Algorithm for Healthcare Data Classification

Samson Mihirette[(⊠)] and Qing Tan

School of Computing and Information Systems, Athabasca University, 1 University Dr, Athabasca, AB T9S3A3, Canada
smihirette1@athabasca.edu, qingt@athabascau.ca

Abstract. To strengthen the healthcare data privacy protecting techniques and ensure the transparency of healthcare data exchange, many data privacy-preserving methods have been introduced. This paper highlights privacy concerns and introduces techniques and research directions towards data privacy in Healthcare Information Systems (HIS). The paper demonstrates the use and the power of the Shapley Additive exPlanations (SHAP) algorithm to identify and classify critical data elements that can put personal privacy at risk within a dataset. A conceptual patient-centric healthcare information system architecture with a data broker is proposed in this paper. The proposed architecture also includes the privacy broker that leverages application programming interface services and integration middleware in safeguarding healthcare data privacy.

Keywords: Data privacy · Data privacy broker · Healthcare Information System (HIS) · SHAP

1 Introduction

Data privacy has deep historical roots and has been discussed and explained among philosophers, sociologists, psychologists, and legal scholars. They all agree that there should be a limit for authorities, governments, or private sectors to use citizens' data. Citizens have the right to know who can access their private or personal information, and for what purpose. With the development of information technology - especially social media, centralized information systems, and surveillance platforms, Big Data Analytics, and data privacy protection has become an urgent issue. Healthcare data is one of the most sensitive and private information type of citizens. It is usually created, transferred, and stored in centralized information systems, Healthcare Information Systems (HIS).

A recent example of the concern on healthcare data privacy is the strategy of using citizens' data to track and trace COVID-19 infections and transmissions. Although leveraging every possible angle in controlling pandemics is beneficial towards the safety and health of the public, and for effective health policy decisions during outbreaks, citizens' healthcare data needs to be accessed in an ethical and secured manner [1].

This paper introduces the use of the SHAP algorithms to identify and classify critical data elements that are collected as a part of personal healthcare information that can put

© Springer Nature Switzerland AG 2022
P. García Bringas et al. (Eds.): HAIS 2022, LNAI 13469, pp. 363–374, 2022.
https://doi.org/10.1007/978-3-031-15471-3_31

personal privacy at risk within a dataset. The identified data elements by the algorithm will be used as an input for a privacy broker to safeguard healthcare data privacy in the HIS and create awareness of the movement of data in a transparent and timely manner.

2 Related Work

A significant number of research papers focus on the following topics or methods in mitigating the risk of healthcare data breaches and creating a healthcare information system with a strong data privacy management system:

- Encryption and decryption methods (AES-256 or RSA)
- Masking, de-identification, randomization, and pseudonymization
- Service level or data-sharing agreement between stakeholders
- Adopting blockchain technology for healthcare information system
- Data perturbation methods and Differential privacy methods
- Detect privacy violations and eliminate inferences.

In their research paper, Sagar Sharma, Keke, Chen, and Amit Sheth discuss the modern healthcare system and its progress towards practical privacy-preserving analytics for IoT and cloud-based healthcare systems [2]. The researchers based their study on kHealth – a personalized digital healthcare information system that is developed and tested for disease monitoring. They discuss potential trade-offs among privacy, efficiency, and model quality. The researchers propose solutions based on the cost of each method. The suggested methods are privacy-preserving computation on untrusted platforms – fully homomorphic encryption (FHE) and Yao's garbled circuit (GC together with oblivious transfer (OT). The researchers indicate that ideal data privacy is almost unachievable throughout the storage, processing, and communication phases. They discuss the potential of AES-256 encryption methods to protect data from potential adversaries that could seriously jeopardize computations and services in providing useful analytics data. Although simpler anonymization techniques can better demonstrate the data use from a modelling or analytics perspective, it is somehow inadequate against data privacy breaches [2]. The researchers also discuss the use of third-party methods such as homomorphic encryption schemes, data perturbation, and differential privacy techniques to mitigate data privacy risks.

Researchers Kingsford Kissi Mireku, Fengli Zhang and Komlan Gbongli, discuss the importance of healthcare data privacy in their research paper "Patient Knowledge and Data Privacy in Healthcare Records System," which focuses on the consent of patients when accessing healthcare data. They used Pearson Correlation and linear regression analysis techniques to determine the relationship of the knowledge with data privacy [3]. In their research, they refer to, "anonymization algorithms such as k-Anonymity, l-diversity, t-closeness mostly deal with the removal of identifiable information in data mining is widely accepted in privacy preservation but unfortunately, the anonymization algorithms do not completely preserve privacy." [3]. Kah Meng Chong presents a compressive survey of the state-of-the-art privacy enhancing methods that ensure a secure data sharing environment, and proposed schemes based on data anonymization and differential privacy approaches in the protection of healthcare data privacy and highlighted the practical strengths and limitations of these two technologies. The paper concludes by suggesting the development of standardization of privacy protection [4].

Researchers Ibrar Yaqoob, Khaled Salah, present projects and case studies to show the practicality of blockchain technology for healthcare applications. They discuss advantages of adopting blockchain technologies for healthcare applications and identify important challenges hindering the successful adoption of blockchain for the healthcare sector. In their conclusion, they point out that leveraging blockchain technology for integration of healthcare systems can lead to some technical challenges such as: blockchain immaturity, scalability and interoperability [5]. Researchers Sascha Welten, Yongli Mou, present the implementation of the Personal Health Train, which is a paradigm that uses distributed analyses methods to protect data and prevent potential privacy breaches. They present an architecture design that has multiple components including monitoring, privacy, and security components. They compare their solution with other privacy enhancing techniques. The main contribution of their work is adding another layer of security in the infrastructure design using learning paradigms which has been the sequential approach (incremental learning.) The asynchronous learning scheme involves an additional layer of security when accessing information [6]. Researchers Amin Aminifar and Martin Shkri present the privacy-preserving distributed using extremely randomized trees algorithm for learning without privacy concerns in the healthcare domain. They evaluate and propose a framework with algorithm extensively using on two popular structured healthcare datasets and two mental health datasets associated with the Norwegian INTROducing Mental health through Adaptive Technology (INTROMAT) project. They conclude by suggesting the possibility of extending the proposed framework [7].

3   Characteristics of Healthcare Data and Its Privacy Sensitivity

Although some data anonymization methods such as data masking, encryption, and tokenization can mitigate the risk of compromising data privacy in any type of infrastructure whether it is a cloud or on-premises platform, patients and all healthcare data stakeholders should have a mechanism to monitor and control the access logs of healthcare data at any given time. The goal of data anonymization and data masking evolves in removing and encrypting personally identifiable and leak of private information but depending on the usage it can reduce the insight of meaningful information. Data masking, and anonymization methods are very useful in providing data access to authorized parties through a mechanism where there is a possibility of guarding or hiding sensitive data from being exposed by creating a version of data that cannot be identified. However, there are situations where the real data is needed without masking. For example, data masking or anonymization can distort health research findings. Day encryption is also extremely important when sending data to the right party. In situations where data is lost during transfer or data at rest, the data cannot be exposed without decryption keys. There is also a risk that unauthorized users can also use de-identification processes inappropriately. To mitigate risk there are some academic researches, for example, the proposal of data anonymization evaluation framework that aims at measuring data utility, privacy metrics, and information loss were used to evaluate data anonymization schemes that can further optimise the data anonymization process to keep risks of re-identification low [8].

As Yuval Noah Harari said - "If corporations and governments start harvesting our biometric data en masse they can… Not just predict our feelings but also manipulate our feelings and sell us anything they want – be it a product or a politician." [9]. Therefore, we need to be mindful of the privacy of citizens' healthcare data.

In this research paper, to enhance the healthcare information system by including a data privacy-preserving component, we introduce the use of SHAP (**SHapley Additive exPlanations**) algorithms to identify and classify important data elements that are very useful for protecting data privacy.

4   Classification of Data Using the SHAP Algorithm

4.1   The SHAP (Shapley Additive Explanation)

SHAP - which stands for SHaply Additive Explanations is a state of art in Machine Learning Explainability. The algorithm is mainly used to understand data elements, features, or variables' contribution towards a prediction or models. SHAP is a game theoretic approach to explain the output of any machine learning model. Machine learning (ML) is a domain within the field of artificial intelligence that leverages algorithms that can provide accurate predictions or models without explicit programming. The fundamental premise of machine learning is to prepare algorithms that can get data and use statistical analysis to predict an outcome while updating the prediction as new data is found.

In this research, by using SHAP algorithms, we demonstrate how to identify the relative importance of each variable in the dataset to the outcome of the model or prediction on healthcare data. SHAP is an interpretability method based on Shapley values and was introduced by Lundberg and Lee [10] to explain individual predictions of any machine learning model. Shapley Values (SHAP) is a method used in game theory to determine how much each player in a collaborative game has contributed to the result of the game. In other words, the Shapley value is the average of all marginal contributions to all possible coalitions or combinations. In our research, Shapley value is used to measure or quantify how much each data element within a healthcare dataset contributes to the final model.

The influence of each data element in the dataset is represented by Shapley values, which reflect the privacy sensitivity of data elements within the dataset. Using SHAP values, we can classify the data elements and identify the critical data elements having high risk in terms of privacy. Once we identify the important data elements that are useful for data privacy protection, the next step is to ensure that we provide a systematic and integrated approach to protect the privacy of the data.

Within Machine learning (ML), Principal Component Analysis (PCA) is a commonly used algorithm to reduce dimensions of a dataset. PCA helps identify relationships among different variables then coupling. Although there are some similarities in terms of reducing variables in a dataset, the use of SHAP in this paper is mainly for identifying the contribution of each variable in a model. The SHAP value will give us the contribution of each individual variable in the dataset towards the final model or prediction. For example, the model can display, what is the contribution of a variable "Income" towards the model that shows the COVID-19 affection rate in each community in a dataset that has a number of variables including "income."

4.2 SHAP for Data Element Classification

In this paper, the model is performed using data science tools for ML prediction and modelling tools (to what degree negative or positive) - SHAP and CATBOOST algorithms [11]. In this experiment, we are explaining the contribution of data elements for a healthcare dataset towards the prediction.

We have conducted multiple experiments using various data sources to verify the result of the SHAP algorithm. The healthcare data source used in this paper comes from the "COVID-19 Community Vulnerability Crosswalk – Crosswalk by Census Tract" provided at the US healthcare research opensource data set. (*COVID-19 Community Vulnerability Crosswalk - Crosswalk by Census Tract\HealthData.Gov*, n.d.) [12]. This healthcare data set is real data collected by the US health officials and is accessible for researchers in the official government website. The data set is used to create a model on which community group is vulnerable to COVID-19 and to identify which community group is hardly hit by COVID-19. The dataset consists of a list of US residents, their state, zip code, income group and other information. The goal of our experiment is which data elements are crucial for the model – ***the prediction of highly COVID-19 affected in the states***. In this paper we are using this dataset to demonstrate how the dataset can be used for predictions and explain how each individual data element contributes to the prediction using SHAP values. For example, to apply health related policy decisions, the dataset can be used to predict the vulnerability of a group of communities being affected by COVID-19. Each record in the dataset consists of nineteen data elements. For this research experiment, to demonstrate how the SHAP algorithm works, we used multiple groups of data elements to measure the contribution to a model. First, we identified data elements that can be useful and significantly affect healthcare policies. We identified the following set of data elements for the SHAP experiment. We defined set A with a member of data elements for the experiment. Each element in set A will be measured in SHAP values in the model it contributes (model is to determine if a particular community is vulnerable towards COVID-19.)

$$A \in \{\text{Low}-\text{income area score, Total score, Low Income Area (Poverty percentage), Hard}-\text{Heat}-\text{Area-Score}\}$$

We established Set A - a group of data elements for our experiment to explain and demonstrate how SHAP works and how much each element contributed to the model. Once we understand the importance of each data element towards the prediction, we will determine which data element is a candidate for data privacy. For example, if an element - Low-income area score, is identified with a high SHAP value, for a given prediction, then we need to determine if data privacy rules can be applied to the data element. We have conducted steps to make the application read the dataset and product models. CATBOOST algorithm [11] is used to handle the categories of data elements. We identified only the following variables – ['Total-Score', 'Max Possible Score', 'Hard-Heat-Area-Score', 'Low-Income-Area-Score', 'Low-Income Area (LIA) Census Tract (Poverty Percentage)', 'Rural', 'Rural - Score']. We dropped data elements that we don't need for the demonstration. This plotting technique of SHAP simplifies the model to understand the influence of each element in the model. It is a non-linear regression model based on classifiers. The data set used in this experiment is to demonstrate how a model can be used to influence health related policies and other measures that can affect healthcare related support assistance given to a given community. As a result of this model, health policies can be established based on the vulnerability of the low-income communities. The question we need to ask is - are stakeholders aware of the decision that was taken because of this model. This data set consists of the income earned in each community which can expose the privacy of the respective individuals. Therefore, SHAP values can help enhance the data privacy protection measures in terms of data classification. With these SHAP values, once we determine the importance of each data element towards a prediction model (in this example COVID-19 venerability model), we can make sure that we apply data privacy protection mechanisms. When elements with a substantial value in determining the outcome (model) are used by authorities or any other public policy organizations, data privacy mechanism will help document and notify the appropriate stakeholders.

To summarize, in this dataset, 'Low-Income Area-Score', the 'Total-Score' is the most important variable. Therefore, in combination with other elements in the dataset Total-Score' related data has a high SHAP value (high in red and low in blue) with a significant impact on the model. The following SHAP summary plot diagram provides information for the selected variables and the impact of each variable on the model (Fig. 1).

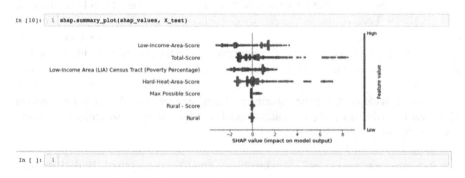

Fig. 1. SHAP values for each player (data element)

The y-axis indicates – the variable name (data element), in order of importance (influence) from top to bottom.

The x-axis (SHAP value) quantifies the contribution of each data element, it can either be a positive or negative value towards the model. Additionally, each plot in the graph is made of all the dots in the train data. This summary diagram demonstrates:

- Data element importance: The horizontal location shows whether the effect of that value is associated with a higher or lower prediction.
- Impact: The horizontal location shows whether that variable is high (in red) or low (in blue) for that observation.
- Original value: Color shows whether that variable is high (in red) or low (in blue) for that observation.

– Correlation: A high level of the "Total-Score" content has a high and positive impact on the quality rating. The "high" comes from the red color, and the "positive" impact is shown on the X-axis.

By identifying the SHAP values for each data element, we are quantifying the contribution of each element to the final prediction. In the demonstration, we provided in this paper, the "Low Income Area Score" has the highest SHAP value as compared to the rest of the seven data elements in the SHAP summary diagram.

1. Blue indicates that, if a community group is classified as a Low-income group in this tribal community data, according to this data, the vulnerability to be exposed to the COVID-19 was low. Note: - In the dataset, tribal communities are located outside urban areas.
2. The sequence of the data elements from top to down indicates that, the contribution of each element to the final prediction. In this SHAP summary, Low-Income-Area-Score has a very important influence on the result of the model as compared to the other data elements.
3. The x-axis, which indicates the SHAP value impact on the model provides a positive or negative influence. The more you go to the right, it will have the more positive outcome. In this example, Low-Income-Area score has a positive, meaning, if a community is residing in a tribal community, the community will not be vulnerable.

Therefore, as indicated in the SHAP summary diagram, we have demonstrated that income has a significant factor towards the vulnerability prediction. This dataset is used as an example to demonstrate how we can use to quantify the relative importance of data elements towards any model. This data set consists of address information of community members and the result can be used for a different purpose therefore data privacy preserving techniques should be applied for those classified data elements. This method of identifying and classifying privacy sensitive data elements can be used for different highly sensitive datasets.

5 Research Design

Once we classify data elements by using SHAP algorithms, we need to implement data privacy protecting mechanisms. One of the options is to introduce a middleware or a data privacy broker. In this paper, we propose a conceptual Secured Data Privacy broker as a middleware and integration component in the healthcare information system architecture to enrich the existing data privacy-preserving techniques, such as encryption, anonymization, differential privacy, and others. The foundation of this proposed HIS architecture is based on a service-oriented architecture that utilizes the application programming interfaces (API) extensively. Using APIs, forward and reverse proxies, we proposed a middleware integration solution that will help enhance the healthcare data privacy capability concerning the cloud environment. This integrated architecture is implemented and will be used among healthcare providers, cloud providers, patients, and other healthcare data stakeholders (Fig. 2).

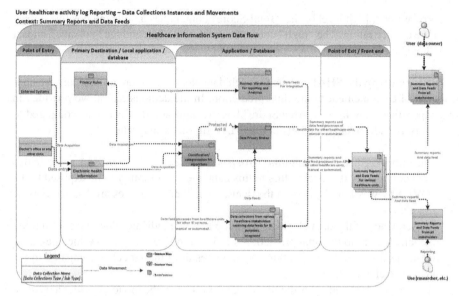

User healthcare activity log Reporting – Data Collections Instances and Movements
Context: Summary Reports and Data Feeds

Fig. 2. Conceptual architecture, data collection instances and movements

5.1   Architecture – Data Movement

Point of Entry (Internal and External Sources)
Internal Sources:
Data entry by clinic receptionist over the phone, online or at the doctor's office. Information such as name, last name, age, health care information of the patient and other similar information.

External Sources:
Clinical data, diagnostic imaging, laboratory results and other health-related information from laboratories, Machine-generated or sensor data such as data from devices monitoring health data and healthcare IoTs.

Primary and Secondary Destination
Electronic healthcare applications reside locally on the computers in the doctor's office. Secondary destinations - Centralized healthcare information system infrastructure, healthcare research centers, universities, government agencies, hospitals, and other third-party healthcare information stakeholders.

Data Classification
The machine learning algorithm categorizes the data into three classifications:

Class A – Data that should be protected and need to pass through the data privacy broker so that all stakeholders are aware when it is accessed by anyone.
Class B – Data with a medium impact on privacy issues but it needs to pass through the data privacy broker if it is not encrypted.

Class C – Data with no impact on privacy therefore these data are sharable with other stakeholders. No need to let Class C data pass through a data privacy broker because if the data is exposed to any stakeholder, it will not expose personal information.

Data Privacy Broker

- A middleware used to make sure that data access is controlled and monitored by authorized body and with proper data classification rules determined by SHAP algorithm. Data movement to a patient in the form of notification or when accessed by the user through a data privacy dashboard. Central healthcare portal system with user login capability providing all healthcare access log information.
- Any data which is vetted by the SHAP algorithm as protected data, or 'Class A' passes through the data privacy broker. The data privacy broker also provides a middleware component therefore it is a health integration access layer that uses API services.

5.2   Architecture – High Level Design

Integration – Mapping
The objective of Integration mapping is to take data format and translate it to a new format to make sure there exists a compatible messaging or data understanding between two or more parties. Data element or database table field filtering to make sure the destination application received the requested and authorized data only. Scripting functionalities can be included if robust messaging or communication is needed. Scripts such as XSL or CSS can be used to package the message (Fig. 3).

Data Privacy Broker (Explained in Sect. 5.1)

Central Advanced Adopter

- Adapter Engine contains a set of communication formats or adapters or protocols such as SOAP, HTTP, SFTP, and FTP

Integration Directory
Integration directory provides answers to multiple questions – information on the receiver and the sender. Are there multiple receivers? What is the mapping program to transform the sender message to the target message format?

Metadata Information Repository
Metadata repository contains useful information that can be leveraged to enable security and governance, data recommendations, and user telemetry. This repository stores technical and operational metadata that allows the logical mapping of physical data stores

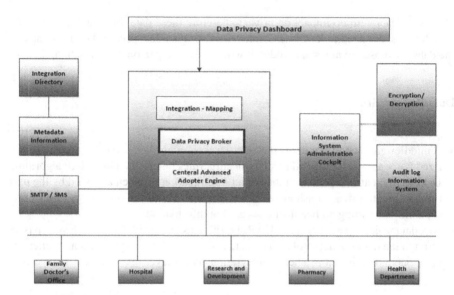

Fig. 3. Healthcare information system including the data privacy broker

to a more application-neutral model. This neutral model assists in representing common keys with different names. Look-up tables to translate keys between systems can also be part of the metadata repository. The translation from a neutral term to a specific data detail enables query planning and execution. Metadata also forms the foundation of data fabric capabilities, which offer AI/ML recommendation engines for various purposes, including data recommendations and performance optimization recommendations.

SMTP/SMS Services
Simple Mail Transfer Protocol (SMTP) is used for sending and receiving emails and SMS is a text messaging solution used on smartphones or any text messaging applications. Both SMTP/SMS services will be used to notify important information to the patient or any stakeholder that need-to-know data access by any party.

Encryption/Decryption
The encryption/decryption component of the healthcare infrastructure is responsible for any type of high-level secured communication between applications.

Audit Log Information System
All access logs will be captured by the audit log capturing component of the infrastructure. It will capture who requested the private data, the date and time the data is accessed, and other information.

Data Privacy Dashboard
This is an element of the healthcare information system that provides users and all other healthcare stakeholders front end services. As a point of entry portal system for any authorized user including the patient, it provides basic information such as a data access activity log for a given period.

6 Conclusion and Future Work

In this research paper, we delivered multiple state-of-the-art contributions to the health-care information system concerning the healthcare data privacy-preserving techniques – a) the use of SHAP for identifying and classifying data elements for a dataset based on their impact in models and b) conceptual architecture that includes the data privacy broker component that provides transparent monitoring and data access awareness solution to all stakeholders including the patient who owns the data.

As technology, healthcare monitoring devices, smartphones, mobile devices and IoTs, continue to grow substantially, further research should continue in keeping the security of data and preserving the privacy of healthcare data. The research needs to close the privacy-preserving gap and the concerns of organizations and users. Several scenarios of the concern include methods in collecting and retaining sensitive personal information; processing personal information in environments, such as the cloud; and information sharing mechanisms, authentication mechanisms, encryption technologies, data masking and other privacy-protecting techniques.

Ultimately, the challenge lies in balancing the concerns in healthcare data privacy and the consumption of healthcare data for the good of society by healthcare policymakers, healthcare providers, and researchers in academic researchers.

References

1. Mihirette, S.: Safeguarding Patients' Healthcare Data: Introducing Data Privacy Broker, 07 January 2022. https://dt.athabascau.ca/jspui/handle/10791/366. Accessed 15 Jan 2022
2. Sharma, S., Chen, K., Sheth, A.: Toward practical privacy-preserving analytics for IoT and cloud-based healthcare systems. IEEE Internet Comput. **22**(2), 42–51 (2018). https://doi.org/10.1109/MIC.2018.112102519
3. Kissi Mireku, K., Zhang, F., Komlan, G.: Patient knowledge and data privacy in healthcare records system. In: 2017 2nd International Conference on Communication Systems, Computing and IT Applications (CSCITA), pp. 154–159, April 2017. https://doi.org/10.1109/CSCITA.2017.8066543
4. Chong, K.M.: Privacy-preserving healthcare informatics: a review. ITM Web Conf. **36**, 04005 (2021). https://doi.org/10.1051/itmconf/20213604005
5. Yaqoob et al.: Blockchain for healthcare data management: opportunities, challenges, and future recommendations (2021). https://link.springer.com/content/pdf/10.1007/s00521-020-05519-w.pdf. Accessed 18 June 2022
6. Welten, S., et al.: A privacy-preserving distributed analytics platform for health care data. Methods Inf. Med. (2022). https://doi.org/10.1055/s-0041-1740564
7. Aminifar, A., Shokri, M., Rabbi, F., Pun, V.K.I., Lamo, Y.: Extremely randomized trees with privacy preservation for distributed structured health data. IEEE Access **10**, 6010–6027 (2022). https://doi.org/10.1109/ACCESS.2022.3141709
8. Ni, C., Cang, L.S., Gope, P., Min, G.: Data anonymization evaluation for big data and IoT environment. Inf. Sci. **605**, 381–392 (2022). https://doi.org/10.1016/j.ins.2022.05.040
9. Frej, M.B.H., Dichter, J., Gupta, N.: Comparison of privacy-preserving models based on a third-party auditor in cloud computing. In: 2019 IEEE Cloud Summit, Washington, DC, USA, pp. 86–91, August 2019. https://doi.org/10.1109/CloudSummit47114.2019.00020

10. Lundberg, S.M., Lee, S.-I.: A unified approach to interpreting model predictions. In: Advances in Neural Information Processing Systems, vol. 30 (2017). https://proceedings.neurips.cc/paper/2017/hash/8a20a8621978632d76c43dfd28b67767-Abstract.html. Accessed 15 Oct 2021

11. CatBoostClassifier. https://catboost.ai/docs/concepts/python-reference_catboostclassifier. Accessed 11 Dec 2021

12. COVID-19 Community Vulnerability Crosswalk - Crosswalk by Census Tract—HealthData.gov. https://healthdata.gov/Health/COVID-19-Community-Vulnerability-Crosswalk-Crosswa/x2y5-9muu. Accessed 11 Dec 2021

Assessment of Creditworthiness Models Privacy-Preserving Training with Synthetic Data

Ricardo Muñoz-Cancino[1], Cristián Bravo[2], Sebastián A. Ríos[1], and Manuel Graña[3(✉)]

[1] Business Intelligence Research Center (CEINE),
Industrial Engineering Department, University of Chile,
Beauchef 851, 8370456 Santiago, Chile
[2] Department of Statistical and Actuarial Sciences,
The University of Western Ontario, 1151 Richmond Street,
London, Ontario N6A 3K7, Canada
[3] Computational Intelligence Group, University of Basque Country,
20018 San Sebastián, Spain
manuelgrana@ehu.es

Abstract. Credit scoring models are the primary instrument used by financial institutions to manage credit risk. The scarcity of research on behavioral scoring is due to the difficult data access. Financial institutions have to maintain the privacy and security of borrowers' information refrain them from collaborating in research initiatives. In this work, we present a methodology that allows us to evaluate the performance of models trained with synthetic data when they are applied to real-world data. Our results show that synthetic data quality is increasingly poor when the number of attributes increases. However, creditworthiness assessment models trained with synthetic data show a reduction of 3% of AUC and 6% of KS when compared with models trained with real data. These results have a significant impact since they encourage credit risk investigation from synthetic data, making it possible to maintain borrowers' privacy and to address problems that until now have been hampered by the availability of information.

Keywords: Credit scoring · Synthetic data · Generative adversarial networks · Variational autoencoders

1 Introduction

For decades financial institutions have used mathematical models to determine borrowers' creditworthiness and consequently manage credit risk. The main objective of these models is to characterize each borrower with the probability of not complying with their contractual obligations [24], avoiding to give loans to applicants that will not be able to pay them back. Despite all the years

P. García Bringas et al. (Eds.): HAIS 2022, LNAI 13469, pp. 375–384, 2022.
https://doi.org/10.1007/978-3-031-15471-3_32

of research on credit scoring, there is still little done on behavioral scoring models, which are the credit scoring models used on those clients who have already been granted a loan, because it requires large volumes of data and a relevant historical depth [8,14]. In addition, financial institutions are often reluctant to collaborate in this type of investigation due to concerns about data security and personal privacy. Until now, the use of synthetic data in credit scoring is mainly restricted to balancing the minority class in classification problems using the traditional SMOTE [7], variational autoencoders [26], and lately generative adversarial networks [4,16,20]. In these studies, synthetic records of the minority class are generated, and the original data set is augmented.

In this paper, we present a framework that allows us to train a model on synthetic data and then apply it to real-world data. We also analyze if the model copes with data drift by applying both models to real-world data representing the same problem but obtaining the dataset one year later.

The main findings of our work are:

- It is possible to train a model on synthetic data that achieves good performance in real situations.
- As the number of features increases, the synthesized data quality gets worse.
- There is a performance cost for working in a privacy-preserving environment. This cost corresponds to a loss of predictive power of approximately 3% if measured in AUC and 6% in KS.

2   Related Work

2.1   Credit Scoring

Credit scoring aims to manage credit risk, defined as the potential for a borrower to default on established contractual obligations [24]. These models intensively use borrower data, demographic information, payment behavior, and even alternative data sources such as social networks [18,21], psychometrics [3], and geolocation [23].

2.2   Generative Models for Synthetic Data Generation

Generative models are a subset of machine learning models whose main objective is to learn the real-data distribution and then to generate consistent samples from the learned distribution. Working with synthetic data allows addressing problems where real-data is expensive to obtain, where a large dataset is needed to train a model, or where the real-data is sensitive or cannot be shared [25]. For years, statistical methods were the most used ones to estimate the real-world data joint distribution. In this group, Gaussian Mixture Models are the most utilized for this task when there are fewer continuous variables. At the same time, Bayesian Networks are commonly used for discrete variables. The main problem of these methods is dealing with datasets containing numerical and categorical variables. They also present problems when the continuous variables have more than one mode and the categorical variables present small categories [27]. During the last years, deep learning models have gained popularity to generate synthetic data due to their performance and because they allow us to deal with the problems mentioned above. The generative adversarial networks and the variational autoencoders stand out within these models.

Generative Adversarial Networks. Generative adversarial networks are a deep learning framework based on a game theory scenario where a generator network $\mathcal{G}(\cdot)$ must compete with a discriminator network $\mathcal{D}(\cdot)$. The generator network produces samples of synthetic data that attempt to emulate real data. In contrast, the discriminator network aims to differentiate between real examples from the training dataset and synthetic samples obtained from the generator [9].

Its most basic form, vanilla GAN, $\mathcal{G}(\cdot)$ maps a vector z from a multivariate Gaussian distribution $\mathcal{N}(\mathbf{0}, \mathbf{I})$ to a vector $\hat{x}$ in the data domain X. While $\mathcal{D}(\cdot)$ outputs a probability that indicates whether $\hat{x}$ is a real training samples or a fake sample drawn from $\mathcal{G}(\cdot)$ [27]. The generator $\mathcal{G}(\cdot)$ and the discriminator $\mathcal{D}(\cdot)$ are alternatively optimized to train a GAN. Vanilla GANs have two main problems, representing unbalanced categorical features and expressing numerical features having multiple modes. To solve this, Xu et al. (2019) [28] present a conditional generator (CTGAN) that samples records from each category according to the log-frequency; this way, the generator can explore all discrete values. Moreover, the multimodal distributions are handled using kernel density estimation to assess the number of modes in each numerical feature.

Variational Autoencoders. Autoencoders (AE) are an unsupervised machine learning method that enables two main objectives: low-dimensional representation and synthetic data generation. Variational Autoencoder [15] interpret the latent space produced by the encoder as a probability distribution modeling the training samples as independent random variables, assuming the posterior distribution defined by the encoder $q_\theta(z|x)$ and generative distribution $p_\phi(x|z)$ defined by the decoder. To accomplish that the encoder produces two vectors as output, one of means and the other of standard deviations, which are the parameters to be optimized in the model. Xu et al. (2019) [28] present TVAE, a variational autoencoder adaption for tabular data, using the same pre-processing as in CTGAN and the evidence lower bound (ELBO) loss.

3 Methodology and Experimental Design

3.1 Dataset

In this work, we use a dataset provided by a financial institution already used for research on credit scoring [18,19]. This dataset includes each borrower financial information and social interactions features over two periods: January 2018 and January 2019; each dataset contains 500,000 individuals. Each borrower is labeled based on their payment behavior in the following 12-month observation period. Each borrower in the 2018 dataset is labeled as a defaulter if it was more than 90 days past due between February 2018 and January 2019 and is labeled as a non-defaulter if it was not more than 90 days past due. Borrowers from the Jan-2019 dataset are similarly tagged. This dataset contains three feature subsets: X_{Fin} corresponds to the borrower's financial information, X_{Degree} corresponds to the number of connections the borrower has in the social interaction network, and X_{SocInt} are the features extracted from the social interactions.

3.2 Synthetic Data Generation

A step to privacy-preserving credit scoring model building is to generate a synthetic dataset that mimics real-world behavior. In order to accomplish this, we compare the performance of two state-of-the-art synthetic data generators, CTGAN and TVAE, defined in Sect. 2. The first experiment (**S01**) only compares these methods using borrowers' features X_{Fin}. The objective of this stage is to find a method to generate synthetic data from real data, and it is not part of this study to find the best way to generate them. Despite not generating an exhaustive search for the best hyper-parameters, we will test two different architectures (Arch) for each synthesizer. Arch A is the default configuration for both methods. In the case of CTGAN, Arch B set up the generator with two linear residual layers and the discriminator with two linear layers, both of size (64, 64). In the case of TVAE Arch B, set hidden layers of (64, 64) for both the encoder and the decoder. Then, in experiment **S02**, we train a new synthesizer using the best architecture from S01. This experiment uses the borrowers' features X_{Fin} and exclusively one feature from the network data, the node degree X_{Degree}. We only include node degree because its feature enables us to reconstruct an entire network using the random graphs generators. Finally, in experiment **S03**, the borrowers and social interaction features $(X_{Fin} + X_{Degree} + X_{SocInt})$ are used to train a synthesizer. This experiment corresponds to the traditional approach to generating synthetic data from a dataset using social interaction features.

3.3 Borrower's Creditworthiness Assessment

The objective of this stage is to have a framework that allows us to estimate the borrower's creditworthiness from a feature set. This modeling framework is based on previous investigations [18,19]. This stage begins by discarding attributes with low or null predictive power and selecting uncorrelated attributes. The correlation-based selection method begins by selecting the attribute with the highest predictive power. It then discards the possible selections if the correlation exceeds a threshold ρ. This step is repeated until no attributes are left to select. To ensure the model generalization capability, we work under a K-fold cross-validation scheme; in this way, the feature selection and the model training use K-1 folds, and the evaluation is carried out with the remaining fold. Additionally, we use two holdout datasets, one generated with information from the same year as the training dataset but not contained. The second contains information from one year later. Both the results of the validation fold and the holdout dataset are stored to use a t-test later to compare different models [5, Ch. 12].

3.4 Evaluation Metrics

In this section, we describe a set of metrics that will help us to evaluate the performance of the synthetic data generators and the classification models used for creditworthiness assessment. The area under the curve (**AUC**) is a performance measure used to evaluate classification models [2]. The AUC is an overall measure of performance that can be interpreted as the average of the true positive rate for all possible values of the false positive rate. A higher AUC indicates a higher overall performance of the classification model [11]. Another classification performance measure is the **F-measure**. This metric is calculated as the harmonic mean between precision and recall. It is beneficial for dichotomous outputs and when there is no preference between maximizing the model's precision or recall [13]. Kolmogorov-Smirnov (**KS**) statistic measures the distance separating two cumulative distributions [12]. The KS statistic ranges between 0 and 1 and is defined as $D = \max_x |F_1(x) - F_2(x)|$, where F_1 and F_2 are two cumulative distributions. In the case of creditworthiness assessment, we are interested in the difference between the cumulative distributions of defaulters and non-defaulters, and a higher D indicates a higher discriminatory power. However, in the case of synthetic data generation, we are interested in the real data distribution and the synthetic data distribution being as similar as possible; in this way, a lower D indicates a better synthetic data generation. In order for all the acceptance criteria to be the same, we define the **KSTest** as $1 - D$; in this way, a higher KSTest indicates a better synthetic data generator. In the synthetic data generation problem, the KS is only valid to measure the performance for continuous features; to handle categorical features, we will use the chi-square test (CS). The CS is a famous test to assess the independence of two events [17]. We will call **CSTest** to the resulting p-value for this test. Therefore a small value indicates we can reject the null hypothesis that synthetic data has the real data distribution. In the synthetic data generation problem, we want to maximize the CSTest.

3.5 Experimental Setup

The parameters of the univariate selection are set at $KS_{min} = 0.01$ and $AUC_{min} = 0.53$, i.e., we discard feature with a univarite performance lower than KS_{min} or AUC_{min}. In the multivariate selection process, we set $\rho = 0.7$ in the process to avoid high correlated features [1]. The N-Fold Cross-Validation stage is carried out considering $N = 10$, and in each iteration, the results of regularized logistic regression and gradient boosting [6] models are displayed.

4 Results and Discussion

In this section, we present the results of our methodology. We start with the implementation details. Then, we compare the synthesizers, and finally, we analyze the creditworthiness assessment performance of the models trained using synthetic data.

4.1 Implementation Details

In this work, we used the Python implementations of Networkx v2.6.3 [10] and
Synthetic Data Vault (SDV) v5.0.0 [22] for networks statistics and synthetic
data generation, respectively. To conduct the experiments, we used a laptop
with 8 CPU cores Intel i7 and 32 GB of RAM.

4.2 Synthetic Data Generation Performance

The first objective is to analyze the performance of the methods to generate
synthetic data presented above, CTGAN and TVAE. Table 1 shows the results
obtained. The features Synthesizer training features corresponds to the training
feature set, while Arq indicates the network architecture defined in Sect. 3.2.
The experiment S01 consisted in comparing both synthesizer using two different
architectures. It is observed that a reduction in the number of layers reduces the
execution times considerably in both cases, being TVAE, the one that presented
the fastest execution times. KSTest show us the performance to synthesize continuous features, where TVAE achieves better performance than CTGAN. The
difference between TVAE architectures is almost negligible when evaluate continuous features performance. The performance to synthesize categorical features
is measured using CSTest. In this case, TVAE obtained higher performance
again, the differences between architectures is slightly higher to architecture A.
Another popular approach to measuring the synthesizer performance is training
a classifier to distinguish between real and synthetic data. The column Logistic
Detection in Table 1 shows the result after training a logistic regression model;
the value displayed corresponds to the complementary F-measure. In this way,
values closer to 1 indicate that the classifier cannot distinguish between real and
synthetic data, and values closer to 0 mean the classifier efficiently detects synthetic data. It can be seen that TVAE achieve the best performance, but this
performance decreases as we include more features to the synthesizer.

Table 1. Synthetic data generators performance

Experiment	Synthesizer training features	Synthesizer	Arch	Exec time (m)	CSTest	KSTest	Logistic detection
S01	X_{Fin}	CTGAN	A	410	0.836	0.864	0.697
			B	260	0.861	0.846	0.749
		TVAE	A	230	0.962	0.868	0.803
			B	130	0.952	0.861	0.756
S02	$X_{Fin} + X_{Degree}$	TVAE	B	140	0.935	0.836	0.644
S03	$X_{Fin} + X_{Degree} + X_{SocInt}$	TVAE	A	400	0.924	0.809	0.539
S03	$X_{Fin} + X_{Degree} + X_{SocInt}$	TVAE	B	320	0.907	0.825	0.542
S03	$X_{Fin} + X_{Degree} + X_{SocInt}$	TVAE	B	465	0.930	0.819	0.513

4.3 Creditworthiness Assessment Performance on Real Data

This section establishes a comparison line for the performance of the models
trained with synthetic data. In order to establish this comparison, we first trained

classifiers using real-world data and tested their performance using the holdout datasets previously defined. Table 2 shows the results of training models according to the methodology described in 3.3. The performance is measured using AUC and KS on the two holdout datasets; the 10-folds mean and its standard deviation are shown for each statistic. For each feature set, we trained two classifiers, logistic regression and gradient boosting. The results show that gradient boosting obtains better results compared to logistic regression. More details of this comparison are shown in Table 3, where we quantify the higher predictive power of gradient boosting.

Table 2. Creditworthiness assessment performance for models trained on real data

Classifier training features	Classifier	Holdout 2018		Holdout 2019	
		AUC	KS	AUC	KS
X_{Fin}	GB	0.88 ± 0.001	0.59 ± 0.002	0.82 ± 0.001	0.50 ± 0.002
X_{Fin}	LR	0.87 ± 0.001	0.58 ± 0.001	0.82 ± 0.001	0.50 ± 0.002
$X_{Fin} + X_{Degree} + X_{SocInt}$	GB	0.88 ± 0.001	0.59 ± 0.002	0.82 ± 0.001	0.50 ± 0.002
$X_{Fin} + X_{Degree} + X_{SocInt}$	LR	0.87 ± 0.001	0.58 ± 0.002	0.83 ± 0.001	0.50 ± 0.002
$X_{Degree} + X_{SocInt}$	GB	0.61 ± 0.002	0.17 ± 0.002	0.62 ± 0.001	0.18 ± 0.002
$X_{Degree} + X_{SocInt}$	LR	0.60 ± 0.001	0.17 ± 0.002	0.61 ± 0.001	0.18 ± 0.002

Based on the results presented above, we will select gradient boosting for the comparisons against the models trained on synthetic data that we will present in the next section.

Table 3. Gradient boosting and logistic regression comparison on real data (holdout 2018)

Classifier training features	AUC diff (%)	KS diff (%)	AUC diff p-value	KS diff p-value
X_{Fin}	0.70%	1.65%	0.000	0.000
$X_{Fin} + X_{Degree} + X_{SocInt}$	0.84%	1.91%	0.000	0.000
$X_{Degree} + X_{SocInt}$	1.65%	2.36%	0.000	0.000

4.4   Creditworthiness Assessment Performance on Synthetic Data

This section aims to know how the performance of a creditworthiness assessment model (the classifier) behaves when trained on synthetic data and applied to real-world data. Table 4 shows the performance indicators on real-world data. Considering all synthesizers are trained with at least the feature set X_{Fin}, the results of training the classifier with X_{Fin} are also displayed for all synthesizers. It is observed that regardless of the synthesizer, training the classifier incorporating at least feature set X_{Fin} produces similar performances in 2018 except in **S02**. However, when we analyze how much the model degrades, the model trained with synthetic X_{Fin} from synthesizer **S01** is the one that suffers a minor discrimination power loss. It can be explained in part that a better synthesizer manages to capture better the proper relationship between the borrower features and the default.

Table 4. Creditworthiness assessment performance on real data for model trained on synthetic data

Synthesizer experiment	Classifier training features	Holdout 2018		Holdout 2019	
		AUC	KS	AUC	KS
S01	X_{Fin}	0.85 ± 0.003	0.53 ± 0.002	0.82 ± 0.002	0.48 ± 0.002
S02	X_{Fin}	0.82 ± 0.001	0.51 ± 0.001	0.80 ± 0.001	0.46 ± 0.002
S03	X_{Fin}	0.85 ± 0.002	0.55 ± 0.002	0.80 ± 0.002	0.46 ± 0.002
S03	$X_{Fin} + X_{Degree} + X_{SocInt}$	0.85 ± 0.002	0.56 ± 0.003	0.80 ± 0.002	0.47 ± 0.003
S03	$X_{Degree} + X_{SocInt}$	0.60 ± 0.002	0.16 ± 0.002	0.61 ± 0.003	0.18 ± 0.002

The comparison of performance obtained by the models trained with synthetic data against the models trained on real-world data is presented in Table 5. We can understand this comparison as the cost of using synthetic data, and it corresponds to the loss of predictive power to preserve the borrower's privacy. We can observe that in the best cases, this decrease in predictive power is approximately 3% and 6% when we measure the performance in AUC and KS, respectively.

Table 5. Comparison between models trained using synthetic data and models trained on real data. ** denotes when the difference is statistically significant using 0.05 as the p-value threshold, while * uses 0.1.

Synthesizer experiment	Classifier training features	Holdout 2018		Holdout 2019	
		AUC diff	KS diff	AUC diff	KS diff
S01	X_{Fin}	−3.59%**	−10.09%**	−0.86%**	−3.92%**
S02	X_{Fin}	−6.24%**	−13.24%**	−3.32%**	−6.48%**
S03	X_{Fin}	−2.81%**	−6.01%**	−3.21%**	−6.70%**
S03	$X_{Fin} + X_{Degree} + X_{SocInt}$	−3.12%**	−5.68%**	−2.54%**	−4.73%**
S03	$X_{Degree} + X_{SocInt}$	−1.85%**	−4.31%**	−0.69%**	1.10%*

5   Conclusions and Future Work

5.1   Conclusions

This work aimed to use synthetic data to train creditworthiness assessment models. We used a massive dataset of 1 million individuals and trained state-of-the-art synthesizer methods to obtain synthetic data and achieve this goal. Then, we presented a training framework that allows us to analyze trained models with synthetic data and observe their performance on real-world data. In addition, we observed their performance one year after being trained to see how susceptible they are to data drift. Our results show that lower quality synthetic data is obtained as we increase the number of attributes in the synthesizer. Despite this, it is possible to use these data to train models that obtain good results in real-world scenarios, with only a reduction in the predictive power of approximately 3% and 6% when we measure the performance in AUC and KS, respectively. These findings are of great relevance since they allow us to train accurate creditworthiness models. At the same time, we keep borrowers' privacy and encourage financial institutions to strengthen ties with academia and foster collaboration and research in credit scoring without the privacy and security restrictions.

5.2 Future Work

Our future work will delve into how to synthesize social interactions' information in the form of graphs and not as added attributes to the training dataset since, as we show, this deteriorates the quality of the synthetic data.

Acknowledgements. This work would not have been accomplished without the financial support of CONICYT-PFCHA/DOCTORADO BECAS CHILE/2019-21190345. The second author acknowledges the support of the Natural Sciences and Engineering Research Council of Canada (NSERC) [Discovery Grant RGPIN-2020-07114]. This research was undertaken, in part, thanks to funding from the Canada Research Chairs program. The last author thanks the support of MICIN project PID2020-116346GB-I00.

References

1. Akoglu, H.: User's guide to correlation coefficients. Turk. J. Emerg. Med. **18**(3), 91–93 (2018)
2. Bradley, A.P.: The use of the area under the roc curve in the evaluation of machine learning algorithms. Pattern Recognit. **30**(7), 1145–1159 (1997)
3. Djeundje, V.B., Crook, J., Calabrese, R., Hamid, M.: Enhancing credit scoring with alternative data. Expert Syst. with Appl. **163**, 113766 (2021)
4. Fiore, U., De Santis, A., Perla, F., Zanetti, P., Palmieri, F.: Using generative adversarial networks for improving classification effectiveness in credit card fraud detection. Inf. Sci. **479**, 448–455 (2019)
5. Flach, P.A.: Machine Learning - The Art and Science of Algorithms that Make Sense of Data. Cambridge University Press, Cambridge (2012)
6. Friedman, J.H.: Greedy function approximation: a gradient boosting machine. Annals Stat. **29**, 1189–1232 (2001)
7. Gicić, A., Subasi, A.: Credit scoring for a microcredit data set using the synthetic minority oversampling technique and ensemble classifiers. Expert Syst. **36**(2), e12363 (2019)
8. Goh, R.Y., Lee, L.S.: Credit scoring: a review on support vector machines and metaheuristic approaches. Adv. Oper. Res. 2019 (2019)

9. Goodfellow, I., Bengio, Y., Courville, A.: Deep Learning. MIT Press, Cambridge (2016)
10. Hagberg, A., Swart, P., SChult, D.: Exploring network structure, dynamics, and function using networkx. In: In Proceedings of the 7th Python in Science Conference (SciPy), pp. 11–15. Citeseer (2008)
11. Ho, P.S., Mo, G.J., Chan-Hee, J.: Receiver operating characteristic (ROC) curve: practical review for radiologists. Korean J. Radiol. **5**(1), 11–18 (2004)
12. Hodges, J.: The significance probability of the smirnov two-sample test. Arkiv för Matematik **3**(5), 469–486 (1958)
13. Hripcsak, G., Rothschild, A.S.: Agreement, the F-measure, and reliability in information retrieval. J. Am. Med. Inform. Assoc. **12**(3), 296–298 (2005)
14. Kennedy, K., Mac Namee, B., Delany, S., O'Sullivan, M., Watson, N.: A window of opportunity: assessing behavioural scoring. Expert Syst. Appl. **40**(4), 1372–1380 (2013)
15. Kingma, D.P., Welling, M.: Auto-encoding variational bayes. arXiv preprint arXiv:1312.6114 (2013)
16. Lei, K., Xie, Y., Zhong, S., Dai, J., Yang, M., Shen, Y.: Generative adversarial fusion network for class imbalance credit scoring. Neural Comput. Appl. **32**(12), 8451–8462 (2019). https://doi.org/10.1007/s00521-019-04335-1
17. McHugh, M.L.: The chi-square test of independence. Biochemia. Med. **23**(2), 143–149 (2013)
18. Muñoz-Cancino, R., Bravo, C., Ríos, S.A., Graña, M.: On the combination of graph data for assessing thin-file borrowers' creditworthiness. arXiv preprint arXiv:2111.13666 (2021)
19. Muñoz-Cancino, R., Bravo, C., Ríos, S.A., Graña, M.: On the dynamics of credit history and social interaction features, and their impact on creditworthiness assessment performance. arXiv preprint arXiv:2204.06122 (2022)
20. Ngwenduna, K.S., Mbuvha, R.: Alleviating class imbalance in actuarial applications using generative adversarial networks. Risks **9**(3), 49 (2021)
21. Óskarsdóttir, M., Bravo, C., Sarraute, C., Vanthienen, J., Baesens, B.: The value of big data for credit scoring: enhancing financial inclusion using mobile phone data and social network analytics. Appl. Soft Comput. **74**, 26–39 (2019)
22. Patki, N., Wedge, R., Veeramachaneni, K.: The synthetic data vault. In: 2016 IEEE International Conference on Data Science and Advanced Analytics (DSAA), pp. 399–410 (2016)
23. Simumba, N., Okami, S., Kodaka, A., Kohtake, N.: Spatiotemporal integration of mobile, satellite, and public geospatial data for enhanced credit scoring. Symmetry **13**(4), 575 (2021)
24. The Basel Committee on Banking Supervision: Principles for the management of credit risk. Basel Committee Publications 75 (2000). www.bis.org/publ/bcbs75.pdf
25. Torres, D.G.: Generation of synthetic data with generative adversarial networks. Ph.D. thesis, Ph. D. Thesis, Royal Institute of Technology, Stockholm, Sweden, 26 November (2018)
26. Wan, Z., Zhang, Y., He, H.: Variational autoencoder based synthetic data generation for imbalanced learning. In: 2017 IEEE Symposium Series on Computational Intelligence (SSCI), pp. 1–7 (2017)
27. Xu, L.: Synthesizing tabular data using conditional GAN. Ph.D. thesis, Massachusetts Institute of Technology (2020)
28. Xu, L., Skoularidou, M., Cuesta-Infante, A., Veeramachaneni, K.: Modeling tabular data using conditional GAN. CoRR abs/1907.00503 (2019)

Combination of Neural Networks and Reinforcement Learning for Wind Turbine Pitch Control

Jesus Enrique Sierra-Garcia[1] and Matilde Santos[2]($\boxtimes$)

[1] University of Burgos, Burgos, Spain
jesierra@ubu.es
[2] Institute of Knowledge Technology, Complutense University of Madrid, Madrid, Spain
msantos@ucm.es

Abstract. In this work, a hybrid proposal that combines reinforcement learning control (RLC) and neural networks for wind turbine pitch control is presented. A reward calculator updates a reward or a punishment depending on the value of the power error derivative. These rewards are used to train a neural network, which can learn the rewards expected to be received if an action is carried out at a given state. In this way, the controller executes the action that will receive the greatest reward. The approach is validated in simulation with the mathematical model of a small onshore 7 kW wind turbine. Results show how the error obtained with the RLC is much smaller than the obtained with a conventional PID regulator. Indeed, the RMSE is 30% lower, the MAE is reduced in a 34%, and the STD is 58% smaller.

Keywords: Wind turbine · Pitch control · Reinforcement learning · Neural networks

1 Introduction

Wind turbines (WT) use the wind resource to extract clean and renewable energy [1]. The rotor with the blades converts the wind energy into rotational torque in the nacelle; the generator transforms this mechanical energy into electricity, and the gearbox adapts the rotor speed to the demanded generator's speed.

However, the final goal is to produce as much energy as possible. Thus, there are different control strategies to maximize the output power at each state of the WT, depending on the wind speed. In general, a turbine may be controlled by adjusting the generator speed, the blade angle, and the rotation of the whole wind turbine. Blade angle and turbine rotation adjustments are also known as pitch and yaw control, respectively. In this work, we are going to focus on the pitch control of the wind turbine, that is, the angle of the blades.

The generation of the pitch control signal is not straightforward due to the nonlinear system dynamic, the coupling of internal variables, unknown parameters, and above all, the wind unpredictable behavior [2]. These are major challenges, particularly for floating

© Springer Nature Switzerland AG 2022
P. García Bringas et al. (Eds.): HAIS 2022, LNAI 13469, pp. 385–392, 2022.
https://doi.org/10.1007/978-3-031-15471-3_33

offshore wind turbines (FOWT), which in addition can experience larger vibration and bigger fatigue due to the harsh environment (waves, current, ice) [3]. To address these problems, different control approaches, including artificial intelligence and ML based ones have been tested on these wind energy devices [4–6]. Among them, reinforcement learning (RL) seems to be a promising strategy due to ability to learn from experience and face non-linear control problems.

Nevertheless, there are only a few recent publications on WT control using RL. To name some, an updated summary of deep RL for power system applications is given in [7]. In [8], RL is used to control variable-speed WTs. It adapts typical variable speed WT controllers to changing wind conditions. The same authors investigated this challenging control issue with large state-action spaces using conditioned RL [9]. In [10], a policy iteration RL model and an adaptive actor-critic approach are used to operate an online doubly-fed induction WT generator. Saénz-Aguirre et al. presents an RL-based artificial neural network for WT yaw control [11].

But the works that use RL for pitch control are scarce. In this work we design a hybrid WT pitch control architecture that combines RL and neural networks. The controller learns how to select the best action to control the blade angle considering the power error and the wind speed. The proposal is tested in simulation with a sawtooth wind profile, and it is shown how outperforms a PID controller.

The rest of the paper is organized as follows. In Sect. 2 the mathematical model of the wind turbine is presented. Section 3 explains the architecture of the intelligent controller. Simulation results are shown and discussed in Sect. 4. The paper ends with the conclusions and future works.

2 Mathematical Model of the Wind Turbine

The WT control solution here presented has been validated with the mathematical model of a small onshore 7 kW wind turbine. The equations that describe the electromechanical performance of this turbine are taken from [1, 4]. Here we summarize the main expressions (1–10).

$$\ddot{\theta} = \frac{1}{T_\theta}\big[K_\theta(\theta_{ref} - \theta) - \dot{\theta}\big], \tag{1}$$

$$f_{blade}(s) = \frac{\beta \cdot s + \sqrt{2}}{\beta^2 \cdot s^2\left(\sqrt{\left(\frac{2}{\alpha}\right)} + \sqrt{\alpha}\right) \cdot \beta \cdot s + \sqrt{2}} \cdot \frac{\gamma \cdot s + \frac{1}{\tau}}{s + \frac{1}{\tau}}, \tag{2}$$

$$v_{ef} = f_{blade}(v_W), \tag{3}$$

$$\lambda = w \cdot R/v_{ef}, \tag{4}$$

$$\lambda_i = \left[\left(\frac{1}{\lambda + c_8}\right) - \left(\frac{c_9}{\theta^3 + 1}\right)\right]^{-1}, \tag{5}$$

$$C_p(\lambda_i, \theta) = c_1 \left[\frac{C_2}{\lambda_i} - c_3\theta - c_4\theta^{c_5} - c_6 \right] e^{-\frac{c_7}{\lambda_i}}, \tag{6}$$

$$\dot{w} = \frac{1}{2 \cdot J \cdot w} \left(C_p(\lambda_i, \theta) \cdot \rho\pi R^2 \cdot v_{ef}^3 \right) - \frac{1}{J} \left(K_g \cdot K_\phi \cdot I_a + K_f \cdot w \right), \tag{7}$$

$$\dot{I}_a = \frac{1}{L_a} \left(K_g \cdot K_\phi \cdot w - (R_a + R_L)I_a \right), \tag{8}$$

$$P_{out} = R_L \cdot I_a^2 \tag{9}$$

where L_a is the armature inductance (H), K_g is a dimensionless constant of the generator, K_ϕ is the magnetic flow coupling constant (V·s/rad), R_a is the armature resistance (Ω), R_L is the resistance of the load (Ω), considered in this study as purely resistive; w is the angular rotor speed (rad/s), I_a is the armature current (A), λ is the tip-speed-ratio (TSR), which is dimensionless, and $[\alpha, \beta, \gamma, \tau]$ is the set of values of the filter that the blades implement.

The power coefficient C_p depends on the characteristics of the WT; J is the rotational inertia (Kg.m^2), R is the radius or blade length (m), ρ is the air density (Kg/m^3), K_f is the friction coefficient (N.m/rad/s), K_θ and T_θ are dimensionless parameters of the pitch actuator, v_{ef} is the effective wind speed in the blades (m/s), and v_W is the wind speed measured by the anemometer sensor.

To simulate the model, the parameters listed in Table 1 has been considered [1].

Table 1. Parameters of the model.

Parameter	Description	Value/units
L_a	Inductance of the armature	13.5 mH
K_g	Constant of the generator	23.31
K_ϕ	Magnetic flow coupling constant	0.264 V/rad/s
R_a	Resistance of the armature	0.275 Ω
R_L	Resistance of the load	8 Ω
J	Inertia	6.53 kg m^2
R	Radius of the rotor	3.2 m
ρ	Density of the air	1.223 kg/m^3
K_f	Friction coefficient	0.025 N m/rad/s
$[c_1, c_2, c_3]$	C_p constants	[0.73, 151, 0.58]
$[c_4, c_5, c_6]$	C_p constants	[0.002, 2.14, 13.2, 18.4]
$[c_7, c_8, c_9]$	C_p constants	[18.4, −0.02, −0.003]
$[K_\theta, T_\theta]$	Pitch actuator parameters	[0.15, 2]
$[\alpha, \beta, \gamma, \tau]$	Blade filter constants	[0.55, 0.832, 1.17, 9]

3   Intelligent Control Loop Architecture

The architecture of the intelligent controller is shown in Fig. 1. By measuring the output power, P_{out} and the power reference, P_{ref}, the power error is calculated as the difference, P_{err}. The power error and its derivative are used by the reward calculator to estimate the reward, r_t. In addition, there is a neural network (NN) that receives as inputs the wind speed, v_w, the power error and its derivative - that is, the state of the system-, and then it estimates the expected rewards, $[N_{er_1}(t_i), \ldots N_{er_n}(t_i)]$, where $N_{er_j}(t_i)$ is the estimated expected reward if the j-action is executed in the time t_i. Considering the values obtained, the action selector choses the action which maximizes the expected reward.

The neural network is trained online with the expected rewards obtained by the Q learning algorithm $[Q_{er_1}(t_i), \ldots Q_{er_n}(t_i)]$, (15–16). These expected rewards are estimated from the real rewards, $r(t_i)$, calculated by the reward calculator while the RL algorithm is working (12). It can be trained each simulation step, or several experiences can be grouped and trained the network with all together. Overall, the later makes the training process faster than training one by one. However, grouping of experiences increases the time between trainings, and thus limits the reaction capacity. Therefore, it is important to correctly select the number of experiences used to train in order to reach a proper balance between computational effort and reactiveness.

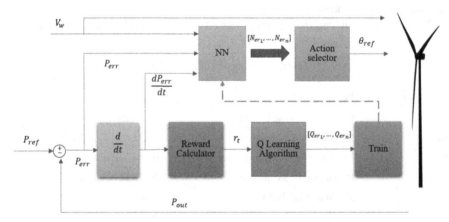

Fig. 1. Architecture of the proposed RL-NN intelligent controller.

In principle, any neural network can be used with this control architecture, but networks with capability to perform regression tasks will work better [12]. To validate this proposal, a multilayer perceptron with two hidden layers and 10 neurons per layer has been implemented. The number of outputs was set to 18. The neural network was trained with the Levenberg-Marquardt algorithm.

The action selector chooses the output of the NN with the maximum expected reward (13–14). This output is transformed to a pitch reference considering the number of actions n_{act} to cover the range $[0, \pi/2]$. Therefore, the number of outputs of the network is equal to the number of actions.

The reward calculator assigns a reward or a punishment to the previous action based on the derivative of the power error. If the power error moves closer the reference it gets a reward, otherwise a punishment is assigned. This is calculated by (12).

The behavior of the controller can be formalized by the following expressions.

$$P_{err}(t_i) = P_{ref}(t_{i-1}) - P_{out}(t_{i-1}) \tag{10}$$

$$[N_{er_1}(t_i), \ldots N_{er_n}(t_i)] = f_{NN}(v_w(t_{i-1}), P_{err}(t_{i-1}), \dot{P}_{err}(t_{i-1}), w_N(t_{i-1})) \tag{11}$$

$$r(t_i) = \begin{cases} -dP_{err_S}(t_i)/dP_{err_{MAX}} & P_{err}(t_{i-1}) > 0 \\ dP_{err_S}(t_i)/dP_{err_{MAX}} & P_{err}(t_{i-1}) < 0 \end{cases} \tag{12}$$

$$a(t_i) = argMAX([N_{er_1}(t_i), \ldots N_{er_n}(t_i)]) \frac{\pi/2}{n_{act}} \tag{13}$$

$$\theta_{ref}(t_i) = a(t_i) \cdot \frac{\pi/2}{n_{act}} \tag{14}$$

$$[Q_{er_1}(t_i), \ldots Q_{er_n}(t_i)] = f_{QL}(r(t_{i-1}), v_w(t_{i-1}), P_{err}(t_{i-1}), \dot{P}_{err}(t_{i-1}), a(t_{i-1})) \tag{15}$$

$$w_N(t_i) = \begin{cases} f_L\left(\begin{bmatrix} v_w(t_{i-1}, \ldots, t_{i-N}) \\ P_{err}(t_{i-1}, \ldots, t_{i-N}) \\ \dot{P}_{err}(t_{i-1}, \ldots, t_{i-N}) \end{bmatrix}, \begin{bmatrix} Q_{er_{1..n}}(t_i, \ldots, t_{i-(N-1)}) \\ a(t_i, \ldots, t_{i-(N-1)}) \end{bmatrix}, w_N(t_{i-1}) \right) & t_i \in T_L \\ w_N(t_{i-1}) & t_i \notin T_L \end{cases} \tag{16}$$

where f_{NN} is the output function of the neural network which depends on its architecture, f_L is the learning algorithm of the NN, f_{QL} represents the QLearning algorithm [13], and T_L is the times when the NN is trained.

4 Discussion of the Results

The results have been obtained using the software Matlab/Simulink. Simulations last 100 s. The simulation step is variable to consider both, how long it takes to run the simulation and how big the discretization error is. The maximum simulation step has been set to 20 ms. The control period is 250 ms.

The performance of the controller has been compared with a PID regulator (17).

$$\theta_{ref}(t_i) = \frac{\pi}{4} - K_p\left[P_{err}(t_i) + K_d \cdot \frac{d}{dt}P_{err}(t_i) + K_i \cdot \int P_{err}\right] \tag{17}$$

where $[K_p, K_d, K_i]$ are set to $[\pi/4000, 0.2, 01]$. These control constants have been tuned by trial and error. A sawtooth wind velocity profile has been used as input signal. This signal has a minimum value of 12.25 m/s and a maximum value of 12.8 m/s. The period of the signal is 30 s.

Figure 2 shows the comparison of the output power when the different controllers are applied. The nominal power is represented with a black dashed line. The blue line indicates the output power obtained when the angle of the blades is fixed to 90°, and the red one when it is maintained at 0°. These lines represent the power boundaries as the power extracted when the angle is 0° is maximum and minimum when it is 90°. The yellow line shows the output power when the PID is applied, and the purple line when the RLC is used.

As it can be observed, the sawtooth profile of the wind is noticeable in the output power. This effect is higher when the blade angle is constant and when the PID is applied. However, the RLC is more robust against wind changes, and it is able to adapt to them, and the sawtooth profile is not perceived in the output power. In general, the RLC provides a better performance than the PID.

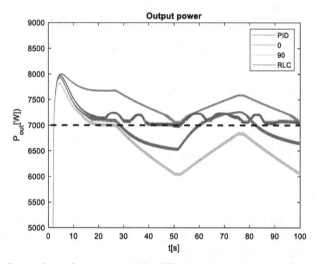

Fig. 2. Comparison of output power for different controllers. (Color figure online)

Figure 3 shows a comparison of the rotor angular speed with different pitch control methods. The color code is the same as in Fig. 2. In general, the shape of the rotor angular velocity is similar to the output power. Again, the RLC follows better the nominal value and gives smaller error with respect to the reference.

The quantitative results of the errors (Table 2) confirm what it has been shown in the previous figures. The mean squared error (RMSE), the mean absolute error (MAE), and the standard deviation (STD) of the output power are calculated. It is possible to see how the error given by RLC is much smaller than the PID's. Specifically, the RMSE is 30% smaller, the MAE is 34% reduced, and the STD is 58% lower.

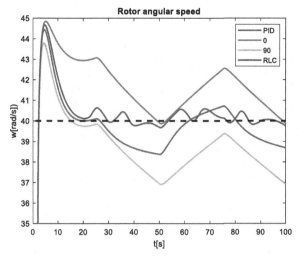

Fig. 3. Comparison of rotor angular velocity for different controllers.

Table 2. Quantitative comparison of the performance of the controllers.

Metric	PID	RLC
RMSE [W]	242.7244	167.7562
MAE [W]	210.8265	139.1034
STD [W]	234.5103	96.9613

5 Conclusions

In this work, an RL control architecture based on neural networks is presented. It is applied to control the pitch angle of a small wind turbine. The state vector of the system includes the wind speed, the power error, and the derivative of the latter. Based on this state, each output of the NN gives an estimation of the expected reward when the action associated to that output is carried out. This way, the controller learns the actions that maximizes the expected reward at each state.

Results show how the power error obtained by RLC is much lower than the one given by the PID regulator. Indeed, the reduction of the error regarding the conventional controller is up to 30% for the RMSE, 34% for the MAE, and 58% for the STD.

As future works, we can highlight the validation of the proposal with a wind turbine prototype and to extend the architecture to consider deep neural networks.

Acknowledgements. This work was partially supported by the Spanish Ministry of Science, Innovation and Universities under MCI/AEI/FEDER Project number RTI2018-094902-B-C21.

References

1. Mikati, M., Santos, M., Armenta, C.: Electric grid dependence on the configuration of a small-scale wind and solar power hybrid system. Renew. Energy **57**, 587–593 (2013)
2. Zhang, C., Plestan, F.: Individual/collective blade pitch control of floating wind turbine based on adaptive second order sliding mode. Ocean Eng. **228**, 108897 (2021)
3. Tomás-Rodríguez, M., Santos, M.: Modelado y control de turbinas eólicas marinas flotan-tes. Revista Iberoamericana de Automática e Informática Industrial **16**(4), 381–390 (2019)
4. Sierra-García, J.E., Santos, M.: Redes neuronales y aprendizaje por refuerzo en el control de turbinas eólicas. Revista Iberoamericana de Automática e Informática industrial **18**(4), 327–335 (2021)
5. Sierra-García, J.E., Santos, M.: Performance analysis of a wind turbine pitch neurocontroller with unsupervised learning. Complexity **2020**, 1–15 (2020)
6. Alzayed, M., Chaoui, H., Farajpour, Y.: Maximum power tracking for a wind energy conversion system using cascade-forward neural networks. IEEE Trans. Sustain. Energy **12**(4), 2367–2377 (2021)
7. Zhang, Z., Zhang, D., Qiu, R.C.: Deep reinforcement learning for power system applications: an overview. CSEE J. Power Energy Syst. **6**(1), 213–225 (2019)
8. Fernandez-Gauna, B., Fernandez-Gamiz, U., Grana, M.: Variable speed wind turbine controller adaptation by reinforcement learning. Integr. Comput.-Aided Eng. **24**(1), 27–39 (2017)
9. Fernandez Gauna, B., Osa, J.L., Graña, M.: Experiments of conditioned reinforcement learning in continuous space control tasks. Neurocomputing **271**, 38–47 (2018)
10. Abouheaf, M., Gueaieb, W., Sharaf, A.: Model-free adaptive learning control scheme for wind turbines with doubly fed induction generators. IET Renew. Power Gener. **12**(14), 1675–1686 (2018)
11. Saénz-Aguirre, A., Zulueta, E., Fernández-Gamiz, U., Lozano, J., Lopez-Guede, J.M.: Artificial neural network based reinforcement learning for wind turbine yaw control. Energies **12**(3), 436 (2019)
12. Sierra-García, J.E., Santos, M.: Switched learning adaptive neuro-control strategy. Neurocomputing **452**, 450–464 (2021)
13. Sutton, R.S., Barto, A.G.: Reinforcement Learning: An Introduction. MIT Press, Cambridge (2018)

Survey for Big Data Platforms and Resources Management for Smart Cities

Carlos Alves⬤, António Chaves⬤, Carla Rodrigues⬤, Eduarda Ribeiro⬤,
António Silva⬤, Dalila Durães^(✉)⬤, José Machado⬤, and Paulo Novais⬤

ALGORITMI Centre, University of Minho, Braga, Portugal
{carlos.alves,antonio.silva,dalila.duraes}@algoritmi.uminho.pt,
{id10053,a85412,a84710}@alunos.uminho.pt
{jmac,pjon}@di.uminho.pt

Abstract. Currently, smart cities are a hot topic and their tendency
will be to optimize resources and promote efficient strategies for the
preservation of the planet as well as to increase the quality of life of its
inhabitants. In this sense, this research presents an initial component
of investigation about Big Data Platforms for Smart Cities in order to
be implemented in integrated and innovative solutions for development
in urban centers. For this, a survey was carried out on "Big Data Platforms", "Data Science Platforms", "Security & Privacy" and "Resources
Management". The extraction of the results of this research was done
through the SCOPUS repository in articles from the last 5 years to conclude what has been done so far and what will be the trends in the
coming years, define proposals for possible solutions for smart cities and
identify the right technologies for the design of a smart city architecture.

Keywords: Big data platform · Data science · Security · Privacy ·
Internet of Things

1 Introduction

The population of urban centers has been increasing in recent decades, and it is
predicted that by 2050, two-thirds of the global population will be metropolitan
inhabitants [1]. Cities are like dynamic living organisms and are constantly evolving, especially under the pressure caused on their infrastructure by the movement
of people to urban centers [1–4]. In this way, a city is a complex system, and new
methods are needed to manage it and use the huge amounts of data that are
generated. Municipal administrations can acquire knowledge that is hidden in
large-scale data to provide better urban management by applying Information
and Communication Technologies (ICT) solutions. Such ICT solutions allow for
better transport planning, efficient water management, new energy efficiency
strategies, better waste management, and effective risk management policies for
city users [1,5,6].

Supported by University of Minho & Algoritmi Centre.

Thus, trends in data acquisition and processing are essential for the communication of Internet of Things (IoT)-based systems in a city. Examples of this are household appliances, homes, hospitals, waste collection systems, lighting, traffic control, etc. [7–10]. To improve this type of integration, solutions to mitigate problems both in terms of data processing and in terms of security of the information that can be generated by the devices will be necessary for the near future. Thereunto, first, it will be essential to analyze some literature in order to obtain some solutions to the problems that may exist either in the creation of a big data platform for Data Science and Artificial Intelligence (AI) or in the implementation of IoT devices or other technologies for smart cities.

After this introductory section, the work is divided into three main sections. First, the methodology used for the systematic review is described, including the research strategy and exclusion criteria. Then, the results found and their discussion are exhibited. Finally, the main conclusions are presented and some directions for future research are provided.

2   Methodology

This survey is based on the PRISMA[1] (Preferred Reporting Items for Systematic Reviews and Meta-Analyses) information and checklist. The reason is justified because PRISMA is commonly accepted by the scientific community in the field of engineering and computer science. The actions considered in this process were: identification of the study's research questions and appropriate keywords; creation of the search string required to query the academic research databases; characterization of the exclusion criteria to filtrate the papers and decrease the sample; a reflection of the selected studies; and finally, exhibition and discussion of the results.

```
1   (("Big Data Platforms") OR ("Data Science Platforms"))
2   AND
3   (("Privacy") AND ("Security")) AND ("Smart Cities") AND (IOT) AND
4   (LIMIT-TO(OA, "all")) AND
5   (LIMIT-TO(PUBYEAR, 2022) OR LIMIT-TO(PUBYEAR, 2021) OR
6   LIMIT-TO(PUBYEAR, 2020)
7   OR LIMIT-TO(PUBYEAR, 2019) OR LIMIT-TO(PUBYEAR, 2018))
8   AND
9   (LIMIT-TO(EXACTKEYWORD, "Big Data") OR LIMIT-TO (EXACTKEYWORD,
10  "Internet Of Things")
11  OR LIMIT-TO(EXACTKEYWORD, "Smart City") OR LIMIT-TO(EXACTKEYWORD,
12  "Data Analytics")
13  OR LIMIT-TO(EXACTKEYWORD, "Decision Making") OR
14  LIMIT-TO(EXACTKEYWORD, "Internet
15  Of Things(IOT)") OR LIMIT-TO(EXACTKEYWORD, "Smart Cities"))
```

---

[1] http://www.prisma-statement.org.

The literature search was conducted on 27 April 2022 in SCOPUS. The keywords are shown in the previous code. Each keyword topic was aggregated with a disjunction and three conjunctions were used to bring the topics together. The query shown above was used for the search in the SCOPUS repository applied to the title, keywords and abstract of the documents.

In order to screening the articles found, some exclusion criteria were defined. Thus, the documents are excluded if fall in one of these:

EC1 Duplicate papers.
EC2 Not freely accessible.
EC3 Do not come from the field of computer science or engineering.
EC4 Have not been peer-reviewed.
EC5 Do not focus on the variables studied or is out of context.
EC6 Has not been produced in the last 5 years (from 2018).

3   Results and Discussion

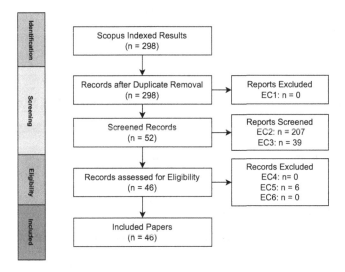

Fig. 1. PRISMA diagram of exclusion criteria.

The search in the SCOPUS repository identified 289 articles and exclusion criteria were subsequently applied. On the search page of the database, the documents that met the first exclusion criterion (EC1) were filtered out, and no paper was removed. Next, the second set of exclusion criteria (EC2 and EC3) were applied and the research was limited to 52 papers. Finally, by applying the last set of exclusion criteria (EC4, EC5, and EC6) 6 papers were removed.

This left 46 documents for a full reading. Figure 1 illustrates, according to the PRISMA methodology, the article selection process.

Next, the papers presented below are divided into two major issues: big data platforms for smart cities and resources management for smart cities. These two topics are presented in Tables 1 and 2, respectively. The topics and the authors that address them were identified in these two tables.

Table 1. Summary of the results for big data platforms for smart cities.

Author(s)/Article	Findings
Big data platforms	[1, 11–27]
Security & Privacy	[8, 15, 22, 24, 25, 27–31]
IoT security	[7, 16, 19, 24, 31–33]
IoT architecture	[7, 23, 25, 30–32, 34]

Table 2. Summary of the results for resources management for smart cities.

Author(s)/Article	Findings
Mobility	[8, 21, 23, 24, 31, 34, 35]
Occupation	[8, 23, 35, 36]
Waste management	[8, 31]
Digital transformation	[8, 15, 16, 37]
Energy	[8, 15, 17, 21–23, 25, 30–32, 38, 39, 49]
Air and water quality control	[8]
Intelligent governance	[21, 35]
Healthcare	[38–45]

3.1 Big Data Platforms for Smart Cities

Soomro et al. [23] claim that distributed storage solutions like Amazon EC2 and HDFS help manage large amounts of smart city data.

Venkatraman et al. [28] approach the current trends in big data, its concepts and dimensions, and the security and privacy challenges associated. This paper presents four realistic solutions derived from current technologies, such as data provenance, encryption and access control, data mining, blockchain, and real-world implementation examples.

Wang et al. [46] proposed a novel fog-computing-based IBDIS approach called Fog-IBDIS. By relocating the integration duty from the cloud to the edge of networks, this technique intends to integrate and distribute industrial big data with high raw data security and low network traffic burdens. Munawar et al. [20] present a review to identify data trends and how the construction industry can benefit from Big Data. The analysis performed by the authors shows that big data research in construction safety, site management, heritage conservation, waste minimization and quality improvements are key areas to the integration of big data in the construction industry. Yousif et al. [48] present a literature review, on the integration of Big Data in the construction industry by addressing the current usage, problems, potential ways of working, and challenges associated with the execution. The authors concluded that the adoption of such technologies in this industry remains at an early stage and lacks deep research. Hassani et al. [49] present a review regarding big data in climate change studies. The authors verified that existing research has applied big data analytics, mainly, in the aspects of energy efficiency, smart agriculture, sustainable urban planning and infrastructure and, natural disaster management. They also concluded that big data in climate change mainly function in four aspects: observing and monitoring, understanding, predicting, and optimizing.

Security and Privacy: Muheidat et al. [14] address emerging concepts that are using blockchain and big data. They conclude that generating big data with blockchain, collects, analyzes and predicts events that consist of device-to-device communication, allowing the technology in blockchain to perform offline operations and more securely.

Rejeb et al. [24] and Sanchez-Corcuera et al. [47], mention that currently current database technologies do not manage effectively and that something that can mitigate this problem will be the blockchain technology.

IoT Security: Yaïci et al. [32] summarise a literature review on IoT applications aiming at lowering greenhouse gas (GHG) emissions and boosting energy efficiency in buildings. The authors desired to understand how further IoT systems and technologies are being developed to improve energy efficiency in residential and commercial facilities.

Thaseen et al. [33] suggested a Hadoop based framework to specify the malicious IoT traffic. So, they used a changed Tomek-link under-sampling integrated with automated Hyper-parameter tuning of Machine Learning (ML) classifiers. This research is unique in that it uses a big data platform to benchmark IoT datasets in order to minimize computational time.

Villamil et al. [29] present a general overview of IoT, based on a review of recently published papers. The authors concluded that IoT technology is a key enabler in increasing commercial and industrial performance, as well as the overall quality of life, and can overcome the challenges in terms of security, processing capacity and data mobility.

Moon et al. [37] propose a collaborative approach based on sample data correlation to select the optimal model for the edge among candidate cloud models. This strategy allows to employ the best model without requiring any training on the edge, and it also reduces privacy concerns.

Saleem and Chishti [21] present a current state-of-the-art in the field of IoT data analytics. The authors investigated the purpose of data analytics in IoT applications, as well as the key enablers, challenges, and future research opportunities for performing desired data analytics in IoT applications.

Zhang et al. [45] present a paper with a new architecture, SafeCity, which highlights the smart city ecosystem and as a result evaluated that the architecture presents precious information about a safe smart city in the context of the sensor-based IoT environment.

IoT Architecture: Belli et al. [8] analyzed the main aspects of an IoT infrastructure for smart cities, where they conclude that it is difficult to define all the viable ways in which city managers can improve the services of their city, but that, citizens will certainly benefit from adopting solutions oriented towards IoT in the urban context in question.

Diaconita et al. [15] proposed numerous architectures based on Apache Hadoop that can provide the ICT backbone for efficient management of a smart city. They conclude that the adoption of Apache Hadoop, together with Spark and the infinity of NoSQL databases and the Apache projects that accompany them, are a mature ecosystem, which is the reason for the adoption of this technology as the most suitable for the construction of a smart city.

Zhang et al. [45] tested reliable datasets using the Apache Hadoop platform. The evaluations revealed that the architecture presents precious information about a safe smart city in the context of the sensor-based IoT environment.

3.2 Resources Management for Smart Cities

Mobility: Baddi et al. [7] presented a solution for mobility in smart cities that consists of a platform capable of configuring heterogeneous and complex scenarios that integrate sensors/actuators such as IoT/Internet of Everything (IoE) in a general scenario of Big Data, Machine Learning and Data Analytics. The authors demonstrated that a configuration flexible and dynamic is possible, supporting security, protection, on-premises, cloud and mixed solutions.

In their state-of-the-art review, Jabbar et al. [34] explain the core concepts surrounding blockchain technology and its historical transformation. The intent of the paper is the classification of the reviewed blockchain applications on the Internet of Vehicles (IoV) systems across different categories. Vehicle capacity management due to increasing levels of urban congestion, IoV, vehicle-to-vehicle emergency messaging systems and smart meters can significantly address energy efficiency in the constitution of a city with intelligent mobility in favour of intelligent life [35].

Soomro et al. [23] report that big data analytics can help solve traffic congestion problems.

Occupation: Parisi et al. [50] present a literature review regarding the tools and strategies of Information and Communication Technologies implemented in intelligent buildings, considering the specific applications in the different phases of the building's life cycle. The authors concluded that the ICT tools in the architecture, engineering, and construction fields are not adequately exploited and need future consideration.

Lawal and Rafsanjani [51] present a review of existing IoT technologies and applications in residential and commercial buildings. The authors identified the trends, current benefits and risks, and future challenges of implementing IoT in built environments. In addition, they conclude that privacy and security risks are the main challenges to IoT implementation.

Kasznar et al. [22] presented a new review framework that analyzes how smart city infrastructure is related to the urbanization process, presenting developments in IoT sensor networks, big data analysis of generated information and green construction and, concluded that future research on the theme should conceptualize smart cities as an emerging socio-technical phenomenon.

Soomro et al. [23] report that big data analytics can support governance by allowing the mining of citizens' preferences and opinions.

Smart cities, through participatory governance, investment in human capital and, traditional and modern communication infrastructures, aim to boost sustainable economic growth in order to improve the quality of life of the people who inhabit them. Overall, the field of smart cities is continually expanding its scope to address urban issues in a more comprehensively way [36].

Waste Management: According to Wang et al. [31], the lack of intelligent waste collection systems was a factor identified as a future challenge to be overcome with regard to smart cities.

Digital Transformation: Wang et al. [31] state that the adoption of AI and IoT in developing countries presents some challenges. In addition, they claim that the lack of infrastructure, insufficient funds, cybersecurity risks and lack of trust in AI, are the factors that are slowing the adoption of AI and IoT in the development of smart cities.

Energy: Bhattarai et al. [17] present a state-of-the-art review regarding big data analytics and its application in power grids. The authors addressed the utility and industry perspectives on the current status of big data implementation in power systems and, the main challenges for deploying big data analytics to smart grids and transforming big data into actionable information. With the focus on the employment of blockchain technology across different sectors (Government, Financial, Manufacturing, among others), [39] presents a double-sided view of the key points of blockchain within these sectors. The core goal of the research highlights the energetic resources required for the continuity of the blockchain environment and whether it is possible to keep it a sustainable process.

Soomro et al. [23] claim that big data analytics can help in energy recycling, lower energy consumption and emissions to reduce pollution and lower energy bills for homes and businesses.

Nguyen et al. [25] and urRehman et al. [13] present a literature review surrounding big data analytics within the context of the oil and gas industry and how recent technology has helped improve operational efficiency. The paper reflects upon the role of international cooperation - both in the industrial and governmental sectors - needed to achieve optimal solutions. There is also a clear concern about cybersecurity issues and the benefits these implementations can bring to legacy systems.

Air and Water Quality Control: Soomro et al. [23] claim that big data analytics can support people's lives in smart cities, providing a cleaner environment. The quality of water and air can be monitored through sensors and aim to improve and prevent risks to human and animal health [8].

Intelligent Governance: Soomro et al. [23] report that big data can help manage large amounts of data and predict future economic scenarios.

Ismagilova et al. [35] mention that the involvement of the population in the planning of smart cities would be of great importance.

Healthcare: Zhang et al. [45], designed and implemented a secure medical big data system based on Hadoop and blockchain. Their goal is to improve the efficiency of traditional medical rehabilitation activities and enable patients to maximize their understanding of their treatment status.

Jouahri et al. [38] propose a design along with an implementation of the next generation of smart homes, where heterogeneous energy and health data from multiple sensors and household equipment is managed, secured and visualized. One goal of the authors is to integrate remote health monitoring in a smart home.

Anisha et al. [43] aimed to develop a smart assistive device that combines sensors and telecommunications technology to enable remote monitoring of disabled patients. The care system is based on IoT, enabling disabled patients to lead independent lives.

Mehmood et al. [3] implemented and tested various active concept drift detection algorithms for time series analysis within a distributed environment. Concept drift detection methods aim to mitigate the problem of poor prediction performance and ineffective decisions, lead by the changes in the statistical features of data streams which have been introduced by the proliferation of IoT deployments across the smart city ecosystem.

Rasool et al. [40] present a review on the classification of the security and privacy main challenges regarding different variants of the Internet of Medical Things (IoMT) based on their actual use in the healthcare domain. This paper also presents a comprehensive attack taxonomy on the overall IoMT infrastructure and the best practices and guidelines for developing security and privacy solutions for the IoMT devices.

Vitabile et al. [41] present a discussion on health monitoring systems. The authors addressed issues in data collection, merger, ownership, and data privacy; models, technologies and solutions for processing and analyzing medical data; big data analytics for remote health status monitoring; and also presented examples of case studies and practical solutions.

4   Conclusions

In the future, more and more intelligent systems based on data analysis and ML will be developed, pushing computing to the cognitive "edge" of the city [3].

Unlike traditional ML models that assume data distribution to be static, real-world streaming data often deviates from the learned distribution, i.e. concept drift. Concept drift can be framed as a change in the latent distribution of a target variable for which predictions are made over a certain period of time due to unforeseen reasons, such as sensor wear or replacement for one with a different calibration [3].

The presence of concept drift can make prediction results inaccurate and therefore lead to poor decisions. Thus, there is an urgent need to improve intelligent systems that operate on real-world data streams with learning methodologies sensitive to deviation from the concept, ensuring the validity of model results. According to Wang et al. [31], the lack of infrastructure, insufficient capital (funds), cybersecurity, intelligent waste collection systems and energy, transport, health and, education management were identified as future challenges to be overcome with regard to smart cities, not forgetting the lack of trust in AI and IoT as they are emerging technologies. In addition, is magilova et al. [35] claim that smart cities have the potential to meet many of the UN's sustainable development goals.

We agree that more research should be carried out in this regard, namely in technologies to improve the quality of life of citizens in smart cities and also, in analyzing platforms such as Home Assistant for the processing of data from homes and companies. In conclusion, the desire to achieve intelligence in cities around the world depends on government will, technological capabilities and availability of funds [12].

Acknowledgment. This work has been supported by FCT-Fundação para a Ciência e Tecnologia within the R&D Units Project Scope: UIDB/00319/2020 and the project "Integrated and Innovative Solutions for the well-being of people in complex urban centers" within the Project Scope NORTE-01-0145-FEDER-000086.

References

1. Kousis, A., Tjortjis, C.: Data mining algorithms for smart cities: a bibliometric analysis. Algorithms **14**(8), 242 (2021)
2. Costa, A., Heras, S., Palanca, J., Novais, P., Julián, V.: A persuasive cognitive assistant system. In: Ambient Intelligence- Software and Applications – 7th International Symposium on Ambient Intelligence (ISAmI 2016). AISC, vol. 476, pp. 151–160. Springer, Cham (2016). https://doi.org/10.1007/978-3-319-40114-0_17
3. Mehmood, H., Kostakos, P., Cortes, M., Anagnostopoulos, T., Pirttikangas, S., Gilman, E.: Concept drift adaptation techniques in distributed environment for real-world data streams. Smart Cities **4**(1), 349–371 (2021)
4. Lavrijssen, S., Vitéz, B.: Good governance and the regulation of the district heating market. In: Weijnen, M.P.C., Lukszo, Z., Farahani, S. (eds.) Shaping an Inclusive Energy Transition, pp. 185–227. Springer, Cham (2021). https://doi.org/10.1007/978-3-030-74586-8_9
5. Bernardes, M. B., de Andrade, F. P., & Novais, P. : Smart cities, data and right to privacy: a look from the Portuguese and Brazilian experience. In Proceedings of the 11th International Conference on Theory and Practice of Electronic Governance, pp. 328–337 (2018)

6. Santos, F., et al.: In-car violence detection based on the audio signal. In: Yin, H., et al. (eds.) IDEAL 2021. LNCS, vol. 13113, pp. 437–445. Springer, Cham (2021). https://doi.org/10.1007/978-3-030-91608-4_43

7. Badii, C., Bellini, P., Difino, A., Nesi, P.: Sii-mobility: an IoT/IoE architecture to enhance smart city mobility and transportation services. Sensors 19(1), 1 (2018)

8. Belli, L., Cilfone, A., Davoli, L., Ferrari, G., Adorni, P., Di Nocera, F., Bertolotti, E.: IoT-enabled smart sustainable cities: challenges and approaches. Smart Cities 3(3), 1039–1071 (2020)

9. Alves, C., Luís Reis, J.: The intention to use e-commerce using augmented reality - the case of IKEA place. In: Rocha, Á., Ferrás, C., Montenegro Marin, C.E., Medina García, V.H. (eds.) ICITS 2020. AISC, vol. 1137, pp. 114–123. Springer, Cham (2020). https://doi.org/10.1007/978-3-030-40690-5_12

10. Machado, J., Abelha, A., Novais, P., Neves, J., Neves, J.: Quality of service in healthcare units. Int. J. Comput. Aided Eng. Technol. 2(4), 436–449 (2010)

11. Zhang, H., Babar, M., Tariq, M.U., Jan, M.A., Menon, V.G., Li, X.: SafeCity: toward safe and secured data management design for IoT-enabled smart city planning. IEEE Access 8, 145256–145267 (2020)

12. Omotayo, T., Awuzie, B., Ajayi, S., Moghayedi, A., & Oyeyipo, O.: A systems thinking model for transitioning smart campuses to cities. Front. Built Environ. 7 (2021)

13. Ur Rehman, M.H., Yaqoob, I., Salah, K., Imran, M., Jayaraman, P.P., Perera, C.: The role of big data analytics in industrial internet of things. Future Gener. Comput. Syst. 99, 247–259 (2019)

14. Muheidat, F., Patel, D., Tammisetty, S., Lo'ai, A.T., Tawalbeh, M.: Emerging concepts using blockchain and big data. Procedia Comput. Sci. 198, 15–22 (2022)

15. Diaconita, V., Bologa, A.R., Bologa, R.: Hadoop oriented smart cities architecture. Sensors 18(4), 1181 (2018)

16. Thasnimol, C.M., Rajathy, R.: The paradigm revolution in the distribution grid: the cutting-edge and enabling technologies. Open Comput. Sci. 10(1), 369–395 (2020)

17. Bhattarai, B.P., et al.: Big data analytics in smart grids: state-of-the-art, challenges, opportunities, and future directions. IET Smart Grid 2(2), 141–154 (2019)

18. Hajjaji, Y., Boulila, W., Farah, I.R., Romdhani, I., Hussain, A.: Big data and IoT-based applications in smart environments: a systematic review. Comput. Sci. Rev. 39, 100318 (2021)

19. Hadi, M.S., Lawey, A.Q., El-Gorashi, T.E., Elmirghani, J.M.: Big data analytics for wireless and wired network design: a survey. Comput. Netw. 132, 180–199 (2018)

20. Munawar, H.S., Ullah, F., Qayyum, S., Shahzad, D.: Big data in construction: current applications and future opportunities. Big Data Cogn. Comput. 6(1), 18 (2022)

21. Saleem, T.J., Chishti, M.A.: Data analytics in the Internet of Things: a survey. Scalable Comput.: Pract. Experience 20(4), 607–630 (2019)

22. Kasznar, A.P.P., Hammad, A.W., Najjar, M., Linhares Qualharini, E., Figueiredo, K., Soares, C.A.P., Haddad, A.N.: Multiple dimensions of smart cities' infrastructure: Rev. Build. 11(2), 73 (2021)

23. Soomro, K., Bhutta, M.N.M., Khan, Z., Tahir, M.A.: Smart city big data analytics: an advanced review. Wiley Interdisc. Rev.: Data Min. Knowl. Discovery 9(5), e1319 (2019)

24. Rejeb, A., Rejeb, K., Simske, S.J., Keogh, J.G.: Blockchain technology in the smart city: a bibliometric review. Qual. Quant. 1–32 (2021). https://doi.org/10.1007/s11135-021-01251-2

25. Nguyen, T., Gosine, R.G., Warrian, P.: A systematic review of big data analytics for oil and gas industry 4.0. IEEE Access **8**, 61183–61201 (2020)
26. Torre-Bastida, A.I., Díaz-de-Arcaya, J., Osaba, E., Muhammad, K., Camacho, D., Del Ser, J.: Bio-inspired computation for big data fusion, storage, processing, learning and visualization: state of the art and future directions. Neural Comput. Appl. 1–31 (2021). https://doi.org/10.1007/s00521-021-06332-9
27. Tang, L., Li, J., Du, H., Li, L., Wu, J., Wang, S.: Big data in forecasting research: a literature review. Big Data Res. **27**, 100289 (2022)
28. Venkatraman, S., Venkatraman, R.: Big data security challenges and strategies. AIMS Math. **4**(3), 860–879 (2019)
29. Villamil, S., Hernández, C., Tarazona, G.: An overview of internet of things. Telkomnika (Telecommun. Comput. Electron. Control) **18**(5), 2320–2327 (2020)
30. Himeur, Y., Ghanem, K., Alsalemi, A., Bensaali, F., Amira, A.: Artificial intelligence based anomaly detection of energy consumption in buildings: a review, current trends and new perspectives. Appl. Energy **287**, 116601 (2021)
31. Wang, K., Zhao, Y., Gangadhari, R.K., Li, Z.: Analyzing the adoption challenges of the Internet of things (IoT) and artificial intelligence (AI) for smart cities in china. Sustainability **13**(19), 10983 (2021)
32. Yaïci, W., Krishnamurthy, K., Entchev, E., Longo, M.: Recent advances in Internet of Things (IoT) infrastructures for building energy systems: a review. Sensors **21**(6), 2152 (2021)
33. Thaseen, I.S., Mohanraj, V., Ramachandran, S., Sanapala, K., Yeo, S.S.: A hadoop based framework integrating machine learning classifiers for anomaly detection in the internet of things. Electronics **10**(16), 1955 (2021)
34. Jabbar, R., et al.: Blockchain technology for intelligent transportation systems: a systematic literature review. IEEE Access **10**, 20995–21031 (2022)
35. Ismagilova, E., Hughes, L., Dwivedi, Y.K., Raman, K.R.: Smart cities: advances in research-an information systems perspective. Int. J. Inform. Manag. **47**, 88–100 (2019)
36. Zhao, L., Tang, Z.Y., Zou, X.: Mapping the knowledge domain of smart-city research: a bibliometric and scientometric analysis. Sustainability **11**(23), 6648 (2019)
37. Moon, J., Kum, S., Lee, S.: A heterogeneous IoT data analysis framework with collaboration of edge-cloud computing: focusing on indoor PM10 and PM2. 5 status prediction. Sensors **19**(14), 3038 (2019)
38. El Jaouhari, S., Jose Palacios-Garcia, E., Anvari-Moghaddam, A., Bouabdallah, A.: Integrated management of energy, wellbeing and health in the next generation of smart homes. Sensors **19**(3), 481 (2019)
39. Ali, O., Jaradat, A., Kulakli, A., Abuhalimeh, A.: A comparative study: blockchain technology utilization benefits, challenges and functionalities. IEEE Access **9**, 12730–12749 (2021)
40. Rasool, R.U., Ahmad, H.F., Rafique, W., Qayyum, A., Qadir, J.: Security and privacy of internet of medical things: a contemporary review in the age of surveillance, botnets, and adversarial ML. J. Netw. Comput. Appl. **201**, 103332 (2022)
41. Vitabile, S., et al.: Medical data processing and analysis for remote health and activities monitoring. In: Kołodziej, J., González-Vélez, H. (eds.) High-Performance Modelling and Simulation for Big Data Applications. LNCS, vol. 11400, pp. 186–220. Springer, Cham (2019). https://doi.org/10.1007/978-3-030-16272-6_7
42. Albahri, A.S., et al.: Based multiple heterogeneous wearable sensors: a smart real-time health monitoring structured for hospitals distributor. IEEE Access **7**, 37269–37323 (2019)

43. Anisha, M., et al.: Automated assistive health care system for disabled patients utilizing internet of things. J. Eng. Sci. Technol. Rev. **13**(4), 206–213 (2020)

44. Albahri, O.S., et al.: Fault-tolerant mHealth framework in the context of IoT-based real-time wearable health data sensors. IEEE Access **7**, 50052–50080 (2019)

45. Zhang, X., Wang, Y.: Research on intelligent medical big data system based on Hadoop and blockchain. EURASIP J. Wirel. Commun. Netw. **2021**(1), 1–21 (2021). https://doi.org/10.1186/s13638-020-01858-3

46. Wang, J., Zheng, P., Lv, Y., Bao, J., Zhang, J.: Fog-IBDIS: industrial big data integration and sharing with fog computing for manufacturing systems. Engineering **5**(4), 662–670 (2019)

47. Sánchez-Corcuera, R., et al.: Smart cities survey: technologies, application domains and challenges for the cities of the future. Int. J. Distrib. Sens. Netw. **15**(6), 1550147719853984 (2019)

48. Yousif, O.S., et al.: Big data integration in the construction industry digitalization. Fronti. Built Environ. 159 (2021)

49. Hassani, H., Huang, X., Silva, E.: Big data and climate change. Big Data Cogn. Comput. **3**(1), 12 (2019)

50. Parisi, F., Fanti, M.P., Mangini, A.M.: Information and communication technologies applied to intelligent buildings: a review. J. Inf. Technol. Constr. **26**, 458–488 (2021)

51. Lawal, K., Rafsanjani, H.N.: Trends, benefits, risks, and challenges of IoT implementation in residential and commercial buildings. Energy Built Environ. **3**, 251–266 (2021)

52. Hassani, H., Huang, X., MacFeely, S., Entezarian, M.R.: Big data and the united nations sustainable development goals (UN SDGs) at a glance. Big Data Cogn. Comput. **5**(3), 28 (2021)

A Proposal for Developing and Deploying Statistical Dialog Management in Commercial Conversational Platforms

Pablo Cañas[ID], David Griol[✉][ID], and Zoraida Callejas[ID]

University of Granada, CITIC-UGR. Pdta. Daniel Saucedo Aranda sn., 18071 Granada, Spain
pcanas@correo.ugr.es, {dgriol,zoraida}@ugr.es

Abstract. Conversational interfaces have recently become ubiquitous in the personal sphere by improving an individual's quality of life and industrial environments by automating services and their corresponding cost savings. However, designing the dialog model used by these interfaces to decide the following response is a hard-to-accomplish task for complex conversational interactions. This paper proposes a statistical-based dialog manager architecture, which provides flexibility to develop and maintain this module. Our proposal has been integrated using DialogFlow, a natural language understanding platform provided by Google to design conversational user interfaces. The proposed hybrid architecture has been assessed with a real use case for a train scheduling domain, proving that the user experience is highly valued and can be integrated into commercial setups.

Keywords: Conversational systems · Dialog management · Statistical approaches

1 Introduction

Conversational Interfaces (CUI) are systems that simulate interactive conversations with humans [10, 11, 18]. These systems are meant to display human-like characteristics and support the use of spontaneous natural language to interact with different purposes, such as performing transactions, responding to questions, or chatting.

These interfaces have become a key research subject for many companies, as they have understood the potential revenue of introducing these devices into society's mainstream. Virtual Personal Assistants (VPAs), such as Google Now, Apple's Siri, Amazon's Alexa, or Microsoft's Cortana, are the most representative examples. These VPAs are used for a wide variety of tasks, from setting the alarm to updating the calendar, passing through getting directions, finding the nearest restaurants or stores, or even planning a recipe or reporting the news. CUis are also used to make their services

The research leading to these results has received funding from the European Union's Horizon 2020 research and innovation program under grant agreement no. 823907 (MENHIR project: https://menhir-project.eu) and the projects supported by the Spanish Ministry of Science and Innovation through the GOMINOLA (PID2020-118112RB-C21 and PID2020-118112RB-C22, funded by MCIN/AEI/10.13039/501100011033), and CAVIAR (TEC2017-84593-C2-1-R, funded by MCIN/AEI/10.13039/501100011033/FEDER).

more engaging and increase customer satisfaction in various applications, such as making appointments and reservations, answering questions, e-government procedures, and support services.

On the other hand, machine learning and deep learning methodologies have become part of many of the current intelligent systems [4, 8, 9]. These methodologies have traditionally been used in the field of conversational interfaces to develop automatic speech recognizers [13, 17] and, more recently, to implement natural language understanding modules, natural language generators, and dialog management processes [1, 3, 9].

The dialog model of the dialog manager defines the conversational system behavior in response to user utterances and environmental states. Thus, the system performance depends significantly on the quality of this model. Early models for dialog strategies were implemented using expert systems, predefined rules, and dialog trees [11]. This methodology consists of manually determining the system's response to each user input. This approach, which is still widely used in most commercial platforms, can be appropriate for very simple use cases. For instance, on systems answering a reduced set of isolated frequently asked questions. However, highly complex dialog systems usually require several user-system turns for a successful interaction, thus making the use of this methodology unfeasible to maintain and scale [10].

Commercial platforms to develop conversational interfaces have integrated new statistical methodologies proposed in the academic settings for natural language understanding (intents and entity recognition). However, they have not yet adopted this perspective for dialog management, which would allow straightforward extension and adjustment of these interfaces to various application domains [10, 11].

To address this problem, this paper proposes a practical framework to develop statistical-based dialog managers that can be easily integrated with toolkits like Google Dialogflow. As proof of concept, we have implemented a practical conversational system for a train scheduling domain. We use the functionalities provided by DialogFlow for natural language understanding, and a statistical dialog manager developed using our proposal with a dialog corpus acquired for the task.

The remainder of the paper is as follows. Section 2 describes related work on statistical data-driven dialog management and the leading available commercial platforms for developing conversational interfaces. Sections 3 and 4 describe an implementation of a practical conversational agent following our proposal and the results of its evaluation. Finally, Sect. 5 presents the main conclusions and future research lines.

2   State of the Art

2.1   SDS Modules

Conversational interfaces and dialog systems have traditionally been developed through the conjunction of several modules that emulate the main processes in human-human speech communication:

- Automatic Speech Recognition (ASR): obtains the sequence of words uttered by a speaker. It is a very complex task that has been a matter of research (see [12, 15]) to overcome the issues of linguistics, transmission channel, or interaction context.

- Spoken Language Understanding (SLU): obtains the semantics from the recognized sentence. It involves morphological, lexical, and syntactical analysis [16].
- Dialog Management (DM): decides the following system's action, interpreting the incoming semantic representation of the user input in the context of the dialog [14].
- Natural Language Generation (NLG): usually considered part of the dialog manager, obtains sentences in natural language from the internal representation of information handled by the dialog system [5].
- Text-To-Speech Synthesis (TTS): transforms the generated sentences into synthesized speech [2].

2.2 Dialog Management

The dialog management process is the main object of study in this paper. The objective of the dialog manager is to consider the information provided by the user during a conversation and the results of accessing the data repositories to decide the subsequent system response. Different statistical techniques have been proposed to automatically learn the dialog model from dialog corpora [10,11]. A widespread methodology to implement statistical dialog management systems is applying reinforcement learning techniques in conjunction with simulated user models to iteratively learn the dialog model by means of the interactions with the simulated user [7]. First, the user model is trained with a real dialog corpus to learn how to respond to a given dialog state. During the second phase, the simulated user interacts with the dialog manager so that it optimizes the dialog strategy based on the feedback given. This allows automatic training with simulated dialogs, and it even enables the exploration of dialog strategies that were not present in the original dialog corpus.

Early models for dialog strategy learning were implemented using Markov Decision Processes (MDPs). MDPs allow forward planning and hence are used as a statistical framework for dialog policy optimization by several authors, such as [6]. However, MDPs assume that the entire space of states is observable, and this assumption does not hold when uncertainty is introduced in the system. As a result, different authors proposed the use of Partially Observable MDPs (POMDPs), which account for uncertainty [19]. Other statistical approaches for dialog management include example-based dialog management, dialog modeling using Hidden Markov Models, stochastic finite-state transducers, and Bayesian networks.

Our proposal for statistical dialog management is described in [1]. The dialog model decides the next system response by employing a classification process that considers the set of pairs (system turn A_i - user turn U_i) preceding the current one to decide the next system response. A data structure (that we called Dialog Register, DR) is used to store the values of the entities provided by the user during the dialog. The classification function can be defined in different ways. In [1], we propose the use of artificial neural networks to improve the results of classifying the information in the DR and the current state of the dialog to select as output the following system response.

2.3 Tools for SDS Implementation

The rise of conversational systems in recent years has boosted the interest of large technology companies, such as Google, Amazon, Microsoft, and IBM, in offering commercial platforms that facilitate the development and deployment of these systems. These platforms provide a set of pre-built components and share key characteristics related to their use and implementation. One of the most important is the use of intents and entities to define the language understanding task.

DialogFlow is a human-computer interaction technology developed by Google to create natural language conversations. It sets the foundation for some Google products, such as Google Assistant or Google Home. Both the speech-to-text and text-to-speech modules are implemented using their technology. The NLU component can be implemented by defining the different possible inputs corresponding to each user's intent, and their pre-trained algorithms will easily identify the intent. The dialog manager can be implemented with predefined answers to each intent, but it also provides a method to add the developer's trained system.

Amazon Lex is a service for building conversational interfaces into any application using voice and text. It is the system powering Alexa, Amazon's VPA. As with DialogFlow, Amazon Lex has already integrated the ASR and TTS modules, utilizing Amazon's deep learning technologies, and the NLU component uses the same intent identification idea. This allows developers to focus on designing the dialog manager response. Besides, it supports easy deployment of the resulting chatbot into mobile applications and other messaging services such as Facebook Messenger or Slack. A very nice feature is that responses for the CUI can be defined directly from the Amazon Web Services (AWS) console, allowing very simple integration of the dialog management system into the SDS.

Microsoft LUis, an acronym for Language Understanding Intelligent Service, is a machine learning-based service to build natural language into apps, bots, and IoT devices. It is used by big multinational companies such as UPS or the actual Microsoft team for their internal and external interactions. As well as its competitors, it allows developers to define user intents in any of the top 15 spoken languages, and it has a predefined machine learning algorithm that predicts the user response very accurately. It also identifies entities and other context variables. This service can be integrated with Microsoft Bot Framework and the Azure Bot Service to develop, deploy, and evaluate intelligent chatbots based on an extensible SDK, tools, templates, and AI services related to speech, such as natural language understanding or question-answering. The developed chatbots can be connected to channels such as Facebook, Messenger, Kik, Slack, Microsoft Teams, Telegram, text/SMS, or Twilio.

As a proof of concept, we have selected Google DialogFlow to show a practical integration of our proposal for statistical dialog management into commercial platforms to develop conversational interfaces. The process to complete this integration does not differ between the described platforms, given the similar descriptions used for the natural language understanding task and the availability of cloud technologies to integrate deep learning frameworks and other services, such as Google Tensorflow and Firebase.

3   Implementation of a Practical Conversational Agent

This section details the implementation of a real conversational system by developing a statistical dialog manager, and its integration with the rest of the functionalities and modules provided by DialogFlow, Firebase Functions, and Firebase Realtime Database. The practical application task that we have selected is to provide information about the Spanish railway system to aid in rail travel planning [3].

3.1 DialogFlow Basic Elements

As described in Sect. 2, a spoken dialog system requires the integration of five main components: the automatic speech recognizer, the language understanding module, the dialog manager, the natural language generator (which can also be integrated as additional functionality of the dialog manager), and the text-to-speech synthesizer. DialogFlow, as part of Google's development toolkit, already provides the ASR and TTS modules. As a result, the only module left to build is the natural language understanding module. For this purpose, DialogFlow has three basic elements that developers can use to build their systems: intents, entities, and contexts.

Intents are the main element in building conversations. Each intent defines examples of user utterances that can trigger the intent, what to extract from the utterance, and how to respond. As a result, the agent will map user inputs to a specific intent to provide a response. This would represent one dialog turn within the conversation. Intents consist of four main components:

1. Intent name. Used to identify the matched intent.
2. Training phrases. Examples of what users can say to match a particular intent. From the ones the developer provides, DialogFlow automatically expands the phrases to match similar user utterances.
3. Actions and parameters. Define the relevant information extracted from user utterances. Examples of this kind of information include dates, times, names, or places.
4. Response. The system output displayed to the user. In our case, responses will not be defined. Instead, user inputs will be sent to a webhook to respond to the user using our statistical model for dialog management.

For the train scheduling task, the conversations gathered in the training corpus were taken into account to build the intent set. Figure 1 shows different examples for the considered intents, along with their translation to English.

Parameters indicate the type of information that should map in each training sentence. For example, although we have trained the intent *Say-Destination* with the city Barcelona, this indicates that it is an object of type *destination*, and could be substituted by any other representative of the same type (Madrid, Seville, Bilbao...). As a result, any time the intent *Say-Destination* is matched, we will have a value for type *destination* that will give us relevant information about the user's query.

It is important to remark that not all parameters necessarily have to appear in each intent. That is because the training phrases were defined based on the corpus set, and parameters were not mentioned in every type of utterance. To give an example, the *departureDate* parameter might appear in the *Say-Destination* intent depending on the information provided by the user.

Intent Name	Training Phrases	Parameters
Change-Departure-Date	¿Y a las 4 de la tarde ? (And for 4 pm ?) ¿Y para el 4 de abril de 2019 ? (And for the 4th of April 2019 ?) ¿Puedes decirme para el 3 de mayo a las 5:00 ? (Could you tell me for May the 3rd at 5:00 ?)	departureDate departureHour
Confirmation	Claro (Of course) Es correcto (That is correct) Correcto (Correct) Sí (Yes)	-
Say-Destination	Mi destino es Barcelona (My destination is Barcelona) Quiero ir a Barcelona en un AVE (I want to go to Barcelona by AVE) A Barcelona (To Barcelona) Ir a Barcelona (Go to Barcelona) A Barcelona el día 7 de abril (To Barcelona the 7th of April) Viajo a Barcelona para mañana (I am travelling to Barcelona tomorrow) El destino es Barcelona (The destination is Barcelona) Voy a Barcelona el día 8 de mayo en AVE (I am going to Barcelona the 8th of May by AVE) Viajo a Barcelona (I am travelling to Barcelona)	destination departureDate trainType
Say-Departure-Date	Para mañana (For tomorrow) Me gustaría salir el 2 de abril (I would like to depart April the 2nd) Para mañana a las 3 (For tomorrow at 3) Salgo el 4 de marzo a las 8 de la tarde (I depart March the 4th at 8 pm) Me gustaría coger el tren a las 5 y cuarto de hoy (I would like to take the train today at quarter past 5) Me gustaría salir el 2 de abril a las 16:00 (I would like to depart April the 2nd at 16:00) Me gustaría coger el tren el 3 de abril (I would like to take the train April the 3rd) Salgo el 4 de marzo (I depart March the 4th)	departureDate departureHour
Say-Arrival-Hour	Sí, quiero que me digas los que lleguen a las 4:00 (Yes, I would like to know the ones that arrive at 4:00) Quiero que llegue a las 4:00 (I would like the train to arrive at 4:00 Me gustaría llegar a las 3:00 (I would like to arrive at 3:00) La hora de llegada debe ser las 3:00 (The arrival hour must be 3:00)	arrivalHour

Fig. 1. Intents defined for the train scheduling SDS, along with their corresponding translation to English

3.2 Entities

Entities are objects used as a mechanism to identify and extract valuable data from natural language. That information can be used and treated as an input into other logic, such as looking up information, carrying out a task, or returning a personalized response.

DialogFlow incorporates a wide variety of predefined system entities, which allow the extraction of information without any additional configuration. Some of these include dates, times, cities, colors, units of measure, or names. However, developers can define custom entities depending on the domain for which the CUI is being built. To do so, two main components are needed:

1. **Entity type**. Defines the type of information to extract from user input.
2. **Entity entry**. These are the elements that belong to the same entity type.

Table 1 shows all the parameters used for this application domain together with their corresponding entity type (system or custom). These parameters correspond to each domain-dependent slot specified to create the deep learning model. As a result, every time one of the parameters is mentioned in the intents, that slot will be denoted as mentioned in order to predict the system response.

Table 1. Parameters defined for the train scheduling SDS

Parameter Name	Entity Type
origin	city (system)
destination	city (system)
departureDate	date (system)
arrivalDate	date (system)
departureHour	time (system)
arrivalHour	time (system)
ticketClass	ticketClass (custom)
trainType	trainType (custom)
services	services (custom)

3.3   Contexts

Contexts represent the current state of a user's request and allow agents to carry information from one intent to another. They can be combined to control the conversational path that the user will take during the dialog. However, the dialog state was not kept using the DialogFlow context element for the train scheduling SDS. In this case, we wanted to keep a considerable number of context variables to make the system work, and using contexts became an arduous task to accomplish. For that reason, using Firebase Realtime Database arose as a feasible alternative. As a result, every time a user input is made, the system will look for the last stored state, and it will update it with the new information gathered, hence always having an updated dialog state. Subsequently, a new database table was created, containing a field for every slot needed for the model prediction as well as the next state and the real values of the users' query parameters.

3.4   DialogFlow Fulfillment

As described, intents have a component to provide a system response once an intent is matched. However, this system is very limited because responses do not include backend logic and can be associated with individual intents. For this reason, we created a service using Firebase Cloud Functions that handle the information extracted by DialogFlow's natural language processor to generate dynamic responses based on the deep learning model created. Figure 2 shows a summary of how all the pieces are assembled, constituting a complete dialog manager for the conversational system.

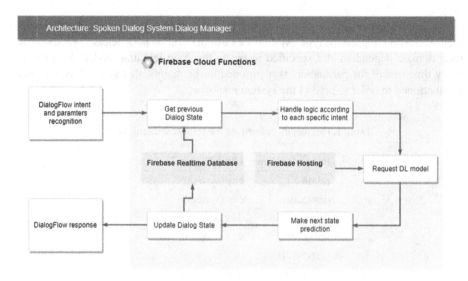

Fig. 2. Dialog manager architecture for the proposed SDS implementation

The architecture goes as follows: first, the model is loaded; later, the input tensor is created; finally, the output is predicted. For this project, the model was uploaded to a server using Firebase Hosting services. Therefore, anytime a request is made to the cloud function, a GET petition is made to the URL where the model is hosted, and the model is loaded for subsequent use. Inside the cloud function, there is a specific handler for each of the different DialogFlow intents previously defined. Such handlers receive as an argument a conversation object that stores all the variables gathered by DialogFlow during the language understanding phase (for example, parameters).

After this, the statistical model predicts the next action based on the current dialog state. Based on this, the system then updates the dialog state, and finally retrieves the appropriate answer to the user.

That is the algorithm followed to retrieve an answer to user utterances: an intent is identified, the handler inside the cloud function is triggered, the model is requested to a hosting service, the database object is updated, the state is sent to the model to make a prediction, and an output message is sent back depending on the prediction made.

There are currently two versions of the Dialogflow platform: Dialogflow Essentials (ES) and Dialogflow Customer Experience (CX). Both versions use the previously described terms of entities, intents, parameters, fulfillment and webhooks. The CX version provides the concept of flows to model more complex conversations by defining conversational paths and grouping them into topics. The concept of pages is also introduced to define the state of the conversation. Each flow is usually composed of several pages and these are in charge of maintaining the dialog with the user. For each page a form can be defined, which contains a list of parameters that are obtained from the users on each page. It is also possible to migrate a project from ES to CX.

The process previously described allows to integrate the statistical dialog model in a conversational system designed with the Dialogflow Essentials version. With regard the Dialogflow Customer Experience version, it is possible to migrate the model created using Dialogflow ES or to define each of the dialog states of the statistical dialog model by means of pages, in which the list of parameters of each page corresponds to the set of entities provided by the user until the current instant of the dialog.

The proposed solution constitutes a hybrid approach for a spoken dialog system implementation. We combine a statistical dialog model with the rule-based response system developed by commercial chatbot development platforms. The statistical model can also follow a hybrid implementation, since it can be made of an ensemble of statistical methods in order to increase the robustness of the system.

4   Evaluation of the Conversational Agent

This section reports the evaluation process followed to validate the conversational agent in an authentic setting. As a result, an evaluation methodology was planned to perform the assessment of the system. After this, real users tested the CUI, out of which the evaluation was made. The assessment process was divided into an objective and a subjective evaluation.

4.1   Experiment Setup

A total of 20 users were interviewed to evaluate the train scheduling chatbot. The selection of the study group was heterogeneous, comprising men and women from an age range between 21 and 60 years old, with the basic knowledge of technology to make them comfortable using a smartphone and who take a medium distance train at least once per month. Furthermore, users were told different details regarding the conversational system:

– Some users were not given any more information and were asked to try the conversational system to solve their requests. With this, we tried to look for possible outliers and classes we had not taken into account during the modeling process.
– Other users were only given specific functionalities of the application. For example, some were explained that the system could solve requests regarding scheduling, while others were asked to request information about prices and services. With this approach, we tested how the conversational system responded to very particular scenarios.
– Finally, another set of users were told about all the system specifications. This population set allowed to test the conversational system as a whole entity.

4.2   Objective Evaluation

For the objective evaluation, we analyzed 9 different metrics extracted from the interactions between the user and the conversational system:

1. Dialog success rate: percentage of dialogs that were finished successfully, with users having their requests resolved.

2. Turn success rate: percentage of turns in which the system responded with a coherent answer to the user's input.
3. Dialog length: average number of turns per dialog. Considering that a turn corresponds to a two-side interaction, a user-system communication will count as one turn.
4. Request length: average number of requests made in one dialog.
5. Number of turns of the shortest dialog.
6. Number of turns of the longest dialog.
7. Percentage of different dialogs.
8. Repetitions of the most observed dialog.
9. Number of turns of the most observed dialog.
10. Number of requests of the most observed dialog.

Table 2 shows the results obtained for the objective evaluation. The results show an 80% success rate and 78% coherent answers rate. Most of the incoherences were found in dialogs ending unsuccessfully. We also found dialogs scoring a perfect 100% turn success rate.

Table 2. Objective evaluation results

Metric	M1	M2	M3	M4	M5	M6	M7	M8	M9	M10
Evaluation value	80%	78%	7	2.89	10	5	55%	6	5	2

4.3 Subjective Evaluation

Besides the objective metrics, users were asked to assess the system's performance subjectively. A total of seven questions were made: 1. How well did the system understand you?; 2. How well did you understand the system messages?; 3. Was it easy for you to get the requested information?; 4. Was the interaction with the system quick enough?; 5. If there were system errors, was it easy for you to correct them?; 6. In general, are you satisfied with the system's performance?; 7. Would you use this system to schedule your future train rides?

Each question was scored from 1 (lowest) to 5 (highest) according to the response given by the user: 1. Never/Not at all; 2. Rarely/Poorly; 3. Sometimes/In some measure; 4. Usually/Well; 5. Always/Very well.

Table 3 shows the subjective evaluation results. It can be observed that results are generally positive, with most answers scoring between 4 and 5. The weakest points of the application have appeared to be related to the system understanding in specific contexts. Users would perceive that the system was coherent usually, but once a mistake was made, it was not easy to recover the conversation thread. There is a total agreement for question 2, where users thought system messages were clear. Also, most of the respondents believed that the interaction was very fast, being a recurrent weak point in spoken dialog systems. In general, users are very satisfied with the performance of the system. Most of them believed that it was easy to get the information they were looking for, and many interviewees affirmed that they would use the system to schedule their future train rides.

Table 3. Subjective evaluation results

Question	Avg. response	Std. deviation	Med. response	Mode response
Q1	3.80	0.83	4	3
Q2	5	0	5	5
Q3	4	1.12	4	5
Q4	4.60	0.71	5	5
Q5	3.30	1.41	4	4
Q6	4.40	0.73	5	5
Q7	4.20	0.68	4	4

5   Conclusions and Future Work

In this paper, we have described a proposal for the integration of statistical methodologies for dialog management. The proposal seamlessly interacts with the rest of the components of an SDS, constituting a hybrid approach in the development of these systems. On the one hand, it simplifies the time and effort required for the development of this module of the conversational system. On the other hand, it allows the development of the rest of its modules using the commercial platforms offered by large technology companies to build chatbots. As a proof of concept, we have integrated a statistical dialog management model in a practical conversational system developed with Google's DialogFlow and Firebase technologies. The results of this system's objective and subjective evaluation show the correct functioning of the system and the overall acceptance by the users. Future work includes proposing an improved statistical management model with additional information that requires error management models and developing techniques for the efficient deployment of this type of system according to the number of users interacting simultaneously with it.

References

1. Canas, P., Griol, D., Callejas, Z.: Towards versatile conversations with data-driven dialog management and its integration in commercial platforms. J. Comput. Sci. **55**, 101443 (2021)
2. Dutoit, T.: An Introduction to Text-to-Speech Synthesis. Kluwer Academic Publishers, New York (1996)
3. Griol, D., Hurtado, L., Segarra, E., Sanchis, E.: A domain-independent statistical methodology for dialog management in spoken dialog systems. Comput. Speech Lang. **28**(3), 743–768 (2014)

4. Kumar, N., Raubal, M.: Applications of deep learning in congestion detection, prediction and alleviation: a survey. Transp. Res. **133**, 103432 (2021)
5. Lemon, O.: Learning what to say and how to say it: joint optimisation of spoken dialogue management and natural language generation. Comput. Speech Lang. **25**, 210–221 (2011)
6. Levin, E., Pieraccini, R., Eckert, W.: Using Markov decision process for learning dialogue strategies. In: Proceedings of ICASSP 1998. vol. 1, pp. 201–204. Seattle, WA, USA (1998)
7. Levin, E., Pieraccini, R., Eckert, W.: A stochastic model of human-machine interaction for learning dialog strategies. IEEE Trans. Speech Audio Process. **8**(1), 11–23 (2000)
8. Li, W., et al.: A perspective survey on deep transfer learning for fault diagnosis in industrial scenarios: theories, applications and challenges. Mech. Syst. Signal Process. **167**, 108487 (2022)
9. Matějů, L., Griol, D., Callejas, Z., Molina, J.M., Sanchis, A.: An empirical assessment of deep learning approaches to task-oriented dialog management. Neurocomputing **439**, 327–339 (2021)
10. McTear, M.: Conversational AI. Dialogue Systems, Conversational Agents, and Chatbots. Morgan and Claypool Publishers (2020)
11. McTear, M., Callejas, Z., Griol, D.: The Conversational Interface. Springer, Cham (2016). https://doi.org/10.1007/978-3-319-32967-3
12. O'Shaughnessy, D.: Invited paper: automatic speech recognition: history, methods and challenges. Pattern Recogn. **41**, 2965–2979 (2008)
13. Rejaibi, E., Komaty, A., Meriaudeau, F., Agrebi, S., Othmani, A.: MFCC-based recurrent neural network for automatic clinical depression recognition and assessment from speech. Biomed. Signal Process. Control **71**, 103107 (2022)
14. Traum, D.R., Larsson, S.: The information state approach to dialogue management. In: van Kuppevelt, J., Smith, R.W. (eds.) Current and New Directions in Discourse and Dialogue. Text, Speech and Language Technology, vol. 22, pp. 325–353. Springer, Netherlands (2003). https://doi.org/10.1007/978-94-010-0019-2_15
15. Tsilfidis, A., Mporas, I., Mourjopoulos, J., Fakotakis, N.: Automatic speech recognition performance in different room acoustic environments with and without dereverberation preprocessing. Comput. Speech Lang. **27**, 380–395 (2013)
16. Wu, W.L., Lu, R.Z., Duan, J.Y., Liu, H., Gao, F., Chen, Y.Q.: Spoken language understanding using weakly supervised learning. Comput. Speech Lang. **24**, 358–382 (2010)
17. Xue, J., Zheng, T., Han, J.: Exploring attention mechanisms based on summary information for end-to-end automatic speech recognition. Neurocomputing **465**, 514–524 (2021)
18. Young, S.: Hey Cyba. The Inner Workings of a Virtual Personal Assistant, Cambridge University Press, Cambridge (2021)
19. Young, S., et al.: The hidden information state model: a practical framework for POMDP-based spoken dialogue management. Comput. Speech Lang. **24**, 150–174 (2010)

Image and Speech Signal Processing

Roadway Detection Using Convolutional Neural Network Through Camera and LiDAR Data

Martín Bayón-Gutiérrez[1]([✉]) [ID], José Alberto Benítez-Andrades[2] [ID],
Sergio Rubio-Martín[2] [ID], Jose Aveleira-Mata[1] [ID], Héctor Alaiz-Moretón[1] [ID],
and María Teresa García-Ordás[1] [ID]

[1] SECOMUCI Research Group, Escuela de Ingenierías Industrial e Informática,
Universidad de León, Campus de Vegazana s/n, 24071 León, Spain
martin.bayon@unileon.es

[2] SALBIS Research Group, Department of Electric, Systems and Automatics
Engineering, Universidad de León, Campus of Vegazana s/n, León, 24071 León, Spain

Abstract. Roadway detection is one of the main topics that autonomous vehicles must face to safety navigate along roads. In this paper, we present the architecture and results of a roadway detection system which uses both camera and LiDAR data to segment the road surface from a Bird's-eye view. Discussion about how camera and LiDAR data has been combined is presented along with example images to later discuss about the neural model that has been developed. The proposed method performs among other state-of-the-art methods on the Kitti-Road dataset. Finally, future research lines are introduced, and it is discussed how the use of the full LiDAR FOV could bring benefits for road detection.

Keywords: Roadway detection · LiDAR · Semantic segmentation

1 Introduction

Autonomous driving is becoming more and more advanced and has become one of the technological challenges of interest to society and science [27,35]. It is important to note the difference between driverless driving, which has existed since 1925, when the company Houdina Radio Control introduced a driverless vehicle remotely controlled by a driver in a nearby vehicle [26], and autonomous driving, which aims to avoid the need for a human being to control the movement of the vehicle [23,40].

In the field of autonomous driving there are different problems that have been solved and some that are still to be solved. Within the problem of roadway detection, it is of interest to obtain systems that are able to detect roadways and their limits. The ultimate interest is to decide whether a region in space is suitable for an autonomous vehicle to drive or not [45].

© Springer Nature Switzerland AG 2022
P. García Bringas et al. (Eds.): HAIS 2022, LNAI 13469, pp. 419–430, 2022.
https://doi.org/10.1007/978-3-031-15471-3_36

One of the first approaches that was taken in this topic was the use of satellite imagery [28]. This method consists of applying image analysis software or artificial intelligence over satellite images of the region of interest (ROI), resulting in a map of the roadway system of such area. Although useful for creating a road network map [46], satellite imagery lacks the necessary precision to detect the bounds of a roadway in which an autonomous vehicle is driving. Another option is detection by cameras installed in the vehicle [1,3]. Although this is a simple and low-cost solution, it has several disadvantages, e.g. when weather conditions are not optimal [29,34]. Some of the methods that seem to offer the best results are those that make use of LiDAR sensors [8]. Although this approach allows a very high level of detail to be obtained [5], it has some drawbacks, for example, the cost of the necessary sensors [24]. However, the fusion of several sensor sources tends to increase the robustness and precision of any system, taking advantage of each sensor strengths and avoiding weaknesses at an individual level [4].

Authors have developed a Neural Network model, which combines RGB images and LiDAR data in a Bird's-eye view (BEV) perspective in order to predict the roadway the vehicle is driving on. The model has been trained and validated using the Kitti-Road benchmark [10], which allows it to be compared along other state of the art methods.

The rest of this paper is organized as follows. Related work is presented in Sect. 2. The methodology of the proposed method is detailed in Sect. 3. In Sect. 4 the setup for the deep learning technique is explained. Finally, results have been discussed in Sect. 5, and we conclude in the same section.

2 Related Work

Within the field of autonomous driving, this research work focuses on the need to develop a roadway detection system that automatically detects the limits of the road on which an autonomous vehicle is driving in order to provide the vehicle with information on whether the ground is suitable for driving or not. As in many other research areas, artificial intelligence and artificial vision seems to have a great impact in terms of precision and computing time [2,7,18].

One approach, is to detect the road lane markings, and consider them as the road boundaries, which the vehicle should never cross to drive safely [22]. This is a well known problem, as this is the way many Advanced Driver Assistance Systems (ADAS) are developed [21,32]. However, this alternative does not take into account the fact that road shoulders are often paved, and thus, they should be considered as an area in which emergency maneuvers could be performed [30]. Considering the road detection system to be installed in an autonomous vehicle, responsible of performing any necessary emergency maneuver, it is of interest to include shoulders into the "driveable" surface of the road.

One of the most common alternatives for road detection, is to perform this as if a semantic segmentation problem was faced. In that case, it is of interest to perform a single class pixel-wise classification, so the result may be a binary map

(Road or not-Road) or a confidence map on about if each pixel corresponds to a road or not [11]. The next important step is to decide which sensors may be used to perform the detection task and the perspective in which the detection will be performed. Here, RGB cameras have traditionally prevail over other types of sensors. Works as [20,42] presented solutions that can perform well in daylight conditions. However, night-time environments tends to negatively impact on camera-based systems as discussed in [13].

Recently, LiDAR-based detection techniques are becoming popular, and novel methods have been published [4,8]. LiDAR sensors are not affected by external light conditions, and therefore, perform well under low-light environments [31]. As other authors have used the Kitti-Road dataset and benchmark for this work [10], it is of interest to note how the results to this benchmark have gone over the years in relation to LiDAR methods.

Even though the Kitti-Road benchmark was made available online in 2013, not many researchers used LiDAR data for theirs work for the few first years, as can be observed in the benchmark website. However, a significant increase in LiDAR-relate works interest can be noted starting in 2019.

In [5,6], authors used just LiDAR data for the road segmentation, while other, e.g. [8,14], combined camera and LiDAR data for this purpose. The use of both types of sensors demonstrated to have a positive impact in comparison with just using LiDAR, leading to higher positions in the leader-board.

Following a revision of the cited methods, authors have developed a neural network model, that takes advantage of both camera and LiDAR data to provide a robust road detection method which is discussed in the following sections.

3   Material and Methods

One of the key points on any research work is to present the methodology that was followed to obtain the results, as well as the data used during this process, so if the work is replicated, the same results should be obtained. In this section, we present the deep learning model we have developed to solve the road detection problem. Discussion about the data used for this work, as well as the preprocessing performed on the data, is presented first. Then, the neural network model architecture is presented to finally discuss about the training performed on the dataset.

3.1   LiDAR and Camera Data

Authors used the Kitti-Road dataset, a public available dataset which includes a leader-board of evaluated methods [12].

The Kitti-Road dataset contains 289 labeled training images, (and 290 unlabeled testing images), from various urban road scenarios with a great variety of circumstances e.g. whether the road is marked or not.

The raw data contains information of several common sensors for autonomous vehicles such as color and gray-scale cameras, 3D LiDAR, GPS and IMU, which are synchronized. Authors have opted to use just the color cameras and 3D LiDAR as the method inputs.

The raw data and labels are organized as follows:

- **Color cameras:** 2 color cameras are installed on top of the vehicle roof facing forward. RGB images are captured at 10 fps and triggered by the 3D LiDAR. Just the left camera is used, as that is the one for which label data is available.
- **3D LiDAR:** A Velodyne HDL-64E 3D LiDAR is installed on top of the vehicle roof. The sensor spins 360° at 10 Hz and generates a 3D pointcloud with a vertical FOV of 26,8°
- **Ground truth Label:** For each color camera image, a ground truth image is provided. The label is encoded as a 3 channel image, in which the third channel contains the road area ground truth.

According to the dataset authors, data has been manually labeled in the perspective view and subsequently refined in the BEV space, which ensures its accuracy [10].

3.2   Data Preprocessing

The evaluation of the images will be performed in the BEV with a ROI of [−10 m, 10 m] laterally and [6 m, 46 m] longitudinally, and a resolution of 0.05 m, which lead to images of 800 × 400 px. The dataset website provides a toolkit which include the code to convert the RGB images to the BEV perspective.

For the 3D LiDAR to adapt to these constraints, a bi-dimensional image, containing the most relevant information, is created from the 3D pointcloud. As in other works, a grid is created from the pointcloud and each point is assigned to a cell according to its coordinates [5]. The result is a 800 × 400 grid that can be encoded as a 3 channel image according to the following schema:

- **Red Channel:** Each pixel is encoded attending to whether any point belongs to the cell (255) or not (0).
- **Green Channel:** Each pixel is encoded attending to the mean intensity of the points belonging to the cell. The intensity is normalized in the [0–255] range.
- **Blue Channel:** Each pixel is encoded attending to the mean height of the points belonging to the cell. The height is first clipped to the [−1.8 m, −1.2 m] range and later normalized in the [0–255] range.

The result is a 3-channel image with dimensions 800 × 400, for which some examples are exposed in Fig. 1. RGB and LiDAR BEV images are then combined to form a 800 × 400 × 6 image that acts the neural network input.

3.3   Neural Network Architecture

Authors have opted to face the problem by means of developing a novel neural network model. The architecture is inspired in the U-Net model presented in [33] from which we have taken the idea of implement two skip connections from the end of convolution blocks 3 and 4 and from VGG-16 model [39], which first layers form the encoder of our method. However, we have substituted the convolutional layers from the middle of the network with two fully connected layers followed by dropout layers. The output of our model is a single channel confidence map of whether each pixel corresponds to a road or not, which is known as road semantic segmentation. The architecture has been implemented on Tensorflow, using the Keras API. A representation of the implemented neural network is presented in Fig. 2.

Fig. 1. Some examples of LiDAR data preprocessing

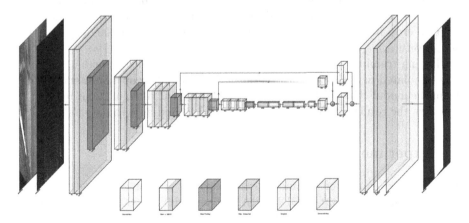

Fig. 2. Neural network architecture for the proposed method

3.4 Training Process

As the Kitti-Road dataset merely consists of 289 labeled images, along with 290 unlabeled testing images, it may be considered as an small dataset, which makes training the model complicated. Authors have divided the dataset in training and validation datasets in a 80/20 rate and have performed data augmentation techniques in the training datasets, as this approach has been demonstrated to have great impacts on artificial intelligence training process [9,38].

The data augmentation process has been performed as follows:

- The training dataset entries have been duplicated, so each instance appears four times in the dataset. This allows for a higher chance that the image suffers from data augmentation, and therefore, the neural network is trained in a wider range of environments.
- Randomly, for each entry:
  - RGB image, LiDAR image and Mask might be flipped in their vertical axis.
  - RGB image brightness might be altered. LiDAR image is keep unchanged as external light conditions does not affect LiDAR output.
  - RGB image contrast might be altered. LiDAR image is keep unchanged as external light conditions does not affect LiDAR output.

Due to the architecture of the neural network developed, in particular the skip connections step, input images are required to have a dimension in each axis that is multiple of 2^5, as there are 5 max pooling layers for which the input shape must be an even number. Authors applied resizing to the dataset from the original 800×400 px to 800×416 px images. During prediction, the output image is resize the opposite way, from 800×416 px to 800×400 px images. This slightly deformation of the image is almost imperceptible, and does not affect the model performance, while it allows to use the full image range, without the need to crop it. Other input sizes could be adapted in a similar way for other sensor setups.

During training, MeanSquaredError was used as the loss function, accuracy was calculated to evaluate model performance and Adam optimizer was used to train the model. The model was trained with different configurations in order to compare the results of those configuration and to find an optimal model.

4 Experiments and Results

For the development of the model, the following environment has been used: Intel Core i9-9900K, 32 GB-DDR4 RAM, Nvidia RTX 2080Ti 11 GB, Ubuntu 20.04, Python 3.9, Tensorflow 2.8.0.

The training was set up to perform up to 200 epochs with an *early-stopping* callback. With the tested configuration, the model converged after 82 epochs, and presented an accuracy of 99,66% on validation split. In Fig. 3, an example of the model prediction on the validation dataset is exposed.

Camera   LiDAR   True mask   Prediction

Fig. 3. Model prediction on a demo sample from the validation dataset

4.1 Results

One of the main advantages of the use of this specific dataset is the fact that it includes an evaluation server that can be used to estimate our method performance on the testing dataset, i.e. data that the model has never seen during training.

Authors submitted their results to the kitti evaluation server and obtained the results exposed in Table 1 for the different categories of the benchmark.

Table 1. Model results from the Kitti-Road evaluation server

Benchmark	MaxF	AP	PRE	REC	FPR	FNR
UM_ROAD	94.06 %	92.13 %	94.32 %	93.80 %	2.57 %	6.21 %
UMM_ROAD	95.41 %	94.83 %	95.23 %	95.59 %	5.26 %	4.41 %
UU_ROAD	92.41 %	90.16 %	92.30 %	02.52 %	2.52 %	7.48 %
URBAN_ROAD	94.20 %	92.66 %	94.25 %	94.14 %	3.16 %	5.86 %

In Fig. 4, some examples for the evaluation process are shown. Here, red denotes false negatives, blue areas correspond to false positives and green represents true positives.

Our method performs among the best works in the Kitti-Road benchmark. In Table 2, authors compare their model along with other state of the art methods published at the dataset leader-board for the category UM_ROAD. Only methods that use LiDAR are included here as it is not in the aim of the authors to compare along camera only methods. The proposed method sits in between the top models for the Kitti-road benchmark, and even performs along the top-3 methods if *Average Precision* is taken as the evaluation metric.

The proposed methods provides a road prediction in 0.05 s (20 fps), which implies that the method can perform under real-time constraints, as the cameras and LiDAR provides information at a 10 fps rate.

Fig. 4. Model prediction on some samples from the Kitti-Road evaluation server

Table 2. LiDAR-based road detection systems on the Kitti-Road dataset

Method	Year	MaxF	AP	LiDAR	Camera	Time	Cite
PLARD	2019	97.05 %	93.53 %	✓	✓	0.16 s	[8]
CLCFNet	2021	95.65 %	89.49 %	✓	✓	0.02 s	[14]
LidCamNet	2018	95.62 %	93.54 %	✓	✓	0.15 s	[4]
CLCFNet (LiDAR)	2021	95.16 %	89.18 %	✓	✗	0.02 s	[14]
TVFNet	2019	94.96 %	89.17 %	✓	✗	0.04 s	[16]
LC-CRF	2019	94.91 %	86.41 %	✓	✓	0.18 s	[15]
CLRD(ours)	2022	94.06 %	92.13 %	✓	✓	0.05 s	
ChipNet	2019	93.73 %	87.62 %	✓	✗	0.01 s	[25]
HID-LS	2019	93.10 %	86.38 %	✓	✓	0.25 s	[17]
LoDNN	2017	92.75 %	89.98 %	✓	✗	0.01 s	[5]
DCF	2021	92.54 %	82.80 %	✓	✓	0.1 s	[44]
MixedCRF	2017	91.57 %	84.68 %	✓	✓	6 s	[19]
HybridCRF	2018	90.99 %	85.26 %	✓	✓	1.5 s	[43]
LidarHisto	2017	89.87 %	83.03 %	✓	✗	0.1 s	[6]
FusedCRF	2015	89.55 %	80.00 %	✓	✓	2 s	[41]
GRES3D+VELO	2015	85.43 %	83.04 %	✓	✗	0.06 s	[37]
RES3D-Velo	2014	83.81 %	73.95 %	✓	✓	0.36 s	[36]

5   Discussion and Conclusions

This paper presented the work carried out to the design of a roadway detection system based on a novel neural network architecture.

Authors have conducted a revision on state of the art road detection methods and techniques, which allowed them to realize that a multi-sensor setup may provide great results in challenging weather conditions. In Sect. 3, authors discussed their data preprocessing procedure along the novel architecture implemented in this work. The developed model performed along the best models of its type on the Kitti-Road benchmark as stated in Sect. 4 and serves as a basis for the authors to continue working in this research line for future work.

The results of the model have been uploaded to the Kitti-Road leader-board, under the name *CLRD*, and are publicly available on the dataset website.

5.1   Future Work

As stated previously, the fact that the dataset only contains 289 labeled images limits how well the model can learn to detect a road. In addition, the fact that LiDAR and cameras are used together limits the ROI in which LiDAR data can be used, as camera images are only available for the front region of the vehicle. Using the publicly available Raw data from the Kitti Dataset project [12], authors have created an alternative dataset composed of 2000 manually labeled images from a wider BEV than the one available on the Kitti-Road dataset. Each label has been annotated as a single channel image from a LiDAR BEV ROI of $[-10\,\mathrm{m}, 10\,\mathrm{m}]$ laterally, $[-20\,\mathrm{m}, 20\,\mathrm{m}]$ longitudinally. The result is a much bigger dataset in which 360° predictions can be performed.

In Fig. 5, we present some of our preliminary results of our model trained in the aforementioned dataset using only LiDAR data. Authors expect to present and release the dataset in a future paper.

Fig. 5. Model predictions on a larger own dataset

References

1. Alvarez, J., Lopez, A., Baldrich, R.: Illuminant-invariant model-based road segmentation, pp. 1175–1180. IEEE (2008). https://doi.org/10.1109/IVS.2008.4621283
2. Bogue, R.: The role of artificial intelligence in robotics. Ind. Robot: Int. J. **41**, 119–123 (2014). https://doi.org/10.1108/IR-01-2014-0300
3. Bolte, J.A., Bar, A., Lipinski, D., Fingscheidt, T.: Towards corner case detection for autonomous driving, pp. 438–445 (2019). https://doi.org/10.1109/IVS.2019.8813817. ISSN 2642-7214
4. Caltagirone, L., Bellone, M., Svensson, L., Wahde, M.: Lidar-camera fusion for road detection using fully convolutional neural networks. Robot. Auton. Syst. **111**, 125–131 (2019). https://doi.org/10.1016/J.ROBOT.2018.11.002
5. Caltagirone, L., Scheidegger, S., Svensson, L., Wahde, M.: Fast lidar-based road detection using fully convolutional neural networks, pp. 1019–1024. IEEE (2017). https://doi.org/10.1109/IVS.2017.7995848
6. Chen, L., Yang, J., Kong, H.: Lidar-histogram for fast road and obstacle detection. In: Proceedings of the IEEE International Conference on Robotics and Automation, pp. 1343–1348 (2017). https://doi.org/10.1109/ICRA.2017.7989159
7. Chen, L., Chen, P., Lin, Z.: Artificial intelligence in education: a review. IEEE Access **8**, 75264–75278 (2020). https://doi.org/10.1109/ACCESS.2020.2988510
8. Chen, Z., Tao, D., Zhang, J.: Progressive lidar adaptation for road detection. IEEE/CAA J. Automatica Sinica **6**, 693–702 (2019)
9. van Dyk, D.A., Meng, X.L.: The art of data augmentation. J. Comput. Graph. Stat. **10**, 1–50 (2001). https://doi.org/10.1198/10618600152418584
10. Fritsch, J., Kühnl, T., Geiger, A.: A new performance measure and evaluation benchmark for road detection algorithms. In: 16th International IEEE Conference on Intelligent Transportation Systems (ITSC 2013) (2013)
11. Garcia-Garcia, A., Orts-Escolano, S., Oprea, S., Villena-Martinez, V., Garcia-Rodriguez, J.: A review on deep learning techniques applied to semantic segmentation. arXiv preprint arXiv:1704.06857 (2017)
12. Geiger, A., Lenz, P., Stiller, C., Urtasun, R.: Vision meets robotics: the KITTI dataset. Int. J. Robot. Res. **32**, 1231–1237 (2013). https://doi.org/10.1177/0278364913491297
13. González, A., et al.: Pedestrian detection at day/night time with visible and fir cameras: a comparison. Sensors **16**, 820 (2016). https://doi.org/10.3390/s16060820
14. Gu, S., Yang, J., Kong, H.: A cascaded lidar-camera fusion network for road detection, pp. 13308–13314. IEEE (2021). https://doi.org/10.1109/ICRA48506.2021.9561935
15. Gu, S., Zhang, Y., Tang, J., Yang, J., Kong, H.: Road detection through CRF based lidar-camera fusion. In: Proceedings of the IEEE International Conference on Robotics and Automation, pp. 3832–3838 (2019). https://doi.org/10.1109/ICRA.2019.8793585
16. Gu, S., Zhang, Y., Yang, J., Alvarez, J.M., Kong, H.: Two-view fusion based convolutional neural network for urban road detection. In: IEEE International Conference on Intelligent Robots and Systems, pp. 6144–6149 (2019). https://doi.org/10.1109/IROS40897.2019.8968054
17. Gu, S., Zhang, Y., Yuan, X., Yang, J., Wu, T., Kong, H.: Histograms of the normalized inverse depth and line scanning for urban road detection. IEEE Trans. Intell. Transp. Syst. **20**, 3070–3080 (2019). https://doi.org/10.1109/TITS.2018.2871945

18. Hamet, P., Tremblay, J.: Artificial intelligence in medicine. Metabolism **69**, S36–S40 (2017). https://doi.org/10.1016/j.metabol.2017.01.011
19. Han, X., Wang, H., Lu, J., Zhao, C.: Road detection based on the fusion of lidar and image data. Int. J. Adv. Robotic Syst. **14**, 172988141773810 (2017). https://doi.org/10.1177/1729881417738102
20. Hu, X., Rodriguez, F.S.A., Gepperth, A.: A multi-modal system for road detection and segmentation, pp. 1365–1370. IEEE (2014). https://doi.org/10.1109/IVS.2014.6856466
21. Jung, C.R., Kelber, C.R.: Lane following and lane departure using a linear-parabolic model. Image Vis. Comput. **23**, 1192–1202 (2005). https://doi.org/10.1016/j.imavis.2005.07.018
22. Kaur, G., Kumar, D.: Lane detection techniques: a review. Int. J. Comput. Appl. (2015)
23. Kröger, F.: Automated driving in its social, historical and cultural contexts. In: Maurer, M., Gerdes, J.C., Lenz, B., Winner, H. (eds.) Autonomous Driving, pp. 41–68. Springer, Heidelberg (2016). https://doi.org/10.1007/978-3-662-48847-8_3
24. Li, Y., Ibanez-Guzman, J.: Lidar for autonomous driving: the principles, challenges, and trends for automotive lidar and perception systems. IEEE Signal Process. Mag. **37**, 50–61 (2020). https://doi.org/10.1109/MSP.2020.2973615. Conference Name: IEEE Signal Processing Magazine
25. Lyu, Y., Bai, L., Huang, X.: ChipNet: real-time lidar processing for drivable region segmentation on an FPGA. IEEE Trans. Circuits Syst. I: Regular Papers **66**, 1769–1779 (2019). https://doi.org/10.1109/TCSI.2018.2881162
26. Magazine, T.: Science: Radio auto. Time Magazine (1925)
27. Martínez-Díaz, M., Soriguera, F., Pérez, I.: Autonomous driving: a bird's eye view. IET Intell. Transp. Syst. **13**, 563–579 (2019). https://doi.org/10.1049/iet-its.2018.5061
28. Mokhtarzade, M., Zoej, M.V.: Road detection from high-resolution satellite images using artificial neural networks. Int. J. Appl. Earth Obs. Geoinf. **9**, 32–40 (2007). https://doi.org/10.1016/j.jag.2006.05.001
29. Nayar, S., Narasimhan, S.: Vision in bad weather. In: Proceedings of the Seventh IEEE International Conference on Computer Vision (1999)
30. Ogden, K.: The effects of paved shoulders on accidents on rural highways. Accid. Anal. Prev. **29**, 353–362 (1997). https://doi.org/10.1016/S0001-4575(97)00001-8
31. Rashed, H., Ramzy, M., Vaquero, V., Sallab, A.E., Sistu, G., Yogamani, S.: Fuse-MODNet: real-time camera and lidar based moving object detection for robust low-light autonomous driving. In: Proceedings of the 2019 International Conference on Computer Vision Workshop, ICCVW 2019, pp. 2393–2402 (2019). https://doi.org/10.1109/ICCVW.2019.00293
32. Raviteja, S., Shanmughasundaram, R.: Advanced driver assistance system (ADAS), pp. 737–740. IEEE (2018). https://doi.org/10.1109/ICCONS.2018.8663146
33. Ronneberger, O., Fischer, P., Brox, T.: U-Net: convolutional networks for biomedical image segmentation. In: Navab, N., Hornegger, J., Wells, W.M., Frangi, A.F. (eds.) MICCAI 2015. LNCS, vol. 9351, pp. 234–241. Springer, Cham (2015). https://doi.org/10.1007/978-3-319-24574-4_28
34. Sakaridis, C., Dai, D., Gool, L.V.: Semantic foggy scene understanding with synthetic data. Int. J. Comput. Vis. **126**, 973–992 (2018)
35. Shima, T., Nagasaki, T., Kuriyama, A., Yoshimura, K., Sobue, T.: Fundamental technologies driving the evolution of autonomous driving. Hitachi Rev. **65**, 427 (2016)

36. Shinzato, P.Y., Wolf, D.F., Stiller, C.: Road terrain detection: avoiding common obstacle detection assumptions using sensor fusion. In: Proceedings of the IEEE Intelligent Vehicles Symposium, pp. 687–692 (2014). https://doi.org/10.1109/IVS. 2014.6856454

37. Shinzato, P.Y.: Estimation of obstacles and road area with sparse 3D points. Universidade de São Paulo (2015). https://doi.org/10.11606/T.55.2015.tde-07082015-100709

38. Shorten, C., Khoshgoftaar, T.M.: A survey on image data augmentation for deep learning. J. Big Data **6**, 60 (2019). https://doi.org/10.1186/s40537-019-0197-0

39. Simonyan, K., Zisserman, A.: Very deep convolutional networks for large-scale image recognition (2015)

40. Wong, K., Gu, Y., Kamijo, S.: Mapping for autonomous driving: opportunities and challenges. IEEE Intell. Transp. Syst. Mag. **13**, 91–106 (2021). https://doi.org/10.1109/MITS.2020.3014152. Conference Name: IEEE Intelligent Transportation Systems Magazine

41. Xiao, L., Dai, B., Liu, D., Hu, T., Wu, T.: CRF based road detection with multi-sensor fusion. In: Proceedings of the IEEE Intelligent Vehicles Symposium, pp. 192–198 (2015). https://doi.org/10.1109/IVS.2015.7225685

42. Xiao, L., Dai, B., Liu, D., Zhao, D., Wu, T.: Monocular road detection using structured random forest. Int. J. Adv. Robot. Syst. **13**, 101 (2016). https://doi.org/10.5772/63561

43. Xiao, L., Wang, R., Dai, B., Fang, Y., Liu, D., Wu, T.: Hybrid conditional random field based camera-lidar fusion for road detection. Inf. Sci. **432**, 543–558 (2018). https://doi.org/10.1016/J.INS.2017.04.048

44. Yang, F., Yang, J., Jin, Z., Wang, H.: A fusion model for road detection based on deep learning and fully connected CRF. In: 2018 13th System of Systems Engineering Conference, SoSE 2018, pp. 29–36 (2018). https://doi.org/10.1109/SYSOSE.2018.8428696

45. Yenikaya, S., Yenikaya, G., Düven, E.: Keeping the vehicle on the road - a survey on on-road lane detection systems. ACM Comput. Surv. **46**, 1–43 (2013). https://doi.org/10.1145/2522968.2522970

46. Zaletelj, J., Burnik, U., Tasic, J.F.: Registration of satellite images based on road network map, pp. 49–53. IEEE (2013). https://doi.org/10.1109/ISPA.2013.6703713

A Conversational Agent for Medical Disclosure of Sexually Transmitted Infections

Joan C. Moreno[1]([⊠])(iD), Victor Sánchez-Anguix[2]([⊠])(iD), Juan M. Alberola[1]([⊠])(iD), Vicente Julián[1]([⊠])(iD), and Vicent Botti[1]([⊠])(iD)

[1] Valencian Research Institute for Artificial Intelligence (VRAIN), Universitat Politècnica de València, Camino de Vera, s/n, 46022 Valencia, Spain
{joamoteo,jalberola,vjulian,vbotti}@upv.es
[2] Instituto Tecnológico de Informática, Grupo de Sistemas de Optimización Aplicada, Ciudad Politécnica de la Innovación, Edificio 8g, Universitat Politècnica de València, Camino de Vera s/n, 46022 Valencia, Spain
vicsana1@upv.es

Abstract. Sexually transmitted infections (STIs) are serious health problems worldwide, increasing the risk of infection by Human Immunodeficiency Virus (HIV)/Acquired Immune Deficiency Syndrome (AIDS). Despite the significant efforts to address the pandemic, especially with sex education programs, STI and HIV remain a significant concern. Meanwhile, conversational agents are becoming popular in healthcare to interact with users and improve their health. This paper reports the design, implementation, and evaluation of VIHrtual-App: an online conversational agent which uses user-centered development and supervised machine learning to offer an engaging sex educational tool to promote awareness and prevention. VIHrtual-App can identify more than 250 STI/HIV-related questions and respond accordingly, attractively providing reliable information.

Keywords: Conversational agent · Sexual education · STI · HIV · Natural language · Machine learning · Chatbot

1 Introduction

Sexually transmitted infections (STIs) represent a major public health problem. The World Health Organization (WHO) reported in 2016 that it was estimated that more than 1 million people become infected daily by chlamydia, gonorrhea, syphilis, and other infections [1]. When left untreated, STIs increase the risk of transmission and the susceptibility to Human Immunodeficiency Virus (HIV); therefore, they might be significant in fighting it [2].

Despite intense efforts to address the pandemic [3], HIV and Acquired Immune Deficiency Syndrome (AIDS) remain an issue. In 2020, the European Centre for Disease Prevention and Control and the WHO Regional Office for Europe published their annual HIV/AIDS surveillance report [4]. The study

© Springer Nature Switzerland AG 2022
P. García Bringas et al. (Eds.): HAIS 2022, LNAI 13469, pp. 431–442, 2022.
https://doi.org/10.1007/978-3-031-15471-3_37

pointed out that, although there have been improvements in the early diagnosis of HIV, there were 24801 new diagnoses and 2772 new AIDS cases in 2019. Besides, stigma and discrimination may have hidden the disease in many societies and continues to do so [5].

In this context, prevention and awareness programs have proved to be a positive approach to reducing new infections [6]. Historically, STI/HIV education programs have been designed to be delivered in schools or community settings, where they can target young people with an elevated risk of infection [7]. Although these strategies have successfully prevented risky sexual behaviors, they present inherent barriers that can reduce efficiency. As exposed by Simon and Daneback [8], sexual health could be a difficult topic for adolescents, who reported that discussing sex with teachers, parents, or even friends could be embarrassing. This perception can prevent teenagers from asking and solving their doubts, and consequently, it can increase the risk of infection. In contrast, the Internet is perceived as a private and anonymous place where they can safely seek information. Nowadays, youth people rely on it as a resource to find information about different health topics, such as physical fitness, diet, and sex [9]. However, the large volume of available information and its unregulated nature pose potential problems for these users, which usually use suboptimal strategies to decide which health sites to trust [10]. As a result, users could adapt their behavior based on false or low-quality health information and, thus, likely suffer a deterioration of their health. Therefore, it is necessary to adapt sex educational programs to this new situation and offer reliable information on the Internet.

Furthermore, conversational agents are gaining momentum in healthcare [11, 12]. Conversational agents are interfaces that enable users to interact with an application using natural language, either by voice or text. The survey by Zeni et al. [13] discussed the current state of the art of these conversational agents in health, which noted that conversational agents' growing popularity had attracted attention as a human-machine interaction way to provide health care services or to automate some medical procedures. Accordingly, such agents could be helpful to help users by providing relevant information about a disease or guiding them through medical routines like taking medication or improving lifestyle habits. In this sense, artificial intelligence allows us to build conversational agents available at any time to offer meaningful and relevant responses about STI/HIV. Using natural language understanding (NLU), conversational agents can engage users and educate them to avoid risky behaviors. Moreover, users can ask and solve their specific doubts without worrying about shaming barriers or privacy.

This paper reports the design, implementation, and evaluation of VIHrtual-App[1]: an online Spanish conversational agent which uses user-centered development and supervised machine learning to offer an engaging and reliable source of information about STI/HIV. Thus, the project aims to offer an innovative educational tool to promote awareness and prevention.

The rest of the paper is organized as follows. In Sect. 2, we detail the architecture of VIHrtual-App. In Sect. 3, we present an evaluation of the proposal. Finally, Sect. 4 draws some concluding remarks and future work lines.

---

[1] http://vihrtualapp.gti-ia.upv.es.

2   Conversational Agent Proposal

VIHrtual-App development has been carried out in three steps: *design, implementation*, and *evaluation* of the model. The first step involved platform and UI design. The conversational agent's requirements and features were also articulated in this step. The second step involved building and deploying the platform with all the features specified in the first step. A conversation-driven development was used during this stage to adapt and improve the conversational agent to the user's needs. The evaluation of the conversational agent model is presented and discussed in the next section.

2.1   Framework Choice

Several technologies can be found in the literature to build conversational agent platforms, such as *Microsoft Bot Framework*[2], *Botkit*[3], *Amazon Lex*[4], or *Rasa*[5]. For this project, we selected the free and open-source *Rasa* framework because it also offers *Rasa X*[6], a tool for collecting and reviewing conversations.

In a *Rasa* project, the NLU pipeline defines the processing steps that convert unstructured user messages into intents and entities. It consists of a series of components, which can be configured and customized [14]. These components shape the NLU pipeline and work sequentially to process user input into a structured output. Different components can be used for entity extraction, intent classification, response selection, and more [15]. As seen in Fig. 1, the processing stage is composed of a tokenizer, featurizer, entity extractor, and intent classifier.

Fig. 1. Rasa NLU pipeline [14]

2.2   Design

Knowledge Base. The definition of the information requirements during the design step was supported by the Infections Unit of Elche Hospital (Spain). Several clinic researchers built a corpus with 132 linked questions and answers (QA) about the virus, including symptoms, transmission, prevention, treatments, and testing. This report formed the initial knowledge database of the conversational agent and was designed to cover the most common topics providing valuable and reliable information to the users.

---

[2] https://docs.microsoft.com/en-us/azure/bot-service/?view=azure-bot-service-4.0.
[3] https://github.com/howdyai/botkit/blob/main/packages/docs/index.md.
[4] https://aws.amazon.com/en/lex/.
[5] https://rasa.com/.
[6] https://rasa.com/docs/rasa-x/.

App Architecture. Due to the high computational and storage requirements, VIHrtual-App is based on traditional client-server architecture. On the one hand, the back-end consists of an HTTP server that exposes the conversational agent's model through an API that can be consumed from different channels, such as a website or an instant messaging platform like *Facebook* or *Telegram*. On the other hand, an additional action server also allows the execution of *Python* code for implementing additional functions into the conversational agent, namely custom text processing and rich responses.

Considering the project's open and informative nature, the conversational agent has a web design. Accordingly, VIHrtual-App uses both servers and a web application as the front-end. As seen in Fig. 2, users interact with the conversational agent using a web page, which communicates with the conversational server. Then, when required, it can request some custom functions from the action server to offer richer responses.

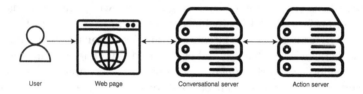

Fig. 2. VIHrtual-App architecture

User Interface. The user interface (UI) is designed by considering different specifications. First, it should be adaptable to multiple resolutions and form factors. Second, it should be easy to use by providing confidence and trust. Finally, the UI should provide a wiki to check information in a more traditional way. To define the conversational agent's colors, it is important considering the colors and their effects on people's feelings. The final result of the design can be seen in Fig. 3 and was approved by experts from the Infections Unit of Elche Hospital.

Social Characteristics. An important issue when designing a conversational agent is the difficulty of engaging users. Usually, conversational agents do not show feelings, affecting their credibility [16]. In this context, the study by Chaves and Gerosa (2020) [17] presents a survey about social characteristics in human-agent interaction design. In their paper, they review the current literature on how a conversational agent should act, what personality traits it should have, and different observed behaviors that improve the user's experience. Specifically, they recommend implementing traits such as proactivity, conscientiousness, communicability, social intelligence, and damage control techniques to recover from failures gracefully.

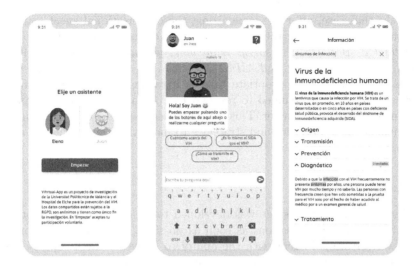

Fig. 3. VIHrtual-App wireframes

During the design step, several conversation paths were designed to engage and inform users about the essential HIV topics. For this task, various techniques were planned to implement the abovementioned features. First, a welcome message was considered to inform users about the conversational agent's purpose and functionalities, which has proven to minimize bad experiences [17]. Second, the use of buttons as quick replies allowed users to quickly dive into the experience with minimum effort. For example, three topics are suggested after the welcome message using easy-to-tap buttons. During the conversation, the agent can use the same technique to provide more topics to keep the conversation alive. All these guiding techniques were used in association with sporadic follow-up questions as a way to introduce new topics and inspire users. Furthermore, conversation paths are designed to follow the conversational flow to demonstrate attentiveness and understanding of the context. This approach is critical to enable the conversational agent to interpret each piece of text as part of a whole conversation, providing the ability to respond considering the previous conversation turns and offering meaningful answers. Thus, the agent conveys conscientiousness and empathy by keeping track of the discussed topic [17]. Additionally, several primary expression pathways are considered to manage typical situations in dialogue (e.g.: *Hi!*, *Good morning*, *How are you?*, *I'm fine, thanks!*). Moreover, some paths are also planned to deal with chitchat and off topics (e.g.: *What's your name?*, *How old are you?*, *Are you a robot?*).

Besides, some damage control techniques are also planned to deal with and recover from possible failures. A graceful path is set to apologize when the conversational agent could not recognize the user's intentions, including asking to rephrase the expression to overcome the lack of knowledge. A similar strategy is set to handle situations where the user could indicate frustration or dissatisfaction. In that case, after the apologizing, the conversational agent could ask some questions to identify the problem and try to solve it (e.g.: *User: I don't understand it. Bot: Oh, I'm sorry! What don't you understand? User: What's the difference between HIV and AIDS?*).

Finally, to handle inappropriate talk or harassing attitudes by users, the conversational agent is planned to respond in a socially acceptable manner to avoid encouraging the behavior. This plan includes responses demonstrating the user's unacceptable behavior and strategies to move the conversation to a relevant topic.

2.3   Implementation

The first implementation stage was developed to create a very first prototype in order to collect conversations as soon as possible. This aim was to provide a basic conversational agent with the ability to answer a small subset of proposed questions while interacting with real users. Then, conversations could be collected and used for further developments.

Initial Dataset. The initial training dataset is the information used to train a conversational system to recognize users' intentions and respond appropriately. This data was structured through files in *YAML* format (*YAML Ain't Markup Language*) containing different entities representing a conversation. This information is represented using several concepts: *Intents* define different example questions that users may ask; *responses* responses contain the messages that the conversational agent will return, and *stories* define conversation patterns by linking *intents* with *responses*.

In order to allow the conversational agent to have broad and rich conversations, *intents*, *responses*, and *stories* belonging to 3 different areas were added to this initial dataset: medical data, basic expressions, and chitchat. Medical data was oriented to answer all the questions about HIV and included all QA from the knowledge database provided by the experts. Basic expressions, as mentioned before, were added to manage sentences that are usually part of a dialogue and chitchat to deal with off topics. In addition, since an intention can be expressed in multiple ways, each one was initially added with 20 different variations to broaden the recognition spectrum. These variations increased as more data was collected from real conversations (see next section).

Responses are crucial to form and transmit the social characteristics discussed in Sect. 2.2. Therefore, when constructing the dataset, different techniques were applied to build responses before adding them:

- Split long answers in multiple messages.
- Replace very formal terminology with more familiar terms.
- Use some emojis to illustrate the information.
- Add answers with video and images to improve understanding.

Additionally, multiple variations of responses were added, which are used randomly to avoid conveying robotic feelings. These alterations were introduced to find a balance between information quality and attraction. Finally, all training data was used to generate a new model and deploy the agent's prototype.

Conversation Driven Development (CDD). One of the big concerns when developing a conversational agent is how training datasets are built. As seen in the last section, the VIHrtual-App's dataset was initially hand-created. Although this is necessary (especially when starting a new project), it presents some long-term problems. Usually, these datasets are created by developers themselves; therefore, they are biased and far from real situations. In this scenario, a conversation-driven development paradigm is vital to solving this issue since it defines a process of listening to real users and using those insights to improve the conversational agent [18]. This development strategy fits with the project aim of deploying a prototype and using its gathered conversational data for machine learning. Therefore, once the VIHrtual-App prototype was ready, it was deployed to collect and supervise conversational agent training.

During this period, the conversational agent entered a CDD cycle where conversations were collected and reviewed to extract useful information. This collected data was added to the dataset, fixing detected problems and performing new trainings. The agent's knowledge database and capacity to chat with users grew by iterating multiple times over this cycle. Over 150 conversations were collected and used throughout this process to improve the agent. In total, 140 new *intents* and 38 *stories* were added to the model.

3 Evaluation

Once the conversational agent is open to the public, it receives and processes information that is not part of its training dataset. To observe how it would behave under these circumstances, it is possible to evaluate the generated NLU model and measure confidence values. In this section several test results are presented, comparing different pipelines and evaluating models.

3.1 Tunning Pipeline

As seen in Sect. 2.1, incoming messages are processed by a sequence of components. Choosing an NLU pipeline allows one to customize and fine-tune the generated model on the dataset. In this work, we experimented with a dataset including 260 intents, 5 entities, and 6040 examples to find the best performance. To this end, three pipeline configurations were tested: default, BERT, and SpaCy. Default configuration was created using the *Suggested Config* [19] feature by Rasa, which chose a sensible default pipeline that includes a white space tokenizer, *RegexFeaturizer*, and the *Dual Intent Entity Transformer* (DIET) [20] classifier for intents and entities. In the BERT pipeline, a white space tokenizer was used along with the *LaBSE* model [21], a language-agnostic model pre-trained on the top 109 languages using BERT architecture [22,23]. The DIET classifier was also used for intent and entities. Finally, the SpaCy pipeline was built using SpaCy Spanish tokenizer and *SpacyFeaturizer*, which provides pre-trained word embeddings for the Spanish language. In particular, this pipeline used the medium news model in its Spanish version[7].

All these three models were evaluated based on three metrics: precision (1), recall (2), and F1-score (3), where TP (true positives) is the number of correctly predicted observations and FN (false negatives) is the number of incorrectly predicted observations.

$$Precision = \frac{TP}{TP + FP} \tag{1}$$

$$Recall = \frac{TP}{TP + FN} \tag{2}$$

$$F1 = 2 \times \frac{Precision \times Recall}{Precision + Recall} \tag{3}$$

The evaluation was performed using cross-validation to measure intent classification and response selection. The results in Table 1 show the macro-average score on the test set over 5-fold cross-validation. As can be seen, the three models obtain similar results, where the BERT model is the best for intent recognition (F1-score = 0.898) and response selection (F1-score = 0.939). In contrast SpaCy gives the worst results for intent recognition (F1-score = 0.859) and response selection (F1-score = 0.863). This can be explained since SpaCy models are trained on Wikipedia's data; thus, they do not fit with the specific terminology of the STI/HIV domain. Hence, the default configuration is more suitable for the training set and offers slightly better results (intent recognition F1-score = 0.879 and response selection F1-score = 0.903).

In addition, a pipeline comparison with different amounts of training data was performed. Figure 4 shows how the default, BERT, and SpaCy models behave using different percentages of training data (0%, 25%, 50%, and 75%). As shown, BERT obtains better scores with higher data exclusions, although this difference becomes lower as the exclusion decreases. As it can be appreciated, the behavior of both the default and SpaCy pipelines is similar. Based on these results, the BERT pipeline is the configuration that generates the model with the best performance. Hence, it is used as the default configuration for training the model.

---

[7] https://spacy.io/models/es.

Table 1. Comparison results of three different NLU pipelines

Model	Metric	Intent prediction	Response selection
Default	Precision	0.879	0.915
	Recall	0.882	0.904
	F1-score	0.879	0.903
BERT	Precision	0.905	0.950
	Recall	0.896	0.934
	F1-score	0.898	0.939
SpaCy	Precision	0.873	0.889
	Recall	0.847	0.855
	F1-score	0.859	0.863

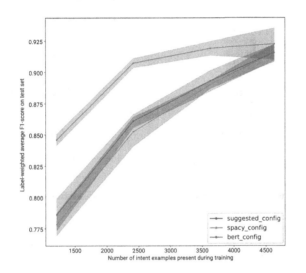

Fig. 4. Pipelines comparison with different training data exclusions

3.2   NLU Model

As mentioned in Sect. 2.3, VIHrtual-App development has been centered in a CDD paradigm, and therefore, the conversational agent's model has been improving its capacities since it was first deployed. This caused the number of intents to grow from 120 to 260 and training examples from 2728 to 6040. Furthermore, 38 stories were added to the initial 43 set. In Table 2, some intent examples are presented with their evolution since the prototype was deployed.

Figure 5 shows the BERT pipeline's intent prediction and response selection confidence distribution. As it can be seen, most examples are accurately selected, and just a few items are not correctly classified or predicted. Accordingly, this experiment, along with the evaluation seen above, shows that the overall performance of the current model is good. Nevertheless, future work can be done

Table 2. Intents evolution

Intent	Init. examples	Init. F1-score	Curr. examples	Curr. F1-score
greet	48	0.72	77	0.81
understand	30	0.69	35	0.91
want_to_ask	39	0.75	72	0.84
affirm	47	0.61	57	0.76
do_i_have_hiv	30	0.88	48	0.91

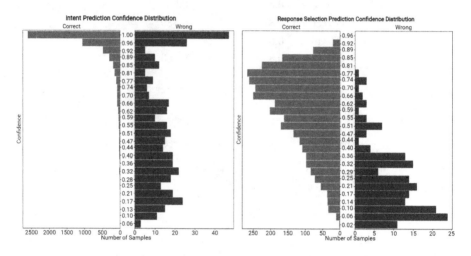

Fig. 5. Intent and response prediction histograms

to fix the remaining errors, such as adding more stories, merging intents, and splitting their meaning using entities.

4 Conclusions

Sex education has been one of the keys to fighting against STIs. Although there have been significant efforts toward ending HIV/AIDS pandemic, it remains a problem. Nowadays, especially adolescents, use the Internet as a source to seek health information, including sexual health. This new situation presents potential issues due to the lack of control of its quality or truthfulness. Therefore, sex education programs should be updated to respond to this new trend.

After reviewing the state-of-the-art conversational agents and the current STI/HIV situation, we presented the design, implementation, and evaluation of VIHrtual-App: an online Spanish conversational agent which offers reliable information to promote STI/HIV awareness and prevention. The development used a CDD paradigm, which allowed us to review and extract useful information from real conversations while building the agent. Thanks to this strategy, the model grew following users' needs, and several questions not initially contemplated were discovered. Finally, three pipeline configurations were evaluated and scored to get the most out of the model. The language-agnostic BERT pipeline generated the best model for intent recognition and response selection, followed by the default configuration and Spanish version of spaCy. This evaluation may show that generic pre-trained word embeddings for the Spanish language do not fit with the domain-specific terminology of the training data. In the future, it would be interesting to train a model using STIs related information and use it to identify relevant entities and improve performance.

The final conversational agent model can recognize more than 250 intentions and respond accordingly, conveying awareness and entertainingly educating users. However, we are aware of the limitations of this preliminary development. Although real users have used the assistant from the beginning, future work is needed to validate and measure user experiences and perceptions, especially across different user profiles. Thus, further studies will focus on studying VIHrtual-App's usability.

In addition, it is required to check how competent is this new tool for fighting against STIs/HIV. Therefore, a future validation should be performed to clarify how effective VIHrtual-App is as a sex educational tool.

Acknowledgements. This work was partially funded by PI2021_25 POLISABIO project of the Universitat Politècnica de València.

References

1. World Health Organization et al.: Report on global sexually transmitted infection surveillance 2018 (2018)
2. Galvin, S.R., Cohen, M.S.: The role of sexually transmitted diseases in HIV transmission. Nat. Rev. Microbiol. **2**(1), 33–42 (2004)
3. Govender, R.D., Hashim, M.J., Khan, M.A., Mustafa, H., Khan, G.: Global epidemiology of HIV/AIDS: a resurgence in North America and Europe. J. Epidemiol. Glob. Health **11**(3), 296 (2021)
4. HIV/AIDS surveillance in Europe 2021–2020 data (2021)
5. Gökengin, D., Doroudi, F., Tohme, J., Collins, B., Madani, N.: HIV/AIDS: trends in the middle east and North Africa Region. Int. J. Infect. Dis. **44**, 66–73 (2016)
6. Kirby, D.B., Laris, B.A., Rolleri, L.A.: Sex and HIV education programs: their impact on sexual behaviors of young people throughout the world. J. Adolesc. Health **40**(3), 206–217 (2007)
7. Haberland, N.A.: The case for addressing gender and power in sexuality and HIV education: a comprehensive review of evaluation studies. Int. Perspect. Sex. Reprod. Health **41**(1), 31–42 (2015)
8. Simon, L., Daneback, K.: Adolescents' use of the internet for sex education: a thematic and critical review of the literature. Int. J. Sex. Health **25**(4), 305–319 (2013)
9. Gray, N.J., Klein, J.D., Noyce, P.R., Sesselberg, T.S., Cantrill, J.A.: Health information-seeking behaviour in adolescence: the place of the internet. Soc. Sci. Med. **60**(7), 1467–1478 (2005)

10. Harris, P.R., Sillence, E., Briggs, P.: Perceived threat and corroboration: key factors that improve a predictive model of trust in internet-based health information and advice. J. Med. Internet Res. **13**(3) (2011)
11. Laranjo, L., et al.: Conversational agents in healthcare: a systematic review. J. Am. Med. Inform. Assoc. **25**(9), 1248–1258 (2018)
12. Car, L.T., et al.: Conversational agents in health care: scoping review and conceptual analysis. J. Med. Internet Res. **22**(8), e17158 (2020)
13. Montenegro, J.L., da Costa, C.A., da Rosa, R.: Survey of conversational agents in health. Expert Syst. Appl. **129**, 56–67 (2019)
14. Warmerdam, V.: Intents & entities: Understanding the rasa NLU pipeline (2021). https://rasa.com/blog/intents-entities-understanding-the-rasa-nlu-pipeline/
15. Rasa pipeline components. https://rasa.com/docs/rasa/components
16. Eisman, E.M., Navarro, M., Castro, J.L.: A multi-agent conversational system with heterogeneous data sources access. Expert Syst. Appl. **53**, 172–191 (2016)
17. Chaves, A.P., Gerosa, M.A.: How should my chatbot interact? A survey on social characteristics in human-chatbot interaction design. Int. J. Hum.-Comput. Interact. **37**(8), 729–758 (2021)
18. Nichol, A.: Conversation-driven development (2020). https://blog.rasa.com/conversation-driven-development-a-better-approach-to-building-ai-assistants/
19. Rasa suggested config. https://rasa.com/docs/rasa/model-configuration/#suggested-config
20. Bunk, T., Varshneya, D., Vlasov, V., Nichol, A.: Diet: lightweight language understanding for dialogue systems. arXiv preprint arXiv:2004.09936 (2020)
21. Feng, F., Yang, Y., Cer, D., Arivazhagan, N., Wang, W.: Language-agnostic BERT sentence embedding. arXiv preprint arXiv:2007.01852 (2020)
22. Warmerdam, V.: Non-English tools for rasa NLU (2021). https://rasa.com/blog/non-english-tools-for-rasa/
23. Devlin, J., Chang, M.-W., Lee, K., Toutanova, K.: BERT: pre-training of deep bidirectional transformers for language understanding. arXiv preprint arXiv:1810.04805 (2018)

A Neuro-Symbolic AI System for Visual Question Answering in Pedestrian Video Sequences

Jaeil Park, Seok-Jun Bu, and Sung-Bae Cho$^{(\boxtimes)}$

Department of Computer Science, Yonsei University, Seoul, South Korea
{wodlf603,sjbuhan,sbcho}@yonsei.ac.kr

Abstract. With the rapid increase in the amount of video data, efficient object recognition is mandatory for a system capable of automatically performing question and answering. In particular, real-world video environments with numerous types of objects and complex relationships require extensive knowledge representation and inference algorithms with the properties and relations of objects. In this paper, we propose a hybrid neuro-symbolic AI system that handles scene-graph of real-world video data. The method combines neural networks that generate scene graphs in consideration of the relationship between objects on real roads and symbol-based inference algorithms for responding to questions. We define object properties, relationships, and question coverage to cover the real-world objects in pedestrian video and traverse a scene-graph to perform complex visual question-answering. We have demonstrated the superiority of the proposed method by confirming that it answered with 99.71% accuracy to 5-types of questions in a pedestrian video environment.

Keywords: Visual question-answering · Neuro-symbolic reasoning · Scene graph · Pedestrian video

1 Introduction

Visual question-answering is an attempt to combine awareness of the relationship between question text and image objects, which is an important topic at the intersection of computer vision and natural language processing. In particular, the attempt to combine neural network approaches for recognition and inference for mapping between recognized symbols is adapted to various tasks like phishing detection [1] and validates the combination of deep learning and inference by obtaining the best accuracy in synthetic photo environments where data collection and object relationships are simple [2].

Meanwhile, although the inference-based question and answering model has been validated in a limited environment, attempts to extend it to more complex real-world data have been pointed out with the expansion of cognitive neural networks and question and answering methods [3]. For example, in a pedestrian video environment, where the types, relationships, and possible situations of objects are more diverse, a broader definition

© Springer Nature Switzerland AG 2022
P. García Bringas et al. (Eds.): HAIS 2022, LNAI 13469, pp. 443–454, 2022.
https://doi.org/10.1007/978-3-031-15471-3_38

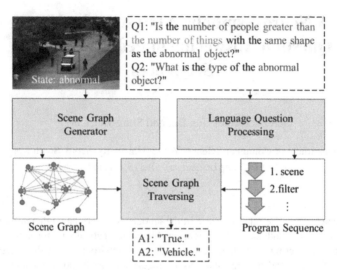

Fig. 1. Combination of reasoning using scene graph traversing and cognition for a visual question and answering in a pedestrian video environment.

of the scope of perceptions and questions is required, and effective combinations with previously developed recognition methods must be considered (Fig. 1).

In addition, the existing methods using pedestrian video data excluding object property and location have a limitation in expressing the relationship between objects. The methods through those data calculate the positional relationship based on the coordinates confirmed through object detection. This has a disadvantage in that the relationship between objects is fragmented with location information and is difficult to applied to a complex relationship of the real-world environment. In consideration of the problem, various methods are being studied. A knowledge expression method that can better express the relationship between objects [4–7], and a method of calculating the relationship between objects in various ways [8–10]. A scene graph is a representative knowledge expression method that expresses the relationship between objects. This has the advantage of being able to clearly express the properties and relationships between objects on the screen.

In this paper, we define the object properties, object relations, and question coverage for the question and answering between the manager and the system when an abnormal situation occurs in the automated monitoring environment of pedestrian video. We propose a neuro-symbolic AI system that handles scene-graph reasoning with scene-graph of pedestrian video and programs of questions. The scene graph reasoning based on the object property and relationship enables complex inference for question-answering method in the real environments. We confirm that a variety of complex queries can be handled. In detail, we demonstrate the superiority of the method compared to the existing visual question-answering methods through graph reasoning. The proposed method is evaluated for five types of questions, and it is found to achieve the best performance compared to the existing methods. According to the experimental results, we argue that the neuro-symbolic AI system of scene-graph reasoning and deep learning-based question-answering provides inference suitable for the realistic environments.

Table 1. Visual question and answering environments and methods with deep learning and neuro-symbolic approach.

Approach	Visual perception	QA processing	Environment
End-to-end neural network	Convolutional neural network	Long short-term memory [11]	Synthetic visual scenes (CLEVR)
		Modular network with encoder-decoder [12]	General objects (MS-COCO)
Neuro-symbolic	Mask R-CNN (Object tables)	Domain-specific language [2]	Synthetic visual scenes (CLEVR)
	Mask R-CNN (Feature vectors)	Quasi-symbolic program execution [8]	
	Faster R-CNN	Differentiable first-order logic [3]	General objects with scene graph (GQA)

2 Related Works

Table 1 summarizes the methods of combining deep learning and inference algorithms for visual question and answering in terms of approach, method, and environment. Attempts to implement both initial image recognition and processing and mapping as neural networks define visual query-answering tasks in a synthetic environment [11]. Methods using modular neural networks demonstrate that tasks of recognition and natural language processing must be distinguished [12].

Meanwhile, in the neuro-symbolic approach to combining inference algorithms with deep learning, symbol grounding and inference methods of objects were studied [2]. Research to develop domain-specific languages and symbolic processes for query and relationship representation [3] implement high-performance question-answering compared to human respondents for synthetic environments [8]. In preparation for previous studies, this paper redefines objects and query range to extend neuro-symbolic VQA to more complex pedestrian video surveillance environments and maximizes and evaluates the practicality by incorporating anomaly detection module with neural networks.

For scene graphs, several generative methods such as conditional random field (CRF), CNN, RNN, LSTM, and graph neural network have been developed. CRF-based scene graph generation models like SG-CRF can model statistical correlation effectively in the visual relationship [13]. With the idea of using the neural model for scene graph generation, the CNN-based model and RNN-based model are investigated. BAR-CNN is a CNN-based model with an attention mechanism, and in the most advanced feature extractors, the receptive region of neurons may still be limited, which means that the model may include a full attention map [14]. Zoom-Net is an RNN-based model that achieves competitive performance by successfully recognizing complex visual relationships through deep message propagation and interaction between local object features and global predicate features without linguistic dictionary [15].

VCTREE is also an RNN-based model proposed for composite dynamic tree structures [16]. However, as GCN has been proven to be highly effective in graph reasoning tasks, many researchers have directly studied the scene graph generation method based on the graph [17, 18]. Factorizable Net is a graph-based model for scene graph generation using GCN with both bottom-up and top-down subgraph-based methods, but unlikely relationships are remained by sub-graph operation [19]. Graph R-CNN is a graph-based model that seeks to trim the original scene graph to generate sparse candidate graph structures [20]. In this paper, considering the complexity of the actual situation, Graph-RCNN is adopted to apply effective and efficient scene graph generation.

3   Proposed Method

The range of questions and answers available in pedestrian video surveillance environments depends on the number of programs defined as a set of question queries written in a domain-specific language, and a program considering the trade-off between practicality and mapping performance should be defined. In this paper, as shown in Fig. 2, a scene graph was generated through the combination of the abnormality detection result and the

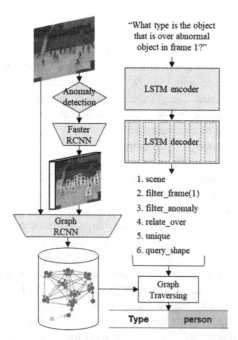

Fig. 2. Combination of reasoning using scene graph traversing and cognition for the visual question and answering in a pedestrian video environment.

object detection result with the pedestrian video data. The set of input questions and programs is mapped to the supervised long short-term memory (LSTM) encoder-decoder structure. The set of programs extracted from the question performs a filter operation implemented by scene graph traversing, which is a method of calculating results by applying a program to a set of nodes in a scene graph.

Table 2. Definition of object property and relation on a pedestrian video environment.

Name	Description
Object property	
Shape	Person, bicycle, car, skateboard, wheelchair, cart, truck, others
Size	Small, large
Position	x-y location
Velocity	Computed by comparing x-y coordinates within frames
Object relation	
Relative position	Left, right, in_front, behind, over, under
Relative size	Larger_than, smaller_than
Relative velocity	Faster_than, slower_than
Numbers	More_than, less_than
Equal	Same_shape, same_size, same_position, same_abnormal

3.1 Scene Graph Generation Using Graph R-CNN

The scene graph describes the properties and relationships of objects. Given a set of object property categories $C = \{C_1, \ldots, C_m\}$ and a set of object relationship categories R, a scene graph is a tuple (O, E) where $O = \{o_1, \ldots, o_n\}$ is a set of objects with each o_i, an object that $o_i = \{c_{i1}, c_{i2}, \ldots, c_{im}\}$ where $c_{ij} \in C_j$, and $E \subseteq O \times R \times O$ is a set of directed edges of the form (o_i, r, o_j) where $o_i, o_j \in O$ and $r \in R$. For this scene graph, object property and relation are defined in this paper as shown in Table 2.

In this paper, images in the pedestrian video sequences are converted into scene graphs through three steps based on the scene graph and object property defined above. First, an autoencoder based on a 3D convolution operation in consideration of a time axis is passed to return whether the corresponding image is normal or not. Second, the result is detected by detecting an object of an image in which an abnormal situation is detected through Faster R-CNN. Finally, the detection result is generated through Graph R-CNN with the original image.

Abnormal detection has been studied in various ways, including GAN [21]. In the proposed model, an autoencoder for anomaly detection operates with the following equations. The lth hidden layer vector h_l and the reconstructed frame X_i for the i-th input frame sequence X_i are defined as shown in Eq. (1). The three-dimensional convolution operation p_i applied at this time learns the filter operation of $(k \times k \times k)$-size on the

m, n, t^{th} pixels, extracts spatial correlation within the frame and inter-frame correlation, and is defined by Eq. (2).

$$h^l = \phi_c\left(W^l x_i + b^l\right); \; \widehat{x_i} = \phi_c(W^{l+1} h^l + b^{l+1}) \tag{1}$$

$$\phi_c^l(x_i) = \sum_{a=0}^{k-1}\sum_{b=0}^{k-1}\sum_{c=0}^{k-1} w_{abc}^l x_{(m+a)(n+b)(t+c)}^{l-1} \tag{2}$$

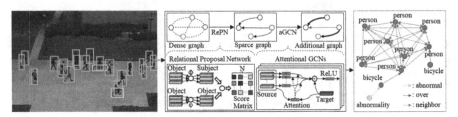

Fig. 3. Scene graph generation for a real pedestrian video.

Faster R-CNN detects objects for scene graph parsing. The proposed model inputs the results of anomaly detection and image to detect objects with a representation of whether objects are normal or not (Fig. 3).

Graph R-CNN, a representative method of scene graph generation algorithms, is a successful algorithm to express the relationships between objects more effectively using a relationship proposition network (RePN) that efficiently handles secondary potential relationships between objects in images and a GCN. In this paper, we input images of pedestrian video frame into corresponding algorithm models learned from the GQA dataset and maintained valid information corresponding to the object properties defined above in the results. Through the corresponding scene graph, the image is expressed in a knowledge expression method that can more clearly express the relationship between objects while maintaining information on objects.

3.2 Scene Graph Traversing and Question-Answering

In the proposed model, the question is translated into the sequence of the program $\{p_1, \ldots, p_s\}$ via the LSTM encoder-decoder structure. Programs are defined as the six classes, as shown in Table 3, and the format of each input and output is also defined. After training the LSTM encoder-decoder model through the predefined questions and program pairs in the dictionary, the questions are translated into programs through the model.

Scene graph traversing is conducted through the corresponding translated program and the resulting scene graph in Sect. 3.1, which turns the program on the set of nodes in the scene graph. For each program, the scene program returns all objects in the current scene, and programs other than relation and scene do not require any relational information, and each of the objects in the set is turned if-else and the resulting output is calculated.

Programs corresponding to relationships require the relation information between objects. This model applies a method of searching through the edge of the scene graph. After confirming whether an edge corresponding to a connection for filtering is connected to each node for a set of nodes that come in as input when the edge is present, the program for connection proceeds by calculating a set of nodes consisting of target nodes and outputting it. This is shown in Algorithm 1. Through this algorithm, logical reasoning for each program stage, including relational programs, becomes possible, and a reasonable result can be obtained accordingly.

Table 3. Definition of domain-specific language set for visual question and answering.

Function	Input	Output	Function	Input	Output
Basic program					
scene	-	Object list	count	Object list	Integer
unique	Object list	Object	exist	Object list	Boolean
relate	Object list	Object list	get_frame	Integer	scene
Filter program					
filter_size	List, size	List	filter_object	List, abnormal	List
filter_shape	List, shape	List	filter_scene	List, integer	List
filter_position	Position	List	filter_frame	List, integer	List
filter_velocity	List, integer	List			
Query program					
query_size	Object	Size	query_velocity	Object	Integer
query_shape	Object	Shape	query_type	Object	List
query_position	Object	Position			
Logic program					
AND	List, List	Object list	OR	List, List	Object list
Sameness program					
same_size	Object	Object list	same_position	Object	Object list
same_shape	Object	Object list	same_velocity	Object	Object list
Compare function					
equal_integer	Int, int	Boolean	equal_shape	Shape, shape	Boolean
equal_size	Size, size	Boolean	less_than	Int, int	Boolean
equal_color	Col, col	Boolean	greater_than	Int, int	Boolean

Algorithm 1. Algorithm for scene graph traversing.

Input: scene graph $G = \{O,E\}$ and program sequence $P=\{p_1,p_2,\ldots,p_n\}$

Output: traversing result – answer to question

1: Initialize $S = []$
2: for p_i in P:
3:   if p_i is "scene":
4:    $S.push(\phi)$
5:   elif p_i is in "relation":
6:    $O_{org} = S.pop()$
7:    $O_{new} = \phi$
8:    for o_j in O_{org}:
9:     for e_{jk} in E where $e_{jk} = \overline{o_j o_k}$:
10:     if e_k is relation in p_i:
11:      $O_{new} = O_{new} + \{e_k\}$
12:   $S.push(O_{new})$
13: else:
14:   $S = p_i(S)$
15: return $S.pop()$

---

4 Experiments and Results

4.1 Real-World Pedestrian Video Dataset

To verify the proposed method, we use UCSD pedestrian datasets collected from fixed CCTV video installed on foot. Pedestrians and objects moving in the left and right directions were photographed and an object attribute table consisting of a combination of two-four-wheel vehicles (bicycles, cars, skateboards, wheelchairs, carts, and trucks) and backgrounds (wood, roads, and grass) was extracted by frame from a total of 98 frames.

4.2 Visual Question and Answering Performance in Pedestrian Video

Table 4 compares the performance of the existing question-answering method by program type. When image recognition using convolutional neural networks and question processing with LSTM are mapped with simple supervised learning, the proposed method is rarely misclassified with an accuracy of 0.9971 as the number of objects in the table of perceived object properties is specified, compared to low accuracy of 0.6457. In complex environments such as pedestrian video sequences, modular neural network-based encoder-decoder approaches show an accuracy of 0.9232 and the proposed method combined with inference capabilities achieves an accuracy of 0.9991.

For each classification, the method presented also records the highest accuracy. In particular, the compare number classification is the most difficult part to record the lowest figure for other algorithms, but the proposed method has a very high accuracy of 0.9996. This shows that the neuro-symbolic approach has strengths in problems that can produce various and complex values, such as numerical comparison, and that the scene graph supports this appropriately. In addition, the query attribute, which can lead to complex results, and the count that must be accurately determined, also show higher figures to support the argument.

Table 4. 10-fold cross-validation of accuracy with other VQA methods by query type.

Method	Count	Exist	Compare number	Compare attribute	Query attribute	Overall accuracy
CNN-LSTM [11]	64.57%	87.44%	53.78%	77.47%	77.47%	72.15%
Mask R-CNN with LSTM	85.23%	92.93%	83.45%	90.68%	92.68%	88.99%
Module network with Encoder-decoder [12]	86.77%	96.61%	86.48%	96.51%	95.27%	92.32%
Scene graph reasoning with NS-VQA (Ours)	99.71%	99.97%	99.96%	99.93%	99.98%	99.91%

Table 5 summarizes that the number, type, and detailed properties of objects are queried, and four cases in which question and answer fails while using data including object property and location but succeeds while using scene graph data. In the case of questions requiring relational information, misrecognition of the relationship frequently occurs based on data with object property and location, but when the proposed method is considered as an image information expression method, the relational information can be more accurately handled with more complex programs.

Table 5. Program representation and scene-graph for each case of correct response.

Scene	Query	Question	Program representation	Answer (P&L) / Answer (graph)
	Count	What number of large normal persons are behind the small man?	scene filter_size[small] unique relate[behind] filter_size[large] filter_anomaly[normal] filter_shape[person] count	14 / 15
	Exist	Are there any things in front of the small normal person?	scene filter_size[small] filter_anomaly[normal] filter_shape[person] unique relate[front] exist	False / True
	Compare number	Are there more humans on the left side of the scene than on the right?	scene filter_position[left] filter_shape[person] scene filter_ position[right] filter_shape[person] greater_than	True / False
	Count	What number of large normal persons are behind the small man?	scene filter_size[small] unique relate[behind] filter_size[large] filter_anomaly[normal] filter_shape[person] count	7 / 8

5 Conclusion

In this paper, for visual question-and-answer tasks in pedestrian video sequences that are close to the real environment, object properties, relations, and question coverage are defined, and a neuro-symbolic visual question-answering methods is proposed by incorporating a scene graph generator and an anomaly detector. The proposed method achieves 0.9978 accuracy for five query types.

On the other hand, the inference algorithm for mapping question and answer in the proposed method is implemented as a simple filter algorithm operation and needs to be verified in the general image field with a relationship of objects more complex than pedestrian video sequences. In particular, when various objects emerge, the scene graph becomes complicated, and the number of computations increases accordingly, a learning method considering the optimization of the computation will be needed.

Acknowledgment. This work was supported by Institute of Information & Communications Technology Planning & Evaluation (IITP) grant funded by the Korean government (MSIT) (No. 2020-0-01361, Artificial Intelligence Graduate School Program (Yonsei University); No. 2021-0-02068, Artificial Intelligence Innovation Hub).

References

1. Park, K.-W., Bu, S.-J., Cho, S.-B.: Evolutionary optimization of neuro-symbolic integration for phishing URL detection. In: International Conference on Hybrid Artificial Intelligent Systems, pp. 88–100 (2021)
2. Yi, K., Wu, J., Gan, C., Torralba, A., Kohli, P., Tenenbaum, J.: Neural-symbolic VQA: disentangling reasoning from vision and language understanding. In: Advances in Neural Information Processing Systems, pp. 1031–1042 (2018)
3. Amizadeh, S., Palangi, H., Polozov, O., Huang, Y., Kishida, K.: Neuro-symbolic visual reasoning: disentangling 'visual' from 'reasoning'. In: International Conference on Machine Learning, pp. 279–290 (2020)
4. Shi, J., Zhang, H., Li, J.: Explainable and Explicit Visual Reasoning over Scene Graphs. In: IEEE Conference on Computer Vision and Pattern Recognition, pp. 8368–8376 (2019)
5. Wang, P., Wu, Q., Shen, C., Dick, A., Van Den Hengel, A.: FVQA: fact-based visual question answering. IEEE Trans. Pattern Anal. Mach. Intell. **40**, 1367–1381 (2018)
6. Teney, D., Liu, L., van Den Hengel, A.: Graph-structured representations for visual question answering. In: IEEE Conference on Computer Vision and Pattern Recognition, pp. 1–9 (2017)
7. Song, Y.-S., Cho, S.-B.: Objects relationship modeling for improving object detection of service robots using Bayesian network integration. In: International Conference on Intelligent Computing, pp. 678–683 (2006)
8. Mao, J., Gan, C., Deepmind, P.K., Tenenbaum, J.B., Wu, J.: The neuro-symbolic concept learner: interpreting scenes, words, and sentences from natural supervision. In: International Conference on Learning Representations (2019)
9. Han, C., Mao, J., Gan, C., Tenenbaum, J.B., Wu, J.: Visual concept metaconcept learning. In: Advances in Neural Information Processing Systems, pp. 5001–5012 (2019)
10. Yu, J., et al.: Reasoning on the relation: enhancing visual representation for visual question answering and cross-modal retrieval. IEEE Trans. Multimedia **22**, 3196–3209 (2020)
11. Agrawal, A., et al.: VQA: visual question answering. Int. J. Comput. Vision **123**(1), 4–31 (2016). https://doi.org/10.1007/s11263-016-0966-6
12. Hu, R., Andreas, J., Rohrbach, M., Darrell, T., Saenko, K.: Learning to reason: end-to-end module networks for visual question answering. In: IEEE International Conference on Computer Vision, pp. 804–813 (2017)
13. Cong, W., Wang, W., Lee, W.-C.: Scene Graph Generation via Conditional Random Fields. arXiv preprint arXiv:1811.08075 (2018)
14. Kolesnikov, A., Kuznetsova, A., Lampert, C., Ferrari, V.: Detecting visual relationships using box attention. In: IEEE International Conference on Computer Vision Workshops, pp. 1749–1753 (2019)

15. Yin, G., et al.: Zoom-net: mining deep feature interactions for visual relationship recognition. In: Ferrari, V., Hebert, M., Sminchisescu, C., Weiss, Y. (eds.) ECCV 2018. LNCS, vol. 11207, pp. 330–347. Springer, Cham (2018). https://doi.org/10.1007/978-3-030-01219-9_20

16. Tang, K., Zhang, H., Wu, B., Luo, W., Liu, W.: Learning to compose dynamic tree structures for visual contexts. In: IEEE Conference on Computer Vision and Pattern Recognition, pp. 6619–6628 (2019)

17. Goller, C., Kuchler, A.: Learning task-dependent distributed representations by backpropagation through structure. In: International Conference on Neural Networks, pp. 347–352 (1996)

18. Gori, M., Monfardini, G., Scarselli, F.: A new model for learning in graph domains. In: IEEE International Joint Conference on Neural Networks, pp.729–734 (2005)

19. Li, Y., Ouyang, W., Zhou, B., Shi, J., Zhang, C., Wang, X.: Factorizable net: an efficient subgraph-based framework for scene graph generation. In: European Conference on Computer Vision, pp. 346–363 (2018)

20. Yang, J., Lu, J., Lee, S., Batra, D., Parikh, D.: Graph R-CNN for scene graph generation. In: Ferrari, V., Hebert, M., Sminchisescu, C., Weiss, Y. (eds.) ECCV 2018. LNCS, vol. 11205, pp. 690–706. Springer, Cham (2018). https://doi.org/10.1007/978-3-030-01246-5_41

21. Shin, W.-S., Bu, S.-J., Cho, S.-B.: 3D-convolutional neural network with generative adversarial network and autoencoder for robust anomaly detection in video surveillance. Int. J. Neural Syst. **40**(6), 2050034 (2020)

Aspects on Image Edge Detection Based on Sensitive Swarm Intelligence

Cristina Ticala[1], Camelia-M. Pintea[1(✉)], Gloria Cerasela Crisan[2],
Oliviu Matei[1], Mara Hajdu-Macelaru[1], and Petrica C. Pop[1]

[1] Technical University of Cluj-Napoca, North University Center of Baia-Mare,
430122 Baia-Mare, Romania
cristina.ticala@cunbm.utcluj.ro, dr.camelia.pintea@ieee.org,
oliviu.matei@holisun.com, {mara.macelaru,petrica.pop}@mi.utcluj.ro
[2] Faculty of Sciences, Vasile Alecsandri University, Bacau, Romania
ceraselacrisan@ub.ro

Abstract. Nowadays, swarm intelligence shows a high accuracy while solving difficult problems, including image processing problem. Image Edge detection is a complex optimization problem due to the high-resolution images involving large matrix of pixels. The current work describes several sensitive to the environment models involving swarm intelligence. The agents' sensitivity is used in order to guide the swarm to obtain the best solution. Both theoretical general guidance and a practical example for a particular swarm are included. The quality of results is measured using several known measures.

Keywords: Swarm intelligence · Image processing · Image Edge Detection

1 Introduction

Nowadays, when images are influential data on many domains, image processing should be very efficient. In particular, the Image Edge Detection problem is used to detect discontinuities in color brightness to find boundaries of objects. Due to their complexity and in order to help sick people, medical images are some of the most used images.

Artificial intelligence, AI, and here in particular, swarm intelligence, is involved to improve edges of images, including medical images obtained from X-rays and tomographic images, due to their complex edges. Other AI techniques includes for example evolutive strategies [1,2] used for the radiograph images segmentation. Detecting tuberculosis, pneumonia, and COVID-19 from chest X-ray images are studied in [3] while using neural networks.

Any combinatorial optimization problem, including Image Edge detection problem, could be represented as a graph. Within Image Edge detection problem each pixel from an image is considered a node in a graph.

© Springer Nature Switzerland AG 2022
P. García Bringas et al. (Eds.): HAIS 2022, LNAI 13469, pp. 455–465, 2022.
https://doi.org/10.1007/978-3-031-15471-3_39

Swarm intelligence are nature inspired techniques. In biology, stigmergy occurs when an action of an insect (in swarm intelligence, an agent) is based on the consequences of the previous action of other insects (i.e. agents) [4]. Overall, stigmergy [5, 6] provides a mechanism correlating individual and swarm behaviors: an individual behavior will modify the environment, and the environment will further modify the behavior of other swarm agents.

Swarm Intelligence (SI) is coined in [7], defining the collective behavior of natural or artificial systems made up by interacting individuals. Each individual can move autonomously, can communicate with peers, can store information on its previous activity, and make decisions based on its myopic view of the environment. These simple characteristics add up to intelligent global behavior, manifested in real life for example by fish schooling, bird flocking, or animal herding. These nature-based models are used for solving problems in many research fields, like Data Analysis [8] or Optimization [9]. As an interdisciplinary field, Computer Vision bridges Optimization and Data Analysis, thus benefiting of various opportunities for using these methodologies.

Among Swarm Intelligence approaches, we mention: Ant Colony Optimization [10], Particle Swarm Optimization [11], Gravitational Search [12], etc.

Ant Colony Optimization (ACO) is inspired by the way ants manage to find the shortest path from nest to food. A colony of artificial ants is deployed in the vertices of a graph and concurrently constructs paths, laying artificial pheromone on edges. Each artificial ant tends to choose the closer vertices and also the edges with higher artificial pheromone quantity [10]. ACO has proven to be efficient when applied in Computer Vision, among many other computational fields [13]. For edge detection, ants move from pixel to pixel, to discover and to maintain a path maximizing the local differences in intensity of the image. Usually, the intensities of pixels at most Manhattan distance 3 are considered when one of the neighbor pixels is chosen. The neighborhood consists of the pixels at Manhattan distance either 1 or 2 [14,15]. In [16] image edge detection was solved using adaptive threshold.

Within Particle Swarm Optimization (PSO) a population of individuals, particles, explores the solution space. Each particle is a solution which moves in its neighborhood based on its own experience and that of its best neighbor [11]. When applied to the edge detection problem, PSO uses grayscale images, and each particle is a set of binary values. Each value is assigned to a pixel from the image: 1 if the pixel is identified with edge/vertex pixel and 0 otherwise. The local edge structure is computed using a 3×3 slide window, and the cost of an individual can have weighted multiple components, associated to edge thickness, fragmentation, curvature [17].

Gravitational Search (GS) is an optimization method inspired by the laws of physics, transcending the broad class of biologically-inspired methods. Edge detection based on GS considers each pixel as an object, with mass corresponding to its grayscale value. The sums of gravitational forces on both dimensions are used to compute the edge intensity and orientation [18].

The current paper details some of the most efficient sensitive techniques involving swarm intelligence. The sensitivity is a parameter or a group of parameters which modifies/modify in time. Sensitivity was proposed for solving different problems including sensor networks security [19, 20].

The next section of the paper shows some of the sensitive approaches within Swarm Intelligence. Section 3 describes the Sensitive Ant Colony Optimization involved in solving the Image Edge Detection problem. Furthermore, the quality of sensitive solutions of the Image Edge Detection problem is for the first time measured with Mean Squared Error (MSE), Peak signal-to-noise ratio (PSNR) and Structural similarity (SSIM) index in Sect. 4. The paper concludes with a section including future possible works involving swarm intelligence with sensitive features.

2   Sensitive Approaches Within Swarm Intelligence

There are multiple way to enhance the combinatorial optimization problem solution while using different techniques, including swarm intelligence.

2.1   Pheromone Sensitivity Levels: PSL

The main role of the Pheromone Sensitivity level factor is to keep an equilibrium between exploration and exploitation of searched spaces. The PSL value is a number in the interval $[0, 1]$; its extremes are the minimal 0 value and 1 as the maximal value: for the minimal value ants ignore information and they have a maximum sensitivity to pheromone trails when $PSL = 1$.

- *Exploration* is made by independent agents, the explorer agents with a low PSL value; in some versions are called small PSL agents ($sPSL$ agents) or sensitive-explorer agents.
- *Exploitation* is made by the most sensitive agents to environments (for example to pheromone trails for ants); the intensively exploiters agents have high PSL values; in some versions are called high PSL agents ($hPSL$ agents) or sensitive-exploiter agents; their role si to intensively exploit already identified regions by the $sPSL$ agents.

During optimization the agents sensitivity (PSL) modifies. In particular, during Sensitive Ant Colony Optimization (SACO) [24] the PSL is globally modified based on the graph's search space topology [25].

2.2   Stigmergic Sensitivity Level (SSL)

A particular version of PSL is the Stigmergic Sensitivity Level (SSL) [37,38] parameter similar with temperature in simulated annealing is used to group the swarm, into two distinct groups of agents. The membership of an agent is based on a random variable uniformly distributed over $[0; 1]$, denoted by q and a constant $q_0 \in (0, 1)$.

The $sSSL$ agents are characterized by the inequality $q > q_0$ while for the $hSSL$ agents $q \leq q_0$ holds.

2.3   Heterogeneous Sensitivity

Based on different sensitivity levels Sensitive Ant Model (SAM) was introduced in [36] and used to solve a complex problem, a dynamic generalized drilling problem on a PCB, a particular class of the Generalized Traveling Salesman Problem (GTSP).

The SAM method uses communication' stigmergy and features that induce heterogeneity within agents. Heterogeneous agents could use specific feature, for example ants use stigmergy, to communicate, while taking decisions based on the environment's changes (i.e. ant pheromone) and the sensitivity levels.

- As for $PSL = 1$ sensitive agents are 'classical' agents and the agent with $PSL = 0$ choose paths similar with the random walk algorithm, while ignoring environment features (e.g. ants pheromone trails).
- For the majority of agents, when $0 < PSL < 1$ the transition probabilities are modified using a renormalization process based on $sp_{iu}(t, k) = p_{iu}(t, k) \cdot PSL(t, k)$.
- If $PSL(t, k) \neq 1$ then for all i graph' nodes $\sum_u sp_{iu}(t, k) < 1$.
- While is associated a standard probability distribution to the system, a *virtual state, vs* related to the 'lost' probability is used. The transition probability associated to the virtual state vs is $sp_{i,vs}(t, k) = 1 - \sum_u sp_{iu}(t, k)$.
- For an agent k at time t

$$sp_{i,vs}(t, k) = 1 - PSL(t, k) \sum_u p_{iu}(t, k).$$

- The renormalized probability $sp_{i,vs}(t, k)$ is the granular heterogeneity of agent k at time t. Therefore, a measure of the system heterogeneity, E, showing the distance from the standard associated system, is defined as:

$$E = \sum_k \sum_i (sp_{i,vs}(t, k))^2. \tag{1}$$

Some of the main observations follows.

- Minimum heterogeneity is associated with normal stigmergic agents, the agents with a maximum sensitivity to environment.
- Maximum heterogeneity is associated to 'environment-blind' agents (agents with zero sensitivity).
- Diverse transition mechanisms, including virtual state could be used. In *SAM* are specified actions associated with the virtual state, a "virtual state decision rule": so, if the selected state is *vs*, an accessible state is random chosen with an uniform probability.

Overall, the model increase diversity based on a random selection while decreasing environment sensitivity level *PSL* for an agent.

Notations used: $p_{iu}(t, k)$ is the probability for agent k to choose the next node u from current node i ; $sp_{iu}(t, k)$ is the renormalized transition probability for agent k and $PSL(t, k)$ is the *PSL* value of agent k at time t.

2.4   Proposal: Human-in-the-Loop Influencing Sensitivity Levels

Future work could include the humans possibility to interact and change parameters values and/or functionality in order to enhance optimization. The pheromone sensitivity to the environment, PSL values, and the heterogenous sensitivity approaches could be interactively modified by the human factor as seen in [33–35].

Keeping the nature-inspired ideas, future work will integrate human decisions in orienting the search. Using a permanent and instant visualization of the implementation evolution, the humans are able to update the search parameters. The goal of this action is to use human knowledge and reasoning in order to improve the quality of the algorithm when executed.

3   Sensitive Ant Colony Optimization for Image Edge Detection

Based on Sensitive Ant Colony Optimization (SACO) [27,28], in our work [24] we made a comparison between ant-based techniques with and without sensitivity factor on the image edge detection problem. The SACO method was applied to complex medical images.

Phases of the Sensitive ACO for solving edge detection problem:

- Input: a colony of K ants, in a graph-space $\mathcal{X}$, with $M_1 \times M_2$ nodes.
- Method:
    - initialize ACO parameters: the ants are randomly placed on the image matrix and each value of the pheromone matrix is a constant τ_0 value; another constant value is L, the number of construction process steps.
    - construct solutions through building a pheromone matrix for all pixels within an image; at the n-th step, a random ant moves from i to j with the transition probability $p_{ij}^n = \dfrac{\left(\tau_{ij}^{(n-1)}\right)^\alpha (\eta_{ij})^\beta}{\sum_{i \in \Omega_i}\left(\tau_{ij}^{(n-1)}\right)^\alpha (\eta_{ij})^\beta}$,   if $j \in \Omega_i$,.
    - the pheromone matrix values are updated during search both locally $\left(\tau_{ij}^{(n)} = \tau_{ij}^{n-1} \cdot (1 - \rho) + \rho \cdot \Delta_{ij}\right)$ and globally.
- Output: the solution is identified as the best graph path, the best edge identified, after a given number of iterations.

Specifications of ACO Sensitivity:

- the pheromone sensitivity level, PSL vector elements are initialized with 1;
- globally updating the best tour's PSL vector in the same time with the globally update the pheromone matrix. The PSL vector, with the ant pheromone sensitivity level is updated based on the formula [27,28]; for the image edge detection problem, the Eq. 2 is influenced by the intensity values of image.

$$PSL = ((1 - \rho) * PSL + \rho * \Delta_{ij} * v(I_{ij})) * \Delta_{ij} + PSL * |1 - \Delta_{ij}|. \quad (2)$$

- The *Sensitive ACO* global update (see Eq. 3):

$$\tau^{(n)} = \max_{k=1:K} PSL(k) \cdot \tau^{(n-1)}. \tag{3}$$

Specifications of Image Edge Detection Problem

- Each pixel, $I_{i,j}$, from an image has an 8-connectivity neighborhood denoted $cI_{i,j}$ in its Local configuration; it is used to compute $V_c(I_{i,j})$ (see Eq. 4) processes the "clique" $cI_{i,j}$ [23, 26].
- Based on the local statistic of the pixel (i,j), $\eta_{i,j} = \frac{1}{Z} \cdot V_c(I_{i,j})$, is computed.
- The $V_c(I_{i,j})$ value at pixel $I_{i,j}$ is based on the image's intensity values on $cI_{i,j}$, and is computed as in [26]:

$$\begin{aligned}
V_c(I_{i,j}) = f\,(&|I_{i-2,j-1} - I_{i+2,j+1}| + |I_{i-2,j+1} - I_{i+2,j-1}| \\
&+ |I_{i-1,j-2} - I_{i+1,j+2}| + |I_{i-1,j-1} - I_{i+1,j+1}| \\
&+ |I_{i-1,j} - I_{i+1,j}| + |I_{i-1,j+1} - I_{i+1,j-1}| \\
&+ |I_{i-1,j+2} - I_{i+1,j-2}| + |I_{i,j-1} - I_{i,j+1}|)\,.
\end{aligned} \tag{4}$$

- An edge-solution' validity is made for images pixels based on a threshold T on the pheromone matrix $\tau^{(N)}$ [29]. The Otsu's method [29] uses the maximization between-class variance. The image I represented by L gray levels; the pixels' probabilities at level i: p_i are positive and the overall sum of these probabilities for $i \in [0, L-1]$ is 1. The cumulative probabilities for classes A_i, $i \in [0, k]$ are:

$$w_0 = \sum_{i=0}^{t_0-1} p_i, \; w_1 = \sum_{i=t_0}^{t_1-1} p_i, \ldots w_k = \sum_{i=L-1}^{t_K-1} p_i. \tag{5}$$

t_j are the thresholds separating A_i classes. For $k+1$ classes A_i, $i \in [0, k]$ the goal is to maximize the objective function

$$f(t_0, t_1, \ldots t_{k-1}) = \sum_{i=0}^{k} \sigma_i$$

where the sigma functions are defined by:

$$\sigma_0 = w_0 \Big(\sum_{i=0}^{t_0-1} \frac{ip_i}{w_0} - \sum_{i=0}^{L-1} ip_i\Big)^2, \sigma_1 = w_1 \Big(\sum_{i=t_0}^{t_1-1} \frac{ip_i}{w_1} - \sum_{i=0}^{L-1} ip_i\Big)^2, \ldots \sigma_k = w_k \Big(\sum_{i=t_k}^{L-1} \frac{ip_i}{w_k} - \sum_{i=0}^{L-1} ip_i\Big)^2. \tag{6}$$

Notations used:

- τ_{ij} is the pheromone value of (i,j);
- η_{ij} the heuristic value connecting nodes i and j;
- α, β are the ACO weighting factors for the pheromone and the heuristic;
- Ω_i includes the neighborhoods nodes of node i;
- ρ is the pheromone evaporation rate and Δ_{ij} is the artificial pheromone laid on edge (ij);
- ψ is the pheromone decay rate;
- $I_{i,j}$ is the intensity value of the image pixel (i,j).
- $Z = \sum_{i=1}^{M_1} \sum_{j=1}^{M_2} V_c(I_{i,j})$ is the image normalization factor;

4   Measuring the Quality of the Sensitivity Approach

In order to quantify and measure the Sensitive swarm intelligence quality versus non-sensitive swarm intelligent method, different metrics could be used. Here we tested and show our results when using Sensitive Ant Colony Optimization versus Ant Colony Optimization.

At first we compare image quality while using the Peak signal-to-noise ratio (PSNR). PSNR is a ratio between the maximum possible power of a signal and the power of noise; the noise is affecting the fidelity of an image. We used the PSNR definition based on the Mean Squared Error (MSE). For a noise-free $m \times n$ black-white image, I, with K its noisy approximation, MSE and PSNR, in decibels (dB):

$$MSE = \frac{1}{mn} \sum_{i=0}^{m-1} \sum_{j=0}^{n-1} [I(i,j) - K(i,j)], \ PSNR = 10 \cdot log_{10}(\frac{MAX_I^2}{MSE}).$$

The PSNR formula is the same for any images, including color ones; the MSE formula includes the sum of all squared value differences, for each color, for example three times for the three RGB colors, and divided by the size of the image and by three.

Structural similarity (SSIM) index it is used also for measuring image quality. The index value closer to 1 indicates better image quality. The mean SSMI includes a division of images into blocks further converted into vectors. Further are calculated two means values, two standard derivations and a covariance. In relation with the statistical values, follows the calculus of contrast, luminance, and structure comparisons. In general, SSIM is more accurate than MSE and PSNR despite their cost. The MatLab version of SSIM was used here.

In order to test image qualities we use the MatLab tools. The highest is the PSNR value, the newest image has a better quality than the original image. Table 1 shows the results of our work including [24] while using the medical images, free of copyright: Brain CT (by request from authors), Hand X-ray [30] with the resolution 128×128 and Head CT [31]. The Denoise Convolutional Neural Network (DnCNN) was used [32]. See detailed data and results also on the GitHub author page[1].

Furthermore, in the context of sensitivity, the solution quality, given by the comparative metrics of Sensitive ACO, SACO versus Ant Colony Optimization, ACO for medical image edge detection are presented in Table 1.

The MSE, PSNR and SSMI results from Table 1 reflects the solutions quality, obtain in [24]; therefore, the quality of solutions are improved by the agents sensitivity level. The standard deviation was also computed; the corresponding

---

[1] https://github.com/cristina-ticala/Sensitive_ACO.

Table 1. Comparative metrics values: SACO vs. ACO for image edge detection, based on the results of [24] with the number of correctly identified pixels on the considered images edges.

Image	Sensitive ACO vs. Ant Colony Optimization (ACO) Solution quality: comparative metrics		
	MSE	PSNR	SSIM
Brain CT	0.1219	131.8669	0.9090
Hand X-ray	0.0169	151.6257	0.9353
Head CT	0.0154	152.5716	0.9449

values are 0.05 for MSE, 9.55 for PSNR and 0.02 for the SSIM index. Overall, the SSRMI accuracy values are close to 1, showing the high quality of Sensitive ACO, and the fact that the Pheromone Sensitivity parameter, PSL, has a good influence over problem solution.

5   Conclusion and Further Work

The current work illustrates some of the main aspects linked with solving combinatorial optimization problem while using swarm intelligence. The agents sensitivity to environment seems to be an influential factor when cope with difficult problems as the image edge detection problem.

The quality of sensitivity is measured with the known indexes as Mean Squared Error (MSE), Peak signal-to-noise ratio (PSNR) and Structural similarity (SSIM) index. For the complex medial images tested, Sensitive Ant Colony Optimization results are encouraging when comparing with ACO results.

Further work could be done in using heterogeneous features related to specific swarm used, as for example ants' stigmergy. Human-in-the-loop techniques is one of the most used worldwide model to further enhance the problem complex solution. Sensitivity was tested also on Fuzzy approaches [22] while using Ant Colony System, and could be further used on similar approaches based on [21].

Sensitivity Level vector could be implemented in extracting the edges from image segments, while using a swarm intelligence method. Post processing stages of image processing, could also involve sensitivity factors in order to obtain continuous/compact and thinner edges.

References

1. Matei, O.: Defining an ontology for the radiograph images segmentation. In: Proceedings International Conference on Development and Application Systems, Suceava, Romania, pp. 266–271 (2008)
2. Matei, O.: Applying evolution strategies for chest radiographs segmentation. Comp. Sci. J. Moldova **14**(3), 324–344 (2006)

3. Marginean, A.N., et al.: Reliable learning with PDE-based CNNs and dense nets for detecting COVID-19, pneumonia, and tuberculosis from chest X-ray images. Mathematics **9**, 1–20 (2021). https://doi.org/10.3390/math9040434

4. Grassé, P.-P.: La Reconstruction du nid et les coordinations interindividuelles chez bellicositermes Natalensis et Cubitermes sp. La theorie de la stigmergie: Essai d'interpretation du comportement des termites constructeurs. Insect Sociaux **6**, 41–80 (1959)

5. Di Caro, G., Dorigo, M.: AntNet: distributed stigmergetic control for communications networks. J. Artif. Intell. Res. **9**, 317–365 (1998)

6. Chira, C., et al.: Stigmergic agent optimization. Romanian J. Inf. Sci. Technol. **9**(3), 175–183 (2006)

7. Beni, G., Wang, J.: Swarm intelligence in cellular robotic systems. In: Dario, P., et al. (eds.) Robots and Biological Systems: Towards a New Bionics? NATO ASI Series, vol 102, pp. 703–712. Springer, Heidelberg (1993). https://doi.org/10.1007/978-3-642-58069-7_38

8. Grosan, C., Abraham, A., Chis, M.: Swarm intelligence in data mining. In: Abraham, A., Grosan, C., Ramos, V. (eds.) Swarm Intelligence in Data Mining. Studies in Computational Intelligence, vol. 34, pp. 1–20. Springer, Heidelberg (2006). https://doi.org/10.1007/978-3-540-34956-3_1

9. Kumar, A., Rathore, P.S., Diaz, V.G., Agrawal, R.: Swarm Intelligence Optimization: Algorithms and Applications. Wiley-Scrivener, Beverly (2021)

10. Dorigo, M., Stützle, T.: Ant Colony Optimization. MIT Press, Cambridge (2004)

11. Kennedy, J., Eberhart, R.: Particle swarm optimization. In: Proceedings of ICNN 1995 International Conference on Neural Networks, Perth, WA, Australia, vol. 4, pp. 1942–1948. IEEE (1995). https://doi.org/10.1109/ICNN.1995.488968

12. Rashedi, E., Nezamabadi-pour, H., Saryazdi, S.: GSA: a gravitational search algorithm. Inf. Sci. **179**(13), 2232–2248 (2009)

13. Meshoul, S., Batouche, M.: Ant colony system with extremal dynamics for point matching and pose estimation. In: Proceedings of the 16th International Conference on Pattern Recognition, Quebec, Canada, vol. 3, pp. 823–826. IEEE (2002)

14. Nezamabadi-Pour, H., Saryazdi, S., Rashedi, E.: Edge detection using ant algorithms. Soft. Comput. **10**(7), 623–628 (2006)

15. Jevtić, A., et al.: Edge detection using ant colony search algorithm and multiscale contrast enhancement. In: International Conference on Systems, Man and Cybernetics, San Antonio, Texas, USA, pp. 2193–2198. IEEE (2009)

16. Verma, O.P., Singhal, P., Garg, S., Chauhan, D.S.: Edge detection using adaptive thresholding and ant colony optimization. In: Proceedings of the World Congress Information and Communication Technologies, WICT 2011, pp. 313–318 (2011)

17. Chaudhary, R., Patel, A., Kumar, S., Tomar, S.: Edge detection using particle swarm optimization technique. In: International Conference on Computing, Communication and Automation (ICCCA), pp. 363–367. IEEE, Gagotias University, India (2017). https://doi.org/10.1109/CCAA.2017.8229843

18. Lopez-Molina, C., Bustince, H., Fernandez, J., Couto, P., De Baets, B.: A gravitational approach to edge detection based on triangular norms. Pattern Recogn. **43**(11), 3730–3741 (2010)

19. Pintea, C.-M., Pop, P.C.: Sensor networks security based on sensitive robots agents. A conceptual model. In: Herrero, Á. (ed.) Advances in Intelligent Systems and Computing, vol. 189, pp. 47–56. Springer, Cham (2013). https://doi.org/10.1007/978-3-642-33018-6

20. Pintea, C.-M., Pop, P.C.: Sensitive ants for denial jamming attack on wireless sensor network. In: Herrero, Á. (ed.) Advances in Intelligent and Soft Computing, vol. 239, pp. 409–418. Springer, Cham (2014). https://doi.org/10.1007/978-3-319-01854-6_42

21. Ticala, C., Pintea, C.-M., Ludwig, S.A., Hajdu-Macelaru, M., Matei, O., Pop, P.C.: Fuzzy index evaluating image edge detection obtained with ant colony optimization. In: FUZZ-IEEE 2022, Padua Italy. IEEE (2022, accepted paper)

22. Pintea, C.-M., Matei, O., Ramadan, R.A., Pavone, M., Niazi, M., Azar, A.T.: A fuzzy approach of sensitivity for multiple colonies on ant colony optimization. In: Balas, V.E., Jain, L.C., Balas, M.M. (eds.) SOFA 2016. AISC, vol. 634, pp. 87–95. Springer, Cham (2018). https://doi.org/10.1007/978-3-319-62524-9_8

23. Ticala, C., Zelina, I., Pintea, C.-M.: Admissible perturbation of demicontractive operators within ant algorithms for medical images edge detection. Mathematics 8(1040), 1–13 (2020). https://doi.org/10.3390/math8061040

24. Ticala, C., Pintea, C.-M., Matei, O.: Sensitive ant algorithm for edge detection in medical images. Appl. Sci. 11(23), 1–10, Article no. 11303 (2021). https://doi.org/10.3390/app112311303

25. Pintea, C.-M., Ticala, C.: Medical image processing: a brief survey and a new theoretical hybrid ACO model. In: Hatzilygeroudis, I., Palade, V., Prentzas, J. (eds.) Combinations of Intelligent Methods and Applications. SIST, vol. 46, pp. 117–134. Springer, Cham (2016). https://doi.org/10.1007/978-3-319-26860-6_7

26. Tian, J., Yu, W., Xie, S.: An ant colony optimization algorithm for image edge detection. In: Congress on Evolutionary Computation, pp. 751–756. IEEE (2008)

27. Chira, C., et al.: Learning sensitive stigmergic agents for solving complex problems. Comput. Inform. 29(3), 337–356 (2010)

28. Pintea, C.-M., Chira, C., Dumitrescu, D., Pop, P.C.: A sensitive metaheuristic for solving a large optimization problem. In: Geffert, V., Karhumäki, J., Bertoni, A., Preneel, B., Návrat, P., Bieliková, M. (eds.) SOFSEM 2008. LNCS, vol. 4910, pp. 551–559. Springer, Heidelberg (2008). https://doi.org/10.1007/978-3-540-77566-9_48

29. Otsu, N.: A threshold selection method from gray-level histograms. IEEE Trans. Syst. Man Cybern. 9(1), 62–66 (1979)

30. X-Ray Hand. Vista Medical pack. License: Free for non commercial use. ID, 236487. https://www.iconspedia.com/. Accessed 5 Aug 2021

31. Head CT. Online medical free image. https://www.libpng.org/pub/png/pngvrml/ct2.9-128x128.png. Accessed 5 Aug 2021

32. Denoise image using Deep Neural Network. MATLAB Central. https://www.mathworks.com/help/images/ref/denoiseimage.html

33. Holzinger, A., Plass, M., Holzinger, K., Crişan, G.C., Pintea, C.-M., Palade, V.: Towards interactive Machine Learning (iML): applying ant colony algorithms to solve the traveling salesman problem with the human-in-the-loop approach. In: Buccafurri, F., Holzinger, A., Kieseberg, P., Tjoa, A.M., Weippl, E. (eds.) CD-ARES 2016. LNCS, vol. 9817, pp. 81–95. Springer, Cham (2016). https://doi.org/10.1007/978-3-319-45507-5_6

34. Holzinger, A., Plass, M., Holzinger, K., Crisan, G.C., Pintea, C.M., Palade, V.: A glass-box interactive machine learning approach for solving NP-hard problems with the human-in-the-loop. Creative Math. Inf. 28(2), 121–134 (2019)

35. Holzinger, A., et al.: Interactive machine learning: experimental evidence for the human in the algorithmic loop: a case study on ant colony optimization. Appl. Intell. 49(7), 2401–2414 (2019). https://doi.org/10.1007/s10489-018-1361-5

36. Chira, C., Pintea, C.-M., Dumitrescu, D.: Heterogeneous sensitive ant model for combinatorial optimization. In: GECCO 2008 Proceedings, Atlanta, Georgia, USA, pp. 163–164 (2008). https://doi.org/10.1145/1389095.1389120
37. Chira, C., Pintea, C.-M., Dumitrescu, D.: Sensitive stigmergic agent systems: a hybrid approach to combinatorial optimization. In: Corchado, E., et al. (eds.) Advances in Soft Computing, vol. 44, pp. 33–39. Springer, Cham (2008). https://doi.org/10.1007/978-3-540-74972-1_6
38. Chira, C., Pintea, C.-M., Dumitrescu, D.: Sensitive stigmergic agent systems. In: Tuyls, K., et al. (eds.) ALAMAS Symposium Proceedings, Maastricht, Netherlands, no. 07-04, pp. 51–57 (2007)

Optimization Techniques

Black Widow Optimization for the Node Location Problem in Localization Wireless Sensor Networks

Paula Verde[(✉)][ID], Javier Díez-González[ID], Alberto Martínez-Gutiérrez[ID], Rubén Ferrero-Guillén[ID], Rubén Álvarez[ID], and Hilde Perez[ID]

Department of Mechanical, Computer and Aerospace Engineering, Universidad de León, 24071 León, Spain
{pverg,javier.diez,amartg,ralvf,hilde.perez}@unileon.es, rferrg00@estudiantes.unileon.es

Abstract. Local Positioning Systems (LPS) present higher performance and more accurate target localization than traditional GNSS systems in harsh environments. However, the Node Location Problem (NLP) stands as one of the most important problems when designing LPS since the achievement of optimized sensor distributions in space requires addressing this NP-Hard problem. Therefore, it is common the employment of metaheuristics to tackle this problem. In this sense, the fundamentals of the no free lunch theorems state that, in order to obtain improved results for a specific problem, an investigation on the heuristic that best suits for the characteristics of the problem must be considered. Therefore, in this paper, we propose the application of the black widow optimization algorithm for the first time in the literature for the NLP. This metaheuristic allows a more diversified search adapting to the discontinuous landscape fitness of the NLP when considering NLOS links among the positioning signals. The results obtained are compared with those by a canonical genetic algorithm (CGA) introduced in our previous research, outperforming the localization error by 15% and 10% the single-point and multipoint crossover CGAs analyzed.

Keywords: Black Widow Optimization · Genetic algorithm · Local Positioning Systems · Time Difference of Arrival · Wireless Sensor Networks

1 Introduction

The employment of Global Navigation Satellite Systems (GNSS) has been widespread for providing localization services all over the Earth's surface. However, these systems face some challenges during the signal transmission from the satellites to the targets and during the measurement of the properties of the positioning signals that allow the determination of the target location.

This research has been funded by the Spanish Research Agency (AEI) grant number PID2019-108277GB-C21/AEI/10.13039/501100011033.

P. García Bringas et al. (Eds.): HAIS 2022, LNAI 13469, pp. 469–480, 2022.
https://doi.org/10.1007/978-3-031-15471-3_40

Some of these challenges include the degradation of the positioning signal due to Non-Line-of-Sight (NLOS) links among satellites and targets [18], the appearance of disruptive phenomena on signals [27], changes in the propagation speed of the radioelectric waves [6], ionospheric scintillation [26] or the improper syn-chronization of the system clocks [29]. These problems discourage the application of GNSS for precision tasks in harsh environments. Local Positioning Systems (LPS) [22] are an active topic of research for mitigating the GNSS limitations. LPS consist of ad-hoc deployments of Wireless Sensor Networks (WSN) that particularly adapt to the characteristics of the environment [9].

LPS are characterized depending on the physical property measured for determining the target location: phase, power, time, angle, frequency or mixed forms of them [5,22]. Among them, time-based systems are the most extended in the literature due to their remarkable trade-off among accuracy, stability, costs and easy-to-implement hardware applications [11].

Among time-based LPS architectures, those based on the measurement of the time difference between the reception of the positioning signal in two different architecture sensor receivers are the most flexible and stable for LPS applications [3]. This is known as Time Difference of Arrival (TDOA) localization [23].

TDOA localization requires the collection of at least four time measurements for determining unequivocally the target location (i.e. a minimum of five sensors involved in the localization process) although some special techniques have been proposed in the literature for reducing this requirement to three measurements (i.e. four-sensor WSN deployments) through the optimization of the sensor distribution in space [10]. The flexibility of the TDOA architecture, its unnecessary synchronization among the clocks of the architecture sensor nodes and the target due to the measurement of time differences and the robustness of the technique for harsh environments typical of LPS applications make this architecture succeed in these contexts, justifying the analysis of the architecture in this paper.

Nevertheless, LPS stand as a robust and precise localization system regardless the architecture analyzed, as their deployment adaptability renders them adequate for varying applications. This requires the addressing of the Node Location Problem (NLP) which looks for defining the optimal location for the architecture sensor nodes in order to maximize the coverage region while attaining a reduction of the localization uncertainties due to the limitation of NLOS links and the avoidance of the multipath effect.

This problem has been categorized as NP-Hard both in accuracy and coverage [25] recommending the employment of metaheuristics for finding optimized solutions in acceptable times. Many different techniques have been proposed for addressing this problem such as elephant herding optimization [30], memetic algorithms [12,33] or simulated annealing [32] being the Genetic Algorithms (GA) the most extended methodology in the literature [8,13,17] since their trade-off between intensification and diversification while looking for promising regions of the space of solutions is optimized for the NLP.

However, the No Free Lunch (NFL) theorems stand that there is not an optimal metaheuristic for all the search problems and that the heuristics must be tailored to the problem at hand by using prior knowledge [20]. In this sense, our previous experience tackling the NLP suggest that an optimized metaheuristic for this problem must allow deep exploration of the space of solutions for finding optimized locations for the architecture sensor nodes since the discontinuity of its fitness function complicates this search leading to premature convergence.

One of the algorithms recently proposed in the literature of continuous optimization, the Black Widow Optimization (BWO) algorithm [19], complements the canonical GA (CGA) through a more diverse search of the space of solutions using novel mating techniques and including a novel operator called cannibalism.

This methodology looks promising for finding the optimal locations of the sensor nodes especially in the harsh application environments of the LPS facing potential NLOS or multipath effects. These contexts are typical for the application of these localization technologies considerably improving the GNSS performance in contexts like the urban scenario proposed in this paper.

Therefore, we propose in this paper an optimization of the NLP following the BWO methodology for the first time in the literature of this problem. This requires the adaptation of the BWO algorithm for discrete optimization and the definition of the cannibalism and mating techniques for the NLP. The results obtained through this methodology are compared with the localization accuracy achieved through the CGA that we previously introduced in the literature of this problem in [7]. These results show the preeminence of this novel technique for the NLP attaining more competitive results both in accuracy and stability of the localization services.

The remainder of the paper is organized as follows: the mathematical characterization of the NLP is presented in Sect. 2, the fundamentals of the CGA for the NLP introduced in [7] are highlighted in Sect. 3, a detailed explanation of the fundamentals and adaptation of the BWO algorithm for the discrete optimization of the NLP is presented in Sect. 4, the results obtained for both methodologies in a urban application scenario are analyzed in Sect. 5 while Sect. 6 concludes the paper.

2   Mathematical Definition of the Node Location Problem

The NLP has been widely analyzed in the literature of the WSN since the deployment of valid and optimized sensor distributions in space requires solving this complex problem. The objective is not only the achievement of a maximized coverage region but also the reduction of the target localization uncertainties when deploying positioning WSN.

This requires addressing a problem which has been categorized as NP-Hard both in accuracy and coverage [25]. Particularly, the NLP in localization WSN looks for finding the optimized location of the Cartesian coordinates of each sensor ($s_i = (x_i, y_i, z_i)$) constituting an optimized subset (S_i) of n sensors that represents the combined sensor distribution in space for the WSN within the set (S) containing every possible valid sensor location within the scenario of application of the LPS analyzed.

Mathematically, this defines the following model for the NLP:

$$Maximize: Z = ff(ffcRB, ff_{pen})$$
(1)

Subject to:

$$x_{lim_l} \leq x_i \leq x_{lim_u} \forall x_i \in s_i; s_i \in S_i; s_i \notin F; x_i \in R$$
(2)

$$y_{lim_l} \leq y_i \leq y_{lim_u} \forall y_i \in s_i; s_i \in S_i; s_i \notin F; y_i \in R$$
(3)

$$z_{lim_l} \leq z_i \leq z_{lim_u} \forall z_i \in s_i; s_i \in S_i; s_i \notin F; z_i \in R$$
(4)

$$n_{min_k} \leq \sum_{i=1}^{n_s} cov_{ki} \qquad \forall k \in K_{TLE}$$
(5)

$$cov_{ki} = \begin{cases} 1 \text{ if } (SNR_{ki} \geq SNR_{threshold}) \text{ and } s_i \text{ fullfills Eq. } (2-4) \\ 0 \qquad\qquad\qquad\qquad otherwise \end{cases}$$
(6)

where $ffcRB$ and ff_{pen} are the fitness functions for minimizing the target location uncertianties and for ensuring the fulfillment of the restrictions of the optimization respectively which sum constitute the fitness function of the optimization ff; x_{lim_l}, y_{lim_l}, z_{lim_l}, x_{lim_u}, y_{lim_u} and z_{lim_u} are the lower and upper bounds for the Cartesian coordinates of every sensor of the WSN deployed, F is the set containing the forbidden location for the sensors due to obstacles (e.g. inside obstacles of the scenario); n_{min} is the minimum number of architecture sensors with effective coverage for enabling the target location calculation (i.e. five in the TDOA architecture analyzed in this paper [10]); cov_{ki} represents whether a particular sensor i of the architecture has effective coverage in the point analyzed k; K_{TLE} defines the number of points discretized in which the target localization uncertainties are analyzed during the optimization; SNR_{ki} denotes the signal-to-noise ratio in the positioning signal emitted from the target in the location k and received in the architecture sensor i and $SNR_{threshold}$ represents the value from which an effective coverage between the target and a particular architecture sensor is considered.

Therefore, the fitness function for addressing the NLP must consider a function for reducing the target localization uncertainties and another for the restrictions of the optimization. In this paper, we make use of the Cramér-Rao Bounds (CRB) for characterizing the localization uncertainties in a particular target location. CRB is a maximum likelihood estimator based on the inverse of the Fisher Information Matrix (FIM) which allows the determination of the minimum variance of an unbiased estimator [34]. In the localization field it is widespread since it allows the achievement of the minimum achievable localization error regardless the algorithm used for determining the target location [21]. Kaune et al. [21] introduced a CRB matrix form allowing the characterization of the localization uncertainties into the covariance matrix of the system (R):

$$FIM_{mn} = \left(\frac{\delta h(TS)}{\delta TS_m}\right)^T R^{-1}(TS)\left(\frac{\delta h(TS)}{\delta TS_n}\right)$$
$$+\frac{1}{2}tr\left\{R^{-1}(TS)\left(\frac{\delta R(TS)}{\delta TS_m}\right)R^{-1}(TS)\left(\frac{\delta R(TS)}{\delta TS_n}\right)\right\}$$
(7)

where FIM_{mn} is the (m, n) element of the FIM; $h(TS)$ is the vector containing the links among the Target Sensor (TS) and the architecture sensors. The characterization of the vector $(h(TS))$ and the covariance matrix $(\boldsymbol{R})$ considers a heteroscedastic noise propagation model and a clock error model following a Monte-Carlo simulation as detailed in our previous research of the field [2–4]. This FIM matrix definition directly allows calculating the minimum achievable error of the positioning architecture deployed as follows [21]:

$$RMSE_k = \sqrt{trace(\boldsymbol{FIM}^{-1})} \tag{8}$$

where $RMSE_k$ represents the root mean squared error in the localization analyzed k. Therefore, this characterization defines the uncertainties in the target position calculation in a particular point of study. Since the LPS architecture must provide localization services in a coverage region, the ff_{CRB} of the optimization must consider the reduction of the target uncertainties in the entire coverage region which is defined as:

$$ff_{CRB} = \cfrac{1}{\cfrac{1}{K_{TLE}} \sum_{k=1}^{K_{TLE}} \cfrac{(RMSE_{ref} - RMSE_k)}{RMSE_{ref}}} \tag{9}$$

where $RMSE_{ref}$ is the reference value from which the uncertainties are lower in the scenario, thus constraining the value of ff_{CRB} between [0,1]. The restrictions for the optimization are ensured through the following function:

$$ff_{pen_k} = \begin{cases} 0 & if \quad n_{min_k} \geq cov_{ki} \\ -\dfrac{R_{ik}}{n_{min_k}} & otherwise \end{cases} \tag{10}$$

$$ff_{pen} = \frac{1}{K_{TLE}} \sum_{k=1}^{K_{TLE}} ff_{pen_k} \tag{11}$$

where R_{ik} denotes the number of sensors under effective coverage in the point k.

3 Canonical Genetic Algorithm

GAs are a research branch of Evolutionary Algorithms (EA) proposed by Holland in the 1970s [15]. These algorithms based on population evolution are powerful and usually generates solutions regardless of the initial parameters [15], so are considered robust. The individuals that form the population represent a solution to the problem and those fitter creatures, as well as their genes, survive through the generations [24]. In order to evaluate the degree of adaptation of each individual, it is necessary to determine the fitness function as detailed in Sect. 2. In addition, it is necessary to perform a proper coding of the individuals, as well as a correct selection of the genetic operators employed for GA to achieve convergence and reach a global optimum [31].

3.1 Codification

The encoding of an individual (i.e., chromosome) is a possible solution within the search space. A binary encoding allows the generation of large and complex solutions by increasing the number of bits in the chromosome. This flexibility allows GA to be applied to a wide number of problems and domains. Moreover, the use of a base two for the coding, the alleles that make up the genes are bounded, simplifying genetics operators. Figure 1 shows the coding followed for the individuals of the population.

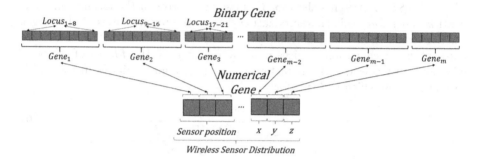

Fig. 1. Proposed genotype coding for GA and BWO.

As shown in Fig. 1, the complete chain of an individual is made up of a combination of n sensors, each of which is composed of different genes. The individual has a total of m genes, which refer to each of the Cartesian Coordinates (i.e., x, y, and z) that represent the position that a sensor of the WSN occupies within the space of solutions (i.e., the 3D scenario characterized in Fig. 3). The coding used allows a discretization of the space into fine meshes [1] that allow the adaptation to the characteristics of the scenario (e.g., obstacles).

3.2 Genetic Operators

In order to ensure the reliability and convergence of the algorithm towards a global optimum, it is necessary to keep those solutions that have the best aptitude and use them in the generation of new individuals [24]. The crossover between two solutions allows exploring potential areas of the space of solutions. The operator generates two new offspring through allele exchange with the aim of reaching solutions that are more optimized than the previous ones.

Different techniques are reported in the literature, such as crossover at a single random point, in which the alleles of an individual are exchanged from a specific point with another individual [14] or multipoint crossover in which there are several exchange points where the solutions are interleaved. In this paper, we call a single crossover (SP) and multipoint crossover (MP) to each of these techniques respectively.

Subsequently, the new individuals generated are subjected to the mutation operator, where each of the alleles forming the individual has a 1% probability [14] of undergoing a mutation and becoming the inverse bit. In addition, the elitism of 10% applied in each of the generations guarantees keeping the best individuals in the population, to ensure the convergence of the algorithm.

4 Black Widow Optimization Algorithm

BWO evolutionary population algorithm was proposed in 2020 [19]. It is based on the unique mating of black widow spiders. As defined in [19] its main advantage is its ability to move away from local optima and to reach a balance between exploration and exploitation compared to other algorithms, which contributes to better quality solutions according to the NFL theorems. In the literature, different applications of this methodology can be found in the WSN field such as energy analysis [28] or data transfer [16]. Therefore, in this paper we propose, for the first time to the best of the authors' knowledge, the use of the BWO for the deployment of a WSN to calculate the minimum associated error in the localization of a target for a TDOA architecture.

The suggested BWO configuration has the same encoding, mutation operator and elitism as those defined for the GA introduced in [7]. The main difference between the algorithms lies in the selection and crossover operators (i.e., for BWO selection and reproduction introduces a new operator called cannibalism) as detailed in Sect. 4.1.

4.1 Operators

The reproductive peculiarities of this type of spider (e.g., sexual cannibalism or consumption between siblings) are interesting aspects that may improve the performance of GA [19].

The selection operator implemented is in charge of choosing the individuals that conceive the individuals of the next population. In this case, as in nature, couples search for the best of their species (i.e., the individuals with the best fitness are selected).

Subsequently, during reproduction, a pair is picked from the previously selected individuals, where the best adapted individual acts as the black widow. A female lays b bags of eggs, each containing u individuals generated with the male. In this paper, we propose a modification on the BWO proposed in [19], where half of the bags are generated from a SP crossover and the other half through the MP crossover taking the benefits from the two techniques, as shown in Fig. 2. From the randomly selected SP, a ρ percentage of the male-forming alleles on the black widow is introduced. The value of ρ is modified for each of the sacs generated by the pair. On the other hand, for the MP cross, two points are randomly generated, from which $\frac{\ell}{2}$ of the alleles of the individual are introduced. Thus, the number of combinations explored for each pair of individuals is much higher, which favors the exploration factor in the search space which is critical for the NLP.

The cannibalism operator applies three types of selections of individuals. First, the black widow, once reproduction is completed, devours her partner, eliminating him from the population. Subsequently, the offspring generated compete within the bag, with the strongest surviving. Thirdly, the offspring individuals compete again with each other and with the mother, with a defined percentage that has the best adaptation surviving. Hence, the operator generates greater elitism within the population than for the GA, leading to a higher exploitation factor as the algorithm converges.

Figure 2 shows the sequence of use of each of the operators and the main parameters used in the reproduction by the BWO algorithm.

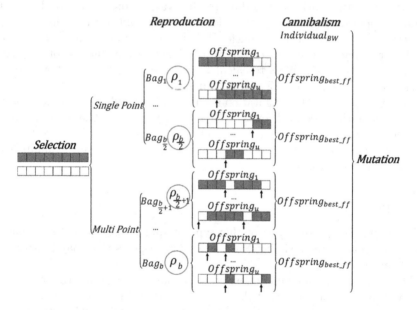

Fig. 2. Operators used by the BWO methodology.

5   Results

The comparison between the different techniques detailed in the previous sections must be performed under the same conditions. For this purpose, an urban scenario has been characterized according to [5] to measure the average error in target localization for a TDOA architecture. In this way, the different sensor networks obtained by the GA-SP, GA-MP and BWO algorithms can be compared. The algorithms have been coded and executed in the MATLAB development environment in an Intel Core i7 2.4 GHz CPU and 16 GB of RAM. The scenario shown in Fig. 3 allows analyzing the degradation suffered by the signals due to the buildings forming this urban area. This application is challenging for the localization of the vehicles moving along the roadways.

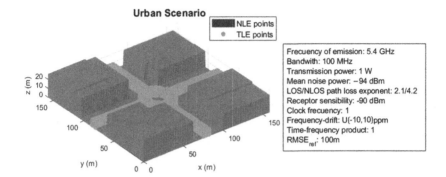

Fig. 3. General characteristics of the urban scenario analyzed. Target Location Environment (TLE) is the location where potential targets can be located, and the Node Location Environment (NLE) is the search space for the algorithms [5,14].

Figure 3 collects those parameters necessary for the TDOA architecture based on real LPS applications (i.e., signal path and sensor characteristics). The designed urban skyline requires at least 8 sensors to achieve coverage at the points studied. We have analyzed the proposed algorithms with the objective of achieving the sensor localization network that improves the accuracy and the standard deviation in target accuracy. Results of the accuracy achieved by the optimizations, as well as the average deviation of the error, are presented in Table 1.

Table 1. Comparison of the RMSE obtained by the GA-SP, GA-MP and BWO algorithms for 8 node sensor networks and 80 individuals from the same initial population.

	GA-SP	GA-MP	BWO
Mean (m)	6.48	6.15	5.53
σ (m)	4.73	3.90	2.91

Therefore, the best methodology is BWO for the deployment of WSN, improving the mean of the RMSE results by 10% over its GA-MP predecessor. The proposed algorithms come through different convergence states along their different generations. Figure 4 shows that the GA-MP achieves better results before than the GA-SP and BWO algorithms. As of execution 70, BWO manages to reach the GA-MP and continues its improvement until it achieves the best individual, while GA-MP and GA-SP have already started its convergence.

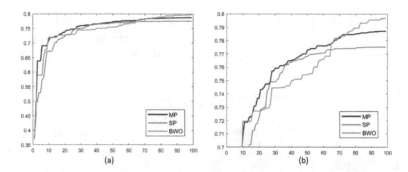

Fig. 4. Comparative of fitness value evolution over generations for GA-SP, GA-MP and BWO in orange, blue and green respectively in (a) while (b) details the fitness interval between 0.7–0.8 for all the algorithms. (Color figure online)

Finally, we present in Fig. 5 the RMSE distribution achieved by the BWO configuration in each of the points of the TLE analyzed in the proposed scenario.

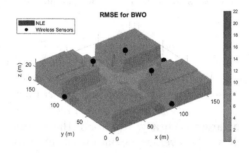

Fig. 5. Accuracy representation in every TLE point of BWO. The use of a base two for the coding

6   Conclusions

The BWO proposed allows the adaptation to the NLP characteristics for the deployment of localization WSNs in harsh environments better than the CGAs implemented in previous works. Therefore, the reached exploration allows decreasing by 15% and 10% the mean of the RMSE obtain for target localization over the GA-SP and GA-MP respectively. Moreover, it improves the uniformity of the error in the area where potential targets can be located by 38% for the GA-SP and 25% for the GA-MP, which increases the stability of the obtained sensor distribution increasing the potential of the TDOA LPS proposed for precision applications such as autonomous navigation or surveillance and rescue operations.

References

1. Al-Qaisi, A., Alhasanat, A.I., Mesleh, A., Sharif, B.S., Tsimenidis, C.C., Neasham, J.A.: Quantized lower bounds on grid-based localization algorithm for wireless sensor networks. Ann. Telecommun. 239–249 (2016). https://doi.org/10.1007/s12243-016-0494-y
2. Álvarez, R., Díez-González, J., Alonso, E., Fernández-Robles, L., Castejón-Limas, M., Perez, H.: Accuracy analysis in sensor networks for asynchronous positioning methods. Sensors **19**(13), 3024 (2019)
3. Álvarez, R., Díez-Gonzalez, J., Sánchez-González, L., Perez, H., et al.: Combined noise and clock CRLB error model for the optimization of node location in time positioning systems. IEEE Access **8**, 31910–31919 (2020)
4. Alvarez, R., Diez-Gonzalez, J., Strisciuglio, N., Perez, H.: Multi-objective optimization for asynchronous positioning systems based on a complete characterization of ranging errors in 3D complex environments. IEEE Access **8**, 43046–43056 (2020)
5. Álvarez, R., Díez-González, J., Verde, P., Perez, H.: Comparative performance analysis of time local positioning architectures in NLOS urban scenarios. IEEE Access **8**, 225258–225271 (2020)
6. Diamant, R., Lampe, L.: Underwater localization with time-synchronization and propagation speed uncertainties. IEEE Trans. Mob. Comput. **12**(7), 1257–1269 (2012)
7. Díez-González, J., Álvarez, R., González-Bárcena, D., Sánchez-González, L., Castejón-Limas, M., Perez, H.: Genetic algorithm approach to the 3D node localization in TDOA systems. Sensors **19**(18), 3880 (2019)
8. Díez-Gonzalez, J., Álvarez, R., Perez, H.: Optimized cost-effective node deployments in asynchronous time local positioning systems. IEEE Access **8**, 154671–154682 (2020)
9. Diez-Gonzalez, J., Alvarez, R., Prieto-Fernandez, N., Perez, H.: Local wireless sensor networks positioning reliability under sensor failure. Sensors **20**(5), 1426 (2020)
10. Díez-González, J., Álvarez, R., Sánchez-González, L., Fernández-Robles, L., Pérez, H., Castejón-Limas, M.: 3D TDOA problem solution with four receiving nodes. Sensors **19**(13), 2892 (2019)
11. Díez-González, J., Álvarez, R., Verde, P., Ferrero-Guillén, R., Perez, H.: Analysis of reliable deployment of TDOA local positioning architectures. Neurocomputing **484**, 149–160 (2022)
12. Díez-González, J., Verde, P., Ferrero-Guillén, R., Álvarez, R., Pérez, H.: Hybrid memetic algorithm for the node location problem in local positioning systems. Sensors **20**(19), 5475 (2020)
13. Domingo-Perez, F., Lazaro-Galilea, J.L., Wieser, A., Martin-Gorostiza, E., Salido-Monzu, D., de la Llana, A.: Sensor placement determination for range-difference positioning using evolutionary multi-objective optimization. Expert Syst. Appl. **47**, 95–105 (2016)
14. Ferrero-Guillén, R., Díez-González, J., Álvarez, R., Pérez, H.: Analysis of the genetic algorithm operators for the node location problem in local positioning systems. In: de la Cal, E.A., Villar Flecha, J.R., Quintián, H., Corchado, E. (eds.) HAIS 2020. LNCS (LNAI), vol. 12344, pp. 273–283. Springer, Cham (2020). https://doi.org/10.1007/978-3-030-61705-9_23
15. Gabriel, P.H.R., Delbem, A.C.B.: Fundamentos de algoritmos evolutivos (2008)
16. Gupta, P., Tripathi, S., Singh, S.: RDA-BWO: hybrid energy efficient data transfer and mobile sink location prediction in heterogeneous WSN. Wirel. Netw. **27**(7), 4421–4440 (2021)

17. Gupta, S.K., Kuila, P., Jana, P.K.: Genetic algorithm approach for k-coverage and m-connected node placement in target based wireless sensor networks. Comput. Electr. Eng. **56**, 544–556 (2016)
18. Guvenc, I., Chong, C.C.: A survey on TOA based wireless localization and NLOS mitigation techniques. IEEE Commun. Surv. Tutor. **11**(3), 107–124 (2009)
19. Hayyolalam, V., Kazem, A.A.P.: Black widow optimization algorithm: a novel meta-heuristic approach for solving engineering optimization problems. Eng. Appl. Artif. Intell. **87**, 103249 (2020)
20. Igel, C.: No free lunch theorems: limitations and perspectives of metaheuristics. In: Borenstein, Y., Moraglio, A. (eds.) Theory and Principled Methods for the Design of Metaheuristics. NCS, pp. 1–23. Springer, Heidelberg (2014). https://doi.org/10.1007/978-3-642-33206-7_1
21. Kaune, R., Hörst, J., Koch, W.: Accuracy analysis for TDOA localization in sensor networks. In: 14th International Conference on Information Fusion, pp. 1–8. IEEE (2011)
22. Kolodziej, K.W., Hjelm, J.: Local Positioning Systems: LBS Applications and Services. CRC Press (2017)
23. Li, Q., Chen, B., Yang, M.: Time difference of arrival passive localization sensor selection method based on Tabu search. Sensors **20**(22), 6547 (2020)
24. Mirjalili, S.: Evolutionary Algorithms and Neural Networks. Studies in Computational Intelligence, vol. 780. Springer, Heidelberg (2019). https://doi.org/10.1007/978-3-319-93025-1
25. Nguyen, N.T., Liu, B.H.: The mobile sensor deployment problem and the target coverage problem in mobile wireless sensor networks are NP-hard. IEEE Syst. J. **13**(2), 1312–1315 (2018)
26. Salles, L.A., Vani, B.C., Moraes, A., Costa, E., de Paula, E.R.: Investigating ionospheric scintillation effects on multifrequency GPS signals. Surv. Geophys. **42**(4), 999–1025 (2021)
27. Seow, C.K., Tan, S.Y.: Non-line-of-sight localization in multipath environments. IEEE Trans. Mob. Comput. **7**(5), 647–660 (2008)
28. Sheriba, S., Rajesh, D.H.: Energy-efficient clustering protocol for WSN based on improved black widow optimization and fuzzy logic. Telecommun. Syst. **77**(1), 213–230 (2021)
29. Skog, I., Handel, P.: Time synchronization errors in loosely coupled GPS-aided inertial navigation systems. IEEE Trans. Intell. Transp. Syst. **12**(4), 1014–1023 (2011)
30. Strumberger, I., Minovic, M., Tuba, M., Bacanin, N.: Performance of elephant herding optimization and tree growth algorithm adapted for node localization in wireless sensor networks. Sensors **19**(11), 2515 (2019)
31. Umbarkar, A.J., Sheth, P.D.: Crossover operators in genetic algorithms: a review. ICTACT J. Soft Comput. **6**(1) (2015)
32. Vecchio, M., López-Valcarce, R., Marcelloni, F.: A two-objective evolutionary approach based on topological constraints for node localization in wireless sensor networks. Appl. Soft Comput. **12**(7), 1891–1901 (2012)
33. Verde, P., Ferrero-Guillén, R., Álvarez, R., Díez-González, J., Perez, H.: Node distribution optimization in positioning sensor networks through memetic algorithms in urban scenarios. In: Engineering Proceedings, vol. 2, p. 73. Multidisciplinary Digital Publishing Institute (2020)
34. Wang, Y., Ho, K.: TDOA source localization in the presence of synchronization clock bias and sensor position errors. IEEE Trans. Signal Process. **61**(18), 4532–4544 (2013)

Hybrid Intelligent Model for Classification of the Boost Converter Switching Operation

Luis-Alfonso Fernandez-Serantes[1]($^{(\boxtimes)}$), José-Luis Casteleiro-Roca[1],
Paulo Novais[2], Dragan Simić[3], and José Luis Calvo-Rolle[1]

[1] CTC, CITIC, Department of Industrial Engineering, University of A Coruña,
Ferrol, A Coruña, Spain
luis.alfonso.fernandez.serantes@udc.es
[2] Algoritmi Center, Department of Informatics, University of Minho, Braga, Portugal
[3] Faculty of Technical Sciences, University of Novi Sad, Trg Dositeja Obradovića 6,
21000 Novi Sad, Serbia

Abstract. The application of a hybrid intelligent model is applied to a boost converter with the aim of detecting the switching operating mode of the converter. Thus, the boost converter has been analyzed and the both operating mode are explained, distinguishing between Hard-switching and Soft-switching modes. Then, the dataset is created out of the data obtained from simulation of the real circuit and, finally, the hybrid intelligent classification model is implemented. The proposed model is able to distinguish between the HS and SS operating modes with high accuracy.

Keywords: Hard-switching · Soft-switching · Boost converter · Power electronics · Classification · Hybrid model

1 Introduction

Nowadays many research are done in the field of power electronics. They focus on increasing the efficiency of the power converter and, therefore, reducing the size of the circuits. The recent works focus on the use of new materials, the so-called wide band-gap materials, and the use of soft-switching techniques [1,5]. The introduction of the Silicon Carbide (SiC) and Gallium Nitride (GaN) in the power converters starts replacing the silicon as manufacturing material of the power transistors [11,28]. They provide better performance and characteristics in comparison with silicon transistors, such as higher switching speeds, higher breakdown voltages, lower on-state resistance, etc. [11,19,21]

Along with the introduction of the Artificial Intelligence (AI) in other fields, the AI starts taking also importance in the power electronics. These techniques are used for supporting the development and design processes, as described in [2,12,31] where the AI are used to design magnetic components. Or, another application, to improve the performance of the power converters with the development of new control schemes, as done in [17,18,30].

© Springer Nature Switzerland AG 2022
P. García Bringas et al. (Eds.): HAIS 2022, LNAI 13469, pp. 481–493, 2022.
https://doi.org/10.1007/978-3-031-15471-3_41

With the aim of improving the efficiency of the power converters, the soft-switching technique can be used. With the aim of verifying that the converter is operated in this mode, and not in hard-switching mode, a classification model based in hybrid techniques has been developed. By assuring the SS mode in the converter, the overall efficiency of it can be improved and a failure condition can be avoided.

The proposed method is a novel use of the AI in power electronics. Until now, the adjustment of the power converter is done by the engineer, who based on own knowledge, determines the operating point of the converter. With the proposed technique, the operating mode of the converter is automatically detected by the algorithm, that aims to increase the efficiency and reduce the switching losses. The AI helps detecting the operation mode of the converter and, thus, reducing the switching losses of the converter [4,23].

After this introduction, this paper continues as follows: first, a detailed description of the analyzed power converter is done. Then, a differentiation between the two operating modes, HS and SS modes, is described. The proposed hybrid intelligent model is defined and the different cluster and classification techniques are described. Afterwards, the obtained results in the classification are shown and, finally, the conclusions are drawn.

2 Case Study

The analysis of a boost converter, as shown in the Fig. 1 is done in this section. The components used in this converter are two transistors, high-side and low-side transistors, which operate in a complementary manner. An input capacitor is used to filter the peak currents drawn by the converter. The transistors generate a pulsed voltage that varies from input voltage and ground, at the switching node (Vsw), which is then filtered. The output filter is made up of an inductor and a capacitor.

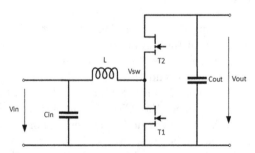

Fig. 1. Circuit topology of a boost converter.

Traditionally, the described converter operates in Hard-Switching (HS) mode: meaning that losses occur during the switching transitions due to the concurrence of current and voltage at the transistor during commutation. When the transistor is turned-off, the voltage is blocked and no current is flowing; in this case, the losses are zero. When the gate signal is applied and the commutation starts, the resistance of the transistor starts to drop as the channel starts conducting. At this point, the voltage drops while the current rises, occurring switching losses as $P = V \cdot I$. This happens during turn-on and turn-off processes. Moreover, when the transistor is switched on, the losses are caused by the on-state resistance times the flowing current: $P_{conduction} = I^2 \cdot R_{on-state}$.

In addition to the losses during the commutation, the parasitic capacitance of the transistor (Coss) is reloaded and discharged through the channel of the switch, causing further switching losses. In the Fig. 2a, the converter losses in HS mode are represented, where the switching losses can be seen.

With the aim of improving the efficiency of the boost power converter, the other operating mode is introduced: soft-switching (SS) mode. In this case, either the current or the voltage through the transistor channels is brought to zero. If the current is zero at the switching instant, it is called Zero-Current-Switching (ZCS); by contrast, if the voltage across the transistor at the switching instant is zero, the SS mode is due to Zero-Voltage-Switching (ZVS).

When the converter operates in SS, the losses during the commutation are nearly zero, as either the voltage or current are zero: $P = V \cdot I = 0 \cdot I$ or $P = V \cdot 0$. The Fig. 2b shows the transitions during this conditions. The condition of SS are achieved thanks to the resonance between the components of the circuit, such as a resonance LC tank. In the proposed converter, the method used for the SS operation is the ZVS and it is achieved during turn-on of the high-side switch.

In the ZVS method, the switches turn-on or -off when the voltage across the transistor is zero. Thus, when the high-side transistor switches off, the current moves to the low-side switch. During the interlock delay, time where both switches are off, the current flows through the antiparallel diode of the low-side transistor, equalizing the voltage across this transistor, causing a ZVS switching of the low-side transistor.

During the time that the low-side transistor is on, the current starts decreasing in the inductor until the current reaches a negative value, flowing from the load to the switches. At this instant, when the current is negative, the low-side switch turns-off. The current at the inductor does not have a path to flow, so it will charge the output capacitance of the transistors, rising the voltage at the switching node.

Once the voltage at the switching node reaches the input voltage, the antiparallel of the high-side switch starts conducting, instant when this switch can turn-on with ZVS. At this point, the high-side transistor can switch lossless, as the voltage across it is just a few volts from the forward diode voltage [27].

When using this topology in SS mode, the design of the inductor is very important, as needs to allow high ripple current.

Traditionally, the designer and developers keep the inductor ripple low, between 10% and 30% of the output current. The definition of the inductance value is done according to the Eq. 1, where the inductor value depends on the switching frequency, output voltage and the allowed current ripple.

$$L = \frac{V_{in} \cdot (V_{out} - V_{in})}{f \cdot I_{ripple} \cdot V_{out}} \tag{1}$$

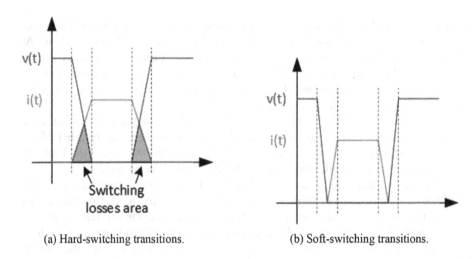

(a) Hard-switching transitions.   (b) Soft-switching transitions.

Fig. 2. Switching transitions.

where L is the inductance value of the inductor, V_{in} is the input voltage to the circuit, V_{out} is the output voltage from the converter, f is the switching frequency and I_{ripple} is the current ripple in the inductor.

With the introduction of this converter operating in SS mode, the design of the inductor needs to be reconsidered. In this mode, the ripple of the current allows the current to drop to zero and beyond, defining the Triangular Current Mode (TCM) [8,10,16].

As mentioned above, when the current ripple is kept low, as shown in the Fig. 3, the converter operates in HS mode, as the ZVS condition is never reached. In opposition, when the current ripple allows the current to drop below zero, the switching losses in the converter can be reduced due to the ZVS mode.

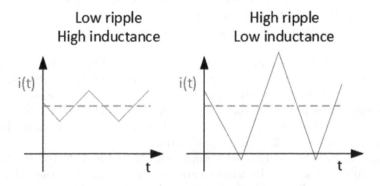

Fig. 3. Current ripple with different filter inductors.

The SS operating mode allows a reduction of the switching losses but with the drawback of increasing the conduction losses, as the Root Means Square (RMS) of the current increases. With the aim of taking advantage of this operating mode, the switching frequency of the converter is increased, reducing in this manner the conduction losses and the filter components; therefore further increasing the power densities of the converter.

3   Model Approach

The development of a hybrid intelligent model able to distinguish and classify the operating mode of the boost converter has been done. This model classifying the switching mode distinguishing between HS and SS mode. The dataset used by this model is obtained from simulation results of the real circuit and a clustering of this data has been done. Afterwards, four different classification algorithms are applied.

The model approach is shown in the Fig. 4, where the input data is obtained and further processed, with the aim of obtaining more representative variables. In a next step, the hybrid intelligent model is applied, first dividing the data into clusters and then applying the classification algorithms.

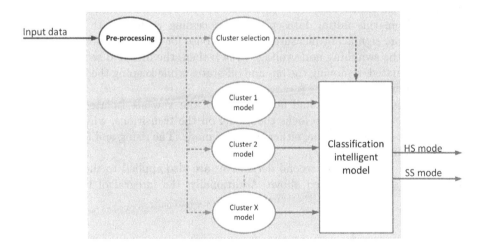

Fig. 4. Model approach.

3.1   Dataset

The dataset is compound of simulation results of a boost converter done in LTSpice. With the aim of obtaining closer results to the real circuit, the transistors models from the manufacturer have been used. In total, the circuit has been simulated 80 times and the obtained simulation curves are used to create the dataset. The simulations combine HS and SS operating mode, being 50% of the data each type. The converter keeps unmodified to allow reproducibility of the experiment, only varying the output load applied to it. The dataset is made up from the following signals measured in the circuit:

- Input voltage: a constant input voltage of 200 V is applied ot the circuit.
- Output voltage: the output voltage of the converter is kept at 400 V, allowing a ripple from 390 V up to 410 V.
- Switching node voltage (Vsw node Fig. 1): at this node, the voltage varies from 0 V when the low-side switch is on up to 400 V when the high-side switch is on. The generated signal is square with a frequency varying from 80 kHz up to 2 MHz.
- Inductor current: is a triangular waveform. The average current depends on the output load and the current ripple depends on the switching frequency. For a constant inductance value, when the switching frequency is higher, the current ripple is lower, while it increases as the switching frequency decreases. In HS mode, the ripple is between 10 % to 30 % while in SS mode, the ripple increases above 200 %.
- Output current: is constant, filtered by the inductor and capacitor, and its value depends on the connected load to the converter.

From previous obtained data, the signal that provides more information about the operating mode is the switching node voltage, as reveals the commutation of the transistors.

Then, from this initial dataset, a pre-processing is done with the aim of obtaining more significant measurements: the base for this pre-processing is the raw data of the switching node voltage (Vsw); then, the first and second derivative are calculated, removing the on- and off-states while keeping the information of the commutations.

Furthermore, the rising and falling edges of the Vsw are isolated, as shown in the Fig. 5. This allows to focus the model on the transitions, which provides details of the operating mode, either HS or SS mode. The rising and falling times are also obtained (tr and tf).

Moreover, the first and second derivatives are also applied to the rising and falling edge signal explained above. Additionally, the integral of the edges is calculated.

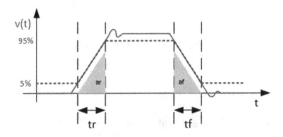

Fig. 5. Rising and falling edge of the switching node voltage, in dashed blue, and the original signal in continuous red. (Color figure online)

As described, 8 signals are obtained from the Vsw: the raw data (red signal in Fig. 5), the first and the second derivatives of the raw data, the rising/falling edge data (dotted blue signal in Fig. 5), the first and second derivatives of rising/falling edge data, the rising edge integral (area at the rising edge, ar, in Fig. 5) and the falling edge integral (area at the falling edge, af, in Fig. 5).

Moreover, the following statistics and indicators are calculated from the 8 obtained signals: average, standard deviation, variance, co-variance, Root Mean Square and Total Harmonic Distortion (THD). Resulting in a matrix of 8×6 for each of the 80 simulations.

3.2   Methods

The classification algorithms used in this research are the Multilayer Perceptron (MLP), the Support Vector Machine (SVM), the Linear Discrimination Analysis (LDA) and the ensemble classifier. These methods are described bellow.

Data Clustering. The K-Means Algorithm. An algorithm commonly uses to divide data is the unsupervised technique called K-Means. This clustering data makes groups according to the similarity of the samples [13, 22, 29]. The unlabeled samples x are compared with the rest of the samples and make the clusters based on the distance between each sample and the centroids. The clusters are defined by their centroids, that are chosen as the geometrical center of the cluster.

This is an iterative algorithm, and the training phase is the calculation of the centroids. To perform this, the algorithm used to start with randomly centroids, chosen from the dataset. In the following steps, all the samples are assigned to each centroids based on the distance from each samples to all centroids; the sample cluster is the one with the smallest distance. Once all the samples are assigned to a cluster, the centroids are re-calculated as the center of the new group of samples. The procedure finish when the centroids do not change its value between two consecutive iterations.

Multilayer Perceptron. A perceptron is an artificial neural network that has only one hidden layer of neurons. When the same structure is used, but with multiple hidden layer, we refer as a multilayer perceptron. The structure is the following: one input layer, which consists of input features to the algorithm, then the multiple hidden layer which have neurons with an activation function, and one output layer, which number of neurons depends on the desired outputs. All these layer are connected in between by weighted connections. These weights are tuned with the aim of decreasing the error of the output [9, 14, 30].

Linear Discrimination Analysis. Another method used for classification is the Linear Discrimination Analysis. This method is based on a dimension reduction, projecting the data from a high dimensional space into a low-dimensional space, where the separation of the classes is done. This method uses a weight vector W, which projects the given set of data vector E in such a way that maximizes the class separation of the data but minimizes the intra-class data [6, 32]. The projection is done accordingly to the Eq. 2. The separation is good when the projections of the class involves exposing long distance along the direction of vector W.

$$P_i = W^T \cdot E_i \tag{2}$$

The LDA provides each sample with its projection and the class label. Two outputs are provided by the analysis, first a gradual decision which is then converted into a binary decision. This method maximizes ratio between the inter-class variance to the intra-class variance, finding the best separation of the classes. The performance of the method increases with the distance between the samples [15,26].

Support Vector Machine. A common used method in classification is the support vector machine, which is a supervised algorithm of machine learning [3,24]. The algorithm tries to find two parallel hyper-planes that maximize the minimum distance between two class of samples [24]. Therefore, the vectors are defined as training instances presented near the hyperplane and a high dimensional feature space is used by a kernel to calculate the projection of the dataset.

Ensemble. The term ensemble is used to define multiple classification methods which are used in combination with the aim of improving the performance over single classifiers [20,25]. They are commonly used for classification tasks. The ensemble does a regularization, process of choosing fewer weak learners in order to increase predictive performance [33].

3.3   Measurement of the Classification Performance

The models are validated with a confusion matrix. The confusion matrix is a commonly used method to asses the quality of a classifier. The entry to this matrix are the true values and the predicted values, where the true values compound the columns while the predicted ones the rows [7].

Usually, the entries to the matrix are two decision classes, positives (P) and negatives (N), and the table entries are called true positives (TP), false positives (FP), true negatives (TN) or false negatives (FN) [15].

Once the confusion matrix have been created, there are 5 indicators that are used to analyzed the performance of the models and compare the results between them. These statistics are the followings: SEnsitivity (SE), SPeCificity (SPC), Positive Prediction Value (PPV), Negative Prediction Value (NPV), and Accuracy (ACC) [15].

3.4   Experiments Description

First, the dataset is divided into two different groups. The first group, that contains 75% of the generated data, is used to train the proposed models; while the rest, 25 % of the dataset, is used to validate the proposed algorithms. It is important to remark that the separation is done randomly. The following experiments are carried out in this research:

– Models without clustering: each of the previously mentioned classification algorithms are tested with the dataset. The aim of this test, is to verify the performance of the models without clustering and later compare the results with the hybrid intelligent model.
– Clustering of the data: the dataset is divided into clusters or groups. The number of them varies from 2 up to 10 clusters, and it is done using the K-means method. Afterwards, to each of the groups, the following algorithms are applied.

The used algorithms to classify the operating mode of the power converter are the nexts:

– Multilayer perceptron (MLP): uses the Levenberg-Marquardt backpropagation algorithm with an hidden layer with 1 up to 10 neurons.
– Linear discrimination analysis (LDA): the used type is the regularized LDA, where all the classes have identical co-variance matrix.
– Support vector machine (SVM): The linear kernel function has been used. The classifier has been trained using the standardized predictors, which centers and scales each predictor variable by the corresponding mean and standard deviation.
– Ensemble: the adaptive logistic regression has been used. This type is commonly applied to binary classification. The number of cycles have been varied in steps of 10 from 10 up to 100. The weak-learners used function is the decision tree.

Finally, once the different models are trained, the models are validated using the previously separated 25 % of the data. This is done to verify the correct functionality and performance of the proposed models. The classification results obtained from the models are then compared with the validation data and by using a confusion matrix, the different statistics are calculated.

4   Results

The results from this research are shown and described in this section. The Table 1 shows a summary of the performance achieved by each model, allowing a comparison between the hybrid intelligent models and the used algorithms without clustering. The performance and accuracy of the models with cluster numbers higher than 4, it is always 100 %.

When the number of clusters is one, so the algorithms are runned without previous clustering, the MLP7 performs the best with an accuracy of 0.97059. By increasing the number of cluster to two, the accuracy is already 100 % and the worst performance is achieved by the ensembles with an accuracy of 0.78947. When using three clusters, most of the classification algorithms performs a 100 % of accuracy and, with four or higher, all the classification models achieve 100 % accuracy.

The Table 1 is organized as follows: first column presents the results of the algorithms without previous clustering of the data. Then, the number of clusters are increased and the results for each of the classification algorithms are preseted.

Table 1. Achieved accuracy with each model and cluster configuration.

Clusters	No cluster	2 clusters		3 clusters			4 clusters			
# of cluster	-	1 of 2	2 of 2	1 of 3	2 of 3	3 of 3	1 of 4	2 of 4	3 of 4	4 of 4
MLP1	0.816	1.0	0.938	0.904	1.0	1.0	1.0	1.0	1.0	1.0
MLP2	0.741	1.0	0.948	1.0	1.0	1.0	1.0	1.0	1.0	1.0
MLP3	0.613	1.0	0.948	1.0	1.0	1.0	1.0	1.0	1.0	1.0
MLP4	0.560	0.935	0.935	0.895	1.0	1.0	1.0	1.0	1.0	1.0
MLP5	0.801	0.891	1.0	1.0	1.0	1.0	1.0	1.0	1.0	1.0
MLP6	0.610	0.946	0.949	1.0	1.0	1.0	1.0	1.0	1.0	1.0
MLP7	0.981	0.946	0.949	0.895	1.0	1.0	1.0	1.0	1.0	1.0
MLP8	0.736	0.946	0.957	1.0	1.0	1.0	1.0	1.0	1.0	1.0
MLP9	0.871	1	0.957	1.0	1.0	1.0	1.0	1.0	1.0	1.0
MLP10	0.519	1	0.947	1.0	1.0	1.0	1.0	1.0	1.0	1.0
SVM	0.789	1	0.951	1.0	1.0	1.0	1.0	1.0	1.0	1.0
LDA	0.603	1	0.939	1.0	1.0	1.0	1.0	1.0	1.0	1.0
Ensem. 10	0.515	0.956	0.798	1.0	1.0	1.0	0.814	1.0	1.0	1.0
Ensem. 20–100	0.509	1.0	0.798	1.0	1.0	1.0	0.814	1.0	1.0	1.0

5   Conclusions and Future Works

This research presents a hybrid intelligent model used for classifying the operating mode of a boost converter. The model detects and distinguish the two possible operating modes of the boost converter, either HS or SS mode. Four different classification algorithms are used for this purpose, previous clustering of the dataset. The data is obtained from the simulation of the boost converter using the models of the transistors, provided by the manufacturer.

The dataset is obtained from the simulation results of the converter, using 8 signals obtained from the switching node voltage. The obtained results of the classification algorithms without clustering of the dataset achieve a maximum performance up to 97 %. In contrast, a hybrid intelligent model is applied, using the classification algorithms in conjunction with the clustering algorithm, the performance reaches 100 % of accuracy.

The hybrid intelligent model presented in this paper is very useful to verify and detect the operating mode of the boost converter, supporting the design of the power converter and increasing the efficiency at the same time.

In future works, the research will focus in applying this technique to other topologies as well as the real implementation of the power converter.

Acknowledgements. CITIC, as a Research Center of the University System of Galicia, is funded by Consellería de Educación, Universidade e Formación Profesional of the Xunta de Galicia through the European Regional Development Fund (ERDF) and the Secretaría Xeral de Universidades (Ref. ED431G 2019/01).

References

1. Al-bayati, A.M.S., Alharbi, S.S., Alharbi, S.S., Matin, M.: A comparative design and performance study of a non-isolated DC-DC buck converter based on Si-MOSFET/Si-Diode, SiC-JFET/SiC-schottky diode, and GaN-transistor/SiC-Schottky diode power devices. In: 2017 North American Power Symposium (NAPS), pp. 1–6 (2017). https://doi.org/10.1109/NAPS.2017.8107192
2. Aláiz-Moretón, H., Castejón-Limas, M., Casteleiro-Roca, J.L., Jove, E., Fernández Robles, L., Calvo-Rolle, J.L.: A fault detection system for a geothermal heat exchanger sensor based on intelligent techniques. Sensors 19(12), 2740 (2019)
3. Basurto, N., Arroyo, Á., Vega, R., Quintián, H., Calvo-Rolle, J.L., Herrero, Á.: A hybrid intelligent system to forecast solar energy production. Comput. Electr. Eng. 78, 373–387 (2019)
4. Casado-Vara, R., et al.: Edge computing and adaptive fault-tolerant tracking control algorithm for smart buildings: a case study. Cybern. Syst. 51(7), 685–697 (2020). https://doi.org/10.1080/01969722.2020.1798643
5. Casteleiro-Roca, J.L., Javier Barragan, A., Segura, F., Luis Calvo-Rolle, J., Manuel Andujar, J.: Intelligent hybrid system for the prediction of the voltage-current characteristic curve of a hydrogen-based fuel cell. Revista Iberoamericana de Automática e Informática industrial 16(4), 492–501 (2019)
6. Crespo-Turrado, C., et al.: Comparative study of imputation algorithms applied to the prediction of student performance. Logic J. IGPL 28(1), 58–70 (2020)
7. Düntsch, I., Gediga, G.: Indices for rough set approximation and the application to confusion matrices. Int. J. Approximate Reasoning 118, 155–172 (2020). https://doi.org/10.1016/j.ijar.2019.12.008
8. Fernandez-Serantes, L.A., Berger, H., Stocksreiter, W., Weis, G.: Ultra-high frequent switching with GaN-HEMTs using the coss-capacitances as non-dissipative snubbers. In: PCIM Europe 2016; International Exhibition and Conference for Power Electronics, Intelligent Motion, Renewable Energy and Energy Management, pp. 1–8. VDE (2016)
9. Fernandez-Serantes, L.A., Casteleiro-Roca, J.L., Berger, H., Calvo-Rolle, J.L.: Hybrid intelligent system for a synchronous rectifier converter control and soft switching ensurement. Eng. Sci. Technol. Int. J. 101189 (2022)
10. Fernandez-Serantes, L.A., Casteleiro-Roca, J.L., Calvo-Rolle, J.L.: Hybrid intelligent system for a half-bridge converter control and soft switching ensurement. Revista Iberoamericana de Automática e Informática industrial (2022). https://doi.org/10.4995/riai.2022.16656
11. Fernández-Serantes, L.A., Estrada Vázquez, R., Casteleiro-Roca, J.L., Calvo-Rolle, J.L., Corchado, E.: Hybrid intelligent model to predict the SOC of a LFP power cell type. In: Polycarpou, M., de Carvalho, A.C.P.L.F., Pan, J.-S., Woźniak, M., Quintian, H., Corchado, E. (eds.) HAIS 2014. LNCS (LNAI), vol. 8480, pp. 561–572. Springer, Cham (2014). https://doi.org/10.1007/978-3-319-07617-1_49
12. García-Ordás, M.T., et al.: Clustering techniques selection for a hybrid regression model: a case study based on a solar thermal system. Cybern. Syst. 1–20 (2022). https://doi.org/10.1080/01969722.2022.2030006
13. Gonzalez-Cava, J.M., et al.: Machine learning techniques for computer-based decision systems in the operating theatre: application to analgesia delivery. Logic J. IGPL 29(2), 236–250 (2020). https://doi.org/10.1093/jigpal/jzaa049
14. Jove, E., Casteleiro-Roca, J., Quintián, H., Méndez-Pérez, J., Calvo-Rolle, J.: Anomaly detection based on intelligent techniques over a bicomponent produc-

tion plant used on wind generator blades manufacturing. Revista Iberoamericana de Automática e Informática industrial **17**(1), 84–93 (2020)

15. Jove, E., et al.: Missing data imputation over academic records of electrical engineering students. Logic J. IGPL **28**(4), 487–501 (2020)

16. Jove, E., Casteleiro-Roca, J.L., Quintián, H., Méndez-Pérez, J.A., Calvo-Rolle, J.L.: A fault detection system based on unsupervised techniques for industrial control loops. Expert. Syst. **36**(4), e12395 (2019)

17. Jove, E., Casteleiro-Roca, J.L., Quintián, H., Simić, D., Méndez-Pérez, J.A., Luis Calvo-Rolle, J.: Anomaly detection based on one-class intelligent techniques over a control level plant. Logic J. IGPL **28**(4), 502–518 (2020)

18. Jove, E., et al.: Comparative study of one-class based anomaly detection techniques for a bicomponent mixing machine monitoring. Cybern. Syst. **51**(7), 649–667 (2020). https://doi.org/10.1080/01969722.2020.1798641

19. Jove, E., Casteleiro-Roca, J.L., Quintián, H., Méndez-Pérez, J.A., Calvo-Rolle, J.L.: A new method for anomaly detection based on non-convex boundaries with random two-dimensional projections. Inf. Fusion **65**, 50–57 (2021). https://doi.org/10.1016/j.inffus.2020.08.011. https://www.sciencedirect.com/science/article/pii/S1566253520303407

20. Jove, E., et al.: Modelling the hypnotic patient response in general anaesthesia using intelligent models. Logic J. IGPL **27**(2), 189–201 (2019)

21. Jove, E., et al.: Hybrid intelligent model to predict the remifentanil infusion rate in patients under general anesthesia. Logic J. IGPL **29**(2), 193–206 (2020). https://doi.org/10.1093/jigpal/jzaa046

22. Kaski, S., Sinkkonen, J., Klami, A.: Discriminative clustering. Neurocomputing **69**(1–3), 18–41 (2005)

23. Leira, A., et al.: One-class-based intelligent classifier for detecting anomalous situations during the anesthetic process. Logic J. IGPL (2020). https://doi.org/10.1093/jigpal/jzaa065

24. Liu, M.Z., Shao, Y.H., Li, C.N., Chen, W.J.: Smooth pinball loss nonparallel support vector machine for robust classification. Appl. Soft Comput. 106840 (2020). https://doi.org/10.1016/j.asoc.2020.106840

25. Luis Casteleiro-Roca, J., Quintián, H., Luis Calvo-Rolle, J., Méndez-Pérez, J.A., Javier Perez-Castelo, F., Corchado, E.: Lithium iron phosphate power cell fault detection system based on hybrid intelligent system. Logic J. IGPL **28**(1), 71–82 (2020)

26. Marchesan, G., Muraro, M., Cardoso, G., Mariotto, L., da Silva, C.: Method for distributed generation anti-islanding protection based on singular value decomposition and linear discrimination analysis. Electric Power Syst. Res. **130**, 124–131 (2016). https://doi.org/10.1016/j.epsr.2015.08.025

27. Mohan, N., Undeland, T.M., Robbins, W.P.: Power Electronics: Converters, Applications, and Design. Wiley, Hoboken (2003)

28. Neumayr, D., Bortis, D., Kolar, J.W.: The essence of the little box challenge-part A: key design challenges solutions. CPSS Trans. Power Electron. Appl. **5**(2), 158–179 (2020). https://doi.org/10.24295/CPSSTPEA.2020.00014

29. Qin, A.K., Suganthan, P.N.: Enhanced neural gas network for prototype-based clustering. Pattern Recogn. **38**(8), 1275–1288 (2005)

30. Tahiliani, S., Sreeni, S., Moorthy, C.B.: A multilayer perceptron approach to track maximum power in wind power generation systems. In: 2019 IEEE Region 10 Conference (TENCON), TENCON 2019, pp. 587–591 (2019). https://doi.org/10.1109/TENCON.2019.8929414

31. Liu, T., Zhang, W., Yu, Z.: Modeling of spiral inductors using artificial neural network. In: Proceedings of the 2005 IEEE International Joint Conference on Neural Networks, vol. 4, pp. 2353–2358 (2005). https://doi.org/10.1109/IJCNN.2005.1556269

32. Thapngam, T., Yu, S., Zhou, W.: DDoS discrimination by linear discriminant analysis (LDA). In: 2012 International Conference on Computing, Networking and Communications (ICNC), pp. 532–536. IEEE (2012)

33. Uysal, I., Gövenir, H.A.: An overview of regression techniques for knowledge discovery. Knowl. Eng. Rev. **14**, 319–340 (1999)

A Comparison of Meta-heuristic Based Optimization Methods Using Standard Benchmarks

Enol García[1(✉)], José R. Villar[1], Camelia Chira[2], and Javier Sedano[3]

[1] Computer Science Department, University of Oviedo, Oviedo, Spain
{garciaenol,villarjose}@uniovi.es
[2] Department of Computer Science, University of Babes Boliay,
Cluj-Napocav, Romania
camelia.chira@ubbcluj.ro
[3] Instituto Tecnológico de Castilla y León, Burgos, Spain
javier.sedano@itcl.es

Abstract. Optimization problems are a type of problem in which multiple solutions satisfy the problem's constraints, so not only must a good solution be found, but the objective is to find the best solution among all those considered valid. Optimization problems can be solved by using deterministic and stochastic algorithms. Those categories can be divided into different kinds of problems. One of the categories inside stochastic algorithms is metaheuristics. This work implements three well-known meta-heuristics –Grey Wolf Optimizer, Whale Optimization Algorithm, and Moth Flame Optimizer–, and compares them using ten mathematical optimization problems that combine non-constrained from other studies and constrained problems from CEC2017 competition. Results show the Grey Wolf Optimizer as the method with faster convergence and best fitness for almost all the problems. This work aims to implement and compare various metaheuristics to carry out future work on solving various real-world problems.

Keywords: Meta-heuristics · Optimization · Benchmarking

This research has been founded by European Union's Horizon 2020 research and innovation programme (project DIH4CPS) under the Grant Agreement no 872548. Furthermore, this research has been funded by the SUDOE Interreg Program -grant INUNDATIO-, by the Spanish Ministry of Economics and Industry, grant PID2020-112726RB-I00, by the Spanish Research Agency (AEI, Spain) under grant agreement RED2018-102312-T (IA-Biomed), by CDTI (Centro para el Desarrollo Tecnológico Industrial) under projects CER-20211003 and CER-20211022, by and Missions Science and Innovation project MIG-20211008 (INMERBOT). Also, by Principado de Asturias, grant SV-PA-21-AYUD/2021/50994 and by ICE (Junta de Castilla y León) under project CCTT3/20/BU/0002.

P. García Bringas et al. (Eds.): HAIS 2022, LNAI 13469, pp. 494–504, 2022.
https://doi.org/10.1007/978-3-031-15471-3_42

1   Introduction

Optimization problems are a type of problem in which multiple solutions satisfy the problem's constraints, so not only must a valid solution be found, but the objective is to find the best solution among all those considered valid. The algorithms that propose to solve this optimization problem are usually grouped into two main categories: i) deterministic and ii) stochastic. On deterministic algorithms, the results are only determined by the input data. So, if the algorithm is run twice with the same input data, it will return the same results. An example of this deterministic algorithm could be the graph search [18]. Stochastic algorithms incorporate an element of randomness, meaning that two runs with the same program and input data do not produce the same execution. In stochastic algorithms, this touch of randomness is used to limit the solutions or guide the search for solutions to where the best solution is believed to be instead of traversing the entire space of valid solutions.

Stochastic algorithms can be divided into two categories: heuristic and meta-heuristic. Heuristic algorithms are given additional information about the problem to guide their search or learn during the search by trial and error. An example of a heuristic algorithm can be A* [6]. In the case of meta-heuristics, a more general strategy is chosen: they try to learn the search space to make the search for the optimal solution more efficient, which means that they can be applied to different types of problems in a more straightforward way. To learn from the problem, meta-heuristics often imitate existing behaviors: music [5], sports [4], mathematics [12], physics [7,9], chemistry [8], biology [3,10,11,13,14], societies [15].

This work aims to implement and compare the performance of some meta-heuristics. Different works can be found in the literature doing this kind of comparison in two ways: using a benchmark with generic mathematical problems [2,16] or applying the meta-heuristic against concrete problems [1,17]. This paper will compare the different meta-heuristics using a benchmark of mathematical problems.

The organization of this paper is as follows. Firstly, a description of the meta-heuristics implemented and the problems used for optimization is in Sect. 2. Then, the experimental results are discussed in Sect. 3. Finally, the conclusions of this paper are in Sect. 4.

2   Materials and Methods

This section outlines each of the three metaheuristic-based optimization methods – Sects. 2.1, 2.2 and 2.3 for the Grey Wolf Optimizer (GWO), the Whale Optimization Algorithm (WOA) and Moth Flame Optimization (MFO), correspondingly. Furthermore, the section ends with the description of the standardized problems for this comparison (see Sect. 2.4) followed by the experimental setup (see Sect. 2.5).

2.1 Grey Wolf Optimizer (GWO)

The GWO meta-heuristic was proposed in [14]; it mimics the hunting behavior of grey wolves. This meta-heuristic employs a series of agents that search for solutions for its operation. These agents in charge of searching for the best solution are the wolves. As in a wolf pack, the solution-seeking agents have different social statuses. The best wolves in the pack are the alpha wolves, followed by the beta, delta, and omega wolves.

Within the pack, the wolves with the highest status –the alpha role–, are in charge of leading the pack and making decisions: when to hunt, when to sleep, where the pack migrates to, and others. The beta role –the second-highest status–, help the alphas in their decision-making and lead the lower-level wolves. The delta role represents the third and lowest social level. The deltas function as a scapegoat within the social structure. They are the last echelon that can give orders. They are in charge of transmitting the orders and directives of the higher echelons to the next level. The lowest echelon is occupied by the omega, who is in charge of carrying out the work within the herd: hunting, fighting and exploring.

The mathematical behavior of meta-heuristics will mimic the hunting process of wolves. The best agent searching for prey will be the alpha, the second the beta, the third the delta, and the rest will be considered omegas. The position and the information of the superior individuals –alpha, beta, and delta roles–, determine the movements that the omegas have to do. The omegas move through the solution space to find new prey or solutions. The meta-heuristic process is iterative. In each iteration, the position of the hunters will be modified to find better prey.

Algorithm 1. Algorithm of the Grey Wolf Optimizer (GWO) metaheuristic

---

Initialize the grey wolf population $X_i (i = 1, 2, ...n))$
Initialize iteration parameters
Calculate the fitness of each search agent
X_α = the best search agent
X_β = the second best search agent
X_δ = the third best search agents
while t < Max number of iterations **do**
    for each search agent **do**
        Update the position of the current search agent
    end for
    Update iteration parameters
    Calculate the fitness of each search agent
    Update X_α, X_β and X_δ
    $t \leftarrow t + 1$
end while

2.2   Whale Optimization Algorithm (WOA)

Whale Optimization Algorithm (WOA) is a meta-heuristic proposed in [13] that mimics the behavior of whales in their search for food. It is based on a series of agents–whales–which move through the solution space searching for food.

The metaheuristic proposes three behaviors for the whale: i) search for prey, ii) envelop the prey, and iii) attack. At all times, a random value determines the behavior of the whales. With a probability of 50%, the method will be considered to be in the phase of attacking the prey. The remaining 50% will determine that we are in one of the other states. Another value will be used to determine whether the prey is searched for or encircled. This second value will be calculated based on our iteration and a random value. Thus, at the beginning of the execution, it is more likely to select the state of searching for a dam, while as the iterations evolve, it will be more likely to go around the dam. These two phases correspond to the exploration and exploitation of the solution space.

Algorithm 2. Algorithm of the Whale Optimization Algorithm (WOA) meta-heuristic

---

Initialize the whales population $X_i (i = 1, 2, ..., n)$
Calculate the fitness of each search agent
X^* = the best search agent
while t < Max number of iterations **do**
    for each search agent **do**
        Update iteration parameters
        $p \leftarrow random \in [0,1]$
        $r \leftarrow random \in [0,1]$
        $A \leftarrow 2 * (2 - t * 2/Max\_iter) * r - (2 - t * 2/Max\_iter)$
        if p<0.5 **then**
            if |A| < 1 **then**
                Update the position of the current search agent by an spiral movement with a radius depending on the current iteration (Attacking)
            else
                Select a random search agent (X_{rand})
                Update the position of the current search agent by moving him on the direction of X_{rand} (Searching for a prey)
            end if
        else
            Update the position of the current search agent by moving him using a small movement (Encircling the prey)
        end if
    end for
    Check if any search agent goes beyond the search space and amend it
    Calculate the fitness of each search agent
    Update X^* if there is a better solution
    $t \leftarrow t + 1$
end while

2.3   Moth Flame Optimizer (MFO)

Moth Flame Optimizer (MFO) [11] is a meta-heuristic that mimics the natural behavior of moths when they fly towards the light of a flame. In this meta-heuristic, the moths are the agents that search for the best solutions, which are candle flames. By observing the behavior of the moths, it is observed how they fly towards the flames in a spiral shape.

To implement this metaheuristic, we will start with a random population of moths evenly distributed in space. It is assumed that there is a flame at the position of the moths, i.e., a solution. The metaheuristic consists of an iterative process. A list of the best flames - best solutions - is used in each iteration. Each moth will select a flame and perform a spiral movement towards it in each iteration. During the spiral movement, it is possible to find other better flames, in which case, the list of best flames will change, and the moth will select a new flame.

Algorithm 3. Algorithm of the Moth Flame Optimizer (MFO) metaheuristic

---
Initialize the moth population $X_i(i = 1, 2, ..., n)$
Calculate the fitness of each search agent
Initialize the best flames population. Initially same as moth population
while t < Max number of iterations **do**
    Update iteration parameters
    for each moth on the population **do**
        Assign a flame
        Update position of the moth moving to the assigned flame
    end for
    $t \leftarrow t + 1$
end while

2.4   Standardized Problems

The experimentation phase uses two different types of problems: unconstrained and constrained. Non-Constrained Optimization problems only use a function $f(x) \in \mathbb{R}$; the goal is to search the value of $x \in \mathbb{R}$ that minimizes the value of $f(x)$. Up to 5 non-constrained problems -extracted from [14]- are used in the comparison; these functions are listed in Table 1.

Besides, a unconstrained function $f(x) \in \mathbb{R}, x \in \mathbb{R}$, includes one or more restrictions and conditions the either the function, the x value or a third function $g(x) \in \mathbb{R}$ must satisfy. For this study, the functions included in Table 2, extracted from the CEC2017 competition [19], are used in this comparison.

2.5   Experimental Setup

Each problem is solved using the three metaheuristics: GWO, WOA, and MFO. Each optimization problem evolved during 1000 iterations. Besides, the methods' parameters were chosen among the best reported in the literature. Each

Table 1. List of unconstrained functions used in this research.

ID	Function
f01	$f(X) = \sum_{i=1}^{n} x_i^2, X \in [-100, 100]$
f02	$f(X) = \sum_{i=1}^{n} |x| + \prod_{i=1}^{n} |x|, X \in [-10, 10]$
f03	$f(X) = \sum_{i=1}^{n} \left(\sum_{j=1}^{i} x_j\right)^2, X \in [-100, 100]$
f08	$f(X) = \sum_{i=1}^{n} -x_i \sin\left(\sqrt{|x_i|}\right), X \in [-500, 500]$
f09	$f(X) = \sum_{i=1}^{n} [x_i^2 - 10\cos(2\pi x_i) + 10], X \in [-5.12, 5.12]$

Table 2. List of contrained functions used in this research.

ID	Function & Constrains
C01	$f(X) = \sum_{i=1}^{n} \left(\sum_{j=1}^{i} x_j\right)^2$ $g(X) = \sum_{i=1}^{n} [x_i^2 - 5000\cos(0.1\pi x) - 4000] \leq 0$ $x \in [-100, 100]$
C04	$f(X) = \sum_{i=1}^{n} [x_i^2 - 10\cos(2\pi x_i) + 10]$ $g_1(X) = \sum_{i=1}^{n} x\sin(2x) \leq 0$ $g_2(X) = \sum_{i=1}^{n} x\sin(x) \leq 0$ $x \in [-10, 10]$
C05	$f(X) = \sum_{i=1}^{n-1} [100(x_i^2 - x_{i+1})^2 + (x_1 - 1)^2]$ $g_1(X) = \sum_{i=1}^{n} [y_i^2 - 50\cos(2\pi y_i) - 40] \leq 0, y_i = M_1 * X$ $g_2(X) = \sum_{i=1}^{n} [w_i^2 - 50\cos(2\pi w_i) - 40] \leq 0, w_i = M_2 * X$ $x \in [-10, 10]$
C13	$f(X) = \sum_{i=1}^{n-1} [100(x_i^2 - x_{i+1})^2 + (x_1 - 1)^2]$ $g_1(X) = \sum_{i=1}^{n} [y_i^2 - 10\cos(2\pi y_i) + 10] \leq 0$ $g_2(X) = \sum_{i=1}^{n} (x_i - 60) \leq 0$ $g_3(X) = \sum_{i=1}^{n} x_i \leq 0$ $x \in [-100, 100]$
C19	$f(X) = \sum_{i=1}^{n} \left(|x_i|^{0.5} + 2\sin x_i^3\right)$ $g_1(X) = \sum_{i=1}^{n-1} \left(-10\exp\left(-0.2\sqrt{x_i^2 + x_{i+1}^2}\right)\right) \leq 0$ $g_2(X) = \sum_{i=1}^{n} 2x_i - 0.5 \leq 0$ $x \in [-50, 50]$

method is run 100 times for each problem to obtain trustworthy metrics about its performance. The following metrics are proposed to measure the performance of the meta-heuristic optimization algorithms:

- Execution time
- Fitness of the best solution found
- Average fitness of the solutions present in the last population.

3 Results and Discussion

Results include both time consumption and fitness values. Table 3 shows the mean and standard deviation of the execution time of each of the algorithms. As

Table 3. Mean and standard deviation –just under the mean value and delimited with parenthesis– of the execution time of each meta-heuristics solving the different unconstrained problems. The best results are marked in bold letters.

Method	f01	f02	f03	f08	f09
GWO	**44.10**	**44.00**	**74.83**	**48.60**	**51.80**
	(0.97)	(1.02)	(0.82)	(1.06)	(1.20)
WOA	258.97	247.73	365.45	261.45	270.82
	(4.22)	(4.93)	(6.34)	(4.78)	(4.05)
MFO	252.71	250.16	371.17	262.84	273.34
	(4.56)	(4.36)	(4.64)	(4.54)	(4.74)
Method	C01	C04	C05	C13	C19
GWO	**86.40**	**56.83**	**74.62**	**62.72**	**68.35**
	(1.71)	(0.97)	(1.41)	(0.85)	(1.40)
WOA	409.90	303.22	431.47	368.10	341.28
	(6.18)	(4.74)	(6.36)	(5.78)	(9.26)
MFO	421.56	313.06	432.06	387.80	308.13
	(5.92)	(5.22)	(5.82)	(4.86)	(6.60)

Table 4. Mean and standard deviation –just below the mean value and delimited with parenthesis–, of the best fitness of each meta-heuristic solving each problem. The best values are remarked in bold letters.

Method	f01	f02	f03	f08	f09
GWO	**0.00**	**0.00**	**0.00**	−743.49	**0.00**
	(0.00)	(0.00)	(0.00)	(129.67)	(0.00)
WOA	**0.00**	**0.00**	**0.00**	**−7313.57**	**0.00**
	(0.00)	(0.00)	(0.00)	(2127.27)	(0.00)
MFO	680.56	0.89	3.95e5	−6510.48	78.57
	(758.83)	(1.94)	(999.61)	(1899.98)	(30.76)
Method	C01	C04	C05	C13	C19
GWO	**0.00**	−43.69	27.21	**70.46**	**−0.75**
	(0.00)	(5.36)	(0.00)	(45.64)	(0.34)
WOA	**0.00**	−235.90	**26.15**	90.89	−0.14
	(0.00)	(76.72)	(0.00)	(186.75)	(0.32)
MFO	3.97e5	**−300**	3.49e5	1.13e9	inf height
	(8.94e4)	(0.00)	(5.76e5)	(1.50e9)	

can be seen, GWO is up to 5 times faster than the rest for all types of problems, while WOA and MFO have very similar execution times to each other.

On the other hand, Table 4 shows the average of the best fitness found in each algorithm run. For all of the problems, it can be observed how MFO always

Table 5. Mean and standard deviation –just below the mean value and delimited with parenthesis–, of the mean fitness of each meta-heuristic solving each problem. The best values are remarked with bold letters

Method	f01	f02	f03	f08	f09
GWO	**0.00**	**0.00**	**0.00**	−743.49	**0.00**
	(0.00)	(0.00)	(0.00)	(129.67)	(0.00)
WOA	**0.00**	**0.00**	**0.00**	**−7313.37**	**0.00**
	(0.00)	(0.00)	(0.00)	(2127.27)	(0.00)
MFO	680.56	0.89	3.95e5	−6510.48	78.57
	(758.83)	(1.94)	(999.61)	(1899.98)	(30.76)

Method	C01	C04	C05	C13	C19
GWO	**0.00**	−43.69	27.21	**70.46**	**−0.75**
	(0.00)	(5.36)	(0.00)	(45.64)	(0.34)
WOA	**0.00**	−235.88	**26.15**	90.90	−0.14
	(0.00)	(76.71)	(0.80)	(186.78)	(0.32)
MFO	3.97e5	**−300**	3.49e5	1.13e9	inf height
	(8.94e4)	(0.00)	(5.76e5)	(1.50e9)	

offers a worse fitness than the other two meta-heuristics. Even for MFO, it is observed that it cannot find a feasible solution in the case of problem C19. Analyzing GWO and WOA, it is seen that both can reach the optimal solution with unconstrained problems. Both meta-heuristics obtain a solution with very similar fitness to the constrained problems.

Besides, Table 5 shows the average fitness of the solutions of the last population of each run. It can be seen that these data are almost identical to those in Table 4, i.e., all meta-heuristics have already stabilized on a solution and are not exploring better solutions.

As seen from the tables, the general tendency is for the GWO meta-heuristic to be the fastest in delivering results, while MFO always takes the longest to run. As for the fitness values, GWO offers very similar values to WOA, but MFO always has a higher value. Figure 1 shows this fact for problem f09, depicting the comparison in times and fitness for all of the runs. There is no doubt that GWO outperforms the other two methods if we consider the two criteria simultaneously.

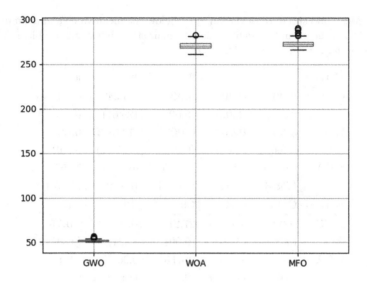

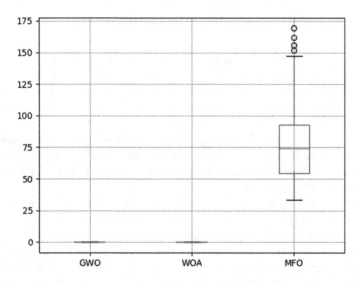

Fig. 1. Comparison box-plots for metrics for function f09. The upper part depicts the time consumption, while the bottom part shows the fitness values.

4   Conclusion

In this study, three known meta-heuristics–Grey Wolf Optimizer, Whale Optimization Algorithm, and Moth Flame Optimizer–were implemented and tested against ten optimization problems, five of which were unconstrained and five of which were constrained. In the analysis of the results, it was observed that the meta-heuristic that offered the fastest results was Grey Wolf Optimizer, but this is not the only criterion to be analyzed. In addition to the execution time, we also examined the minimum value found for optimizing the function. In this case, the values of the Grey Wolf Optimizer were very close to those of the Whale Optimization Algorithm. The meta-heuristic that yielded the worst data for the minimum value of the function was Moth Flame Optimizer, which, even for some more complicated problems, could not find any solution that satisfies the constraints of the function.

In this work, we have tried to collect the most common metaheuristics found in the literature, implement them and conduct a comparative study between them to analyze their performance. This work aims to be the starting point of a line of research in which we will try to solve other problems with real-world applications.

Future work in this line of research will adapt the implementations made in this study to solve problems such as multi-robot path planning, time slot allocation, or collision-free multi-robot trajectory determination. Future work also aims to propose some improvements to improve the performance of these algorithms.

References

1. Azimi, Z.N.: Comparison of metaheuristic algorithms for examination timetabling problem. J. Appl. Math. Comput. **16**(1), 337 (2004). https://doi.org/10.1007/BF02936173
2. Bloomfield, M.W., Herencia, J.E., Weaver, P.M.: Analysis and benchmarking of meta-heuristic techniques for lay-up optimization. Comput. Struct. **88**(5), 272–282 (2010). https://doi.org/10.1016/j.compstruc.2009.10.007
3. Dorigo, M., Di Caro, G.: Ant colony optimization: a new meta-heuristic. In: Proceedings of the 1999 Congress on Evolutionary Computation-CEC99 (Cat. No. 99TH8406), vol. 2, pp. 1470–1477 (1999). https://doi.org/10.1109/CEC.1999.782657
4. Fadakar, E., Ebrahimi, M.: A new metaheuristic football game inspired algorithm. In: 2016 1st Conference on Swarm Intelligence and Evolutionary Computation (CSIEC), pp. 6–11 (2016). https://doi.org/10.1109/CSIEC.2016.7482120
5. Geem, Z.W., Kim, J.H., Loganathan, G.: A new heuristic optimization algorithm: harmony search. Simulation **76**(2), 60–68 (2001). https://doi.org/10.1177/003754970107600201
6. Hart, P.E., Nilsson, N.J., Raphael, B.: A formal basis for the heuristic determination of minimum cost paths. IEEE Trans. Syst. Sci. Cybern. **4**(2), 100–107 (1968). https://doi.org/10.1109/TSSC.1968.300136
7. Hatamlou, A.: Black hole: a new heuristic optimization approach for data clustering. Inf. Sci. **222**, 175–184 (2013). https://doi.org/10.1016/j.ins.2012.08.023. Including Special Section on New Trends in Ambient Intelligence and Bio-inspired Systems
8. Irizarry, R.: A generalized framework for solving dynamic optimization problems using the artificial chemical process paradigm: applications to particulate processes and discrete dynamic systems. Chem. Eng. Sci. **60**(21), 5663–5681 (2005). https://doi.org/10.1016/j.ces.2005.05.028

9. Kirkpatrick, S., Gelatt, C.D., Vecchi, M.P.: Optimization by simulated annealing. Science **220**(4598), 671–680 (1983). https://doi.org/10.1126/science.220.4598.671

10. Mirjalili, S.: The ant lion optimizer. Adv. Eng. Softw. **83**, 80–98 (2015). https://doi.org/10.1016/j.advengsoft.2015.01.010

11. Mirjalili, S.: Moth-flame optimization algorithm: a novel nature-inspired heuristic paradigm. Knowl.-Based Syst. **89**, 228–249 (2015). https://doi.org/10.1016/j.knosys.2015.07.006

12. Mirjalili, S.: SCA: a sine cosine algorithm for solving optimization problems. Knowl.-Based Syst. **96**, 120–133 (2016). https://doi.org/10.1016/j.knosys.2015.12.022

13. Mirjalili, S., Lewis, A.: The whale optimization algorithm. Adv. Eng. Softw. **95**, 51–67 (2016). https://doi.org/10.1016/j.advengsoft.2016.01.008

14. Mirjalili, S., Mirjalili, S.M., Lewis, A.: Grey wolf optimizer. Adv. Eng. Softw. **69**, 46–61 (2014). https://doi.org/10.1016/j.advengsoft.2013.12.007

15. Mousavirad, S.J., Ebrahimpour-Komleh, H.: Human mental search: a new population-based metaheuristic optimization algorithm. Appl. Intell. **47**(3), 850–887 (2017). https://doi.org/10.1007/s10489-017-0903-6

16. Parejo, J.A., Ruiz-Cortés, A., Lozano, S., Fernandez, P.: Metaheuristic optimization frameworks: a survey and benchmarking. Soft. Comput. **16**(3), 527–561 (2012). https://doi.org/10.1007/s00500-011-0754-8

17. Sonmez, M.: Performance comparison of metaheuristic algorithms for the optimal design of space trusses. Arab. J. Sci. Eng. **43**(10), 5265–5281 (2018). https://doi.org/10.1007/s13369-018-3080-y

18. Williams, M.L., Wilson, R.C., Hancock, E.R.: Deterministic search for relational graph matching. Pattern Recogn. **32**(7), 1255–1271 (1999). https://doi.org/10.1016/S0031-3203(98)00152-6

19. Wu, G., Mallipeddi, R., Suganthan, P.: Problem definitions and evaluation criteria for the CEC 2017 competition and special session on constrained single objective real-parameter optimization. Technical report, IEEE Congress on Evolutionary Computation (2016)

An Analysis on Hybrid Brain Storm Optimisation Algorithms

Dragan Simić[1]([✉]) [ID], Zorana Banković[2], José R. Villar[3] [ID], José Luis Calvo-Rolle[4] [ID], Svetislav D. Simić[1] [ID], and Svetlana Simić[5] [ID]

[1] Faculty of Technical Sciences, University of Novi Sad, Trg Dositeja Obradovića 6, 21000 Novi Sad, Serbia
dsimic@eunet.rs, {dsimic,simicsvetislav}@uns.ac.rs
[2] Frontiers Media SA, Paseo de Castellana 77, Madrid, Spain
[3] University of Oviedo, Campus de Llamaquique, 33005 Oviedo, Spain
villarjose@uniovi.es
[4] Department of Industrial Engineering, University of A Coruña, 15405 Ferrol-A Coruña, Spain
jlcalvo@udc.es
[5] Faculty of Medicine, University of Novi Sad, Hajduk Veljkova 1–9, 21000 Novi Sad, Serbia
svetlana.simic@mf.uns.ac.rs

Abstract. Optimisation can be described as the process of finding optimal values for the variables of a given problem in order to minimise or maximise one or more objective function(s). Brain storm optimisation (BSO) algorithm is relatively new swarm intelligence algorithm that mimics the brainstorming process in which a group of people solves a problem together. The aim of this paper is to present hybrid BSO algorithm solutions in general, and particularly: (i) a hybrid BSO for improving the performances of the original BSO algorithm; (ii) a hybrid BSO for the flexible job-shop scheduling problem; and (iii) a feature selection by a hybrid BSO algorithm for the COVID-19 classification. The hybrid BSO algorithm overcomes the lack of exploitation in the original BSO algorithm, and simultaneously, the obtained better results prove their efficiency and robustness.

Keywords: Brain storm optimisation · Dynamic parameters adjustment · Job-shop scheduling problem · Feature selection · Classification

1 Introduction

Optimisation can be described as the process of finding optimal values for the variables of a given problem in order to minimise or maximise one or more objective function(s). Conventional optimisation techniques are mostly based on a gradient descent to find the optimum for a given optimisation problem. This makes them highly dependent on initial solutions and most of the time it results in local optima stagnation of the algorithm. Local optima stagnation occurs when an optimisation mistakenly assumes that a local solution is a global solution. To alleviate the drawbacks of conventional optimisation techniques, stochastic algorithms were proposed. In such approaches, random abrupt or gradual change of the solutions results in a better local optima avoidance.

© Springer Nature Switzerland AG 2022
P. García Bringas et al. (Eds.): HAIS 2022, LNAI 13469, pp. 505–516, 2022.
https://doi.org/10.1007/978-3-031-15471-3_43

A metaheuristic algorithm is a search procedure designed to find a good solution to an optimisation problem that is complex and difficult to solve by optimality. It is imperative to find a near-optimal solution based on imperfect or incomplete information in this real-world of limited resources; it takes computational power and time. The emergence of metaheuristics for solving such optimisation problems is one of the most notable achievements in the last two decades in operations research. Optimisation algorithms can be divided into two classes: single-objective and multi-objective.

In 2011, a metaheuristic-based algorithm called brain storm optimisation (BSO) algorithm was developed [1]. The algorithm is motivated by the brainstorming process of humans. Brainstorming is a creative way of solving a specific problem by a group of people. In the brainstorming process, several people share their ideas with each other related to the problem that should be solved, where any idea is acceptable and criticism is not allowed. In the end, from all suggested ideas, the best possible solution is selected. Similarly, in the BSO algorithm, initially, random solutions are generated; as in any other swarm intelligence algorithm, each solution is analogous to an idea in the brainstorming process. At every iteration the idea is modified; in other words, the solution's position is updated according to the previous knowledge.

Numerous research papers [2] develop and apply the BSO algorithm. In BSO, the solutions are diverged into several clusters. One individual or more individuals are selected to generate new solutions by some genetic operators. Some multi-objective BSO algorithms [3, 4] are proposed to solve MOPs. In these multi-objective BSO algorithms, population is updated by new solutions after the solutions are clustered, which may decrease the speed of convergence.

Therefore, BSO is not only an optimisation method but it could also be viewed as a framework for the optimisation technique. The motivations for this research and new challenges for future research are interested in combining BSO with some other heuristics and metaheuristics methods to create an efficient hybrid BSO optimisation, classification, clustering, and feature selection systems. This research directly continues and expands the authors' previous research on optimisation [5]. Also, this paper, in general, continues the authors' previous research in optimisation of supply chain management, and optimisation in inventory management presented in [6–9].

The rest of the paper is organized in the following way: Sect. 2 provides an overview of the basic idea on brain storm optimisation algorithm. Modelling the bio-inspired hybrid systems combining BSO algorithm and metaheuristics-based algorithm is presented in Sect. 3. Section 4 provides conclusions and some directions for future work.

2 Brain Storm Optimisation Algorithm

Brain storm optimisation (BSO) is relatively new swarm intelligence algorithm. It is inspired by collective behaviour of human beings. It has attracted a number of researchers and has good performance in its applications for complex problems. Brainstorming is a process of collecting new ideas about a specific problem from a group of people without any prejudicing or ordering. Then, these ideas are evaluated and filtered one by one to select the best idea.

The general description of BSO flowchart is displayed in Fig. 1. The basic steps of BSO are summarized by the pseudo code revealed in Algorithm 1.

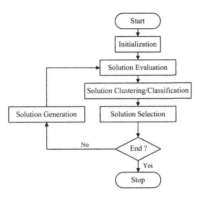

Fig. 1. The process of brain storm optimisation algorithms

Algorithm 1 *The basic procedure of the **brain storm** optimisation algorithm*

Begin

Step 1:   ***Initialization.*** Randomly generate ***n*** individuals (potential solutions), and evaluate the ***n*** individuals;

Step 2:   **While** *not find "good enough" solution or not reach the pre-determined maximum number of iterations* **do**

Step 3:   **Solution clustering/classification** *operation: Diverge n individuals into m groups by a clustering/classification algorithm;*

Step 4:   **New solution generation** *operation: Select solution(s) from one or two groups randomly to generate new individual (solution);*

Step 5:   **Solution selection** *operation: Compare the newly generated individual (solution) and the existing individual (solution) with the same individual index; the better one is kept and recorded as the new individual;*

Step 6:   *Evaluate n individuals (solutions);*

Step 6:   **end While**

Step 7:   *Post-processing the results and visualization;*

End.

In a BSO algorithm, the solutions are separated into several clusters. The best solution of the population will be kept if the new generated solution is not better. New individual can be generated based on one or two individuals in clusters. The exploitation ability is enhanced when the new individual is close to the best solution found till that moment. The exploration ability is enhanced when the new individual is randomly generated, or generated by individuals in two clusters.

2.1 Solution Clustering

The aim of solution clustering/classification is to converge the solutions into small regions. Different clustering algorithms can be utilized in the brain storm optimisation algorithm. In the original BSO algorithm, the basic k-means clustering algorithm is utilized. The clustering strategy has been replaced by other convergence methods.

2.2 New Solution Generation

A new individual generation can be generated based on one or several individuals or clusters. In the original brain storm optimisation algorithm, a probability value is utilized to determine a new individual being generated by one or two "old" individuals. Generating an individual from one cluster could refine a search region, and it enhances the exploitation ability. On the contrast, an individual, which is generated from two or more clusters, may be far from these clusters.

2.3 Selection

Selection strategy plays an important role in an optimisation algorithm. The aim of the selection strategy is to keep good or more representative solutions in all individuals. The better solution is kept by the selection strategy after each new individual generation, while clustering strategy and generation strategy add new solutions into the swarm to keep the diversity for the whole population. The selection strategy determines the lifecycle of individuals. Individuals can only do one of the three things in selection strategy: be chosen, be kept, and be replaced. In other words, individuals are born, live, and die in the optimisation process by the selection strategy. Different selection strategies in genetic algorithms, such as ranking selection, tournament selection, and other selection schemes, are analysed in [10].

2.4 Variants of BSO Algorithms

The brain storm optimisation algorithm is not only an optimisation method but also can be viewed as a framework for the optimisation technique. The process of BSO algorithm could be simplified as a framework with two basic operations: the converging operation and the diverging operation. These two basic operations in BSO algorithms are shown in Fig. 2.

Some variants of BSO algorithms have been proposed to improve the search ability of the original BSO algorithm. The solutions get clustered after a few iterations and the population diversity decreases quickly during the search, which are common phenomena in swarm intelligence. A definition of population diversity in BSO algorithm is introduced to measure the change of solutions' distribution. According to the analysis, different kinds of partial re-initialization strategies are utilized to improve the population diversity in the BSO algorithm [11].

Under similar consideration, chaotic predator-prey brain storm optimisation was proposed to improve its ability for continuous optimisation problems. A chaotic operation is further added to increase the diversity of population.

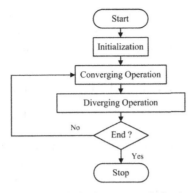

Fig. 2. Two basic operations in brain storm optimisation algorithms [11]

3 Modelling the Hybrid Brain Storm Optimisation Algorithms

Two main processes in any metaheuristic algorithms are intensification and diversification. Often, the algorithm has one of these two processes more enhanced: one of the approaches for making the right balance between these two phases is by hybridizing two or more algorithms.

3.1 A Hybrid BSO for Improving the Performances of the Original BSO

Intelligence algorithms play an increasingly important role in the field of intelligent control. Brain storm optimisation (BSO) is a new kind of swarm intelligence algorithm inspired by emulating the collective behaviour of human beings in the problem-solving process. To improve the performance of the original BSO, many variants of BSO are proposed.

In the research paper [12], an improved BSO algorithm with a dynamic clustering strategy (BSO-DCS) is proposed as a variant of BSO for global optimisation problems. To reduce the time complexity of the original BSO, a new grouping method named dynamic clustering strategy (DCS) is proposed. The aim was to improve the clustering method in the original BSO. To verify the effectiveness of the proposed BSO-DCS, it was tested on 12 benchmark functions of CEC 2005 with 30 dimensions. Experimental results demonstrated that DCS was an effective strategy to reduce time complexity, and the improved BSO-DCS performed greatly better than the original BSO algorithm.

As a novel swarm intelligence optimisation algorithm, brain storm optimisation (BSO) has its own unique capabilities in solving optimisation problems. However, the performance of a traditional BSO strategy in balancing exploitation and exploration is inadequate, reducing the convergence performance of BSO. To overcome these problems, in the research paper [13], a multi-strategy BSO with dynamic parameters adjustment (MSBSO) is presented. In MSBSO, four competitive strategies based on the improved individual selection rules are designed to adapt to different search scopes, thus obtaining more diverse and effective individuals. In addition, a simple adaptive parameter that can dynamically regulate search scopes is designed as the basis for selecting strategies. The proposed MSBSO algorithm and other state-of-the-art algorithms are tested on CEC 2013 benchmark functions and CEC 2015 large scale global optimisation (LSGO) benchmark functions, and the experimental results prove that the MSBSO algorithm is more competitive than other related algorithms.

3.2 A Hybrid BSO for the Flexible Job-Shop Scheduling Problem

The Job-Shop Scheduling Problem (JSSP) is a common and difficult problem in resource allocation whereby various jobs are assigned to physical resources or machines at a given time. The research paper [14] presents the hybrid BSO algorithm with an updating strategy as a solution to the flexible job-shop scheduling problem (FJSSP). With the aim to improve the global search of the BSO algorithm, a new updating strategy is proposed to adaptively perform several selection methods and neighbourhood search operations (BSO_US) (Fig. 3).

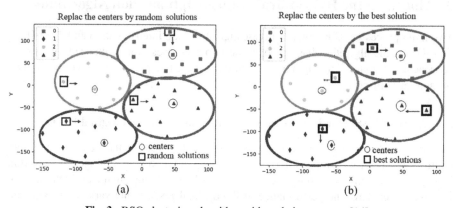

Fig. 3. BSO clustering algorithm with updating strategy [14]

The Flexible Job-Shop Scheduling Problem (FJSSP) is a generalisation of the traditional JSSP whereby every operation is processed by several machines which are chosen from a candidate machines subset. This has been identified as an NP-hard problem [15, 16]. However, FJSSP is considered to be more challenging and complex, since it needs to select the machines properly from a group of machines, as well as balance the machines' workload.

The FJSSP is divided into two problems. The routing sub-problem is the first, in which every operation is assigned to a machine. The second sub-problem is sequencing, which consists of defining a sequence for the assigned operations - resulting from the first step - on each machine. This step aims to achieve a practical schedule to minimize the time needed for the task to be accomplished – makespan. Therefore, two sub-problems present further complexity in solving the problem.

The idea of selecting the solution to be updated comes from the original BSO algorithm. The idea is to utilize the benefit of the clustering of the population, which leads to not searching similar solutions. The selection strategies are listed below: (i) select the centre of a random cluster; (ii) select a random solution from a randomly selected cluster; (iii) select two centres from two random clusters; and (iv) select two random solutions from two random clusters.

The *neighbourhood search operators* are described as the following: (i) *Neighbourhood N1* – where a selection of jobs is random and the operations on each machine are interchanged with other machines without constraints' breaking; (ii) *Neighbourhood N2* – where an operation is selected randomly, and it is taken into account for exchange with another operation or moved to a new random and possible location; (iii) *Neighbourhood N3* – this neighbourhood is based on the critical path. The sequence of two critical operations handled by the same machine is known as the critical path, which presents the longest schedule.

Computational experience proves that the following values are more effective for the proposed hybrid method: *number of iterations* = 200; and *population size* = 50. The number of clusters was chosen based on [17].

Finally, it can be concluded that the new updating strategy is also proposed for the BSO algorithm to enhance the global search by adaptively applying different selection and neighbourhood methods. The experimental results demonstrate that the BSO_US overcomes the basic BSO algorithm. Moreover, when compared with other algorithms, it is obvious that the BSO_US algorithm displayed to be effective for the FJSSP, obtaining comparable results [14].

3.3   Feature Selection for the Hybrid BSO Algorithm for COVID-19 Classification

A large number of features lead to very high-dimensional data. The total possible number of attribute subsets is 2^n in datasets with n number of features. As the dimension of data increases, the possible feature subset increases highly. Hence, the goal in the feature selection problem is to minimize the dimension and at the same time maximize the classification accuracy of the given dataset, which is considered as an optimisation task.

The feature selection method reduces the dimension of data, increases the performance of prediction, and reduces the computation time. Feature selection is the process of selecting the optimal set of input features from a given dataset in order to reduce the noise in data and keep the relevant features. The optimal feature subset contains all useful and relevant features and excludes any irrelevant feature that allows machine learning models to understand better and differentiate efficiently the patterns in datasets. In this article, a binary hybrid metaheuristic-based algorithm for selecting the optimal feature subset is proposed. In feature selection problems, due to the exponential increase in the number of features, this kind of problem belongs to nonpolynomial-hard (NP-hard) optimisation problems, where traditional exact optimisation algorithms would fail. However, stochastic approximation algorithms, such as metaheuristic algorithms, are very successful in tackling such problems.

In the research paper [18], the brain storm optimisation (BSO) algorithm by the firefly algorithm (FA) [19] to achieve a better trade-off between the exploration and exploitation and apply it for feature selection problem by using a wrapper-based method is applied. Concretely, the BSO algorithm is hybridized by the FA and adopted as a wrapper method for feature selection problems on classification datasets. The control parameters for the proposed BSO-FA feature selection system are presented in Table 1. At the end of the BSO-FA algorithm procedure, the Sigmoid function squeezes the solution's value between 0 and 1, afterward based on the threshold value, to decide whether 0 or 1 will be assigned to the corresponding feature. If the solution's value is less than the threshold, which is set to 0.5, after applying the transfer function, the solution will be 0; otherwise, it will be 1.

Table 1. Control parameters for the BSO-FA feature selection system

Parameter	Value
BSO parameters	
One cluster selection probability p1	0.8
Total number of clusters $cluster_{number}$	5
Replacing operator probability $p_{replace}$	0.2
Probability of choosing the centre of cluster 1 $p_{1center}$	0.4
Probability of choosing the centres of clusters 2 $p_{2center}$	0.5
Step size k	20
Parameter Ω_1	0.5
Parameter Ω_2	0.5
FA parameters	
Randomization parameter α	1.0
Attractiveness parameter β_0	1.0
Bright intensity parameter γ	1.0

The proposed algorithm in [18] is evaluated on 21 datasets from UCI data repository [20], which are presented in Table 2, and compared with 11 metaheuristic algorithms. In addition, the proposed method is adopted for the coronavirus disease dataset [21]. That dataset is applied for COVID-19 patient health prediction, where the dataset has 15 different attributes (features), including the patient's location, country, gender, age, and different symptoms [22].

The experimental results of the hybrid BSO-FA system are compared with other approaches, as well as with: (i) the original FA; (ii) the original BSO algorithm; (iii) whale optimisation algorithm (WOA); (iv) binary whale optimisation algorithm with Sigmoid transfer function (bWOA-S); (v) binary whale optimisation algorithm with hyperbolic tangent transfer function (bWOA-V); (vi) three variants of binary ant lion (bALO1) (bALO2) (bALO3); (vii) particle swarm optimisation (PSO); (viii) binary grey wolf optimisation (bGWO); and (ix) binary dragonfly algorithm (bDA). In addition, the proposed method is compared with three gaining-sharing knowledge-based algorithm (GSK) variants [23]: (x) V-shaped GSK-based algorithm (bGSK-V4) [24]; (xi) chaotic GSK-based optimisation algorithm (CBi-GSK1) [25]; and (xii) binary GSK-based optimisation (FS-pBGSK) [26].

The experiment is repeated in 20 runs, the maximum number of iterations = 70, and the population size is set to 8. Moreover, the binary BSO-FA approach is employed

Table 2. Datasets

Dataset name	No. of features	No. of samples
Breast Cancer	9	699
Tic Tac Toe	9	958
Zoo	16	101
Wine EW	13	178
Spect EW	22	267
Sonar EW	60	208
Ionosphere EW	34	351
Heart EW	13	270
Congress EW	16	435
Krvskp EW	36	3196
Waveform EW	40	5000
Exactly	13	1000
Exactly 2	13	1000
M of N	13	1000
Vote	16	300
Breast EW	30	569
Semeion	265	1593
Clean 1	166	476
Clean 2	166	6598
Lymphography	18	148
Penghung EW	325	73

for COVID-19 and compared with other state-of-the art methods, where the proposed hybrid binary BSO-FA method is over performed by other approaches. The accuracy of bBSOFA is 93.57%, whereas the second-best performing approach is hyper learning binary dragonfly algorithm (HLBDA), with the classification accuracy of 92.21% [27].

In the case of the total selected features, both approaches selected two to three features on average. Based on the analysis of the selected features, we can draw a conclusion that specific features are not important for the prediction, and symptom4, symptom5, and symptom6 are never selected by the algorithm in the experiments.

The hybrid BSO algorithm with the FA algorithm overcomes the lack of exploitation in the original BSO algorithm. The obtained experimental results substantiate the robustness of the proposed hybrid binary BSO-FA algorithm. It efficiently reduces and selects the feature subset and at the same time results in higher classification accuracy than other methods in literature. Moreover, some other methods are employed for COVID-19 classification, such as: Gradient-based grey wolf optimiser with Gaussian walks in modelling and prediction of the COVID-19 pandemic [28], COVID-19 diagnosis on

CT images with Bayes optimization-based [29], and a proficient approach to forecast COVID-19 spread via optimized dynamic machine learning [30]. However, the comparison of different traditional methods and novelty heuristics and metaheuristics methods with different datasets is very difficult, maybe even impossible.

4 Conclusion and Future Work

The brain storm optimisation (BSO) algorithm has its own unique capabilities in solving optimisation problems, but the performance of traditional BSO strategy in balancing exploitation and exploration is inadequate. Therefore, an improved BSO algorithm with a dynamic clustering strategy, and a multi-strategy BSO with a dynamic parameter adjustment, intended to reduce the time complexity and overcome problems of the original BSO algorithm are presented in this paper.

However, the focus and the aim of this paper were to propose the hybrid new brain storm optimisation algorithms. First, the hybrid BSO for the flexible job-shop scheduling problem which combined new strategy to enhance the global search by adaptively applying different selection and neighbourhood methods in BSO algorithm was shown. Second, a feature selection by a new hybrid BSO algorithm for COVID-19 classification was presented. The hybrid BSO algorithm for feature selection combines classical BSO and firefly optimisation algorithms in classification problem. Both hybrid BSO algorithms presented in this paper have proved the better efficiency and robustness in comparison with the original BSO algorithm.

Regarding future work, it would be of interest to investigate the effect of all these modifications and improvements of the hybrid BSO algorithms for other optimisations, prediction classifications, and clustering problems, testing their performance and behaviour on broader domains, as well as experimenting with more different datasets.

References

1. Shi, Y.: An optimization algorithm based on brainstorming process. Int. J. Swarm Intell. Res. **2**(4), 35–62 (2011). https://doi.org/10.4018/ijsir.2011100103
2. Cheng, S., Qin, Q., Chen, J., Shi, Y.: Brain storm optimization algorithm: a review. Artif. Intell. Rev. **46**(4), 445–458 (2016). https://doi.org/10.1007/s10462-016-9471-0
3. Guo, X., Wu, Y., Xie, L., Cheng, S., Xin, J.: An adaptive brain storm optimization algorithm for multiobjective optimization problems. In: Tan, Y., Shi, Y., Buarque, F., Gelbukh, A., Das, S., Engelbrecht, A. (eds.) ICSI 2015. LNCS, vol. 9140, pp. 365–372. Springer, Cham (2015). https://doi.org/10.1007/978-3-319-94120-2_41
4. Shi, Y., Xue, J., Wu, Y.: Multi-objective optimization based on brain storm optimization algorithm. Int. Swarm Intell. Res. **4**(3), 1–21 (2013). https://doi.org/10.4018/ijsir.2013070101
5. Simić, D., Ilin, V., Simić, S.D., Simić, S.: Swarm intelligence methods on inventory management. In: Graña, M., et al. (eds.) SOCO'18-CISIS'18-ICEUTE'18 2018. AISC, vol. 771, pp. 426–435. Springer, Cham (2019). https://doi.org/10.1007/978-3-319-94120-2_41
6. Simić, D., Ilin, V., Svirčević, V., Simić, S.: A hybrid clustering and ranking method for best positioned logistics distribution centre in Balkan Peninsula. Logic J. IGPL **25**(6), 991–1005 (2017). https://doi.org/10.1093/jigpal/jzx047

7. Simić, D., Svirčević, V., Ilin, V., Simić, S.D., Simić, S.: Particle swarm optimization and pure adaptive search in finish goods' inventory management. Cybern. Syst. **50**(1), 58–77 (2019). https://doi.org/10.1080/01969722.2018.1558014

8. Simić, D., Svirčević, V., Corchado, E., Calvo-Rolle, J.L., Simić, S.D., Simić, S.: Modelling material flow using the Milk run and Kanban systems in the automotive industry. Expert. Syst. **38**(1), e12546 (2021). https://doi.org/10.1111/exsy.12546

9. Zayas-Gato, F., et al.: A hybrid one - class approach for detecting anomalies in industrial systems. Expert Syst. e12990 (2022). https://doi.org/10.1111/exsy.12990

10. Goldberg, D.E., Deb, K.: A comparative analysis of selection schemes used in genetic algorithms. Found. Genet. Algorithms **1**, 69–93 (1991). https://doi.org/10.1016/B978-0-08-050 684-5.50008-2

11. Cheng, S., Shi, Y., Qin, Q., Ting, T.O., Bai, R.: Maintaining population diversity in brain storm optimization algorithm. In: Proceedings of 2014 IEEE Congress on Evolutionary Computation (CEC 2014), pp. 3230–3237. IEEE, Beijing (2014)

12. Cao, Z., Rong, X., Du, Z.: An improved brain storm optimization with dynamic clustering strategy. MATEC Web Conf. **95**, 19002 (2017). https://doi.org/10.1051/matecconf/201795 19002

13. Liu, J., Peng, H., Wu, Z., Chen, J., Deng, C.: Multi-strategy brain storm optimization algorithm with dynamic parameters adjustment. Appl. Intell. **50**(4), 1289–1315 (2020). https://doi.org/10.1007/s10489-019-01600-7

14. Alzaqebah, M., Jawarneh, S., Alwohaibi, M., Alsmadi, M.K., Almarashdeh, I., Mohammad, R.M.A.: Hybrid brain storm optimization algorithm and late acceptance hill climbing to solve the flexible job-shop scheduling problem. J. King Saud Univ. – Comput. Inf. Sci. (2020). https://doi.org/10.1016/j.jksuci.2020.09.004

15. Garey, M.R., Johnson, D.S., Sethi, R.: The complexity of flowshop and jobshop scheduling. Math. Oper. Res. **1**(2), 117–129 (1976)

16. Genova, K., Kirilov, L., Guliashki, V.: A survey of solving approaches for multiple objective flexible job shop scheduling problems. Cybern. Inf. Technol. **15**(2), 3–22 (2015). https://doi.org/10.1515/cait-2015-0025

17. Bholowalia, P., Kumar, A.: EBK-means: a clustering technique based on elbow method and k-means in WSN. Int. J. Comput. Appl. **105**(9) (2014). https://doi.org/10.5120/18405-9674

18. Bezdan, T., Živković, M., Bacanin, N., Chhabra, A., Suresh, M.: Feature selection by hybrid brain storm optimization algorithm for COVID-19 classification. J. Comput. Biol. **29**(6), 1–15 (2022). https://doi.org/10.1089/cmb.2021.0256

19. Yang, X. S.: Firefly algorithms for multimodal optimization. In: Watanabe, O., Zeugmann, T. (eds.) SAGA 2009. LNCS, vol. 5792, pp. 169–178. Springer, Heidelberg (2009). https://doi.org/10.1007/978-3-642-04944-6_14

20. Dua, D., Graff, C: UCI Machine Learning Repository. http://archive.ics.uci.edu/ml. Accessed 28 Sept 2020

21. https://github.com/Atharva-Peshkar/Covid-19-Patient-Health-Analytics. Accessed 25 Sept 2020

22. Iwendi, C., et al.: COVID-19 patient health prediction using boosted random forest algorithm. Front. Public Health **8**, 357 (2020). https://doi.org/10.3389/fpubh.2020.00357

23. Mohamed, A.W., Hadi, A.A., Mohamed, A.K.: Gaining-sharing knowledge based algorithm for solving optimization problems: a novel nature-inspired algorithm. Int. J. Mach. Learn. Cybern. **11**(5), 1501–1529 (2020). https://doi.org/10.1007/s13042-019-01053-x

24. Agrawal, P., Ganesh, T., Mohamed, A.W.: Chaotic gaining sharing knowledge-based optimization algorithm: an improved metaheuristic algorithm for feature selection. Soft. Comput. **25**(14), 9505–9528 (2021). https://doi.org/10.1007/s00500-021-05874-3

25. Agrawal, P., Ganesh, T., Mohamed, A.W.: A novel binary gaining-sharing knowledge-based optimization algorithm for feature selection. Neural Comput. Appl. **33**(11), 5989–6008 (2021). https://doi.org/10.1007/s00521-020-05375-8

26. Agrawal, P., Ganesh, T., Oliva, D., Mohamed, A.W.: S-shaped and v-shaped gaining-sharing knowledge-based algorithm for feature selection. Appl. Intell. **52**(1), 81–112 (2022). https://doi.org/10.1007/s10489-021-02233-5

27. Too, J., Mirjalili, S.: A hyper learning binary dragonfly algorithm for feature selection: a COVID-19 case study. Knowl. Based Syst. **212**, 106553 (2021). https://doi.org/10.1016/j.kno sys.2020.106553

28. Khalilpourazari, S., Doulabi, H.H., Çiftçioğluc, A.O., Weber, G.-W.: Gradient-based grey wolf optimizer with Gaussian walk: application in modelling and prediction of the COVID-19 pandemic. Expert Syst. Appl. **177**, 114920 (2021)

29. Canayaz, M., Şehribanoğlu, S., Özdağ, R., Demir, M.: COVID-19 diagnosis on CT images with Bayes optimization-based deep neural networks and machine learning algorithms. Neural Comput. Appl. **34**(7) (2022). https://doi.org/10.1007/s00521-022-07052-4

30. Alali, Y., Harrou, F., Sun, Y.: A proficient approach to forecast COVID-19 spread via optimized dynamic machine learning models. Sci. Rep. **12**(1), 2467 (2022). https://doi.org/10.1038/s41 598-022-06218-3

Author Index

Printed in the United States
by Baker & Taylor Publisher Services